Periodic, Quasi-Periodic and Chaotic Motions in Celestial Mechanics: Theory and Applications

Periodic, Quasi-Periodic and Chaotic Motions in Celestial Mechanics: Theory and Applications

Selected papers from the Fourth Meeting on Celestial Mechanics, CELMEC IV
San Martino al Cimino (Italy), 11–16 September 2005

Edited by
A. Celletti
Dipartimento di Matematica, Università di Roma "Tor Vergata", Italy
and
S. Ferraz-Mello
Instituto de Astronomia, Geofísica e Ciências Atmosféricas, Universidade de São Paulo, Brasil

Reprinted from *Celestial Mechanics and Dynamical Astronomy*, Volume 95(1–4), 2006

Springer

A C.I.P. catalogue record for this book is available from the Library of Congress

ISBN-1-4020-5324-X

Published by Springer,
P.O. Box 17, 3300 AA Dordrecht, The Netherlands

Sold and distributed in North, Central and South America
By Springer,
101 Philip Drive, Norwell, MA 02061, USA

In all other countries, sold and distributed
By Springer,
P.O. Box 322, 3300 AH Dordrecht, The Netherlands

Printed on acid-free paper

Cover image: Resonances, tori and chaotic motions in a pendulum-like system through Fast Lyapunov Indicators. ***Courtesy of:*** A. Celletti, C. Froeschlé, E. Lega.

CELESTIAL MECHANICS AND DYNAMICAL ASTRONOMY

Vol. 95 Nos. 1–4 2006

Celestial Mechanics and Dynamical Astronomy (2006) 95:1–2
DOI 10.1007/s10569-006-9040-y

Preface

Alessandra Celletti · Sylvio Ferraz-Mello

The Fourth International Meeting of Celestial Mechanics—CELMEC IV—took place in the welcoming landscape of San Martino al Cimino, about 100 km north of Roma, during the period 11–16 September 2005. Following the tradition of the previous CELMEC meetings (taking place every 4 years), the goal was to gather together scientists from the different communities involved in Celestial Mechanics (such as universities, astronomical observatories, space agencies, research institutes and industries). This synergy was aimed to satisfy the demand of comparing complementary ideas and techniques on the recent advances in Celestial Mechanics. The meeting was particularly crowded with respect to the previous editions, thanks to the participation of about 130 people coming from all over the world.

The strong interdisciplinary character of modern Celestial Mechanics is witnessed by the different contributions presented in the current publication, ranging from advanced mathematical theories to sophisticated numerical investigations of the solar system dynamics. Each section is opened by review papers, which introduce to leading subjects, like the variational approaches to find periodic orbits, the stability theory of the N-body problem, the spin-orbit resonances and chaotic dynamics, the space debris polluting the circumterrestrial space. The subsequent research papers encompass many key topics of Celestial Mechanics, often bridging from theory to applications, from dynamical system theory to planetary science, from natural to artificial satellite theory. This nice intermingling of subjects was made possible by the enthusiastic presentations and discussions of the participants.

A. Celletti (✉)
Dipartimento Di Matematica, Universita' Di Roma Tor Vergata,
Via Della Ricerca Scientifica, Roma 133, Italy
e-mail: celletti@mat.uniroma2.it

S. Ferraz-Mello
IAG-Universidade de Sao Paulo, Rua do Matao,
1226, Cidade Universitária, CEP 05508-900
Sao Paulo, Brazil
e-mail: sylvio@usp.br

The CELMEC IV organizing committee was composed by Alessandra Celletti (Università di Roma "Tor Vergata"), Andrea Milani (Università di Pisa), Ettore Perozzi (Telespazio, Roma) and Giovanni B. Valsecchi (Istituto Nazionale di Astrofisica—IASF, Roma). The realization of the meeting was made possible thanks to the financial supports provided by the following institutions: University of Roma "Tor Vergata" and its Department of Mathematics, Gruppo Nazionale per la Fisica Matematica (GNFM), Telespazio, European Space Agency (ESA), University of Pisa, National Institute for Astrophysics (INAF), Balletti Park Hotel; a special thank to Antonio Giorgilli for his financial contribution through the University of Milano Bicocca. These sponsors allowed many young people and researchers from developing countries to attend the meeting. CELMEC IV was promoted by the Italian Society of Celestial Mechanics and Astrodynamics (SIMCA) and it was hosted by the Balletti Park Hotel (San Martino al Cimino, Viterbo), which provided a very handsome atmosphere.

All authors were encouraged to write papers of a length that they considered suitable for the presentation of their results. The editors and the editorial board of "Celestial Mechanics and Dynamical Astronomy" arranged for competent and fast refereeing so that all papers could be reviewed and, when necessary, revised before publication.

Alessandra Celletti and Sylvio Ferraz-Mello

Celestial Mechanics and Dynamical Astronomy (2006) 95:3–25
DOI 10.1007/s10569-006-9025-x

REVIEW ARTICLE

On the variational approach to the periodic *n*-body problem

S. Terracini

Received: 17 November 2005 /Revised: 6 March 2006 /
Accepted: 12 April 2006 / Published online: 7 July 2006

Abstract This expository paper gathers some of the results obtained by the author in recent works in collaboration with Davide Ferrario and Vivina Barutello, focusing on the periodic n-body problem from the perspective of the calculus of variations and minimax theory. These researches were aimed at developing a systematic variational approach to the equivariant periodic n-body problem in the two and three-dimensional space. The purpose of this paper is to expose the main problems and achievements of this approach. The material here was exposed in the talk that given at the Meeting CELMEC IV promoted by SIMCA (Società italiana di Meccanica Celeste).

Keywords Symmetric periodic orbits · 3-body problem · Collisions · Minimizers of the Lagrangian action

1 Introduction

Among all periodic solutions of the planar 3-body problem, the relative equilibrium motions—the equilateral Lagrange and the collinear Euler-Moulton solutions—are definitely the simplest and most known. They both show an evident symmetry ($SO(2)$ and $O(2)$ respectively), that is, they are equivariant with respect to the symmetry group of dimension 1 acting as $SO(2)$ (resp. $O(2)$) on the time circle and on the plane, and trivially on the set of indexes $\{1, 2, 3\}$. Furthermore, they are minimizers of the Lagrangian action functional in the space of all loops having their same symmetry group. Hence, for given a symmetry group G, G-equivariant minimizers for the action functional can be thought as the natural generalization of relative equilibrium motions.

S. Terracini (✉)
Dipartimento di Matematica e Applicazioni,
Università di Milano Bicocca, Via Cozzi 53,
20125 Milano, Italy
e-mail: susanna.terracini@unimib.it

This perspective has known a wide popularity in the recent literature and has produced a new boost in the study of periodic trajectories to the n-body problem; the recent discovery of the Chenciner and Montgomery eight-shaped orbit is emblematic of this renewed interest (see, for instance Chenciner and Montgomery 2000; Chenciner 2002; Ferrario and Terracini 2004 and the major part of the bibliographical references here). Indeed, by Palais principle of symmetric criticality, periodic and quasi-periodic solutions of the n-body problem can be found as critical points of the Lagrangian action functional restricted to suitable spaces of symmetric paths.

Let us consider n point particles with masses $m_1, m_2, \ldots, m_n$ and positions $x_1, x_2, \ldots, x_n \in \mathbb{R}^d$, in dimension $d \geq 2$. We denote by $\mathcal{X}$ the space of configurations with center of mass in 0, and by $\hat{\mathcal{X}} = \mathcal{X} \setminus \Delta$ the set of collision-free configurations (collision means $x_i = x_j$ for some $i \neq j$). On the configuration space we define the homogeneous (Newton) potential of degree $-\alpha < 0$:

$$U(x) = \sum_{i<j} U_{i,j}(|x_i - x_j|) \qquad U_{i,j}(|x_i - x_j|) = \frac{m_i m_j}{|x_i - x_j|^\alpha}.$$

In many cases we shall simply require the $U_{i,j}$'s be asymptotically homogeneus only at the singularity. It is worthwhile noticing that the major part of our analysis can be extended to logarithmic potentials.

The potential U has an actractive singularity on collisions. We are interested in (relative) periodic solutions to the system of differential equations:

$$m_i \ddot{x}_i = \frac{\partial U}{\partial x_i}.$$

We associate with the equation the Lagrangian integrand

$$L(x, \dot{x}) = L = K + U = \sum_i \frac{1}{2} m_i |\dot{x}_i|^2 + \sum_{i<j} U_{i,j}(|x_i - x_j|)$$

and the action functional:

$$\mathcal{A}(x) = \int_0^T L(x(t), \dot{x}(t))dt.$$

Sometimes it will be preferable to consider the problem in a frame rotating uniformly about the vertical axis, with an angular speed ω; the corresponding action $\mathcal{A}^\omega$ then contains a gyroscopic term, associated with Coriolis force. *Relative equilibria* are stationary solutions in this rotating system; they correspond to those configurations (termed *central*) which are critical for the restriction of the potential U to the ellipsoid $I = \lambda$ where I denotes the momentum of inertia:

$$I(x) = \sum_i m_i |x_i|^2.$$

We shall seek periodic solutions as critical points of the action functional on the Sobolev space of T-periodic trajectories: $\Lambda = H^1(\mathbb{T}, \mathcal{X})$ or, to be more precise, of the action constrained on suitable linear subspaces $\Lambda_0 \subset \Lambda$.

Two are the major difficulties in following the variational approach; the first is due to the *lack of coercivity* (or of Palais–Smale) due to the vanishing at infinity of the force fields: indeed sequences of almost–critical points (such as minimizing sequences) may very well diverge. Furthermore, as the potential U is singular on collisions, miminizers or other critical points can a priori be *collision trajectories*. Many strategies were proposed in the literature in order to face these problems. The development of a suitable Critical Point theory taking into account of the contribution of fake periodic solutions (the critical points at infinity) was proposed by some authors (Bahri and Rabinowitz 1991; Majer and Terracini 1993, 1995a,b; Riahi 1999) and returned a good estimate of the number of periodic trajectories satisfying an appropriate bound on the length, with the main disadvantage of requiring a stronger order of infinity at the collisions (the *strong force condition*). Another suitable strategy in order to recover coercivity of action functional consists in imposing a symmetry constraint on the loop space. Surprisingly enough, once coercivity is recovered, also the problem of collisions looses part of its dramatic character. This fact was remarked for the first time in Degiovanni et al. (1987) and Degiovanni and Giannoni, (1988) for two–body problems, in Serra and Terracini (1994) for three–body problems and, since then, widly exploited in the literature. This is indeed a general fact: in many cases, *minimizing trajectories are free of collisions*; a recent breakthrough in this direction is due of the neat idea, due to C. Marchal, of averaging over all possible variations (generalized and exposed in section 3; Marchal's idea was first exposed in Chenciner 2002). This argument can be used in most of the known cases to prove the absence of collisions for minimizing trajecories.

2 Symmetry groups and equivariant orbits

Let us start by introducing some basics concepts and definitions from Ferrario and Terracini (2004). Let G be a finite group endowed with:

- an orthogonal representation of dimension 2, say, $\tau: G \to O(2)$ (on cyclic time $\mathbb{T} \cong S^1$);
- an orthogonal representation (on the euclidean space $\mathbb{R}^d$), say, $\rho: G \to O(d)$;
- and an homomorphism on the symmetric group on n elements $\mathbf{n} = \{1, 2, \ldots, n\}$, say, $\sigma: G \to \Sigma_n$.

Then G acts on time (translation and reversal) $\mathbb{T}$ via τ and it acts on the configuration space $\mathcal{X}$ via ρ and σ in the following way

$$\forall i = 1 \ldots n : (gx)_i = \rho(g) x_{\sigma(g)^{-1}(i)}.$$

As a consequence we have an action on the space of trajectories:

Definition 2.1 A continuous function $x(t) = (x_1(t), \ldots, x_n(t))$ is *G–equivariant* if

$$\forall g \in G : x(gt) = (gx)(t).$$

The linear subspace $\Lambda_0 = \Lambda^G \subset \Lambda$ denotes the set of periodic trajectories in Λ which are equivariant with respect to the G-action:

The Palais Principle of symmetric criticality (Palais 1979) guarantees that the critical points of an invariant functional restricted to the space of equivariant trajectories are free critical points. We stress that by the special form of the interaction potentials, in our setting invariance is simply implied by equality of those masses which are interchanged by the action of G on the set of the indices. We will always implicitly make this assumption.

2.1 Cyclic and dihedral actions

Consider the normal subgroup $\ker\tau \lhd G$ and the quotient $\bar{G} = G/\ker\tau$. Since $\bar{G}$ acts effectively on $\mathbb{T}$, it is either a *cyclic* group or a *dihedral* group.

- If the group $\bar{G}$ acts trivially on the orientation of $\mathbb{T}$, then $\bar{G}$ is cyclic and we say that the action of G on Λ is of cyclic type

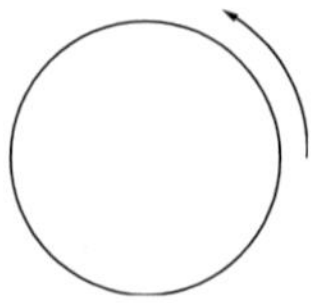

- If the group $\bar{G}$ consists of a single reflection on $\mathbb{T}$, then we say that action of G on Λ is of brake type.

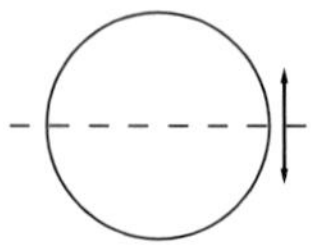

- Otherwise, we say that the action of G on Λ is of dihedral type.

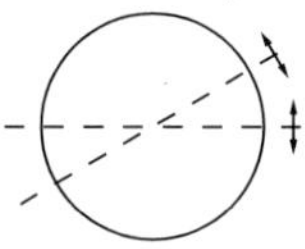

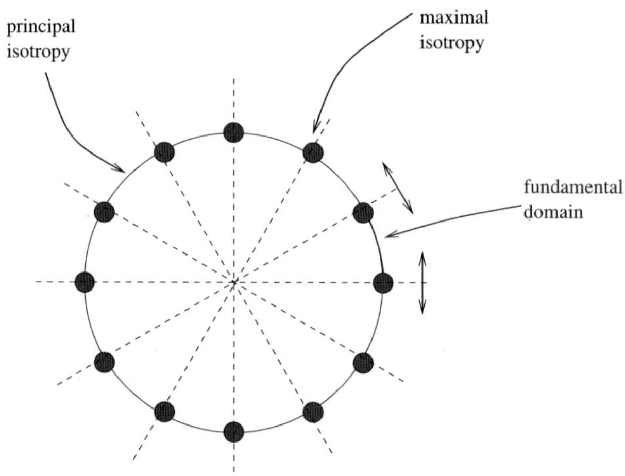

Proposition 2.2 *Let* $\mathbb{I}$ *be the* fundamental domain *(for a dihedral type), or any interval having as length the minimal angle of time–rotation in* $\bar{G}$ *(for a cyclic type). Then the G-equivariant minimization problem is equivalent to problem of minimizing the action over all paths* $x\colon \mathbb{I} \to \mathcal{X}^{\ker\tau}$ *subject to the boundary conditions* $x(0) \in \mathcal{X}^{H_0}$ *and* $x(1) \in \mathcal{X}^{H_1}$*, where* H_0 *and* H_1 *are the maximal isotropy subgroups of the boundary of* $\mathbb{I}$.

2.2 Coercivity

Let us consider the $\mathcal{A}$ restricted to the space of symmetric loops Λ^G.

Proposition 2.3 *The action functional $\mathcal{A}$ is coercive in Λ^G if and only if $\mathcal{X}^G = 0$. Consequently, if $\mathcal{X}^G = 0$ then a minimizer of $\mathcal{A}^G$ in Λ^G exists.*

Given ρ and σ, we can compute $\dim \mathcal{X}^G$ as:

$$\dim \mathcal{X}^G = \frac{1}{|G|} \sum_{g \in G} \mathrm{Tr}(\rho(g)) \# \mathrm{Fix}(\sigma(g)) - d.$$

In the frame rotating with constant angular speed ω the action $\mathcal{A}^\omega$ is generally coercive, except for a (possible) discrete set values of ω.

2.3 The symmetries of the Chenciner and Montgomery eight-shaped trajectory

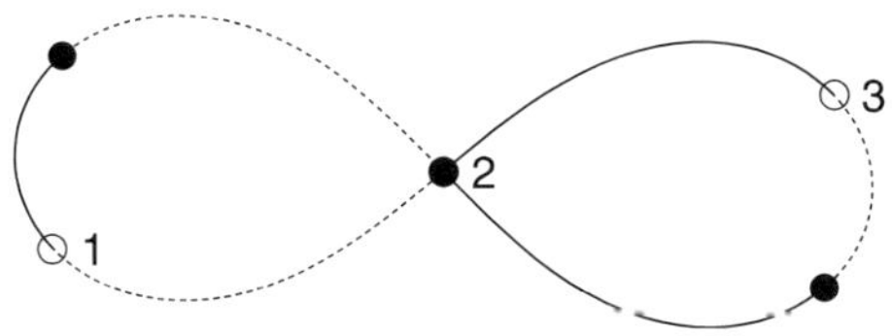

In the celebrated paper (Chenciner and Montgomery 2000), a variational argument was performed to show the existence of a new periodic trajectory for the three body problem, where the three particles move one a single eight-shaped curve, interchanging their positions after a fixed time. One of the the simplest symmetry giving rise to the Chenciner and Montgomery solution is described below. Denote

$$x(t) = (x_1(t), x_2(t), x_3(t)) \in \mathbb{R}^6$$

and impose that

$$\begin{array}{lll} x_1(-t) = -x_1(t), & x_2(-t) = -x_3(t), & x_3(-t) = -x_2(t) \\ x_1(1-t) = -x_2(t), & x_2(1-t) = -x_1(t), & x_3(1-t) = -x_3(t) \end{array}$$

Here $G = D_6$ be the dihedral group generated by two following reflections: $g_1 = (\tau_1, \rho_1, \sigma_1)$ is $\tau_1(t) = -t$, $\rho_1(x) = -x$ and $\sigma_1(1,2,3) = (1,3,2)$, and $g_2 = (\tau_2, \rho_2, \sigma_2)$ is $\tau_2(t) = 1 - t$, $\rho_2(x) = -x$ and $\sigma_2(1,2,3) = (2,1,3)$.

Chenciner and Montgomery minimized the action on the space of symmetric trajectories and they showed, by a level estimate that minimals are free of collisions. Actually, for technical reasons, they used the supergroup D_{12} of D_6. There are indeed three possible groups yielding an eight–shaped trajectory. First, we consider the group of cyclic action type C_6 (the *cyclic eight* having order 6, which acts cyclically on $\mathbb{T}$ (i.e. by a rotation of angle $\pi/3$), by a reflection in the plane E, and by the cyclic permutation $(1, 2, 3)$ in the index set.

The second group, which we denote by D_{12}, is the group of order 12 obtained by extending C_6 with the element h defined as follows: $\tau(h)$ is a reflection in $\mathbb{T}$, $\rho(h)$ is the antipodal map in E (thus, the rotation of angle π), and $\sigma(h)$ is the permutation $(1, 2)$. This is the symmetry group used by Chenciner and Montgomery (2000).

The third group is the dihedral group of order $6D_6$ we described in this section. This can be seen as the subgroup of D_{12} generated by h and the subgroup C_3 of order 3 of $C_6 \subset D_{12}$.

The symmetry groups D_{12} and D_6 are of dihedral type. The choreography group C_3 is a subgroup of all the three groups, thus the action is coercive on G-equivariant loops.

3 Generalized orbits and singularities

The existence of a G-equivariant minimizer of the action is a simple consequence of the direct method of the Calculus of Variations, whenever the action is coercive on the space of equivariant loops. These trajectories, however, solve the associated differential equations only where they are free of collisions. Generally speaking, G-equivariant minimizers may present collision, even though the collision set must clearly be of vanishing Lebesgue measure. A first natural question concerns the number of possible collision instants. Here below we follow Barutello et al. (2005).

Definition 3.1 A path $x\colon(a,b) \to \mathcal{X}$ is called a generalized solution of the n–body problem if, for every $t_0 \in (a,b)$, there exists $\delta > 0$ such that the restriction of x to $[t_0-\delta, t_0+\delta]$ is a local minimizer for the action with respect to G–equivariant compactly supported variations.

We remark that:

- Equivariant minimizers are generalized solutions.
- If the potential is of class $\mathcal{C}^2$ outside collisions then every non collision solution is a generalized solution.
- The generalized solutions possess an index: the minimal number of intervals I_j needed to cover (a,b) such that the restriction of x to I_j is a local minimizer for the action.
- There is a natural notion of maximal existence interval for generalized solutions, even without the unique extension property.

Let us consider the following assumptions on the interaction potentials:

(A1)

$$\left\| \frac{\partial U_{i,j}}{\partial t} \right\| \le C U_{i,j}(t,x);$$

(A2) there exist $\alpha \in (0,2), \gamma, C > 0$ such that

$$\nabla U_{i,j}(t,x)\cdot x + \alpha U_{i,j}(t,x) \ge -C|x|^{\gamma} U_{i,j}(t,x);$$

(A3) there exists a continuous positive function on the sphere $\tilde{U}_{i,j}$ such that

$$\lim_{r\to 0} r^{\alpha} U_{i,j}(t,rs) = \tilde{U}_{i,j}(t,s);$$

uniformly in t and $s \in S^{d-1}$.

The following theorem is proved in Barutello et al. (2005) (see also Barutello (2004) Ferrario and Terracini (2004), and is based on a suitable variant of Sundman's inequality and the asymptotic analysis of possible collisions outlined in the sequel.

Theorem 3.2 *Let $x\colon(a,b) \to \mathcal{X}$ be a generalized solution of the n-body problem. Then collision instants are isolated in (a,b). Furthermore, if (a,b) is the finite maximal extension interval of x and no escape in finite time occurs then the number of collision instants is finite.*

The theorem extends to the logarithmic potentials, under the following hypotheses:

$$\nabla U_{i,j}(t,x)\cdot x \geq -m_{i,j}(t) - C|x|^{\gamma}U_{i,j}(t,x)$$

with $m_{i,j}(t) > 0$ and continuous, $C > 0$, and

$$\lim_{r\to 0} U_{i,j}(t,rs) + m_{i,j}(t)\log(r) = \tilde{U}_{i,j}(t,s)$$

uniformly in t and $s \in S^{d-1}$, and $\tilde{U}_{i,j}$ is a continuous positive function on the sphere.

3.1 Generalized Sundman–Sperling estimates revisited

Theorem 3.2 can be interpreted as a regularity result. The proof relies on the extension, given in Barutello et al. (2005), to the framework of generalized solutions of the classical asymptotic estimates on collisions by Sundman and Sperling (1970) (see also Ferrario and Terracini (2004)).

To start with, let assume $t = 0$ be a collision instant (possibly involving more than one colliding cluster). We denote by $\mathbf{k} \subset \mathbf{n}$ the colliding cluster, by ($x_0 = \sum_{i\in\mathbf{k}} m_i x_i/m_0$, $m_0 = \sum_{i\in\mathbf{k}} m_i$):

$$I_{\mathbf{k}} = \sum_{i\in\mathbf{k}} m_i(x_i - x_0)^2$$

the momentum of inertia with respect to the center of mass (all the bodies in $\mathbf{k}$ collide in x_0 if and only if $I_{\mathbf{k}} = 0$). Moreover, we introduce a system of polar coordinates

$$r = \sqrt{I_{\mathbf{k}}} \qquad s = x/r.$$

Finally we denote the *partial kinetic energy* $K_{\mathbf{k}}$:

$$K_{\mathbf{k}} = \sum_{i\in\mathbf{k}} \frac{m_i}{2}|\dot{x}_i|^2,$$

the *partial potential function*

$$U_{\mathbf{k}} = \sum_{i,j\in\mathbf{k}, i<j} U_{i,j}(t,x),$$

together with the *partial energy*

$$E_{\mathbf{k}} = K_{\mathbf{k}} - U_{\mathbf{k}}.$$

and the *partial Lagrangian*: $L_{\mathbf{k}} = K_{\mathbf{k}} + U_{\mathbf{k}}$.

Theorem 3.3

- *For every maximal (in the number of bodies) colliding cluster* $\mathbf{k} \subset \mathbf{n}$ *the partial energy* $E_{\mathbf{k}}$ *is bounded.*
- *There is* $\kappa > 0$ *such that the following asymptotic estimates hold*:

$$r \sim (\kappa t)^{\frac{2}{2+\alpha}} \qquad \left(r \sim \kappa|t|\sqrt{-\log(|t|)}\right)$$

$$K_{\mathbf{k}} \sim U_{\mathbf{k}} \sim \frac{1}{4-2\alpha}\ddot{I}_{\mathbf{k}} \sim \frac{2}{(2+\alpha)^2}\kappa^2(\kappa t)^{\frac{-2\alpha}{2+\alpha}} \qquad (\sim -\log|t|)\,.$$

– *Let s be the* normalized configuration *of a colliding cluster* $s = (x - x_0)/r$. *Then*

$$\lim_{t \to 0} r^{2+\alpha} |\dot{s}|^2 = 0$$
$$\lim_{t \to 0} U(s(t)) = b < +\infty$$

– *Moreover there exists the angular blow–up, that is angular scaled family* $(s(\lambda t))_\lambda$ *is precompact for the topology of uniform convergence on compact sets of* $\mathbb{R} \backslash \{0\}$.

As a consequence we have the vanishing of the total angular momentum and the absence of partial collisions in a neighbourhood of the maximal collision.

3.2 Dissipation (Mc Gehee revisited)

While binary collisions admit the Levi–Civita regularization, simultaneous collisions involving three bodies or more can not be fully regularized, for the possible occurrence of accumulations of partial collisions. In order to study the motion close to a collision, (McGehee 1974) attached to the phase space a manifold (named the collision manifold) and performed a suitable change of coordinates, showing how some of the motions could be extended also through the singularity. The system written in McGehee coordinates in no longer conservative: on the contrary, it possesses a Lyapounov function.

It is interesting to give a new look at this dissipation phenomenon, from the lagrangian point of view. Here we follow (Barutello 2004); Barutello et al. 2005). To fix our minds, let us think to a homogeneous potential U_α and perform the following change of variables

$$\rho = r^{\frac{2-\alpha}{4}}, \qquad \rho' = \frac{2-\alpha}{4} r^{-\frac{2+\alpha}{4}} r'.$$

In this way, we obtain a new action functional depending on (ρ, s)

$$\mathcal{A}(\rho, s) = \int_0^{\tau^*} \frac{1}{2} \left(\frac{4}{2-\alpha} \right)^2 (\rho')^2 + \rho^2 \left(\frac{1}{2} |s'|^2 + U(s) \right),$$

where $\beta := 2(2+\alpha)/(2-\alpha) > 2$. Here, similarly to Mc Gehee, we have reparametrized the time as

$$dt = r^{\frac{2+\alpha}{2}} d\tau,$$

and hence we have an extra constraint:

$$\int_0^{\tau^*} \rho^\beta d\tau = T,$$

while τ^* is a free parameter. On the other hand, one can prove that at a collision there holds

$$\tau^* = \int_0^T r^{-\frac{2+\alpha}{2}} dt = +\infty.$$

In these coordinates the action integral corresponds to the action associated with the homoclinic problem of a Duffing equation, coupled with the n-body angular system. In particular, since the function ρ acts as viscosity time–varying (decreasing) parameter for the angular lagrangian, we find that the energy of the angular system increases when ρ decreases (and vice-versa). This monotonicity property is indeed equivalent to the Sundman inequality.

3.3 Blow-ups and parabolic collision trajectories

For simplicity, from now on we shall consider only the case of α–homogeneus potentials: the logarithmic case requires some different arguments and can be found in Barutello (2004) and Barutello et al. (2005). Having understood the behaviour of the radial variable close to a collision, we turn to the asymptotic analysis of the angle $s(t)$ as t goes to the collision instant. The convergence of the angular variable is not at all obvious and is the object of a classical problem in Celestial Mechanics (the infinite-spin problem). In general, there is no hope of proving the full convergence of the variable s. Even more, it has to be noticed that our assumptions on the interaction potentials are so weak that there does not even exists a limiting problem for the angular variable. However, thanks to Theorem 3.3, there is enough compactness for completing the asymptotic analysis through a blow–up argument. For every $\lambda > 0$ let

$$x^{\lambda}(t) = \lambda^{-2/(2+\alpha)} x(\lambda t)$$

If $\{\lambda_n\}_n$ is a sequence of positive real numbers such that $s(\lambda_n)$ converges to a normalized configuration $\bar{s}$, then $\forall t \in (0,1) : \lim_{n\to\infty} s(\lambda_n t) = \lim_{n\to\infty} s(\lambda_n) = \bar{s}$. Hence the rescaled sequence will converge uniformly to the *blow-up* of $x(t)$ relative to the colliding cluster $\mathbf{k} \subset \mathbf{n}$ (in $t = 0$). Moreover, the blow-up $\bar{x}$ is parabolic: here a *parabolic collision trajectory* for the cluster $\mathbf{k}$ is the path

$$\bar{x}_i(t) = |t|^{2/(2+\alpha)} \xi_i, \qquad i \in \mathbf{k},\ t \in \mathbb{R}.$$

Proposition 3.4 *The sequences x^{λ_n} and dx^{λ_n}/dt converge to the blow-up $\bar{x}$ and its derivative $\dot{\bar{x}}$ respectively, in the H^1-topology. Moreover $\bar{x}$ is a minimizing trajectory in the sense of Definition* 3.1.

$$\int_0^T \left[\mathcal{L}(\bar{x}+\varphi) - \mathcal{L}(\bar{q}) \geq 0\right] dt.$$

for any compactly supported variation φ, G_0-equivariant, where G_0 is the isotropy group at $t = 0$.

Let us assume, in addition to our assumptions, that the total force is asymptotically homogeneous:

$$\lim_{r\to 0} r^{\alpha+1} \nabla U(t, rs) = \nabla \tilde{U}(t, s),$$

uniformly in t and in s on compact subset of $S^{nd-1} \setminus \Delta$.

In this case, the parabolic motions has a configuration $\xi = (\xi_i)_{i\in\mathbf{k}}$ which is *central* for the limiting force field, namely it annihilates the tangential part of $\nabla U(0, \bar{s})$.

Hence we are left with the problem of collision parabolic solutions in the generalized sense we gave in Definition 3.1. Next step consist in proving by a suitable variation that parabolic solutions can not be local minimizer.

4 Avoiding collisions

4.1 The standard variation and the displacement differential potential

Let G_0 be the isotropy group at the collision time, then the blow–up procedure implies the existence of q, a G_0-equivariant minimizing parabolic collision trajectory. Following Ferrario and Terracini (2004), we first define a class of suitable variations as follows:

Definition 4.1 The *standard variation* associated to δ and T is defined as

$$v^{\delta}(t) = \begin{cases} \delta & \text{if } \ 0 \le |t| \le T - |\delta| \\ (T-t)\frac{\delta}{|\delta|} & \text{if } \ T - |\delta| \le |t| \le T \\ 0 & \text{if } \ |t| \ge T, \end{cases}$$

By chosing δ to be fixed by G_0, we can always manage to have v^{δ} G_0-equivariant. Our next goal is to find a suitable δ such that its associated standard variation v^{δ} decreases the action, or, in other words, such that

$$\Delta\mathcal{A} := \int_{-\infty}^{+\infty} [\mathcal{L}_{\mathbf{k}}(q+v^{\delta}) - \mathcal{L}_{\mathbf{k}}(q)]dt < 0.$$

Definition 4.2 Let us define the *displacement potential differential* associated to the standard variation v^{δ}:

$$S(\xi,\delta) = \int_0^{+\infty} \left(\frac{1}{\left|\xi t^{2/(2+\alpha)} - \delta\right|^{\alpha}} - \frac{1}{\left|\xi t^{2/(2+\alpha)}\right|^{\alpha}} \right) dt$$

where $\xi, \delta \in \mathbb{R}^2$.

It can proven that the function S measures the limiting (as $T \to \infty$) differential of the potential energy. Thus we obtain the fundamental estimate:

Theorem 4.3 *Let $q = \{q\}_i = \{t^{2/(2+\alpha)}\xi_i\}$, $i = 1, \ldots, k$ be a parabolic collision trajectory and v^{δ} any standard variation. Then, as $\delta \to 0$*

$$\Delta\mathcal{A} = 2|\delta|^{1-\alpha/2} \sum_{\substack{i<j \\ i,j\in\mathbf{k}}} m_i m_j S\left(\xi_i - \xi_j, \frac{\delta_i - \delta_j}{|\delta|}\right) + O(|\delta|).$$

We observe that

$$S(\lambda\xi, \mu\delta) = |\lambda|^{-1-\alpha/2}\,|\mu|^{1-\alpha/2} S(\xi,\delta)$$

and hence the sign of S depends on the angle between ξ and δ. Let

$$\Phi(\vartheta) = \int_0^{+\infty} \frac{1}{\left(t^{\frac{4}{\alpha+2}} - 2\cos\vartheta\, t^{\frac{2}{\alpha+2}} + 1\right)^{\alpha/2}} - \frac{1}{t^{\frac{2\alpha}{\alpha+2}}} dt, \quad \alpha \in (0,2)$$

$\Phi(\theta)$ represents the *potential differential* needed for displacing the colliding particle originarily traveling on the x–axis to the point $e^{i\theta}$.

4.2 Averaging properties involving Φ and Marchal's Principle

The value of $\Phi(0)$ ranges from decrease to $+\infty$, in a way also depending on the exponent α. Hence it can take positive and negative values. However, thanks to some harmonic analysis one can prove that suitable *averages* are always negative: the first inequality is particularity useful when dealing with reflected triple collisions from the Lagrange central configuration:

$$\Phi\left(\frac{2\pi}{3}+\gamma\right)+\Phi\left(\frac{2\pi}{3}-\gamma\right)<0, \quad \forall\gamma\in[0,\pi/2].$$

A crucial estimate was proved in Ferrario and Terracini (2002) about the averages of Φ on circles:

Theorem 4.4 *For every $\alpha>0$, $\xi\in\mathbb{R}^3\setminus\{0\}$ and for every circle $\mathbb{S}\subset\mathbb{R}^d$ with center in* 0,

$$\tilde{S}(\xi,\mathbb{S})=\frac{1}{|\mathbb{S}|}\int_{\mathbb{S}}S(\xi,\delta)d\delta=|\xi|^{-1-\alpha/2}\,|\delta|^{1-\alpha/2}\frac{1}{2\pi}\int_0^{2\pi}\Phi(\theta)d\theta<0.$$

Let q be a parabolic collision solution associated to the configuration ξ. Consider $\xi=x_i-x_j$ and δ ranging in a circle. Then the above inequality allows us to state the existence of at least one δ for which the associated standard variation lowers the action.

The first who conjectured that the method of averaged variations could be used to avoid collisions on minimizers was Marchal (2002), who remarked that, being the Newton potential harmonic on $\mathbb{R}^3$, averaging it on a sphere results in a truncation in the interior. In fact, is not so much a matter of harmonicity rather than a subtle balance between the averaged potential and the speed of collisions, as already Marchal observed by considering the case of the planar Keplerian potential generated by a plate, which expression is known since the 19th century. Our estimates on the averages of $S(\xi,\delta)$ show that this is a general fact and Marchal's principle can be extended in the following way (see Ferrario and Terracini (2004) for homogeneous potentials and Barutello (2004); Barutello et al. (2005) for logarithmic potentials):

Theorem 4.5 *(The rotating circle principle) For homogeneus and logarithmic potentials it is always more convenient (from the point of view of the integral of the potential on the time line) to replace one of the colliding particles with a homogeneous circle of same mass and fixed radius which is moving keeping its center in the position of the original particle*

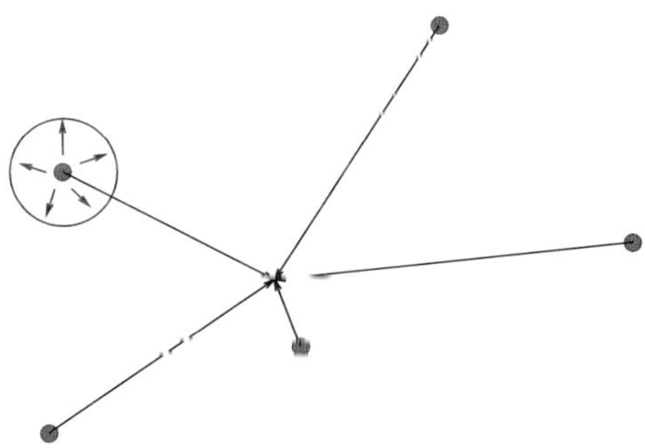

Of course by a circle we mean the intersection of the sphere with any two-dimensional plane.

5 The rotating circle property

As a consequence of the analysis of collisions we find that if the action of G on $\mathbb{T}$ and $\mathcal{X}$ fulfills some conditions (computable) then (local) minimizers of the action functional $\mathcal{A}$ in $\Lambda^G \subset \Lambda$ do not have collisions. This is the rotating circle condition, introduced in Ferrario and Terracini (1977).

For a group H acting orthogonally on $\mathbb{R}^d$, a circle $\mathbb{S} \subset \mathbb{R}^d$ (with center in 0) is termed *rotating under* H if $\mathbb{S}$ is *invariant* under H (that is, for every $g \in Hg\mathbb{S} = \mathbb{S}$) and for every $g \in H$ the restriction $g|\mathbb{S}\colon \mathbb{S} \to \mathbb{S}$ is a *rotation* (the identity is meant as a rotation of angle 0).

Let $i \in \mathbf{n}$ be an index and $H \subset G$ a subgroup. A circle $\mathbb{S} \subset \mathbb{R}^d = V$ (with center in 0) is termed *rotating for i under H* if $\mathbb{S}$ is *rotating under H* and

$$\mathbb{S} \subset V^{H_i} \subset V = \mathbb{R}^d,$$

where $H_i \subset H$ denotes the *isotropy subgroup* of the index i in H relative to the action of H on the index set $\mathbf{n}$ induced by restriction (that is, the isotropy $H_i = \{g \in H \mid gi = i\}$).

Definition 5.1 A group G acts with the rotating circle property if for every $\mathbb{T}$-isotropy subgroup $G_t \subset G$ and for at least $n-1$ indexes $i \in \mathbf{n}$ there exists in $\mathbb{R}^d$ a rotating circle $\mathbb{S}$ under G_t for i.

If the action has the rotating circle property, then for every $g \in G$ the linear map $1 - g$ sends the rotating circle into another circle (thus we can use the rotating circle principle). In most of the known examples the property is fulfilled. In Ferrario and Terracini (2004) we proved the following results.

Theorem 5.2 *Consider a finite group K acting on Λ with the rotating circle property. Then a minimizer of the K-equivariant fixed–ends (Bolza) problem is free of collisions.*

Corollary 5.3 *For every $\alpha > 0$, minimizers of the fixed-ends (Bolza) problem are free of interior collisions.*

Corollary 5.4 *If the action of G on Λ is of cyclic type and* $\ker\tau$ *has the rotating circle property then any local minimizer of $\mathcal{A}^G$ in Λ^G is collisionless.*

Corollary 5.5 *If the action of G on Λ is of cyclic type and* $\ker\tau = 1$ *is trivial then any local minimizer of $\mathcal{A}^G$ in Λ^G is collisionless.*

Theorem 5.6 *Consider a finite group G acting on Λ so that every maximal $\mathbb{T}$-isotropy subgroup of G either has the rotating circle property or acts trivially on the index set* $\mathbf{n}$. *Then any local minimizer of $\mathcal{A}^G$ yields a collision-free periodic solution of the Newton equations for the n-body problem in $\mathbb{R}^d$.*

6 The 3-body problem

The major achievement of Barutello et al. (2004) is to give the complete description of the outcome of the equivariant minimization procedure for the planar three-body problem. First we can ensure that minimizers are always collisionless.

Theorem 6.1 *Let G a symmetry group of the Lagrangian in the* 3*-body problem (in a rotating frame or not). If G is not bound to collision (i.e. every equivariant loop has collisions), then any (possible) local minimizer is collisionless.*

A symmetry group G of the Lagrangian functional $\mathcal{A}$ is termed

- *bound to collisions* if all G-equivariant loops actually have collisions,
- *fully uncoercive* if for every possible rotation vector ω the action functional $\mathcal{A}_\omega^G$ in the frame rotating around ω with angular speed $|\omega|$ is not coercive in the space of G-equivariant loops (that is, its global minimum escapes to infinity);
- *homographic* if all G-equivariant loops are constant up to orthogonal motions and rescaling.
- The *core* of the group G is the subgroup of all the elements which do not move the time $t \in \mathbb{T}$.

If, for every angular velocity, G is a symmetry group for the Lagrangian functional in the rotating frame, then we will say that G is of type R. This is a fundamental property for symmetry groups. In fact, if G is *not* of type R, it turns out that the angular momentum of all G-equivariant trajectories vanishes.

6.1 The classification of planar symmetry groups for 3-body

Theorem 6.2 *Let G be a symmetry group of the Lagrangian action functional in the planar* 3*-body problem. Then, up to a change of rotating frame, G is either bound to collisions, fully uncoercive, homographic, or conjugated to one of the symmetry groups listed in* Table 1 *(RCS stands for Rotating Circle Property and HGM for Homographic Global Minimizer)* (Fig.1).

6.2 Planar symmetry groups

- *The trivial symmetry.* Let G be the trivial subgroup of order 1. It is clear that it is of type R, it has the rotating circle property. It yields a coercive functional on $\Lambda^G = \Lambda$ only when ω is not an integer. If $\omega = \frac{1}{2} \mod 1$ then the minimizers are minimizers for the anti-symmetric symmetry group (also known as *Italian* symmetry) $x(at) = ax(t)$, where a is the antipodal map on $\mathbb{T}$ and E. The masses can be different.

Table 1 Planar symmetry groups with trivial core

Name	$\|G\|$	Type R	Action type	Trans. Dec.	RCP	HGM
Trivial	1	yes		$1+1+1$	yes	yes
Line	2	yes	brake	$1+1+1$	(no)	no
2-1-choreography	2	yes	cyclic	$2+1$	yes	no
Isosceles	2	yes	brake	$2+1$	no	yes
Hill	4	yes	dihedral	$2+1$	no	no
3-choreography	3	yes	cyclic	3	yes	yes
Lagrange	6	yes	dihedral	3	no	yes
C_6	6	no	cyclic	3	yes	no
D_6	6	no	dihedral	3	yes	no
D_{12}	12	no	dihedral	3	no	no

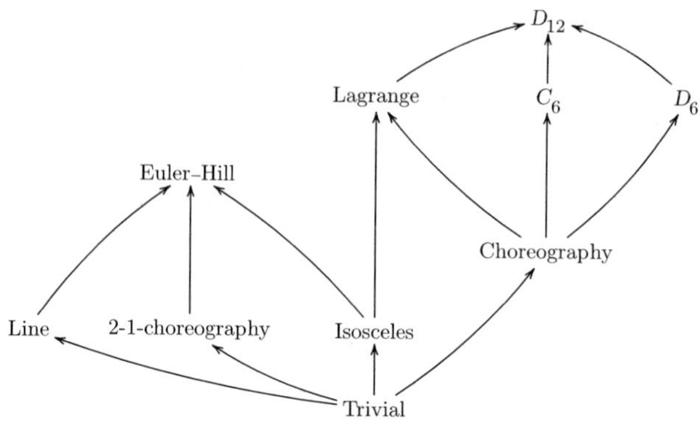

Fig. 1 The poset of symmetry groups for the planar 3-body problem

Proposition 6.3 *For every $\omega \notin \mathbb{Z}$ and every choice of masses the minimum for the trivial symmetry occurs in the relative equilibrium motion associated to the Lagrange central configuration.*

- *The line symmetry.* Another case of symmetry group that can be extended to rotating frames with arbitrary masses is the line symmetry: the group is a group of order 2 acting by a reflection on the time circle $\mathbb{T}$, by a reflection on the plane E, and trivially on the set of indexes. That means, at time 0 and π the masses are collinear, on a fixed line $l \subset E$. It is coercive only when $\omega \notin \mathbf{Z}$. In this case the Lagrangian solution cannot be a minimum, while the relative equilibrium associated with the Euler configuration can be (Fig. 2).
- *The* 2-1-*choreography symmetry.* Consider the group of order 2 acting as follows: $\rho(g) = 1$, $\tau(g) = -1$ (that is, the translation of half-period) and $\sigma(g) = (1,2)$ (that is, $\sigma(g)(1) = 2$, $\sigma(g)(2) = 1$, and $\sigma(g)(3) = 3$. That is, it is a half-period choreography for the bodies 1 and 2. It can be extended to rotating frames and coercive for a suitable choice of $\omega \neq 0, 1 \mod 2$.

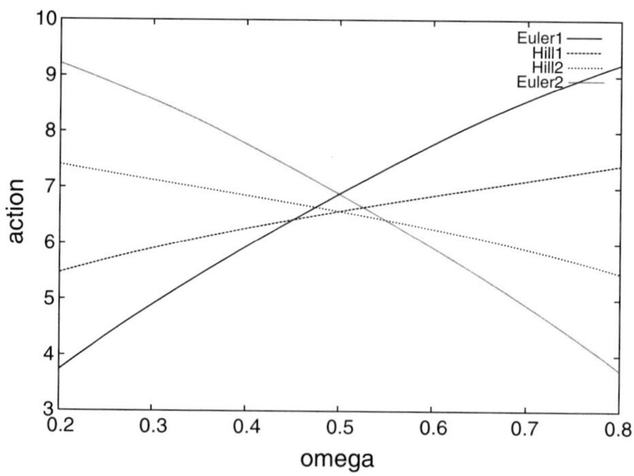

Fig. 2 Action levels for the line symmetry

The Euler's orbit with $k = 1$ and the Hill's orbits with $k = \pm 1$ are equivariant for the 2-1-choreography symmetry, while the Euler's orbit with $k = 0$ is not equivariant for this symmetry. In Fig. 3 Euler 1 represents the action levels on the Euler's orbit with $k = 1$, Hill 1–2 the ones on the Hill's with $k = \pm 1$.

– *The isosceles symmetry.* The isosceles symmetry can be obtained as follows: the group is of order 2, generated by h; $\tau(h)$ is a reflection in the time circle $\mathbb{T}$, $\rho(h)$ is a reflection along a line l in E, and $\sigma(h) = (1,2)$ as above. The constraint is therefore that at time 0 and π the 3-body configuration is an isosceles triangle with one vertex on l (the third).

Proposition 6.4 *For every $\omega \notin \mathbb{Z}$ and every choice of masses the minimum for the isosceles symmetry occurs in the relative equilibrium motion associated to the Lagrange configuration.*

– *The Euler–Hill symmetry.* Now consider the symmetry group with a cyclic generator r of order 2 (i.e. $\tau(r) = -1$) and a time reflection h (i.e. $\tau(h)$ is a reflection of $\mathbb{T}$) given by $\rho(r) = 1$, $\sigma(r) = (1,2)$, $\rho(h)$ is a reflection and σ is the identity in $\{1,2,3\}$. It contains the 2-1-choreography (as the subgroup ker det(τ)), the isosceles symmetry (as the isotropy of $\pi/2 \in \mathbb{T}$) and the line symmetry (as the isotropy of $0 \in \mathbb{T}$) as subgroups.

Proposition 6.5 *The minimum of the Euler–Hill symmetry is not homographic, provided that the angular velocity ω is close to 0.5 and the values of the masses are close to 1.*

– *The choreography symmetry.* The choreography symmetry is given by the group C_3 of order 3 acting trivially on the plane E, by a rotation of order 3 in the time circle $\mathbb{T}$ and by the cyclic permutation $(1,2,3)$ of the indices.

Proposition 6.6 *For every ω the minimal choreography of the 3-body problem is a rotating Lagrange configuration.*

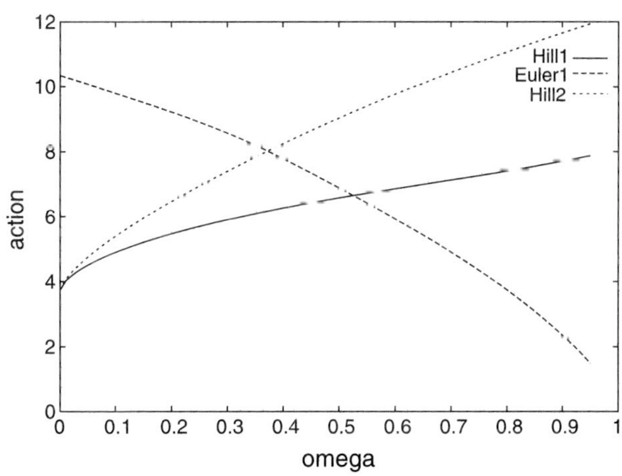

Fig. 3 Action levels for the 2-1-choreography symmetry

- *The Lagrange symmetry.* The Lagrange symmetry group is the extension of the choreography symmetry group by the isosceles symmetry group. Thus, it is a dihedral group of order 6, the action is of type R. Hence, the relative equilibrium motions associated to the Lagrange configuration are admissible motions for this symmetry and, again, the minimizer occurs in the relative equilibrium motion associated to the Lagrange configuration.
- *The Chenciner–Montgomery symmetry group and the eights.* There are three symmetry groups (up to change of coordinates) that yield the Chenciner–Montgomery *figure eight* orbit: they are the only symmetry groups which do not extend to the rotating frame and we have already described them in Sec. 2.3. One can prove that all G-equivariant trajectories have vanishing angular momentum, whenever the group is not of type R. Moreover, we were able to partially answer to the open question (posed by Chenciner) whether their minimizers coincide or not: for two of them (D_6 and D_{12}) the minimizer is necessarily the same.

6.3 Space three-body problem

Based on the classification of planar groups, by introducing a natural notion of space extension of a planar group, Ferrario (2004) gave a complete answer to the classification problem for the three-body problem in the space and at the same time to determine the resulting minimizers and describe its more relevant properties.

Theorem 6.7 *Symmetry groups not bound to collisions, not fully uncoercive and not homographic are, up to a change of rotating frame, either the three-dimensional extensions of planar groups (if trivial core) listed in* Table 2 *or the vertical isosceles triangle (if non-trivial core).*

The next theorem is the answer to the natural questions about collisions and description of some main features of minimizers.

Theorem 6.8 *Let G be a symmetry group not bound to collisions and not fully uncoercive. Then*

(i) *Local minima of $\mathcal{A}^{\omega}$ do not have collisions.*
(ii) *In the following cases minimizers are planar trajectories:*
    (a) *If G is not of type R: $D_6^{+,-}$, $D_6^{-,+}$ and $D_{12}^{-,+}$ (and then G-equivariant minimizers are Chenciner–Montgomery eights).*

Table 2 Space extensions of planar symmetry groups with trivial core

Name	Extensions
Trivial	C_1^-
Line	$L_2^{+,-}, L_2^{-,+}$
Isosceles	$H_2^{+,-}, H_2^{-,+}$
Hill	$H_4^{+,-}, H_4^{-,+}$
3-choreography	C_3^+, C_3^-
Lagrange	$L_6^{+,+}, L_6^{+,-}, L_6^{-,+}$
D_6	$D_6^{+,-}, D_6^{-,+}$
D_{12}	$D_{12}^{-,+}$

(b) *If there is a G-equivariant minimal Lagrange rotating solution:* C_1^-, $H_2^{+,-}$, $C_3^|$, $L_6^{+,+}$, $L_6^{+,-}$ *(and then the Lagrange solution is of course the minimizer).*
(c) *If the core is non-trivial and it is not the vertical isosceles (and then minimizers are homographic).*

(iii) *In the following cases minimizers are always non-planar* (Fig. 4):
(a) *The groups* $L_6^{-,+}$ *and* C_3^- *for all* $\omega \in (-1,1)+6\mathbb{Z}$, $\omega \neq 0$ *(the minimizers for* $L_6^{-,+}$ *are the elements of Marchal family* P_{12}*, and minimizers of* C_3^- *are a less-symmetric family* P'_{12} [1]*).*
(b) *The extensions of line and Hill-Euler type groups, for on open subset of mass distributions and angular speeds* ω*:* $L_2^{+,-}$, $L_2^{-,+}$, $H_4^{+,-}$ *and* $H_4^{-,+}$ *(for* $L_2^{-,+}$ *this happens also with equal masses).*
(c) *The vertical isosceles for suitable choices of masses and* ω*.*

7 Is the choreographical minimizer a homografic motion?

The results of this sections are contained in Barutello and Terracini (2004) and Arioli et al. (2005).

Definition 7.1 A simple choreography is a trajectory where the bodies lie on the same curve and exchange their mutual positions after a fixed time, namely,

$$q_i(t) = x(t + (i-1)\tau), \quad i = 1,\ldots,n,\ t \in \mathbb{R},$$

where $\tau = 2\pi/n$.

In the space of symmetric (choregraphical) loops, the action takes the form

$$\mathcal{A}(x) = \frac{1}{2}\sum_{h=0}^{n-1}\int_0^\tau |\dot{x}(t+h\tau)|^2 dt + \frac{1}{2}\sum_{\substack{h,l=0\\ h\neq l}}^{n-1}\int_0^\tau \frac{dt}{|x(t+l\tau)-x(t+h\tau)|^\alpha}$$

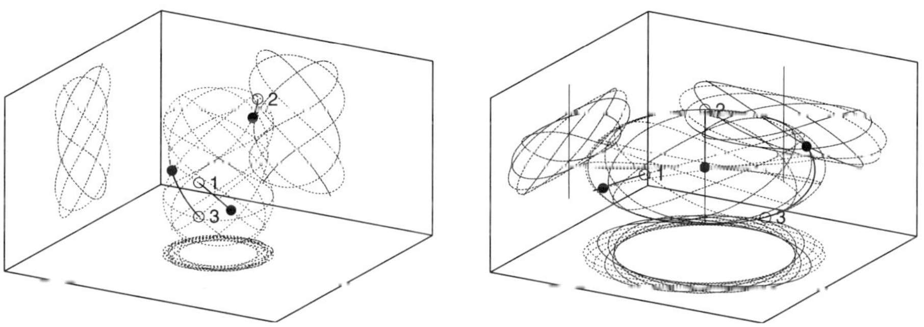

Fig. 4 Non planar minimizers for the groups $L_6^{-,+}$ and $L_2^{+,-}$, plotted in the fixed frame

[1] Highly likely they are not distinct families: this is the recurring phenomenon of "more symmetries than expected" in n-body problems.

Unfortunately, the bare minimization among choreographical loops returns only trivial motions:

Theorem 7.2 *For every $\alpha \in \mathbb{R}_+$ and $d \geq 2$, the absolute minimum of $\mathcal{A}$ on Λ is attained on a relative equilibrium motion associated to the regular n-gon.*

This theorem extends some related result for the italian symmetry by Chenciner and Desolneux (1998), and the results in section 6. The proof is based on a (quite involved) convexity argument together with the analysis of some spectral properties related to the choreographical constraint. Now, in order to find nontrivial minimizers, we look at the same problem in a rotating frame. In order to take into account of the Coriolis force, the new action functional has to contain a gyroscopic term:

$$\mathcal{A}(y) = \frac{1}{2}\int_0^{2\pi} |\dot{y}(t) + J\omega y(t)|^2 dt + \frac{1}{2}\sum_{h=1}^{n-1}\int_0^{2\pi} \frac{dt}{|y(t) - y(t+h\tau)|^\alpha}.$$

Consider the function $h\colon \mathbb{R}_+^* \to \mathbb{N}$, $h(\omega) = \min_{n\in\mathbb{N}^*} \frac{(\omega-n)^2}{n^2}$ and let $\omega^* = \frac{4}{3}$. The same technique used for the inertial system extends to rotating systems having small angular velocity; this gives the following result.

Theorem 7.3 *If $\omega \in (0, \omega^*)\setminus\{1\}$, then the action attains its minimum on a circle with minimal period 2π and radius depending on n, α and ω.*

7.1 When ω is close to an integer

The situation changes dramatically when ω is close to some integer. To understand this phenomenum, let us first check the result of the minimization procedure when ω is an integer:

Proposition 7.4

(1) *If $\omega = n$, then the action has a continuum of minimizers.*
(2) *If $\omega = k$, coprime with n, then the action does not achieve its infimum (escape of minimizing sequences).*
(3) *If $\omega = k$ and k divides n, then the action does not achieve its infimum (clustering of minimizing sequences).*

As a consequence, we have the following result:

Theorem 7.5 *Suppose that n and k are coprime. Then there exist $\epsilon = \epsilon(\alpha, n, k)$ such that if $\omega \in (k-\epsilon, k+\epsilon)$ the minimum of the action is attained on a circle with minimal period $2\pi/k$ that lies in the rotating plane with radius depending on n, α and ω.*

An interesting situation appears when the integer closest to the angular velocity is not coprime with the number of bodies. In this case we prove that the minimal orbit it is not circle anymore, as the following theorem states.

Theorem 7.6 *Take $k \in \mathbb{N}$ and g.c.d.$(k,n) = \tilde{k} > 1$, $\tilde{k} \neq n$. Then there exists $\epsilon = \epsilon(\alpha, n, k) > 0$ such that if $\omega \in (k-\epsilon, k+\epsilon)\setminus\{k\}$ the minimum of the action is attained on a planar 2π-periodic orbit with winding number k which is not a relative equilibrium motion* (Fig. 5).

Also, it has be noticed that, for large number of bodies and angular velocities close to the half on an integer, the minimizer apparently is not anymore planar (Fig. 6).

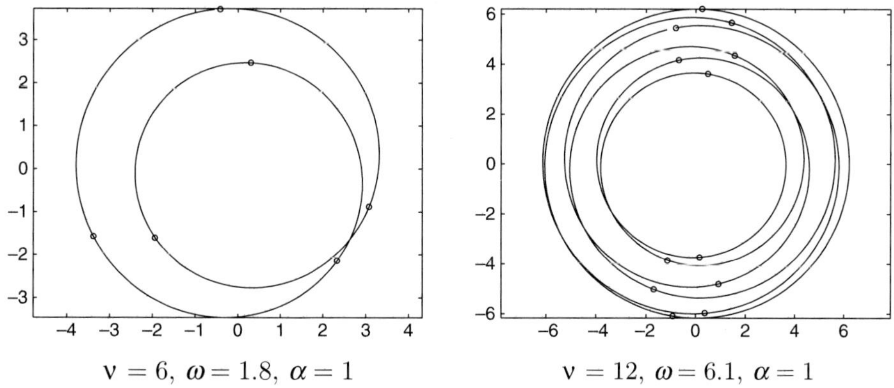

Fig. 5 Examples for Theorem 7.6 close to an integer that divides n.

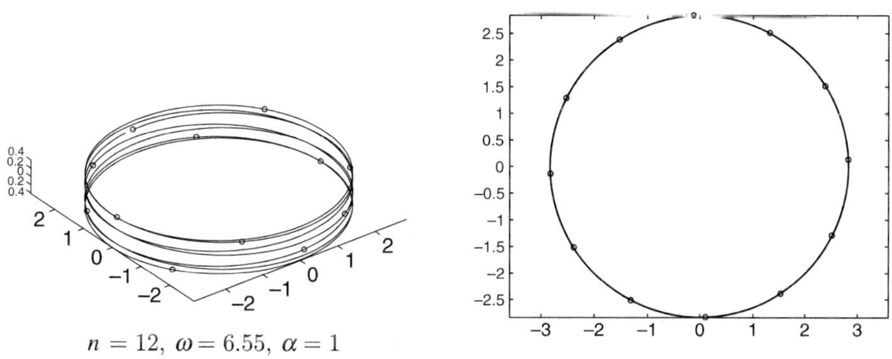

Fig. 6 Non planar minimizers of the action with angular velocities close to the half on an integer

7.2 Mountain pass solutions for the choreographical 3-body problem

The discussion carried in the previous section shows that, as the angular velocity varies, the minimizer's shape must undergoes some transitions (for example it has to pass from relative equilibrium having different winding numbers). This scenario suggests the presence of other critical points, such as local minimizers or mountain pass. This was indeed discovered numerically in Barutello and Terracini (2004) and then proved by a computer assisted proof in Arioli et al. (2005).

To begin with, let us look at Fig. 7, where the values of the action functional $\mathcal{A}^{\omega}$ on the branches of circular orbits L_k^{ω} are plotted:

The analysis of this picture suggests the presence of critical points different from the Lagrange motions. Indeed, let us take the angular velocity $\omega = 1.5$: in this case there are two distinct global minimizers, the uniform circular motions with minimal period 2π and π, lying in the plane orthogonal to the rotation direction. This is a well known structure in Critical Point Theory, referred as the Mountain Pass geometry and gives the existence of a third critical point, provided the Palais-Smale condition

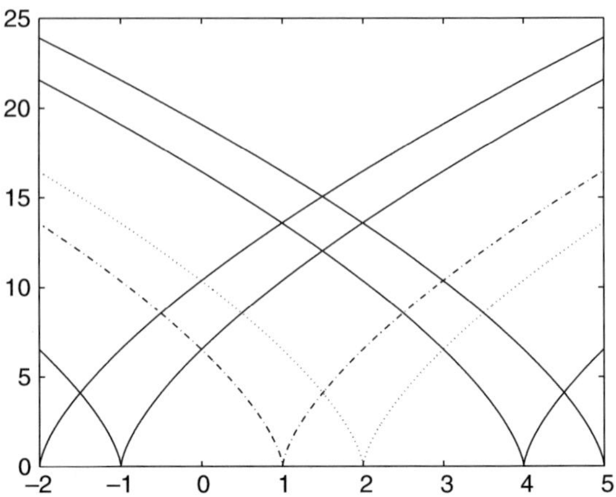

Fig. 7 Action levels for the Lagrange solutions

is fulfilled, with an additional information on the Morse index. Next theorem follows from the application of the Mountain Pass Theorem to the action functional $\mathcal{A}^{3/2}$:

Theorem 7.7 *There exists a (possibly collision) critical point for the action functional* $\mathcal{A}^{3/2}$ *with Morse index smaller then* 1 *and distinct from any Lagrange motion.*

Once the existence of a Mountain Pass critical point was theoretically established, we studied its main properties in order to understand whether it belonged to some known families of periodic trajectories. To this aim we applied the bisection algorithm proposed in Barutello and Terracini (2004) to approximate the maximal of a locally optimal path joining the two strict global minimizers, finding in this a good numerical candidate. Of course, there could be a gap between the mountain pass solution whose existence is ensured by Theorem 7.7 and the numerical candidate found by applying the bisection algorithm. In order to fill this gap we proved the existence of an actual solution very close to the numerical output of the Mountain Pass algorithm (Fig. 8). The argument was based upon a fixed point principle and involved a rigorous computer assisted proof. As a consequence, we obtained the existence of a new branch of solution for the spatial 3-body problem (see Fig. 9).

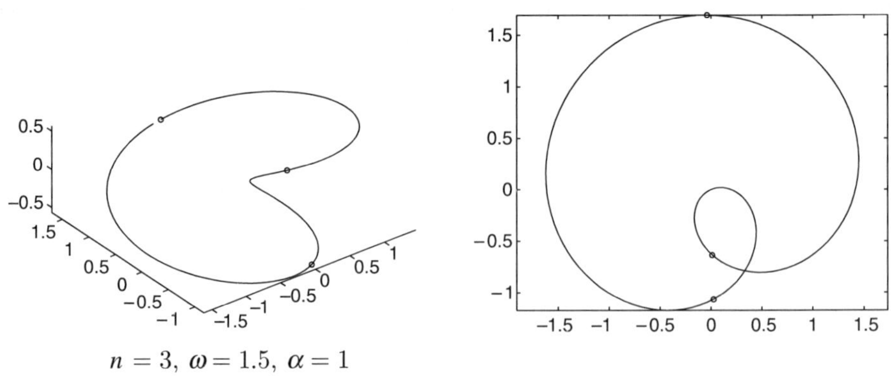

Fig. 8 Mountain pass solution with angular velocities close to the half on an integer

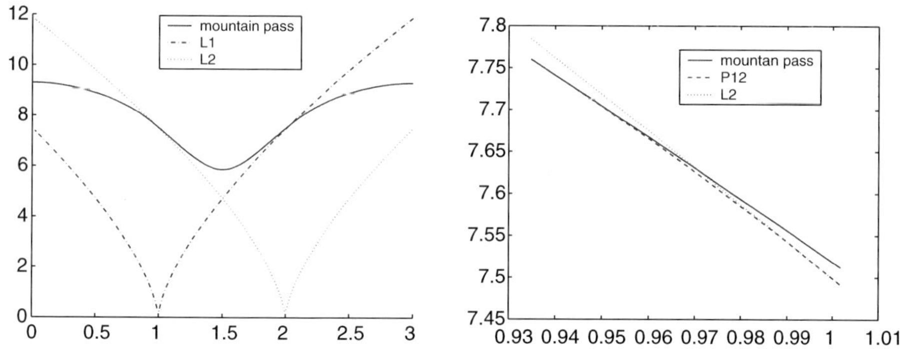

Fig. 9 Action levels for the Lagrange and the mountain pass solution in the 3-body problem. On the x-axes the angular velocity varies in the interval $[0,3)$. The left picture focuses on the biburcation from the P_{12} family

Here are some relevant features of the new solution: the orbit is not planar, its winding number with respect, for instance, to the line $x = -0.2$, $y = 0$ is 2 and it does not intersect itself. A natural question is whether this solution can be continued as a function of the parameter ω. We were able to extend the numerical-rigorous argument to cover a full interval of values of the angular velocity, providing the existence of a full branch of solution.

Theorem 7.8 *There exists a smooth map $B(\omega)$ giving the (locally unique, up to symmetries) branch of solution of the choreographical 3-body problem for all $\omega \in [1,2]$, starting at the Mountain Pass solutions for $\omega = 1.5$.*

A natural question is whether the mountain pass branch meets one of the known branches of choreographical periodic orbits: either one of the Lagrange or Marchal's P_{12} (described by Marchal, (2000)) families. Lead from a wrong intuition, looking at Fig. 9 (left), we first conjectured that the branch should bifurcate from L_1 at the angular velocity value $\omega = 1$. It was A. Chenciner who brough this mistake to our attention, pointing out how this fact would contradict the numerical computation on the local bifurcation structure at $\omega = 1$ that he and J. Féjoz were carrying on Chenciner and Féjoz (2005), see also the results in Chenciner et al. (2005). He also suggested us that very likely, this new branch was bifurcating, by symmetry breaking, from the P_{12}–family. Indeed, this was shown by our computations, both numerical and rigorous. The details are depicted in Fig. 9 (right).

References

Albouy, A.: The symmetric central configurations of four equal masses. Contemp. Math. **198**, 131–135 (1996)

Albouy, A., Chenciner, A.: Le problème des n corps et les distances mutuelles. Invent. Math. **131**, 151–184 (1998)

Ambrosetti, A., Coti Zelati, V.: Periodic solutions of singular Lagrangian systems. In: of Progress in Nonlinear Differential Equations and their Applications, vol. 10, Birkhäuser Boston Inc., Boston, MA (1993)

Ambrosetti, A., Rabinowitz, P.H.: Dual variational methods in critical point theory and applications. J. Funct. Anal. **14**, 349–381 (1973)

Arioli, G., Gazzola, F., Terracini, S.: Minimization properties of Hill's orbits and applications to some N-body problems. Ann. Inst. H. Poincaré Anal. Non Linéaire **17**(5), 617–650 (2000)
Arioli, G., Barutello, V., and Terracini, S.: A new branch of mountain pass solutions for the choreographical 3-body problem. Comm. Math. Phys., (2005) to appear.
Bahri, A., Rabinowitz, P.H.: Periodic solutions of Hamiltonian systems of 3-body type. Ann. Inst. H. Poincaré Anal. Non Linéaire **8**(6), 561–649 (1991)
Barutello, V., Terracini, S.: Action minimizing orbits in the n-body problem with choreography constraint. Nonlinearity **17**, 2015–2039 (2004)
Barutello, V.: *On the n-body problem*, Ph. D. Thesis, Università di Milano-Bicocca, (2004) available at www.matapp.unimib.it/dottorato/.
Barutello, V., Ferrario, D.L., Terracini, S.: Symmetry groups of the planar 3-body problem and action–minimizing trajectories. arxiv:math.DS/0404514, preprint (2004).
Barutello, V., Ferrario, D.L., Terracini, S.: On the singularities of generalized solutions to the N-body problem. in preparation (2005).
Barutello, V., Terracini, S.: A bisection algorithm for the numerical Mountain Pass, NoDEA, (2004) to appear (avaliable at arxiv:math.CA/0410284).
Bessi, U., Coti Zelati, V.: Symmetries and noncollision closed orbits for planar N-body-type problems. Nonlinear Anal. **16**(6), 587–598 (1991)
Chen, K.-C.: On Chenciner-Montgomery's orbit in the three-body problem. Discrete Contin. Dynam. Systems **7**(1), 85–90 (2001a)
Chen, K.-C.: Action-minimizing orbits in the parallelogram four-body problem with equal masses. Arch. Rat. Mech. Anal. **158**, 293–318 (2001b)
Chen, K.-C.: Binary decompositions for planar n-body problems and symmetric periodic solutions. Arch. Ration. Mech. Anal. **170**(3), 247–276 (2003a)
Chen, K.-C.: Variational methods on periodic and quasi-periodic solutions for the N-body problem. Ergodic Theory Dynam. Sys. **23**(6), 1691–1715 (2003)
Chenciner, A.: Action minimizing orbits in the newtonian n-body problem. In: Celestial mechanics (Evanston, IL, 1999), Amer. Math. Soc. Providence, pp. 71–90 (2002)
Chenciner, A.: Action minimizing solutions of the newtonian n-body problem: From homology to symmetry. August 2002, ICM, Peking.
Chenciner, A.: Simple non-planar periodic solutions of the n-body problem. In: Proceedings of the NDDS Conference, Kyoto. (2002)
Chenciner, A., Desolneux, N.: Minima de l'intégrale d'action et équilibre relatifs de n corps. C. R. Acad Sci. Paris, Sér. I **326**, 1209–1212 (1998). Correction in C. R. Acad Sci. Paris, Sér. I **327**, 193 (1998)
Chenciner, A., Féjoz, J.: L'équation aux variations verticales d'un equilibre relatif comme source de nouvelles solutions periodiques du probleme des N corps. C. R. Math. Acad. Sci. Paris **340**(8), 593–598 (2005)
Chenciner, A., Fejoz, J., Montgomery, R.: Rotating eights. I. The three Γ_i families. Nonlinearity **18**(3), 1407–1424 (2005)
Chenciner, A., Gerver, J., Montgomery, R., Simó C.: Simple choreographies of N bodies: a preliminary study. Geometry, Mechanics and Dynamics Springer (2001)
Chenciner, A., Montgomery, R.: A remarkable periodic solution of the three body problem in the case of equal masses. Ann. Math. **152**(3), 881–901 (2000)
Chenciner, A., Venturelli, A.: Minima de l'intégrale d'action du problème Newtonien de 4 corps de masses égales dans $\mathbb{R}^3$: orbites "hip-hop". Celestial Mechanics **77**, 139–152 (2000)
Degiovanni, M., Giannoni, F.: Dynamical systems with newtonian type potentials. Ann. Scuola Norm. Sup. Pisa, Ser. IV **15**, 467–494 (1988)
Degiovanni, M., Giannoni, F., Marino, A.: Dynamical systems with newtonian type potentials. Atti Accad. Naz. Lincei Rend. Cl. Sci. Fis. Mat. Natur. Ser. 8 **81**, 271–278 (1987)
Dell'Antonio, G.: Non-collision periodic solutions of the N-body system. Nonlinear Differ. Equ. Appl. **5**(1), 117–136 (1998)
Ferrario, D.L.: Symmetry groups and non-planar collisionless action-minimizing solutions of the 3-body problem in the 3-dimensional space. ariv:math.DS/0407461, preprint (2004).
Ferrario, D., Terracini, S.: On the existence of collisionless equivariant. Invent. Math. **155**(2), 305–362 (2004)
Gordon, W.B.: A minimizing property of Keplerian orbits. Amer. J. Math. **99**(5), 961–971 (1977)
Gordon, W.B.: Conservative dynamical systems involving strong forces. Trans. Amer. Math. Soc. **204**, 113–135 (1975).

Majer, P., Terracini, S.: Periodic solutions to some problems of n-body type. Arch. Rational Mech. Anal. **124**(4), 381–404 (1993)

Majer, P., Terracini, S.: On the existence of infinitely many periodic solutions to some problems of n-body type. Comm. Pure Appl. Math. **48**(4), 449–470 (1995).

Majer, P., Terracini, S.: Multiple periodic solutions to some n-body type problems via a collision index Variational methods in nonlinear analysis (Erice, 1992), 245–262, Gordon and Breach, Basel, (1995).

Marchal, C.: The three-body problem, vol. 4 of Studies in Astronautics. Elsevier Science Publishers B.V., Amsterdam, 1990.

Marchal, C.: The family P_{12} of the three-body problem— the simplest family of periodic orbits, with twelve symmetries per period, Celest. Mech. Dynam. Astron. **78**(1–4), 279–298 (2000).

Marchal, C.: How the method of minimization of action avoids singularities. Cele. Mech. Dyn. Astron. **83**, 325–353 (2002)

McGehee, R.: Triple collision in the collinear three-body problem. Invent. Math. **27**, 191–227 (1974).

Moeckel, R.: On central configurations. Math. Zeit. **205**, 499–517 (1990)

Montgomery, R.: The N-body problem, the braid group, and action-minimizing periodic solutions. Nonlinearity **11**(2), 363–376 (1998)

Montgomery, R.: Action spectrum and collisions in the planar three-body problem. In: Celestial Mechanics (Evanston, IL, 1999), vol. 292 of Contemp. Math. Amer. Math. Soc., Providence, RI, (2002), pp. 173–184

Moore, C.: Braids in classical dynamics. Phys. Rev. Lett. **70**(24), 3675–3679 (1993)

Pacella F.: Central configurations and the equivariant Morse theory. Arch. Rat. Mech. **97**, 5 9–74 (1987)

Palais, R.S.: The principle of symmetric criticality. Commun. Math. Phys. **69**, 19–30 (1979)

Poincaré, H.: Sur les solutions périodiques et le principe de moindre action. C. R. Acad. Sci. Paris Sér. I Math. **123**, 915–918 (1896)

Riahi, H.: Study of the critical points at infinity arising from the failure of the Palais-Smale condition for n-body type problems. Mem. Amer. Math. Soc. **138**, 658 (1999), viii+112.

Sbano, L.: The topology of the planar three-body problem with zero total angular momentum and the existence of periodic orbits. Nonlinearity **11**(3), 641–658 (1998)

Serra, E., Terracini, S.: Collisionless periodic solutions to some three-body problems. Arch. Rational Mech. Anal. **120**(4), 305–325 (1992)

Serra, E., Terracini, S.: Noncollision solutions to some singular minimization problems with Keplerian-like potentials. Nonlinear Anal. **22**(1), 45–62 (1994)

Sperling, H.J.: On the real singularities of the N-body problem. J. Reine Angew. Math. **245**, 15–40 (1970)

Terracini, S., Venturelli, A.: Symmetric trajectories for the $2N$-body problem with equal masses preprint (2004).

Venturelli, A.: Application de la minimisation de l'action au Problème des N corps dans le plan et dans l'espace. Thesis. University Paris VII, December 2002.

Venturelli, A.: Une caracterisation variationnelle des solutions de Lagrange du probleme plan des trois corps. C. R. Acad. Sci. Paris Ser. I Math. **332**(7), 641–644 (2001)

Wintner, A.: The Analytical Foundations of Celestial Mechanics. Princeton Mathematical Series, vol. 5. Princeton University Press, Princeton, N.J. (1941)

Celestial Mechanics and Dynamical Astronomy (2006) 95:27–54
DOI 10.1007/s10569-006-9021-1

ORIGINAL ARTICLE

On families of periodic solutions of the restricted three-body problem

Alexander D. Bruno · Victor P. Varin

Received: 17 November 2005 / Revised: 20 February 2006 /
Accepted: 5 April 2006 / Published online: 15 August 2006

Abstract We consider the plane circular restricted three-body problem. It is described by an autonomous Hamiltonian system with two degrees of freedom and one parameter $\mu \in [0, 1/2]$ which is the mass ratio of the two massive bodies. Periodic solutions of this problem form two-parameter families. We propose methods of computation of symmetric periodic solutions for all values of the parameter μ. Each solution has a period and two traces, namely, the plane and the vertical one. Two characteristics of a family, i.e., its intersection with the symmetry plane, are plotted in the four coordinate systems used in the investigations: two global and two local ones related to the two massive bodies. We also describe generating families, i.e., the limits of families as $\mu \to 0$, which are known almost explicitly. As an example, we consider the family h, which begins with retrograde circular orbits of infinitely small radius around the primary P_1 of bigger mass. For this family, we cite detailed data for $\mu = 0$ and $\mu \approx 10^{-3}$ and give a brief description of its evolution as μ increases up to $\mu = 1/2$.

Keywords Restricted three-body problem · Periodic solutions · Two-parameter families

1 Introduction

Let three point bodies P_1, P_2, and P_3 move in a single plane under the Newton law of gravitation. The bodies P_1 and P_2 have masses m and m_2 respectively, and the mass of the body P_3 is so small that its influence on the bodies P_1 and P_2 can be ignored. We say that the mass of the body P_3 is equal to zero. Then the body P_2 moves along

A. D. Bruno · V. P. Varin (✉)
Mathematical Department, Keldysh Institute of Applied Mathematics,
Miusskaja sq. 4,
Moscow, 125047, Russia
e-mail: varin@keldysh.ru

a Keplerian orbit with respect to the body P_1. If the body P_2 moves along a circle, then the problem on movements of the body P_3 is called the plane circular restricted three-body problem (briefly: the restricted problem).

We assume that mass, time, and distance units are chosen in such a way that the sum $m+m_2$, the gravitational constant, the distance P_1P_2, and the angular velocity P_2 with respect to P_1 are equal to one. Then there is a unique parameter $\mu = m_2/(m+m_2) \in [0,1/2]$. If we introduce a coordinate system rotating with P_2, then in this (synodical) coordinate system with the center at P_1, the position x_1, x_2 of the body P_3 is described by the Hamiltonian system with two degrees of freedom having one parameter μ (see (Bruno 1990, Chap. III, Sect. 1)):

$$\dot{x}_j = \partial H/\partial y_j, \qquad \dot{y}_j = -\partial H/\partial x_j, \quad j = 1,2, \quad \cdot \stackrel{\text{def}}{=} d/dt, \tag{1.1}$$

where

$$H = H_0 + \mu R, \quad H_0 = \frac{1}{2}(y_1^2 + y_2^2) + x_2 y_1 - x_1 y_2 - r^{-1},$$
$$R = r^{-1} + x_1 - r_2^{-1}, \quad r = \sqrt{x_1^2 + x_2^2}, \quad r_2 = \sqrt{(x_1-1)^2 + x_2^2}. \tag{1.2}$$

For $\mu \neq 0$, the problem is not integrable. For $\mu = 0$, the problem is integrable, and all its solutions can be described (Bruno 1990, Chapts. III–VI). The phase space of this problem for $\mu = 0$ has a complicated structure due to collisions of the body P_3 with the body P_2. For $\mu > 0$, these collisions cause singular perturbations leading to further complication of the structure of the phase space. Here the families of periodic solutions are of most importance, since they form, in a sense, the skeleton of some part of the phase space. For a fixed value of the parameter μ, periodic solutions of the Hamiltonian system constitute one-parametric families; and two-parametric ones for varying μ.

An orbit is the projection of the solution $x_j(t), y_j(t), j = 1,2$ of the system (1.1) onto the plane x_1, x_2. The number of revolutions of a synodical periodic solution around the body P_1 is called the *global multiplicity* of the solution. Although the multiplicity may change along the family when we encounter a collision orbit of P_3 with P_1, still the multiplicity is useful for classification of families. If two families of periodic solutions intersect, and the periods on one family are q-times larger than periods on the other family, then we say that the first family has a *local multiplicity* q.

The system (1.1) is invariant under the substitution

$$t, x_1, x_2, y_1, y_2 \longrightarrow -t, x_1, -x_2, -y_1, y_2, \tag{1.3}$$

which is its *symmetry*. Under the symmetry (1.3), the plane $x_2 = y_1 = 0$ is invariant; it is called (Bruno 1990, Chap. III) the *symmetry plane* Π. Solutions of the system (1.1) invariant under the substitution (1.3) are *symmetric*. A symmetric periodic solution crosses the symmetry plane twice. These intersections are convenient for tracking mutual disposition of such solutions.

A family of periodic solutions of the system (1.1) for a fixed value of the parameter μ is called *natural*, if the solution continues to both sides until the natural endings, which may be the ending at a stationary point or on another family of periodic solutions; contraction of the orbit to a point or extension to infinity; convergence of the period to zero or to infinity, etc.

We begin a new cycle of articles on the study and computation of the principal families of periodic solutions for all values of the parameter $\mu \in [0,1/2]$. Until recently, these families were studied and computed either for fixed values of the parameter μ, or for small μ. For $\mu = 1/2$, it was done in works of Strömgren (1935) and Bartlett (1964). For $\mu = \mu_M = 0.1215585$, corresponding to the Earth (P_1) – Moon (P_2) case, by Broucke (1968). For $\mu = \mu_J = 0.00095388$, corresponding to the Sun (P_1) – Jupiter (P_2) case, by Bruno (1993, 1996). For $\mu = \mu_N = 5.178 \times 10^{-5}$, corresponding to the Sun (P_1) – Neptune (P_2) case, by Kotoulas and Voyatzis (2004), Voyatzis and Kotoulas (2005), and Voyatzis et al. (2005). For $\mu = 0$ (generating families), by Bruno in the book (1990) and preprints (Bruno 1993, 1996), and by Hénon in the books (1997, 2001). Some special families were also studied for other values of μ.

The interest in these studies was stimulated by the need to describe the motion of small bodies in the Solar System (asteroids, Kuiper belt objects (Kotoulas et al. 2004), satellites and particles in rings of the giant planets etc.), and in binary star systems (planets, dust particles), as well as for needs of navigation in the Solar System.

Periodic solutions of the restricted three-body problem were considered in the books (Szebehely 1967), (Bruno 1990), (Hénon 1997, 2001). In (Szebehely 1967, Chapts. 8, 9), are given the data on periodic orbits known up to 1967. This data pertains mainly to solutions of the first and the second kinds, i.e., to the perturbations of circular and elliptic orbits. Solutions of the second species, i.e., passing close to the primary P_2, are not considered there, although there are some examples of computed orbits with this property for small μ. In the book by Bruno (1990), are studied all regular perturbations of periodic solutions, present for $\mu = 0$, including the bifurcations of families. There, it is given a systematic study of the families of arc-solutions, which provides a basis for the study of singular perturbations of periodic orbits, i.e., periodic solutions of the second species. In the book by Hénon (1997), with the help of the Broucke principle, are studied simple bifurcations in generating families, which composed from the parts of families of circular orbits, elliptic orbits, and arc-solutions. There are also described some generating families or their initial parts. In the book by Hénon (2001), using Power Geometry (Bruno 1998), are studied complicated bifurcations in generating families, but there are no new examples of generating families.

In the present work, we propose a program for the study of two-parametric families of periodic solutions starting with $\mu = 0$, i.e., with generating families, and concluding with $\mu = 1/2$. For this purpose, we use the results of the books (Bruno 1990) and (Hénon 1997, 2001) as a starting ground. In addition, we use here our method for a compact representation of a large amount of data.

We intend to compute each family for

$$\mu = 0,\ \mu_J,\ 0.1,\ 0.2,\ 0.3,\ 0.4,\ 0.5 \tag{1.4}$$

and describe its disposition in the phase space in form of curves in the symmetry plane Π (in four different coordinate systems), and also periods of solutions, their plane and vertical traces (i.e., the indices of stability). All this in form of figures. If necessary, we will also consider some other interesting values of μ (especially those where bifurcations occur). Besides, we intend to present the most interesting orbits in the family, intersections of the family with other families, and self-bifurcations of the family. But first, we briefly recall well-known properties of periodic solutions of the restricted three-body problem.

In Sects. 6–8, we give as an example the family h, which begins with retrograde circular orbits of infinitely small radius around the primary P_1 of bigger mass. For $\mu = 0$

and $\mu = \mu_J \approx 0.00095$, we give the detailed data, and we give a brief description of the results for other values of μ.

2 On families of periodic solutions

General properties of families of periodic solutions to a Hamiltonian system with two degrees of freedom are cited in (Bruno 1972) (in detail), and briefly in (Bruno 1990, Chap. II, Sect. 4). For information on families of symmetric periodic solutions refer to (Bruno 1972, Sect. 4). Each symmetric solution M in a family F has *period* T, *trace* Tr of the monodromy matrix of the variational system, and two points of intersection with the symmetry plane Π (after half-period). As a parameter on a family, we can take either the value of the Hamiltonian function, or one of the coordinates of intersections with the symmetry plane.

For $\mu \in [0, 1/2]$, the families of symmetric periodic solutions to the system (1.1) are two-parametric, hence they can have singularities of co-dimension 2, which are studied in (Bruno 1972, 1990, Chapts. I, II). But for $\mu = 0$, the system (1.1) is degenerate (see Sect. 3).

For a fixed $\mu \neq 0$, periodic solutions for the non-integrable Hamiltonian system (1.1) constitute one-parametric families. The limits of these families as $\mu \to 0$ are called *generating families* of periodic solutions. In (Bruno 1990, Introduction), a program was suggested for studying families of periodic solutions of the restricted problem for small $\mu > 0$ by means of their generating families. For this, we need to describe all generating families. Further, we will only be interested in the symmetric periodic solutions, which are invariant under the substitution (1.3), and whose orbits in the plane x_1, x_2 are symmetric with respect to the x_1 axis. Each symmetric periodic solution is represented by the two points in the symmetry plane Π. For fixed μ, each family of these solutions is represented in Π by the two curves, i.e., by characteristics of the family.

To plot characteristics of families in the plane Π using coordinates x_1, y_2 is not convenient, since these characteristics form very close curves (see Broucke 1968). Hence we use in this plane four coordinate systems, two global: system I: x_1, y_2, and system II: $x_1, C = -2H$; and two local, system III: $\tilde{a}, \tilde{e}$ related to the body P_1, and system IV: $\tilde{a}'_*, \tilde{e}_*$ related to the body P_2.

System I. Natural coordinates x_1, y_2. In figures, straight lines $x_1 = 0$ and $x_1 = 1$ corresponding to the bodies P_1 and P_2 are plotted as thick lines. Dotted lines represent the curves $|x_1|y_2^2 = 2(1-\mu)$ which are the boundaries of domains of elliptic motion of the body P_3 under the gravitational force of the body P_1 of mass $1-\mu$.

System II. The mapping of the symmetry plane $\Pi = (x_1, y_2)$ onto the plane x_1, $C = -2H$ is two-sheeted: two different values of y_2 correspond to one value of $C = -2H$; the boundary of the mapping is the line where the discriminant of the polynomial in y_2

$$-y_2^2 + 2x_1y_2 + \frac{2(1-\mu)}{|x_1|} - 2\mu x_1 + \frac{2\mu}{|x_1 - 1|} - C = 0,$$

linking the Jacobi constant C and coordinates x_1, y_2 in the plane Π, vanishes, i.e.,

$$x_1^2 + \frac{2(1-\mu)}{|x_1|} - 2\mu x_1 + \frac{2\mu}{|x_1 - 1|} - C = 0. \qquad (2.1)$$

Figures show components of this curve as dotted lines; the straight lines $x_1 = 0$ and $x_1 = 1$ corresponding to the bodies P_1 and P_2 are plotted as thick lines.

System III. Astronomical coordinates close to the body P_1

$$\tilde{e} = \frac{x_1 y_2 |y_2|}{(1-\mu)}, \quad \tilde{a} = \frac{x_1}{|2 - |\tilde{e}||}. \tag{2.2}$$

Then $a = |\tilde{a}|$ and $e = |1 - |\tilde{e}||$ are the semi-major axis and the eccentricity of the corresponding sidereal orbit of the two-body problem P_1 (with mass $1 - \mu$) and P_3. This is true for elliptic orbits (for $|\tilde{e}| \leq 2$) and hyperbolic orbits (for $|\tilde{e}| \geq 2$). Parabolic orbits are omitted here. But we restrict ourselves by parts where $|\tilde{e}| \leq 2$. The body P_1, i.e., $x_1 = 0$, is represented by the straight line $\tilde{a} = 0$; the body P_2, i.e., $x_1 = 1$, is represented by the curves $\tilde{a}|1-|\tilde{e}|| = 1$. They both plotted as thick lines. The inversion of the transformation (2.2) is

$$x_1 = \tilde{a}\,|2 - |\tilde{e}||, \quad y_2 = \operatorname{sign}(\tilde{a}\,\tilde{e}) \sqrt{\frac{|\tilde{e}|(1-\mu)}{|x_1|}}. \tag{2.3}$$

System IV. Astronomical coordinates for the body P_2

$$\tilde{e}_* = \frac{\xi_1 \eta_2 |\eta_2|}{\mu}, \quad \tilde{a}_* = \frac{\xi_1}{||\tilde{e}_*| - 2|}, \quad \tilde{a}'_* = \frac{\tilde{a}_*}{\mu}, \tag{2.4}$$

where

$$\xi_1 = x_1 - 1, \quad \eta_2 = y_2 - 1 \tag{2.5}$$

are local coordinates for the body P_2. Here the values x_1 and y_2 are taken close to the body P_2 and only for $|\tilde{e}_*| \geq 2$. Here the values $|\tilde{e}_*| - 1$ and $|\tilde{a}_*| = |\tilde{a}'_*|\mu$ are the eccentricity and the real semi-axis of the hyperbolic fly-by orbit of the body P_3 close to the body P_2 of mass μ (ignoring the body P_1).

This system is analogous to the system III, i.e., (2.2), but close to the body P_2. Both systems correspond to two-body problems: P_1, P_3 (system III), and P_2, P_3 (system IV). Since for small μ the value ξ_1 is proportional to μ, then according to (2.4), the same is true for $\tilde{a}_*$. Hence we introduce the coordinate $\tilde{a}'_* = \tilde{a}_*/\mu$, which has a finite limit as $\mu \to 0$. In figures, the straight line $\xi_1 = 0$ corresponding to the body P_2 is plotted as a thick line.

The coordinate systems III and IV are introduced for the purpose of comparison of the computed families for $\mu > 0$ with the generating families (i.e., for $\mu = 0$), which correspond to solutions of the two-body problem (P_3, P_1 for system III, and P_3, P_2 for system IV).

For a fixed μ, every family in the plane Π in coordinate systems I and II is represented by two characteristics. For one of them, we cite the values of the normalized period $\tilde{T} \stackrel{\text{def}}{=} T/2\pi$. The values of traces Tr and Tr_ν are given as functions of $\tilde{T}$. Both traces Tr and Tr_ν may take very big values, hence outside the interval $[-2, 2]$, we plot them in logarithmic scale

$$\widetilde{\text{Tr}} = \begin{cases} \text{Tr}, & \text{if } |\text{Tr}| \leq 2, \\ (1 + \log_2 |\text{Tr}|)\operatorname{sign}(\text{Tr}), & \text{if } |\text{Tr}| > 2, \end{cases} \tag{2.6}$$

and similarly for $\widetilde{\text{Tr}}_\nu$.

The planar circular restricted problem is a part of the space circular restricted problem (see Szebehely 1967, Sect. 10.2). For a solution of the planar problem, the vertical components are separated in the variational system into the subsystem, which computes the vertical trace Tr_v (Hénon 1973). If $|\mathrm{Tr}_v| \leq 2$, then from the plane periodic solution there may bifurcate some families of space periodic solutions having symmetric orbits with respect to the plane of rotation of the bodies P_1 and P_2. To isolate periodic solutions that are stable in linear approximation we need to know both traces, since this stability is only possible for

$$|\mathrm{Tr}| \leq 2 \quad \text{and} \quad |\mathrm{Tr}_v| \leq 2. \tag{2.7}$$

Otherwise periodic solutions are unstable.

Usually, a family of periodic solutions includes some collision orbits with the bodies P_1, P_2. For a family of symmetric periodic solutions for fixed μ, a sequence of parts of orbits near collision with the body P_i is shown in Figure 1. Here a loop is formed around the body P_i. Figure 1. shows one possible sequence; the reverse sequence is also possible. Besides, there are possible situations corresponding to bilateral symmetry of pictures in Fig. 1 with respect to the vertical axis. We emphasize that the point on the axis x_1 where orthogonal crossing of the axis by the orbit takes place remains on the same side of the body P_i before and after the collision. This is a special case of the *Broucke principle* (Broucke 1968): on families of symmetric periodic solutions, the points of orthogonal crossing of the axis x_1 do not change their disposition with respect to the bodies P_1, P_2.

3 Generating families

If a solution $x(t, \mu)$ existing for some $\mu = \mu_0 > 0$ is continual by μ for arbitrary small $\mu > 0$, then its limit as $\mu \to 0$ is called the *generating periodic solution*. Obviously, the generating solution consists of parts of solutions of the problem for $\mu = 0$. These solutions subdivide into two species: the first species consist of solutions where the body P_3 does not collide with the body P_2; to the second species there belong solutions where the body P_3 collides with the body P_2. Solutions of the first species are the solutions of the two-body problem of P_1 and P_3 in the synodic (rotating) coordinate system. Solutions of the second species are formed from several parts of the solutions of the two-body problem of P_1 and P_3, each part beginning and ending with a collision of P_3 with P_2; and all parts have the same value of the Hamiltonian function H (or the Jacobi constant $C = -2H$). All these parts, i.e., arc-solutions, constitute denumerable set of one-parameter families A_i, B_j, C_{kl} (united into families S), and one-parameter families T_N, studied in detail in (Bruno 1990, Chapts. III–V). Families S of symmetric arc-solutions were found by Hénon (1968).

Here we briefly cite those properties of the families of circular orbits, of the families of symmetric elliptical orbits, and of the families of arc-solutions, that we use for description of generating families. For more detailed description see (Bruno 1990, Chapts. III and IV), which are shorter versions of articles (Bruno 1978a, b). Here we keep the notation of the articles.

Generating families of symmetric periodic solutions consist of parts of families of symmetric periodic solutions for $\mu = 0$, of parts of families S, and of parts of pairs of families T_N with bifurcations between these parts. According to (Bruno 1990, Chapts. VII and VIII), for $\mu = 0$ all symmetric periodic solutions without collisions of P_3 with P_2 are generating ones. They form two families *Id* and *Ir* with circular orbits, and denumerable many families $E_N^{\pm}$ of elliptic orbits with fixed semi-major axis (one

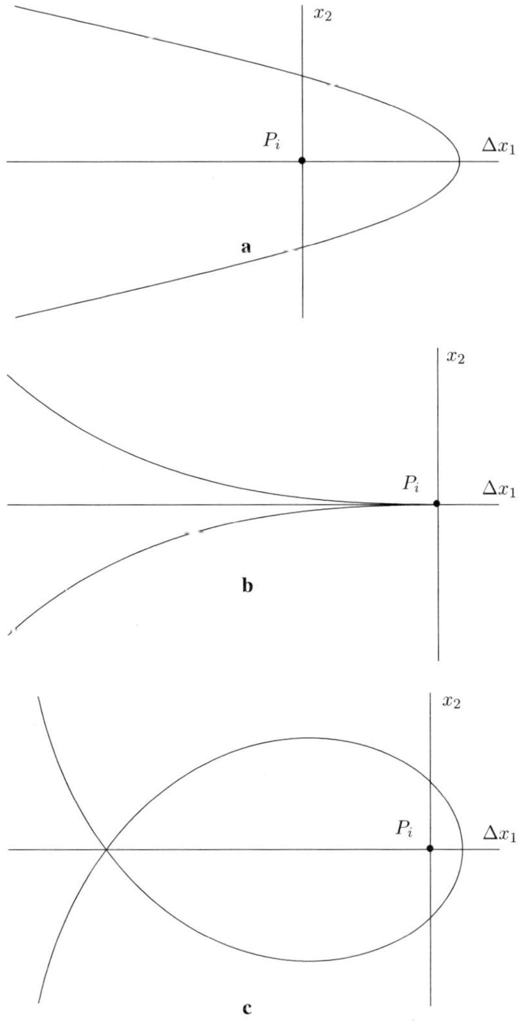

Fig. 1 Parts of orbits of a family near collision (b) with the body P_i

or two families for every rational $N = a^{-3/2} > 0$) (Bruno 1990, Fig. 11). Bifurcations between these families take place at the intersections of the families *Id* with the families $E_{(p+1)/p}$ for $p = 1, \pm 2, \pm 3, \ldots$ We denote these intersections as $Id(N) = Id \cap E_N$. On the plane Π, these intersections of families correspond to the points $N = (p+1)/p$ for $p = 1, \pm 2, \pm 3, \ldots$, where $\tilde{a} = \pm N^{-2/3}$, $\tilde{e} = 1$. They divide the family *Id* into the parts Id_p with $p/(p-1) > N > (p+1)/p$ for $p = 1, \pm 2, \pm 3, \ldots$, and the families $E_{(p+1)/p}$ into two parts $E^+_{(p+1)/p}$ (with $\theta = 0$) and $E^-_{(p+1)/p}$ (with $\theta = \pi$). Perturbations and bifurcations of families *Id*, *Ir*, and $E^\pm_N$ are studied in (Bruno 1990, Chapts. VII and VIII). In particular, the character of bifurcations of the parts Id_p with $E^\pm_N$ is shown in (Bruno 1990, Fig. 90). Perturbations of the period T and the trace Tr at some families E_N are cited in Table 2 in (Bruno 1990, Appendix).

Let a and e be the semi-major axis and the eccentricity of an elliptic orbit of the two-body problem P_2 and P_3 for $\mu = 0$. Collisions of this orbit with the body P_2 are possible only for $a(1-e) \leq 1 \leq a(1+e)$. On the plane Π, these inequalities isolate four domains ω_i, $i = 1,2,3,4$. Each symmetric arc-solution, i.e., belonging to the family S, has one point of intersection with the plane Π, and this point lies in one of the domains ω_i represented by dotted lines. On the plane Π, the families S are represented by their characteristics.

On the plane Π, in the third coordinate system $\tilde{a}, \tilde{e}$, characteristics of the families Id and Ir are straight lines $\tilde{e} = 1$ and $\tilde{e} = -1$, respectively; characteristics of the families E_N are two segments $\tilde{a} = N^{-2/3}$; and characteristics of the families of symmetric arc-solutions A_i, B_j, C_{kl} are shown in (Bruno 1990, Figs. 20 and 40). These figures also show boundaries of domains ω_i by dotted lines.

According to (Bruno 1990, Chapts. III and IV), for $\mu = 0$, every symmetric periodic solution with collision of P_3 with P_2 is composed of arc-solutions from the families S and of even number of symmetrically placed arc-solutions from the families T_N with the same Jacobi constant C for all arc-solutions. It seems that every such combination is generating. But it is proved only for the simplest of them consisting of one or two arc-solutions (Perko 1981). The families S, T_N, and their characteristics in Π are studied in (Bruno 1990, Chapts. III–V). Bifurcations between families of symmetric periodic solutions take place at the intersections of the families Id, Ir, $E_N^{\pm}$, S. On the plane Π, they correspond to the intersections of the characteristics of these families. All of them have rational values of N. For rational $N = (p+q)/p, N \neq 1$, intersections of the family E_N with families S take place at the orbits with

$$e^* = \cos(k\pi/(2|p+q|)),\ k = 0,1,\dots,2|p+q|,\ \text{if } a > 1;$$
$$e^* = \delta\cos(k\pi/(2q)),\ k = 0,1,\dots,[(x - z_0 + \varphi(x))q],\ \text{if } a < 1,$$

where the coordinate e^* was introduced in (Bruno 1990, Chap. IV, Formula (2.17)), $\delta = \pm 1, x = p/q, z_0 = 1/(2\sqrt{2}-1) = 0.5469\dots$, values of the function $\varphi(x)$ are cited in Table 2 (Bruno 1990, Chap. IV), and $[y]$ means the integer part of the number y (see Bruno 1990, Chap. IV, Sect. 2). According to (Bruno 1990, Chap. VI, Sect. 2) and by Theorem 2.4 in (Bruno 1990, Chap. IV), these values of e^* imply certain values of $\tilde{e}$, and certain orbits of intersection, which we denote as $E_N(k)$ for $a > 1$, and $E_N(\delta k)$ for $a < 1$. For $a > 1$, the number of orbits of intersection $E_N(\delta k)$ is odd, and for $a < 1$, the number of orbits $E_N(k)$ is even.

We remark that for $k = 0$ and $2|p+q|$, one of the two conjugate points of the orbit $E_N(k)$ lies at $P_2^{**} \stackrel{\text{def}}{=} P_2 \cap \Pi$, i.e., $x_1 = 1$, and the other lies on the boundary of one of the domains ω_i (Bruno 1990, Chap. III, Sect. 3.D).

If $k \neq 0,\ 2|p+q|$, then the trajectory from $E_N(k)$ is represented in Π by two conjugate points of intersection lying inside the domains ω_i.

According to (Bruno 1990, Chap. III, Sect. 3), there are two other orbits of intersection:

$$E_1(-1) = \{N = 1, \tilde{e} = -1\},\ \text{here families } Ir, E_1^-,\ \text{and } S \text{ intersect;}$$
$$E_1(+1) = \{N = 1, \tilde{e} = 1\}\ \text{here families } Id, E_1^+,\ \text{and } S \text{ intersect.}$$

Thus, on the mentioned orbits $Id(1 + 1/p)$ and $E_N(k)$, bifurcations take place between parts of the generating families of symmetric periodic solutions.

Extremal orbits of families S where the Jacobi constant C attains an extremum on a family also play an important part in the formation of generating families. Their theory is developed in (Bruno 1990, Chapts. IV and V), and numerical values are published in (Hitzl and Hénon 1977a). Every family S of the type A_i and B_j has denumerable number of extremal orbits. In order to numerate them sequentially, we choose an initial extremal orbit $S(0)$ on the family S and the direction of increasing number m of an extremal orbit. Then all extremal orbits of the family S are denoted as $S(m)$. For families A_i, the orbit $A_i(0)$ is multiple of the orbit $E_1(-1)$, i.e., $\tilde{a} = \pm 1$, $\tilde{e} = -1$. For families B_j, the orbit $B_j(0)$ is $E_1(1)$, i.e., $\tilde{a} = 1$, $\tilde{e} = 1$. On families C_{kl}, there are only two extremal points $C_{kl}(\delta)$: $C_{kl}(1)$ for $\tilde{e} > 0$, and $C_{kl}(-1)$ for $\tilde{e} < 0$. Tables of families S are cited in (Hénon, 1968).

Basic generating families Id, Ir, $E_N^{\pm}$, A_i, B_j, C_{kl}, and T_N are described in the book (Bruno 1990, Chapts. III and IV). Points of intersection of their characteristics in the symmetry plane Π are also described there. To these points there correspond some special solutions on which basic families intersect:

$$Id(N) = Id \cap E_N, \qquad Ir(N) = Ir \cap E_N, \qquad E_N(k). \tag{3.1}$$

Besides, in (Bruno 1990, 1993) there were isolated extremal orbits mentioned earlier,

$$A_i(m), \quad B_j(m), \quad C_{kl}(\delta), \tag{3.2}$$

where the corresponding characteristics of a generating family can have a point of return (fold), or a point of the closure. At these special solutions, the bifurcation of intersecting basic families takes place, i.e., the generating family consists of parts of the basic families restricted by special solutions. The character of these bifurcations is discussed in (Bruno 1990), (Strömgren 1935), (Hénon 1997, 2001), but it is not studied yet in all cases.

On the family Id (i.e., $\tilde{e} = 1$), the period T, and traces Tr and Tr_v are

$$T = 2\pi |a^{-3/2} - 1|^{-1}, \quad \mathrm{Tr} = 2\cos T = \mathrm{Tr}_v. \tag{3.3}$$

On the family Ir (i.e., $\tilde{e} = -1$), we have

$$T = 2\pi |a^{-3/2} + 1|^{-1}, \quad \mathrm{Tr} = 2\cos T = \mathrm{Tr}_v. \tag{3.4}$$

On the families $E_N^{\pm}$ with rational $N = a^{-3/2} = (p+q)/p$ (i.e., $\tilde{a}^{-3/2} = N$), we have

$$T = 2\pi p, \quad \mathrm{Tr} = 2 = \mathrm{Tr}_v. \tag{3.5}$$

On the generating families of periodic solutions of the second species (i.e., with collisions of the bodies P_3 and P_2), both traces Tr and $\mathrm{Tr}_v = \pm\infty$. The change of sign of Tr takes place only at special and extremal solutions for the families having periodic solutions consisting of one arc-solution of the family S (i.e., A_i, B_j, or C_{kl}) (Hitzl and Hénon 1977b). For families of periodic solutions consisting of more than one arc-solution; other places of the change of sign of Tr are possible as well, but we know nothing about them as yet. But a regular Hamiltonian system at the intersection of two families of periodic solutions has $\mathrm{Tr} = 2$ on one family and $\mathrm{Tr} = 2\cos(2\pi/q)$ on the other, where the natural number q is the order of the resonance and the multiplicity of the intersection. Hence one-arc-solution generating families may intersect only at the special solutions (3.1) and at the extremal solutions (3.2), and many-arc-solution generating families may intersect in addition at some unknown solutions. The places of the change of sign of Tr_v at solutions of the second species are yet unknown.

Hénon (1997, 2001) developed a very complicated theory of bifurcations of families of generating solutions of the second species. The generating families can be computed by more or less explicit formulas (see Bruno 1990, Chap. IV, and Chap. VI, Sect. 3.3).

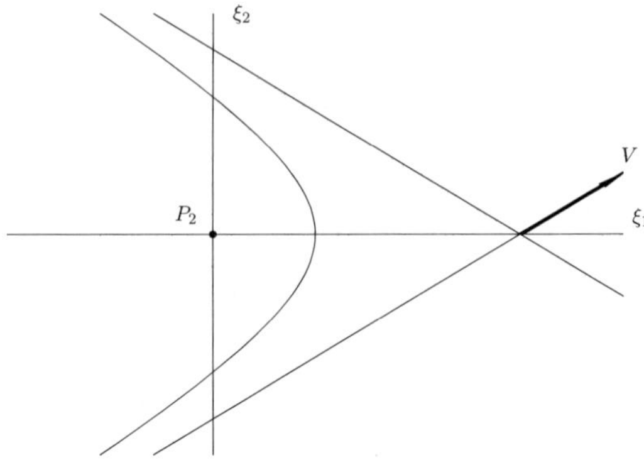

Fig. 2 Hyperbolic orbit near the body P_2

Special care must be taken for characteristics of generating families of the second species near the body P_2. The character of the fly-by of the body P_3 near the body P_2 is described in (Bruno 1978c, 1981). Approximately, the fly-by is described by a symmetric hyperbolic orbit of the two-body problem of P_2 (of mass μ) and P_3 (see Fig. 2).

According to Abalakin et al. (1971, Part II, Sect. 2.03), the motion of the body P_3 in the local coordinates $\xi_1 = x_1 - 1$, $\xi_2 = x_2$ is described by the formulas

$$r = \frac{a_*(e_*^2 - 1)}{1 + e_* \cos \nu}, \quad \xi_1 = r \cos \nu, \ \xi_2 = r \sin \nu,$$

where a_*, e_* are the real semi-axis and the eccentricity of the hyperbolic orbit, ν is the true anomaly. Here the component of the velocity transversal to the radius-vector is

$$v_n = \sqrt{\frac{\mu}{a_*(e_*^2 - 1)}}(1 + e_* \cos \nu),$$

and the energy integral is

$$V^2 = \mu \left(\frac{2}{r} + \frac{1}{a_*} \right),$$

where

$$V^2 = v_1^2 + v_2^2 = \dot{\xi}_1^2 + \dot{\xi}_2^2.$$

At the intersection with the axis ξ_1, we obtain the smallest distance to the body P_2:

$$|\xi_1^0| = a_*(e_* - 1) \quad \text{and} \quad \dot{\xi}_2 = v_n = \sqrt{\frac{\mu(e_* + 1)}{a_*(e_* - 1)}}.$$

At an infinitely remote point of the hyperbolic orbit $r = \infty$, i.e., $1 + e_* \cos \nu_\infty = 0$. Consequently,

$$e_* = -\frac{1}{\cos \nu_\infty} > 1.$$

There, the energy integral gives $V^2 = \mu/a_*$, where $V^2 = v_1^2 + v_2^2$, and (v_1, v_2) is the arrival velocity from infinity. Consequently,

$$a_* = \mu/V^2.$$

Since $|\cos \nu_\infty| = |v_1|/V$, then according to the previous formulas, the intersection of the hyperbolic orbit with the axis ξ_1 takes place at the point

$$\xi_1^0 = \frac{\mu}{V^2}\left(\frac{V}{|v_1|} - 1\right)\operatorname{sign} v_1 \tag{3.6}$$

with the velocity

$$\dot{\xi}_2 = \sqrt{\frac{\mu(1 + V/|v_1|)}{|\xi_1^0|}}\operatorname{sign} v_2 = V\sqrt{\frac{V + |v_1|}{V - |v_1|}}\operatorname{sign} v_2. \tag{3.7}$$

We make the matching of the hyperbolic solution with the arc-solution of the two-body problem of P_1 and P_3 in the most simple way: we assume that the arrival velocity from infinity (v_1, v_2) for the hyperbolic orbit is equal to the velocity of collision (v_1, v_2) of the body P_3 with the body P_2 for solutions of families S. More accurate matchings is the subject of numerous studies (see, for example, Breakwell and Perko 1974).

Since in (2.5)

$$\eta_2 = \xi_1^0 + \dot{\xi}_2,$$

then as $\mu \to 0$, we have $\eta_2 \to \dot{\xi}_2$. If now for the coordinates ξ_1, η_2, we introduce the corresponding astronomical coordinates (2.4) and replace there η_2 by $\dot{\xi}_2$, then, according to (3.6) and (3.7), we obtain

$$\begin{aligned}\tilde{e}_* &= \frac{\xi_1^0\dot{\xi}_2|\dot{\xi}_2|}{\mu} = \frac{V + |v_1|}{|v_1|}\operatorname{sign}(v_1 v_2),\\ \tilde{a}'_* &= \frac{(V - |v_1|)\operatorname{sign} v_1}{V^2|v_1||2 - |\tilde{e}_*||} = \frac{(V - |v_1|)\operatorname{sign} v_1}{V^2|2|v_1| - V - |v_1||} = \frac{\operatorname{sign} v_1}{3 - C},\end{aligned} \tag{3.8}$$

where (v_1, v_2) is the velocity of entry of the orbit at the point P_2, $V^2 = v_1^2 + v_2^2 = 3 - C$, and $\tilde{e}_* - 1$ and $\tilde{a}'_*$ are the limit values of the eccentricity $\tilde{e}_*$ and the normalized real semi-axis $\tilde{a}_*/\mu$ of the local hyperbolic orbit of the fly-by of the body P_3 near the body P_2 for small μ; C is the value of the Jacobi constant on the arc-solution. By the formula (3.8), we compute characteristics of the generating families in the coordinate system IV.

Arc-solutions have velocities of entry $V = (v_1, v_2)$ and exit $(-v_1, v_2)$ at the point $P_2 = (1, 0)$. The corresponding generating and generated orbits have orthogonal intersection with the axis ξ_1 taking place on the left of the point P_1 if $\operatorname{sign} v_1 < 0$, and on the right if $\operatorname{sign} v_1 > 0$. If $v_2 \neq 0$, then the fly-by occurs as described above. If on the other hand $v_2 = 0$, then the generating orbit is the collision orbit. For it, $\xi_1^0 = 0$ and $\dot{\xi}_2 = \infty$. The transition of such generated orbits through the collision is shown in Fig. 1. For generating families, the orbits of collision have $v_2 = 0$. Acording to (Bruno 1990, Chap. III, Sect. 3.7), this takes place when $\sqrt{a(1 - e^2)} = 1$, i.e., when

$$|\tilde{a}|\,\tilde{e}\,(2 - \tilde{e}) = 1. \tag{3.9}$$

4 On computations of families

We intend to compute nine principal families, which, in the notation of Strömgren (1935), Bartlett (1964), and Hénon (1965–2001) are as follows.

Family a goes from the Lagrange stationary point L_2.

Family b goes from the point L_3.

Family c goes from the point L_1.

Family f begins as retrograde circular orbits of infinitely small radius around the body P_2.

Family g begins as direct circular orbits of infinitely small radius around the body P_2.

Family h begins as retrograde circular orbits of infinitely small radius around the body P_1.

Family i begins as direct circular orbits of infinitely small radius around the body P_1.

Family l begins as retrograde circular orbits of infinitely large radius with positive value of C (in sidereal coordinates, they have direct direction of motion).

Family m begins as retrograde circular orbits of infinitely large radius with negative value of C (in sidereal coordinates, they keep retrograde direction of motion).

Strömgren (1935), Broucke (1968), and Bruno (1993, 1996), respectively, used the following notation for these families.

a	b	c	f	g	h	i	l	m
I	J_1	G	C	H_1, H_2	A_1	BD	E_1	F
$L2$	$L3$	$L1$	$E+1/1$	$2T_1$	$IR+$	$ID1$	$ID-1$	$IR-$

For each family, we will start with the description and computation of the generating family (i.e., for $\mu = 0$); then we will compute this family for the values (1.4) of μ. For each value of μ, we will give graphs of the characteristics of the family in the symmetry plane Π in the four coordinate systems described in Sect. 2, as well as graphs of the normalized period $\tilde{T} \stackrel{\text{def}}{=} T/2\pi$, of the trace $\tilde{\text{Tr}}$, and trace $\tilde{\text{Tr}}_\nu$. We will also give tables of critical orbits and figures of some of them. The orbit is *critical* if it is either a collision orbit (with P_1 or P_2), or it has $\text{Tr} = \pm 2$, or $\text{Tr}_\nu = \pm 2$. Then for some computed orbits, we intend to compare their parameters with those predicted by them by the perturbation theory according to (Bruno 1990).

We will also mark the places of intersection of families with other symmetric and asymmetric planar and space families of periodic solutions.

We remark that these basic nine families undergo the following transitions under the substitution $\mu \to 1 - \mu$:

$$a, b, c, f, g, h, i, l, m \longrightarrow c, b, a, h, i, f, g, l, m.$$

It is remarkable that for the description of generating families we need a fairly developed theory and relatively small amount of computations. For the description of families for $\mu > 0$ the situation is the reverse: we need a bit of theory but very extensive computations.

5 Methods of computation

A typical family of periodic solutions can be loosely split into several parts with different behavior of solutions as far as computations are concerned. In some parts, there is no problem with computing solutions, but in other places precision may decrease dramatically if no special care is taken. And finally, there are parts near the collisions of the body P_3 with the body P_1 or P_2, where computations are impossible without regularization.

So rather than one, we need a series of programs developed for optimal performance at various parts of the families.

The precision of an individual trajectory may be determined comparing the Jacobi constant C at the beginning and at the end of the trajectory after half-period. In our computations, this difference is less than 10^{-10} for all trajectories, but usually is less than 10^{-13}.

The bulk of the computations is performed by a Runge-Kutta method of 5th order (requiring 6 evaluations of the right-hand side of the system of ODEs at each step) with fixed stepsize. Individual trajectories are computed by the shooting method and Newton iterations.

Near the collisions, we use the Thiele-Burrau and Levi-Civita regularizations described in (Szebehely 1967).

Finally, at some problematic parts of families, we use expansions of the periodic solutions in Fourier series and fast Fourier transformation to obtain parameters of the trajectory. This is a very powerful tool, but fairly expensive. It can also be used for accuracy control, since it is completely independent of any numerical integration: solutions of initial value problems depend on initial data, but Fourier series treat the whole trajectory as a point in some functional space.

In addition, we have an option of computation with an arbitrary precision using a computer algebra system.

After the trajectory of a periodic solution is computed, we solve the variational system of equations over the period for the trace Tr of the monodromy matrix, and over the half-period for the trace Tr_v according to Hénon and Guyot (1970, Formula (18), p. 416), i.e., we compute the stability of the periodic solution in linear approximation.

Parameters of the orbit are written in a database, which is used to extrapolate the next orbit. Then the extrapolated orbit is refined by Newton iterations. Thus we trace the family until completion (or as far as we need).

6 Generating family h ($\mu = 0$)

6.1 Description of the family

This family begins with retrograde circular orbits of infinitely small radius around the primary P_1 of bigger mass. The generating family h was described, first, in (Bruno 1996, Sect. 3) as the family $IR+$, then it was described in (Hénon 1997, Chap. 10, Sect. 10.2.6).

Tables 1–3 cite the data on 16 critical orbits of computed part of this family. In each column of these Tables, we cite the values according to the heading of the table, namely, the number of the orbit k; coordinates $x_1(0)$ and $y_2(0)$ on the symmetry plane Π; then coordinates $v_1(T/2)$ and $v_2(T/2)$ of the velocity of entry of the orbit into the point $P_2 = (1, 0)$ (if the orbit actually enters there); then the *normalized period* of the orbit $\tilde{T} \stackrel{\mathrm{def}}{=} T/(2\pi)$; the value of the Jacobi constant C; the traces (plane $\widetilde{\mathrm{Tr}}$ and vertical $\widetilde{\mathrm{Tr}}_v$); initial points of the orbit in astronomical coordinates $\tilde{a}(0)$ and $\tilde{e}(0)$ according to (2.2); then the point of the orbit after half-period $\tilde{a}(T/2)$, $\tilde{e}(T/2)$ only for the orbits 4 and 11, and in the same columns, the values $\tilde{a}'_*(T/2)$, $\tilde{e}_*(T/2)$ in

Table 1 Critical orbits of the family h for $\mu = 0$ (a)

k	$x_1(0)$	$y_2(0)$	$v_1(T/2)$	$v_2(T/2)$	$\tilde{T}$
1	−1.00000	1.00000	0	−2.00000	0.50000
2	−2.30499	−0.43384	−0.56616	0	1.50000
3	−2.17441	−0.53828	−0.00258	0.17045	1.99740
4	−2.17480	−0.53820	0	0.17049	2.00000
5	−3.17480	0			2.00000
6	−2.17480	0.53820	0	−2.17049	2.00000
7	−2.57909	0.45608	−0.22116	−2.17628	2.19007
8	−3.34552	0.29891	−0.70109	−2.00000	2.50000
9	−4.25811	−0.23485	−0.76515	0.00000	3.50000
10	−4.03954	−0.31344	−0.00124	0.26615	3.99917
11	−4.03968	−0.31343	0	0.26616	4.00000
12	−5.03968	0			4.00000
13	−4.03968	0.31343	0	−2.26616	4.00000
14	−4.12806	0.30666	−0.08366	−2.26592	4.05653
15	−5.08933	0.19649	−0.80351	−2.00000	4.50000
16	−5.86227	−0.17058	−0.82942	0	5.50000

Table 2 Critical orbits of the family h for $\mu = 0$ (b)

k	C	$\tilde{\text{Tr}}$	$\tilde{\text{Tr}}_v$
1	−1.000000	$[-2, +\infty]$	$[-2, -\infty]$
2	2.679465	$+\infty$	$[-\infty, +\infty]$
3	2.970940	$[+\infty, -\infty]$	$+\infty$
4	2.970934	$[-\infty, 2]$	$[+\infty, 2]$
5	0.629961	2	2
6	−1.711013	$[2, -\infty]$	$[2, +\infty]$
7	−1.785103	$[-\infty, +\infty]$	$+\infty$
8	−1.491531	$+\infty$	$[+\infty, -\infty]$
9	2.414539	$+\infty$	$[-\infty, +\infty]$
10	2.929162	$[+\infty, -\infty]$	$+\infty$
11	2.929161	$[-\infty, 2]$	$[+\infty, 2]$
12	0.396850	2	2
13	−2.135461	$[2, -\infty]$	$[2, +\infty]$
14	−2.141393	$[-\infty, +\infty]$	$+\infty$
15	−1.645629	$+\infty$	$[+\infty, -\infty]$
16	2.312067	$+\infty$	$[-\infty, +\infty]$

coordinates (3.8) for other orbits (coordinates $\tilde{a}(T/2), \tilde{e}(T/2)$ are cited only for those orbits where $|\tilde{e}(T/2)| \leq 2$, and coordinates $\tilde{a}'_*(T/2), \tilde{e}_*(T/2)$ for those orbits where $|\tilde{e}_*(T/2)| \geq 2$). In each Table 1–3, the number of the orbit k is cited again. Figure 3 shows 6 critical orbits in coordinates x_1, x_2; each orbit is marked by its number k from Table 1. Characteristics of the family in coordinates x_1, y_2 are shown in Fig. 4, and in coordinates x_1, C, they are shown in Fig. 5. Figure 6 shows characteristics of the family in coordinates $\tilde{a}, \tilde{e}$. In all Figures, crosses mark the critical orbits from Tables 1–3.

The generating family h begins as a part of the family Ir of retrograde circular orbits of small radius around the primary P_1 of mass 1. This part ends with the orbit 1, where the family h transforms into the part of the family A_0 with $\tilde{e} > -1$, which is continued up to the orbit 4. Here the orbit 2 has collision with the body P_2, i.e., at the point P_2 the velocity coordinate $v_2 = 0$ (see (3.9)), and this collision is preserved

Table 3 Critical orbits of the family h for $\mu = 0$ (c)

k	$\tilde{a}(0)$	$\tilde{e}(0)$	$\tilde{a}'_*(T/2)$	$\tilde{e}_*(T/2)$
1	−1.00000	−1.00000		
2	−1.47175	0.43384	−3.11978	2.00000
3	−1.58720	0.63003	−34.41170	−66.96938
4	−1.58740	0.62996	1.58740	1.37004
5	−1.58740	0		
6	−1.58740	−0.62996		
7	−1.76225	−0.53648	−0.20898	10.89091
8	−1.96669	−0.29891	−0.22264	4.02289
9	−2.41232	0.23485	−1.70806	2.00000
10	−2.51977	0.39686	−14.11681	−215.81601
11	−2.51984	0.39685	2.51984	1.60315
12	−2.51984	0		
13	−2.51984	−0.39685		
14	−2.56117	−0.38821	−0.19450	28.10268
15	−2.82190	−0.19649	−0.21526	3.68244
16	−3.20444	0.17058	−1.45363	2.00000

for $\mu > 0$. The Jacobi constant C attains its maximum at the orbit 3. At the orbit 4, the family h becomes the family $E^+_{1/2}$ up to the orbit 6. Here the orbit 5 has collision with the body P_1. From the orbit 6, the family h continues as the family A_1 up to the orbit 11. Here the Jacobi constant C attains its minimum at the orbit 7. At the orbit 8, there is a jump of the vertical trace $\tilde{\mathrm{Tr}}_v$ from $+\infty$ to $-\infty$. At the orbit 9, there is a collision with the body P_2 of mass zero (see (3.9)). At the orbit 10, the Jacobi constant C attains its maximum. From the orbit 11, the family h continues as the family $E^+_{1/4}$ up to the orbit 13. Here the orbit 12 has collision with the body P_1. From the orbit 13, the family h continues as the family A_0 up to intersection with the family $E^+_{1/6}$. Here the Jacobi constant C attains its minimum at the orbit 14; at the orbit 15, the vertical trace $\tilde{\mathrm{Tr}}_v$ jumps fromâ $+\infty$ to $-\infty$, and at the orbit 16, there is a collision with the body P_2 (see (3.9)). As a whole, the family h consists of the parts

$$\{A_0, E^+_{1/(4k-2)}, A_1, E^+_{1/(4k)}\}, \quad k = 1, 2, \ldots$$

Above we described the first such part ($k = 1$) and the beginning of the second ($k = 2$).

6.2 Characteristics of the family

Characteristics of the family h are shown in Figs. 4–7 in the coordinate systems I–IV, respectively. Thick lines denote there the primaries P_1 and P_2, and for every part, we cite the families A_0, A_1, E^+_N, where that part belongs to. Crosses on a characteristic mark the critical orbits 1–16 from Tables 1–3. In Fig. 4, the dotted line isolates the domain $|\tilde{e}| \leq 2$ that goes to Fig. 6. Figure 7 consists of two fragments with different scales in order to avoid the parts of characteristics near $|\tilde{e}_*| = 2$. Here the first part of the family A_0 corresponds to the upper characteristics of both fragments of Fig. 7. According to (3.8), they have vertical asymptotes at $\tilde{a}'_* \approx -34.4$ ($\tilde{e}_* \to -\infty$, orbit 4), and at $\tilde{a}'_* \approx -0.25$ ($\tilde{e}_* \to +\infty$, orbit 1). The part of the family A_1 corresponds to middle characteristics with the vertical asymptotes at $\tilde{a}'_* \approx -14.1$ ($\tilde{e}_* \to -\infty$, orbit 11), and at $\tilde{a}'_* \approx -0.21227$ ($\tilde{e}_* \to +\infty$, orbit 6). The second part of the family A_0

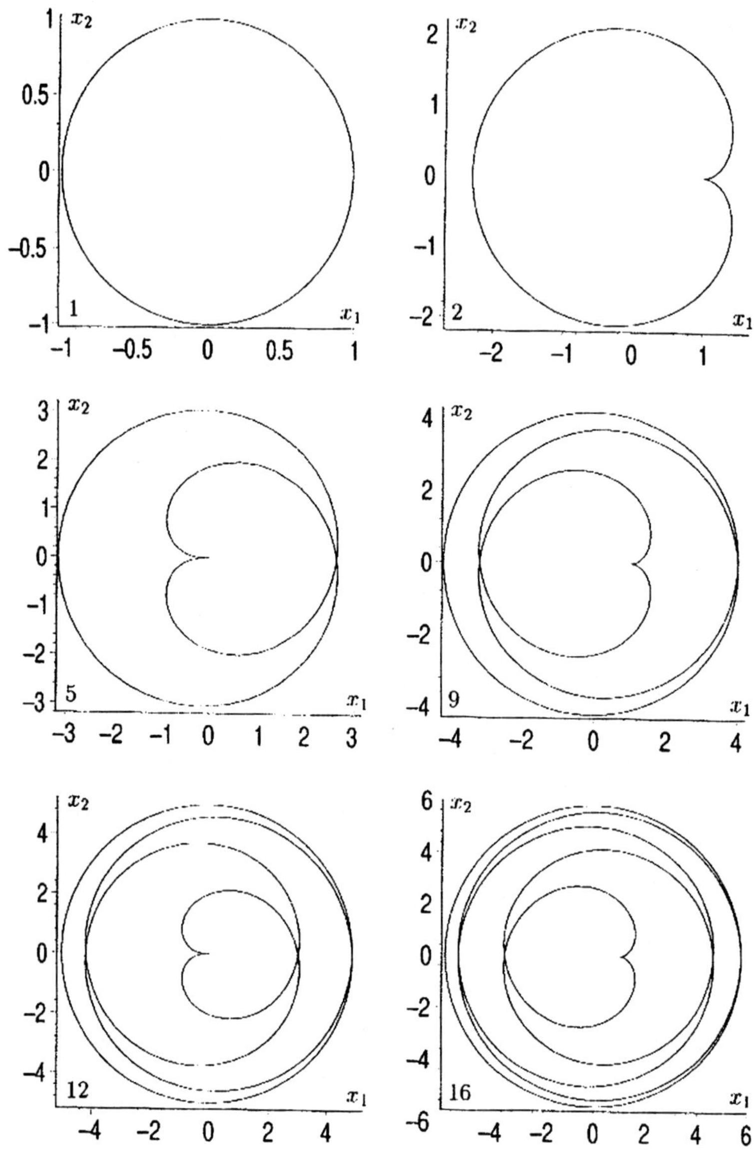

Fig. 3 Some orbits of the family h for $\mu = 0$

corresponds to the lower characteristics of both fragments of Fig. 7 with the vertical asymptotes at $\tilde{a}'_* \approx -10.9$ ($\tilde{e}_* \to -\infty$, no critical orbit), and at $\tilde{a}'_* \approx -0.19472$ ($\tilde{e}_* \to +\infty$, orbit 13).

6.3 Period and traces

Figure 8 shows the dependence of the normalized period $\tilde{T}$ on the coordinate $x_1(0)$ of the left characteristics of the family h. Horizontal parts $\tilde{T} = 2$ and $\tilde{T} = 4$ are passed

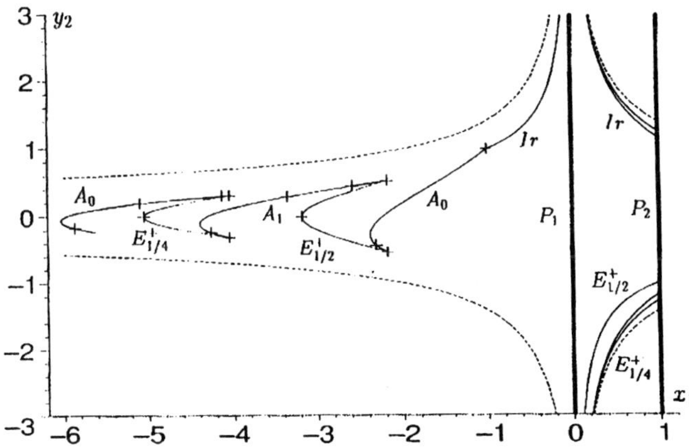

Fig. 4 Characteristics of the family h in coordinates x_1, y_2; $\mu = 0$

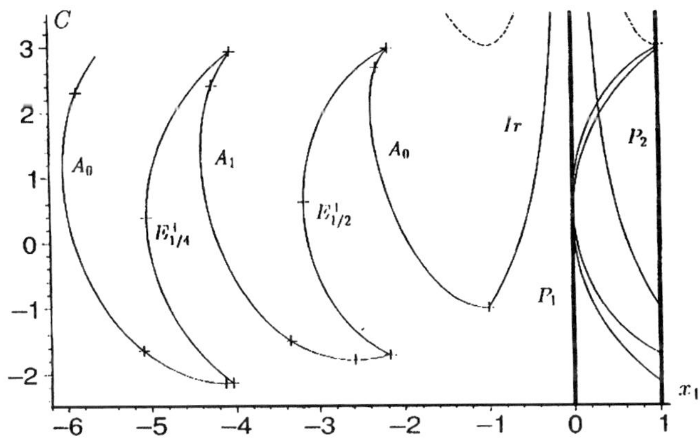

Fig. 5 Characteristics of the family h in coordinates x_1, C; $\mu = 0$

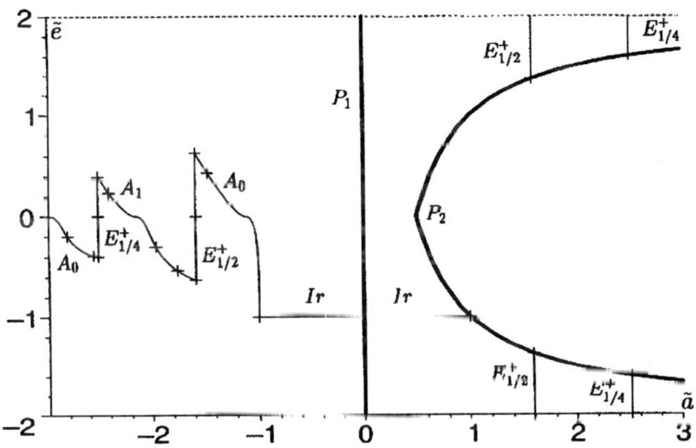

Fig. 6 Characteristics of the family h in coordinates $\tilde{a}, \tilde{e}$; $\mu = 0$

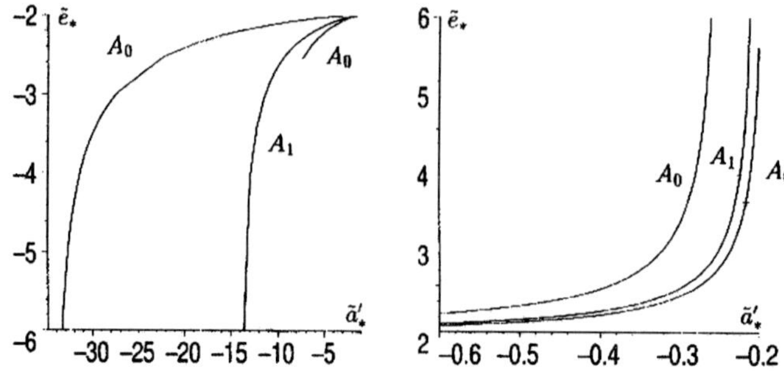

Fig. 7 Characteristics of the family h in coordinates $\tilde{a}'_*, \tilde{e}_*$; $\mu = 0$

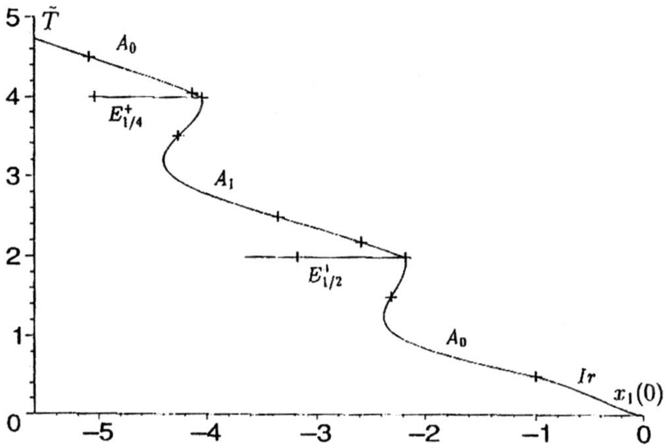

Fig. 8 The dependence of the normalized period $\tilde{T}$ on $x_1(0)$; $\mu = 0$

twice: first, from right to left (from the orbit 4 to the orbit 5, and from the orbit 11 to the orbit 12), and then backwards (from the orbit 5 to the orbit 6, and from the orbit 12 to the orbit 13).

Figure 9 shows in sketch the dependence of the plane trace $\tilde{\mathrm{Tr}}$ on the normalized period $\tilde{T}$. On the parts A_0, A_1, A_0, the trace $\tilde{\mathrm{Tr}} = +\infty$. The jump occurs on the orbits where the Jacobi constant C attains an extremum. According to Formulas (3.4) and (3.5), on the parts Ir and $E^+_{1/2}$, $E^+_{1/4}$, the trace $\tilde{\mathrm{Tr}} = 2\cos T$ and $\tilde{\mathrm{Tr}} = 2$, respectively. The families $E^+_{1/2}$ and $E^+_{1/4}$ have constant values $\tilde{T} = 2$ and 4, respectively. In Fig. 7, the points $\tilde{\mathrm{Tr}} = 2$ and 4 are blown into small intervals for clarity. In matching places of parts of different families, the trace $\tilde{\mathrm{Tr}}$ jumps from a finite value to infinity (see Table 2, column $\tilde{\mathrm{Tr}}$). The relation of $\tilde{\mathrm{Tr}}$ with Tr is in Formula (2.6).

Figure 10 shows in sketch the dependence of the vertical trace $\tilde{\mathrm{Tr}}_v$ on the normalized period $\tilde{T}$. On the parts A_0, A_1, A_0, the trace $\tilde{\mathrm{Tr}}_v = \pm\infty$. According to Table 2, the jump occurs on the orbits with half-integer $\tilde{T} > 1$, i.e., $\tilde{T} = (2k+1)/2$, $k = 1, 2, \ldots$, where either $v_2 = -2$ or $v_2 = 0$. According to Formulas (3.4) and (3.5), on the parts $Ir, E^+_{1/2}, E^+_{1/4}$, the vertical and the plane traces are equal: $\tilde{\mathrm{Tr}} = \tilde{\mathrm{Tr}}_v$. In Fig. 10, the parts corresponding to the families $E^+_{1/2}$ and $E^+_{1/4}$ are also blown into intervals. In matching

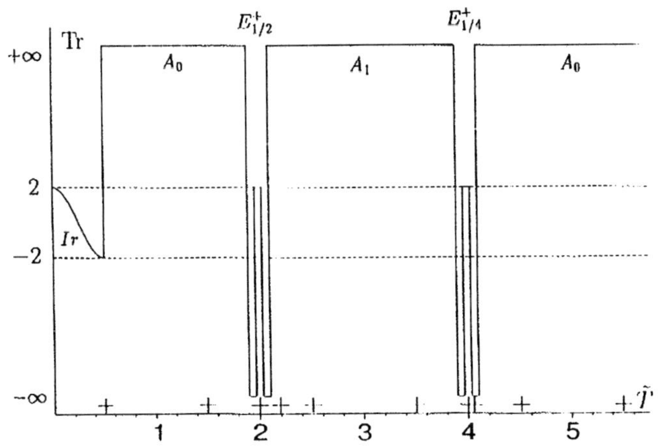

Fig. 9 The dependence of the plane trace $\tilde{\text{Tr}}$ on the normalized period $\tilde{T}$; $\mu = 0$

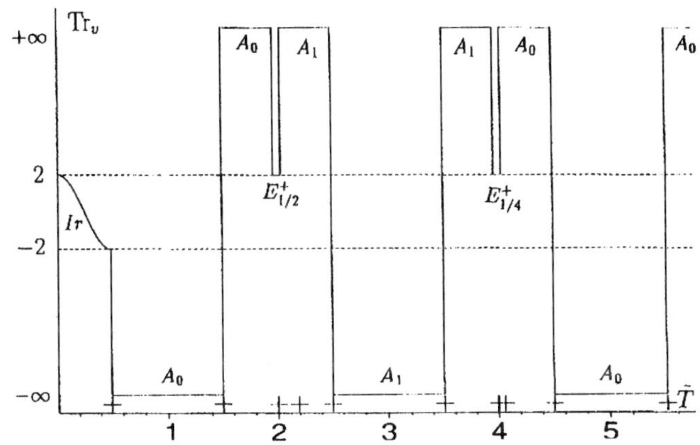

Fig. 10 The dependence of the vertical trace $\tilde{\text{Tr}}$ on the normalized period $\tilde{T}$; $\mu = 0$

places of parts of different families, the trace $\tilde{\text{Tr}}_v$ jumps from a finite value to infinity (see Table 2, column $\tilde{\text{Tr}}_v$).

We remark that the linear stability in both directions (plane and vertical), i.e., inequalities (3.4), is present only on parts of families Ir, $E^+_{1/2}$, $E^+_{1/4}$ included in the generating family h.

6.4 Intersections with other families

On the part Ir, the family h intersects with families $E^{\pm}_N$ with $1 < N < \infty$ (see (Bruno, 1990, Chapts. VII and VIII)). In particular, for $N = 2$, the family h intersects the family E_2 as locally three-multiple, since at the orbit of intersection for the family h, the trace $\tilde{\text{Tr}} = -1$. The family E_2 is a part of the generating family i. Besides, every family E_N with $N = (p+1)/p$ is included in the generating family i and intersects with the family h as locally $(2p+1)$-multiple. At the orbit 1, the trace $\tilde{\text{Tr}} = -2$, and for $\mu > 0$, here forms an interval with $\tilde{\text{Tr}} \leq -2$. Hence at the orbit 1, two locally two-multiple families b and c terminate. In addition, at the orbits where the trace $\tilde{\text{Tr}}$ jumps over the

Table 4 Critical orbits of the family h for $\mu = \mu_J$ (a)

k	$x_1(0)$	$y_2(0)$	$x_1(T/2)$	$y_2(T/2)$	$\tilde{T}$
1	−0.93273	1.03212	0.92705	−1.04392	0.47360
2	−1.00332	0.98645	0.98033	−1.03241	0.50001
3	−1.00405	0.98590	0.98066	−1.03262	0.50026
4	−1.07156	0.92757	0.99413	−1.07961	0.52256
5	−2.30488	−0.43191	1.00000	−55.59079	1.49208
6	−2.30416	−0.43268	1.00000	−63.08153	1.49367
7	−2.29900	−0.43812	1.00000	∞	1.50506
8	−2.19052	−0.52696	0.98811	1.44489	1.84046
9	−2.19074	−0.52691	0.98782	1.44042	1.84168
10	−2.31420	−0.48135	0.86285	1.31876	1.97199
11	−2.35330	−0.46640	0.82329	1.35455	1.98025
12	−3.17449	0.00113	0.00000	∞	2.00034
13	−2.37247	0.46316	0.80700	−1.36105	2.00249
14	−2.29034	0.51365	0.99229	−1.23978	2.05359
15	−2.57854	0.45657	0.99821	−1.41905	2.18864
16	−2.58341	0.45560	0.99824	−1.42222	2.19078
17	−3.34472	0.29965	0.99957	−1.98579	2.49962
18	−3.34546	0.29949	0.99957	−1.98643	2.49991
19	−4.25797	−0.23353	1.00000	−164.34632	3.49615
20	−4.25750	−0.23375	1.00000	−184.15620	3.49681
21	−4.25350	−0.23556	1.00000	∞	3.50241
22	−4.05387	−0.30759	0.98743	1.48502	3.89752
23	−4.05394	−0.30758	0.98734	1.48387	3.89787
24	−4.21783	−0.27652	0.82396	1.43621	3.98844
25	−4.32339	−0.25490	0.71769	1.55118	3.99379
26	−5.03923	0.00075	0	∞	4.00034
27	−4.25718	0.27118	0.78493	−1.46756	4.00165
28	−4.11287	0.30792	0.99351	−1.33511	4.04224
29	−4.13223	0.30653	0.99507	−1.35369	4.05482
30	−4.15581	0.30468	0.99615	−1.37579	4.06927
31	−5.08865	0.19730	0.99965	−2.18556	4.49966
32	−5.08916	0.19723	0.99966	−2.18606	4.49988
33	−5.86212	−0.16939	1.00000	−221.04204	5.49685
34	−5.86169	−0.16952	1.00000	−249.66849	5.49737
35	−5.85779	−0.17069	1.00000	∞	5.50206

interval $[-2,2]$, there are also intersections with other families. At the orbits where the vertical trace $\tilde{\mathrm{Tr}}_v$ jumps over the interval $[-2,2]$, there are intersections of the family h with the space families of doubly symmetric periodic solutions of the space circular restricted three-body problem (Bray and Goudas 1967).

7 Sun-Jupiter case ($\mu = 0.00095388$)

7.1 Description of the family

Here all computations for the family h were done for $\mu = \mu_J \stackrel{\text{def}}{=} 0.00095$ for technical reasons. For this case, the part of the family h was computed in (Bruno 1996, Sect. 4) as the family $IR+J$. Tables 4–6 cite the data on 35 critical orbits of the computed part of the family h. Tables 4–6 are organized in the same way as Tables 1–3, only instead of $v_1(T/2)$ and $v_2(T/2)$, we give the values $x_1(T/2)$ and $y_2(T/2)$. In addition, the pairs

Table 5 Critical orbits of the family h for $\mu = \mu_J$ (b)

k	C	$\tilde{\text{Tr}}$	$\tilde{\text{Tr}}_v$	$\tilde{a}(0)$	$\tilde{e}(0)$
1	−0.844734	−2.00	−1.94	−0.928	−0.995
2	−0.957252	−2.00	−1.99	−0.981	−0.977
3	−0.957916	−1.99	−2.00	−0.981	−0.977
4	−0.979715	2.00	−3.47	−0.995	−0.923
5	2.677267	15.08	−2.00	−1.468	0.430
6	2.679804	15.06	2.00	−1.469	0.432
7	2.697543	14.96	5.79	−1.475	0.442
8	2.948799	2.00	3.57	−1.575	0.609
9	2.948795	−2.00	3.54	−1.575	0.609
10	2.865520	−3.28	2.00	−1.582	0.537
11	2.832678	−2.00	1.99	−1.582	0.512
12	0.629660	1.85	2.00	−1.587	0.000
13	−1.563962	1.95	2.00	−1.592	−0.509
14	−1.738441	−2.00	3.52	−1.642	−0.605
15	−1.781763	−2.00	8.44	−1.764	−0.538
16	−1.781774	2.00	8.49	−1.766	−0.537
17	−1.489176	15.51	2.00	−1.968	−0.301
18	−1.488569	15.52	−2.00	−1.968	−0.300
19	2.412851	18.26	−2.00	−2.409	0.232
20	2.414421	18.25	2.00	−2.409	0.233
21	2.427563	18.21	6.18	−2.412	0.236
22	2.901138	2.00	3.61	−2.508	0.384
23	2.901137	−2.00	3.60	−2.508	0.384
24	2.739222	−4.08	2.00	−2.515	0.323
25	2.610757	−2.00	1.99	−2.515	0.281
26	0.399788	1.50	2.00	−2.520	0.000
27	−1.903701	1.81	2.00	−2.524	−0.313
28	−2.132758	−2.00	3.52	−2.555	−0.390
29	−2.134579	−2.00	4.36	−2.564	−0.389
30	−2.135207	2.00	5.24	−2.575	−0.386
31	−1.643370	17.62	2.00	−2.824	−0.198
32	−1.642830	17.62	−2.00	−2.824	−0.198
33	2.310540	19.64	−2.00	−3.200	0.168
34	2.311900	19.64	2.00	−3.201	0.169
35	2.324027	19.60	6.42	−3.202	0.171

of values $\tilde{a}(T/2), \tilde{e}(T/2)$ and $\tilde{a}'_*(T/2), \tilde{e}_*(T/2)$ are places in different pairs of columns. More accurate values $x_1(T/2)$ for some orbits are

k	5	6	19
$x_1(T/2)$	0.999999407087	0.999999537683	0.999999930558
k	20	33	34
$x_1(T/2)$	0.999999944993	0.999999961994	0.999999969812

and more accurate values Tr_v for the orbits 12 and 26 are

$$\text{Tr}_v(12) = 1.99999842294, \quad \text{Tr}_v(26) = 1.99999646745. \tag{7.1}$$

Table 6 in columns $\tilde{a}(T/2), \tilde{e}(T/2)$ cites only the values where $|\tilde{e}(T/2)| \leq 2$. Similarly, $\tilde{a}'_*$ and $\tilde{e}_*$ are given only for $|\tilde{e}_*| \geq 2$. Increasing number of critical orbits for $\mu = \mu_J$ in comparison with $\mu = 0$ is caused by the fact that sometimes one critical orbit for $\mu = 0$ corresponds to several critical orbits for $\mu = \mu_J$. In particular, it is true for critical orbits where the traces Tr and Tr_v are equal to ± 2. Indeed, if for $\mu = 0$ there is

Table 6 Critical orbits of the family h for $\mu = \mu_J$ (c)

k	$\tilde{a}(T/2)$	$\tilde{e}(T/2)$	$\tilde{a}'_*(T/2)$	$\tilde{e}_*(T/2)$
1	0.938	−1.011	−76.371	320.787
2	1.027	−1.046	−20.245	85.530
3	1.029	−1.047	−19.893	84.089
4	1.183	−1.160	−5.741	26.744
8			−8.993	−2.476
9			−9.213	−2.486
10	1.733	1.502	−98.658	−14.668
11	1.687	1.512	−125.037	−23.382
12	0	0	−526.317	∞
13	1.602	−1.496	−136.297	∞
14	2.096	−1.527	−5.819	40.724
15			−1.285	10.997
16			−1.266	10.868
17			−0.266	4.028
18			−0.266	4.026
22			−8.190	−3.114
23			−8.248	−3.121
24	2.758	1.701	−110.485	−35.260
25	2.644	1.729	−172.891	−90.281
26	0	0	−526.317	∞
27	2.549	−1.692	−134.228	∞
28	4.369	−1.773	−4.245	37.258
29	5.692	−1.825	−3.221	28.754
30	8.840	−1.887	−2.510	22.864
31			−0.202	3.687
32			−0.201	3.685

Table 7 Correspondence between critical orbits of Tables 1–3 (k_1) and Tables 4–6 (k_2)

k_1	1	2	3	4	5	6	7	8
k_2	1–4	5–7	8, 9	10, 11	12	13, 14	15, 16	17, 18
k_1	9	10	11	12	13	14	15	16
k_2	19–21	22, 23	24, 25	26	27, 28	29, 30	31, 32	33, 34

a jump in values of one or both traces at one orbit including the values $\mathrm{Tr} = \pm 2$, then for $\mu > 0$, the values of traces ± 2 are attained at different orbits. Table 7 shows the correspondence between critical orbits of Tables 1–3 and 4–6. There k_1 is the number of a critical orbit in Tables 1–3, and k_2 is the number of the critical orbit in Tables 4–6. The orbits for $\mu = \mu_J$ are similar to orbits for $\mu = 0$ shown in Fig. 3. Using Table 7, we can determine the numbers of orbits for $\mu = \mu_J$, which correspond to the orbits in Fig. 3.

7.2 Characteristics of the family

Characteristics of the family are shown in Figs. 11–14 in various coordinate systems. In comparison to the corresponding figures of Sect. 1, the similarity is evident; some figures seem almost identical. The crosses on characteristics mark the critical orbits from Table 4. Figure 14 shows characteristics of the family near the body P_2 in astronomical coordinates $\tilde{a}'_*$, $\tilde{e}_*$ (see (2.4)). Here the values x_1 and y_2 are taken near the

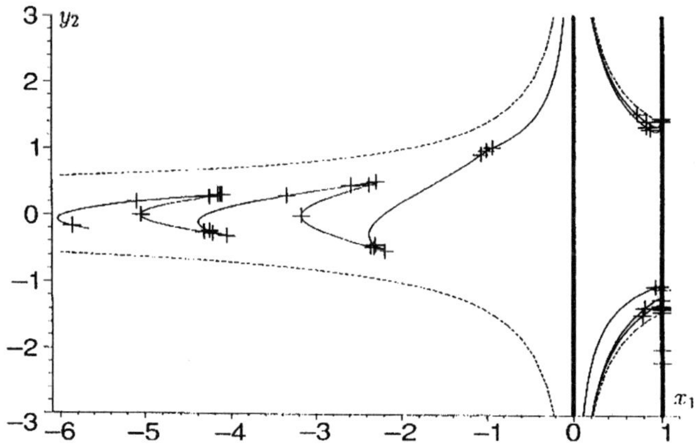

Fig. 11 Characteristics of the family h in coordinates x_1, y_2; $\mu = \mu_J$

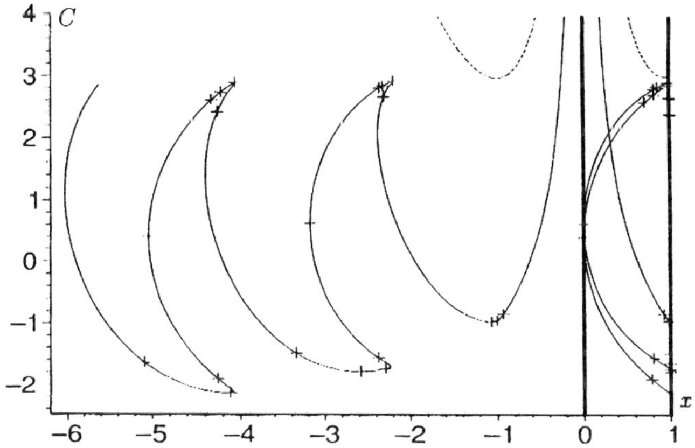

Fig. 12 Characteristics of the family h in coordinates x_1, C; $\mu = \mu_J$

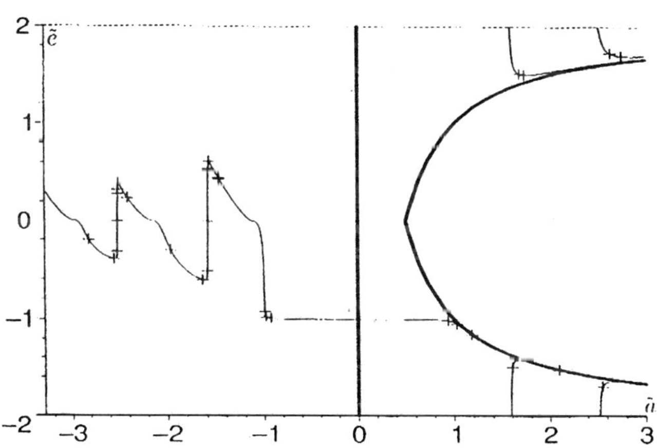

Fig. 13 Characteristics of the family h in coordinates $\tilde{a}, \tilde{e}$; $\mu = \mu_J$

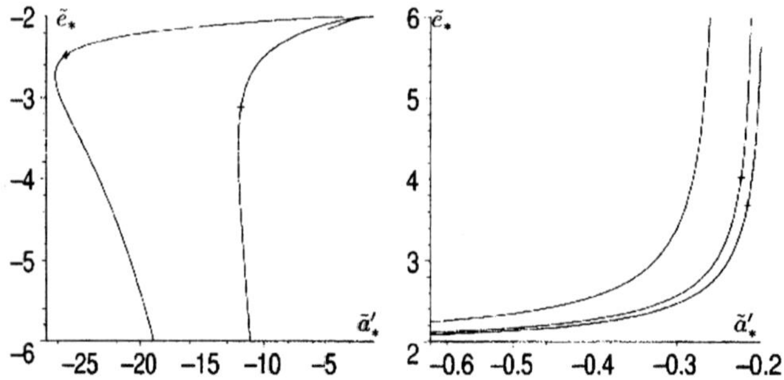

Fig. 14 Characteristics of the family h in coordinates $\tilde{a}'_*, \tilde{e}_*$; $\mu = \mu_J$

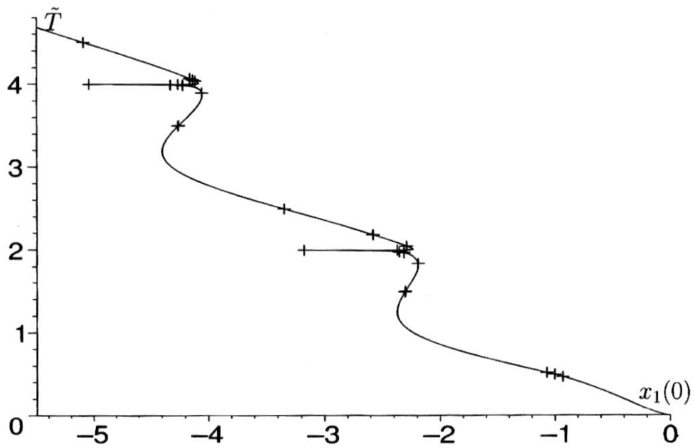

Fig. 15 The dependence of the normalized period $\tilde{T}$ on $x_1(0)$; $\mu = \mu_J$

body P_2, i.e., $x_1(T/2)$, $y_2(T/2)$, and only for $|\tilde{e}_*| \geq 2$. Here the values $|\tilde{e}_*| - 1$ and $|\tilde{a}_*| = |\tilde{a}'_*|\mu$ are the eccentricity and the real semi-axis of the hyperbolic fly-by orbit near the body P_2. Figs. 7 and 14 are almost identical. Especially at the right fragment. At the left fragment, the middle branch has almost the same vertical asymptote as for $\mu = 0$; but vertical asymptote of the left branch is shifted to the right. This means that the more the value $\tilde{a}'_*$, i.e., the farther the body P_3 fly-bys around the body P_2, the greater the perturbation caused by the change of mass of the body P_1 from 1 to $1 - \mu$.

7.3 Periods and traces

Figure 15 shows the dependence of the normalized period $\tilde{T}$ on coordinate $x_1(0)$. Crosses mark critical orbits from Table 4. Analogy with Fig. 8 is evident. Only now the period slightly increases on the parts where for $\mu = 0$ it was constant. Now the period monotonously increases on the whole family.

Figure 16 shows the dependence of the modified plane trace $\tilde{\mathrm{Tr}}$ (see (2.6)) on the normalized period $\tilde{T}$. Comparing with Fig. 9, we see that for $\mu = \mu_J$, big values of $|\tilde{\mathrm{Tr}}|$ are in places where for $\mu = 0$ there were extended intervals of infinite values. Narrow intervals of negative infinite values $\tilde{\mathrm{Tr}}$ for $\mu = 0$ are now turned into narrow intervals of small negative values $\tilde{\mathrm{Tr}}$ for μ_J. We remark that

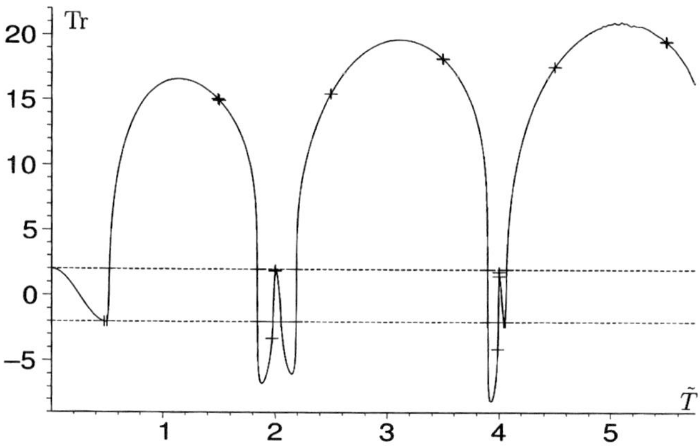

Fig. 16 The dependence of the plane trace $\tilde{\mathrm{Tr}}$ on the normalized period $\tilde{T}$; $\mu = \mu_J$

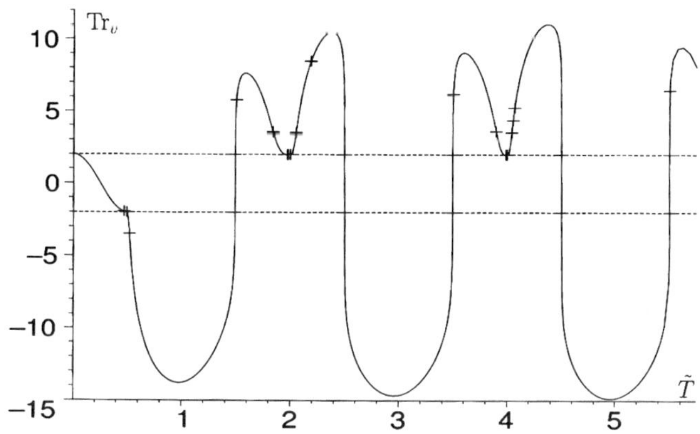

Fig. 17 The dependence of the vertical trace $\tilde{\mathrm{Tr}}$ on the normalized period $\tilde{T}$; $\mu = \mu_J$

near $\tilde{T} = 2$ and 4, the values of the trace $\tilde{\mathrm{Tr}}$ remain slightly less than 2 (see (2.1)). This is in accordance with computations of perturbations of the trace $\tilde{\mathrm{Tr}}$ on families E_N cited in (Bruno 1990, Chap. VII and Table 2 in Appendix). In Table 2 in Appendix in (Bruno 1990), there cited the perturbations Tr_1 of the trace $\tilde{\mathrm{Tr}} = 2 + \mu \mathrm{Tr}_1 + \dots$ on families $E^{\pm}_{1/2}$ and $E^{\pm}_{1/4}$. In particular, for the families $E^{+}_{1/2}$ and $E^{+}_{1/4}$, the values of the trace Tr_1 are cited as Tr^{+}_1. On intervals from the orbit 4 to the orbit 6 for $E^{+}_{1/2}$, i.e., for $|e\varepsilon'| > 1 - |\tilde{e}| \approx 0.37$, and from the orbit 11 to the orbit 13 for $E^{+}_{1/4}$, i.e., for $|e\varepsilon'| > 1 - |\tilde{e}| \approx 0.6$, the values Tr^{+}_1 are negative and tend at the ends to $-\infty$, but for $\varepsilon' = 1$ the tendency to $-\infty$ is stronger than for $\varepsilon' = -1$. Correspondingly, in Fig. 16 near $\tilde{T} = 2$ and 4, the values of the trace $\tilde{\mathrm{Tr}}$ are less on the left, i.e., for smaller values $\tilde{T}$, and they remain less than 2.

Figure 17 shows the dependence of the vertical trace $\tilde{\mathrm{Tr}}_v$ on the normalized period $\tilde{T}$. Analogy with Fig. 10 is evident. Only for the vertical trace $\tilde{\mathrm{Tr}}_v$, the theory of its regular perturbations on families E_N is absent, although it can be developed. In order to clarify the behavior of $\tilde{\mathrm{Tr}}_v$ near $\tilde{T} = 2$, Fig. 18 shows the graphs of $\tilde{\mathrm{Tr}}_v$ in two enlarged

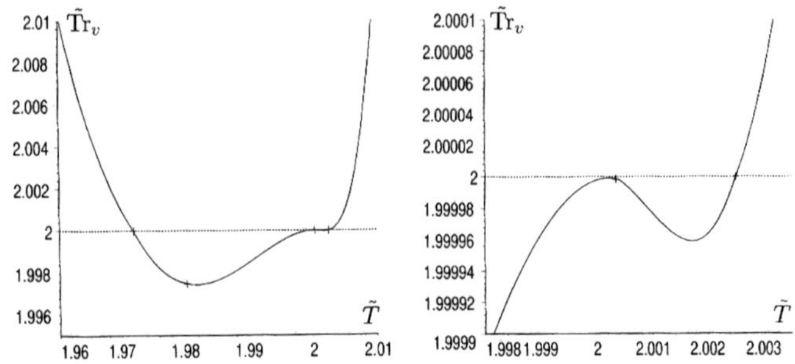

Fig. 18 Fragments of Fig. 17 at $\tilde{T} = 2$ in larger scales

scales. From there we see that between the orbits 10 and 13, everywhere $\tilde{\mathrm{Tr}}_v < 2$, and there are two minima (one is at the orbit 12). Similarly for $\tilde{T} = 4$ between the orbits 24 and 27.

From Figs. 16 and 17, and from Table 5, we see that linear stability on the family h is present only in three intervals: from the beginning to the orbit 3, between the orbits 11 and 13, and between the orbits 25 and 27. These intervals correspond to parts Ir, $E^+_{1/2}$, and $E^+_{1/4}$ of the generating family.

7.4 Intersections with other families

On the parts where $|\mathrm{Tr}| \leq 2$, the family h intersects with other families of plane periodic solutions. In particular, at the orbit with $x_1(0) \approx 0.6$, the trace $\mathrm{Tr} = -1$, and there is an intersection with the locally three-multiple family i. At the orbits 1 and 2, the family h intersects with the families a and c respectively, which are terminated at these orbits as locally two-multiple. Since $\mathrm{Tr} = 2$ is only at the orbits where the Jacobi constant C has an extremum (i.e., at the orbits 4, 8, 16, 22, 30), so the family h does not intersect with the families of asymmetric periodic solutions.

On the parts where $|\tilde{\mathrm{Tr}}_v| \leq 2$, the family h intersects with families of space doubly symmetric periodic solutions (Bray and Goudas 1967), which are as yet barely studied.

7.5 The Earth-Moon case

The Earth-Moon case $\mu = \mu_M \stackrel{\mathrm{def}}{=} 0.012155$. For this case Broucke (1968) computed 10 families of symmetric periodic solutions and published them in great detail. In particular, the family h, denoted there as A_1, was computed from the beginning to slightly further than the orbit 12 in Table 4. In (Broucke 1968) there are detailed tables of orbits of the family, as well as figures of many orbits, characteristics of the family in coordinates $x_1, H = -C/2$, and the graph of the plane trace Tr as a function of $x_1(0)$. These results may be considered as perturbations of the generating family. They were used in (Hénon 1997, Chap. 10, Sect. 10.4.7) as an example of good agreement with the generating family.

8 Evolution of the family h as μ increases

Similarly, the family h is given in detail in Preprint (Bruno and Varin 2005b) for $\mu = 0.1, 0.2$, and in Preprint (Bruno and Varin 2005c) for $\mu = 0.3, 0.4, 0.5$. The family

h and its stability for $\mu = 0.5$ were computed also in works (Strömgren 1935), (Bartlett 1964), (Hénon 1965a, b 1973).

As μ increases from zero, the family h differs from the generating one the more the further the orbit from the primaries P_1 or P_2 (i.e., the more $|\tilde{a}(0)|$ or $|\tilde{a}'_*(T/2)|$). This is valid both, for coordinates of the orbit, and for their traces. For regular perturbation, this is in agreement with Tables 1 and 2 in Appendix in (Bruno 1990), where the cited perturbations for families G_p and E_N increase in absolute value as a increases.

As figures in (Bruno and Varin 2005b, c) show, characteristics of the family h does not have self-bifurcations as μ increases, i.e., it is mapped one-to-one onto the band

$$T > 0, \quad \mu \in [0, 0.5]$$

and unified by these two parameters as two-parametric family. Here the critical orbits form its subfamilies. Some of these subfamilies for $\mathrm{Tr} = \pm 2$ were computed in (Hénon and Guyot 1970, Sect. 3.3) and shown there in Figs. 6 and 7, and in tables on pages 369–371. By the orbits with $k = 1$ and 2, for all $\mu > 0$, there are intersections with families a and e. In addition, as μ increases from 0 to 0.5, the family becomes more uniform. If for $\mu = 0$, it is composed from the parts of various families Ir, E_N, A_0, and A_1 with different behavior of period and traces, so for $\mu = 0.5$, it cannot be divided in parts with different behavior of these values, and its characteristics look more similar as well. We remark that characteristics in System IV undergo great changes for $\tilde{a}'_* < -1$, and change little for $\tilde{a}'_* > -1$.

We remark also that for $\mu > 0.3$, the intervals of complete linear stability coincide with the intervals of plane linear stability, i.e., the vertical component does not introduce additional instability.

Preliminary version of this article was published in preprints (Bruno and Varin 2005a, b, c).

Acknowledgements The authors are grateful to the Associate Editor and Reviewer 1 for useful remarks. This work was supported by RFBR, Grant 05-01-00050.

References

Abalakin, V.K., Aksenov, E.P., Grebennikov, E.A., Riabov, Yu.A.: Reference Guide on Celestial Mechanics and Astrodynamics, p. 584 Nauka, Moscow, (Russian) (1971)

Bartlett, J.H.: The restricted problem of three bodies (1). Kong. Dan. Vidensk, Selsk. Mat.-Fys. Skr. **2**(7), 48 (1964)

Bray, T.A., Goudas, C.L., Three-dimensional periodic oscillations about L_1, L_2, and L_3. In: Kopal, Z. (ed.), Advances in Astronomy and Astrophysics. Academic Press, New York and London, **5**: 71–130 (1967)

Breakwell, J.V., Perko, L.M.: Second order matching in the restricted three-body problem (small μ). Celest. Mech. **9**(4), 437–450 (1974)

Broucke, M.R.: Periodic orbits in the restricted three-body problem with earth–moon masses. NASA Technical Report 32–1168, p. 92, Pasadena, (1968)

Bruno, A.D.: Researches on the restricted three-body problem. I. Periodic solutions to Hamiltonian system. Preprint of KIAM, **18**, p. 44, Moscow, (Russian) (1972). Cosmic Research, **44**(3) (English) (2006)

Bruno, A.D.: Researches on the restricted three-body problem. II. Periodic solutions and arcs for $\mu = 0$. Celest. Mech. **18**(1), 9–50 (1978a)

Bruno, A.D.: Researches on the restricted three-body problem. III. Properties of solutions for $\mu = 0$. Celest. Mech. **18**(1), 51–101 (1978b)

Bruno, A.D.: On periodic flybys of the moon. Preprint of KIAM, **91**, p. 25, Moscow, (Russian) (1978c)

Bruno, A.D.: On periodic flybys of the moon. Celest. Mech. **24**(3), 255–268 (1981)

Bruno, A.D.: The Restricted 3-Body Problem. Plane Periodic Orbits, Nauka, p. 296, Moscow, (Russian) (1990), Walter de Gruyter, p. 368, Berlin, (English) (1994)

Bruno, A.D.: Periodic solutions to the restricted three-body problem of multiplicity one in Sun-Jupiter case. Preprint of KIAM, **66**, p. 27, Moscow, (Russian) (1993)

Bruno, A.D.: Zero-multiple and retrograde periodic solutions to the restricted three-body problem. Preprint of KIAM, **93**, p. 32, Moscow, (Russian) (1996)

Bruno, A.D.: Power Geometry in Algebraic and Differential Equations, p. 288, Fizmatlit, Nauka, (Russian) (1998), p. 385, Elsevier, Amsterdam, (English) (2000)

Bruno, A.D., Varin, V.P.: On families of periodic solutions to the restricted three-body problem. Preprint of KIAM, **10**, p. 20, Moscow, (Russian) (2005a)

Bruno, A.D., Varin, V.P.: Family *h* of periodic solutions of the restricted problem for small μ. Preprint of KIAM, **67**, p. 32, Moscow, (Russian) (2005b)

Bruno, A.D., Varin, V.P.: Family *h* of periodic solutions of the restricted problem for big μ. Preprint of KIAM, **64**, p. 31, Moscow, (Russian) (2005c)

Hénon, M.: Exploration Numérique du Problème Restreint. I – Masses égales, orbites périodique. Ann. Astrophys. **28**(3), 499–511 (1965a)

Hénon, M.: Exploration Numérique du Problème Restreint. II – Masses égales, stabilité des orbites périodique. Ann. Astrophys. **28**(6), 992–1007 (1965b)

Hénon, M.: Sur les orbits interplanetaires qui rencontrent deux fois la terre. Bull. Astron. Ser. 3. **3**(3), 377–402 (1968)

Hénon, M., Guyot M.: Stability of periodic orbits in the restricted problem. In: Giacaglia G.E.O. (ed.) Periodic Orbits, Stability and Resonances, pp. 349–374. Dordrecht-Holland: D Reidel P.C. (1970)

Hénon, M.: Vertical stability of periodic orbits in the restricted problem. Astron. Astrophys. **28**, 415–426 (1973)

Hénon, M.: Generating Families of the Restricted Three-Body Problem, Lect. Notes in Phys. monographs 52, Springer, Berlin, (1997).

Hénon, M.: Generating Families of the Restricted Three-Body Problem. II. Quantitative Study of Bifurcations, Lect. Notes in Phys. monographs 65, Springer, Berlin (2001).

Hitzl, D.L., Hénon, M.: Critical generating orbits for second species periodic solutions of the restricted problem. Celest. Mech. **15**(4), 421–452 (1977a)

Hitzl, D.L., Hénon, M.: The stability of second species periodic orbits in the restricted problem ($\mu = 0$). Acta Astronaut. **4**, 1019–1039 (1977b)

Kotoulas, T., Voyatzis, G.: Comparative study of the 2:3 and 3:4 resonant motion with Neptune: an application of symplectic mappings and low frequency analysis. Celest. Mech. Dynam. Astron. **88**(4), 343–363 (2004)

Perko, L.M.: Second species solutions with an $O(\mu^\nu)$, $0 < \nu < 1$, near Moon passage. Celest. Mech. **24**(2), 155–171 (1981)

Strömgren, E.: Connaissance Actuelle des Orbits dans le Problème des Trois Corps, In: Publications and Minor Communications of Copenhagen Observatory, Publication 100. Copenhagen University, Astronomical Observatory, Denmark (1935)

Szebehely, V.: *Theory of Orbits. The Restricted Problem of Three Bodies*, p. 668, Academic Press, New York and London (1967), Russian translation: p. 656, Nauka, Moscow (1982).

Voyatzis, G., Kotoulas, T.: Planar periodic orbits in exterior resonances with Neptune. Planetary and Space Science. **53**, 1189–1199 (2005)

Voyatzis, G., Kotoulas, T., Hadjidemetriou, J.D.: Symmetric and nonsymmetric periodic orbits in the exterior mean motion resonances with Neptune. Celest. Mech. Dynam. Astron. **91**(4), 191–202 (2005)

Celestial Mechanics and Dynamical Astronomy (2006) 95:55–66
DOI 10.1007/s10569-006-9016-y

ORIGINAL ARTICLE

Hip-hop solutions of the 2N-body problem

Esther Barrabés · Josep Maria Cors ·
Conxita Pinyol · Jaume Soler

Received: 15 November 2005 / Revised: 29 January 2006 /
Accepted: 26 March 2006 / Published online: 17 August 2006

Abstract Hip-hop solutions of the 2N-body problem with equal masses are shown to exist using an analytic continuation argument. These solutions are close to planar regular 2N-gon relative equilibria with small vertical oscillations. For fixed N, an infinity of these solutions are three-dimensional choreographies, with all the bodies moving along the same closed curve in the inertial frame.

Keywords N-body problem · Analytic continuation · Hip-hop · Choreographies

1 Introduction

The equal-mass n-body problem has recently attracted much attention thanks to the work of Chenciner and other authors on the type of orbits called hip-hop solutions, and on the solutions that have eventually been called choreographies.

In a hip-hop solution, 2N bodies of equal mass stay for all time in the vertices of a regular rotating anti-prism whose basis, i.e. the regular polygons that define it, perform an oscillatory motion separating, reaching a maximum distance, approaching,

E. Barrabés (✉)
Departament d'Informàtica i Matemàtica Aplicada, Universitat de Girona, Girona, Spain
e-mail: barrabes@ima.udg.es

J. Maria Cors
Departament de Matemàtica Aplicada III, Universitat Politècnica de Catalunya, Barcelona, Spain
e-mail: cors@eupm.upc.es

C. Pinyol
Departament d'Economia i Història Econòmica, Universitat Autònoma de Barcelona, Barcelona, Spain
e-mail: conxita.pinyol@uab.es

J. Soler
Departament d'Informàtica i Matemàtica Aplicada, Universitat de Girona, Girona, Spain
e-mail: jaume.soler@ima.udg.es

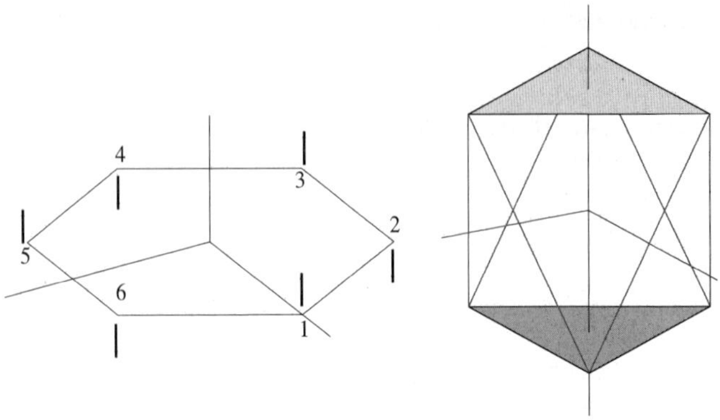

Fig. 1 Qualitative representation of a hip-hop motion in the case of 6 bodies

crossing each other, and so on, as sketched in Fig. 1 for $N = 3$. The $2N$ bodies can be arranged in two groups of N, each group moving on its plane on a rotating regular N-gon configuration (homographic), while the planes are always perpendicular to the z-axis, oscillate along this axis, and coincide (with opposite velocities) at regular intervals when they cross the origin. The orthogonal projection of both N-gons on the $z = 0$ plane is always a regular rotating $2N$-gon.

On the other hand, a choreography is a solution in which n bodies move along the same closed line in the inertial frame, chasing each other at equi-spaced intervals of time. It is well known the 'figure-eight' choreography in the three–body problem, shown by Chenciner and Montgomery (2000) in a most celebrated paper. A great many choreographies with $n > 3$ have been shown numerically to exist by Simo (2001).

The above results were obtained mostly by means of variational methods, which make it possible to find solutions that do not depend on a small parameter, i.e. far from solutions of an integrable problem (see Chenciner and Venturelli 2000; Chenciner et al. 2002; Chenciner and Fèjoz 2005 and references therein for details). In the case of hip-hop solutions, the question arises whether in some simple cases they could be obtained through the traditional analytic continuation method of Poincaré, which would give families (differentiable with respect to a parameter) of periodic solutions, at least in a rotating frame. In this respect, mention should be made of a result by Meyer and Schmidt Meyer and Schmidt (1993) on a similar solution with a large central mass and $2n$ very small, equal masses around it which was suggested as a model for the braided structure of some of Saturn's rings.

In this paper, we show that Poincaré's argument of analytic continuation can be used to add vertical oscillations to the circular motion of $2N$ bodies of equal mass occupying the vertices of a regular $2N$-gon. In this way, a family of three-dimensional orbits, periodic in a rotating frame, can be shown to exist. This is a Lyapunov family of orbits whose periods tend to the period of the vertical oscillations of the linearized system around the relative equilibrium solution. These solutions were found numerically by Davies et al. (1983). Infinitely many of this solutions are periodic in the inertial frame, provided that the quantity $H(N)$ given by (27) does not vanish, and are three-dimensional choreographies, in the sense that all bodies move at equi-spaced time intervals along a closed twisted curve in the inertial frame.

Some solutions found in our article may coincide with the generalized hip-hop solutions obtained by Chenciner (2002). Terracini and Venturelli (2005) recently showed the existence of hip-hop solutions in the same problem using variational methods, adding vertical variations to the planar relative equilibrium in order to reduce the value of the action functional. The variational approach does not depend on any small parameter and yields global existence, while continuation methods give explicit approximations to solutions in a small neighbourhood of the relative equilibrium. A precise comparison of both methods from a purely analytic point of view would involve either estimating the distance from the variational solutions to the relative equilibrium or estimating the size of the neighbourhood in which the family can be continued, but both questions seem far from easy.

2 Equations of motion

Consider 2N bodies with equal mass m moving under their mutual gravitational attraction and let $(\mathbf{r}_i, \dot{\mathbf{r}}_i), i = 1, \ldots, 2N$, be their positions and velocities. The equations of motion of the 2N-body problem are

$$\ddot{\mathbf{r}}_i = Gm \sum_{k=1, k\neq i}^{2N} \frac{\mathbf{r}_k - \mathbf{r}_i}{r_{ki}^3}, \tag{1}$$

where $r_{ki} = |\mathbf{r}_k - \mathbf{r}_i|$. Scaling the time t by $\sqrt{Gm}\,t$ the Lagrangian function associated to the problem becomes

$$\mathcal{L} = \sum_{i=1}^{2N} \frac{1}{2} |\dot{\mathbf{r}}_i|^2 + \sum_{1\leq i<j\leq 2N} \frac{1}{|\mathbf{r}_i - \mathbf{r}_j|}.$$

As we are looking for solutions of the 2N-body problem such that all the bodies stay for all time on the vertices of an anti-prism, it will suffice to know the position and velocity of one of the 2N bodies. Given $\mathbf{r}_1 = \mathbf{r}_1(t)$, we define

$$\mathbf{r}_i = R^{i-1}\mathbf{r}_1, \qquad \dot{\mathbf{r}}_i = R^{i-1}\dot{\mathbf{r}}_1$$

for $i = 2, \ldots, 2N$, where R is a rotation plus a reflection in such a way that all the bodies are on the vertices of an anti-prism. Since in this configuration N of the bodies on a plane and the other N on a parallel plane we can assume, without loss of generality, that both planes are perpendicular to the z axis. In this case, the matrix R can be written as

$$R = \begin{pmatrix} \cos(\frac{\pi}{N}) & -\sin(\frac{\pi}{N}) & 0 \\ \sin(\frac{\pi}{N}) & \cos(\frac{\pi}{N}) & 0 \\ 0 & 0 & -1 \end{pmatrix}.$$

Proposition 1 *$(\mathbf{r}_1, \mathbf{r}_2, \ldots, \mathbf{r}_{2N}) = (\mathbf{r}, R\mathbf{r}, \ldots, R^{2N-1}\mathbf{r})$ is a solution of the 2N-body problem given by (1) if and only if $\mathbf{r}(t)$ satisfies the equation*

$$\ddot{\mathbf{r}} = \sum_{k=1}^{2N-1} \frac{(R^k - I)\mathbf{r}}{|(R^k - I)\mathbf{r}|^3}. \tag{2}$$

Proof Substituting $(\mathbf{r}_1, \mathbf{r}_2, \ldots, \mathbf{r}_{2N}) = (\mathbf{r}_1, R\mathbf{r}_1, \ldots, R^{2N-1}\mathbf{r}_1)$ in (1) we have

$$R^{i-1}\ddot{\mathbf{r}}_1 = \sum_{j=1, j\neq i}^{2N} \frac{(R^{j-1} - R^{i-1})\mathbf{r}_1}{|(R^{j-1} - R^{i-1})\mathbf{r}_1|^3} = \sum_{j=1, j\neq i}^{2N} \frac{R^{i-1}(R^{j-i} - I)\mathbf{r}_1}{|(R^{j-i} - I)\mathbf{r}_1|^3}$$

for $i = 1, \ldots, 2N$. Using the fact that $R^{-l} = R^{2N-l}$, for $l = 1, \ldots, 2N$, we get

$$\begin{aligned}
\ddot{\mathbf{r}}_1 &= \sum_{j=1}^{i-1} \frac{(R^{j-i} - I)\mathbf{r}_1}{|(R^{j-i} - I)\mathbf{r}_1|^3} + \sum_{j=i+1}^{2N} \frac{(R^{j-i} - I)\mathbf{r}_1}{|(R^{j-i} - I)\mathbf{r}_1|^3} \\
&= \sum_{l=1}^{i-1} \frac{(R^{2N-l} - I)\mathbf{r}_1}{|(R^{2N-l} - I)\mathbf{r}_1|^3} + \sum_{l=1}^{2N-i} \frac{(R^{l} - I)\mathbf{r}_1}{|(R^{l} - I)\mathbf{r}_1|^3} \\
&= \sum_{k=1}^{2N-1} \frac{(R^{k} - I)\mathbf{r}_1}{|(R^{k} - I)\mathbf{r}_1|^3}.
\end{aligned}$$

That is, we get the same equation from the initial $2N$ Eq. 1. □

If $\mathbf{r} = (x, y, z)$ is the position of the first body, then the equations of motion (2) can be written as the following differential system of order two

$$\ddot{x} = \frac{\partial U(x,y,z)}{\partial x}, \qquad \ddot{y} = \frac{\partial U(x,y,z)}{\partial y}, \qquad \ddot{z} = \frac{\partial U(x,y,z)}{\partial z}, \tag{3}$$

where $U(x, y, z)$ is the potential function

$$U(x,y,z) = \frac{1}{2} \sum_{k=1}^{2N-1} \frac{1}{\sqrt{4(x^2+y^2)\sin^2\left(\frac{k\pi}{2N}\right) + ((-1)^k - 1)^2 z^2}}.$$

The problem stated by system (3) can be formulated in Hamiltonian terms by the Hamiltonian function

$$H = \frac{1}{2}(p_x^2 + p_y^2 + p_z^2) - U(x,y,z), \tag{4}$$

where p_x, p_y and p_z are the momenta associated to the x, y, z coordinates. Introducing cylindrical coordinates by means of the canonical change

$$\begin{aligned}
x &= r\cos\theta, & p_x &= p_r\cos\theta - \frac{p_\theta}{r}\sin\theta, \\
y &= r\sin\theta, & p_y &= p_r\sin\theta + \frac{p_\theta}{r}\cos\theta, \\
z &= d, & p_z &= p_d
\end{aligned}$$

the Hamiltonian (4) becomes

$$\mathcal{H} = \frac{1}{2}\left(p_r^2 + \frac{p_\theta^2}{r^2} + p_d^2\right) - \frac{1}{2} \sum_{k=1}^{2N-1} \frac{1}{\sqrt{4r^2 \sin^2\left(\frac{k\pi}{2N}\right) + \left((-1)^k - 1\right)^2 d^2}}. \tag{5}$$

Then the equations of motion (2) for the first body are

$$
\begin{aligned}
\dot{r} &= p_r, \\
\dot{\theta} &= \frac{p_\theta}{r^2}, \\
\dot{d} &= p_d, \\
\dot{p_r} &= \frac{p_\theta^2}{r^3} - 2r \sum_{k=1}^{2N-1} \frac{\sin^2\left(\frac{k\pi}{2N}\right)}{\left(4r^2\sin^2\left(\frac{k\pi}{2N}\right) + \left((-1)^k - 1\right)^2 d^2\right)^{3/2}}, \\
\dot{p_\theta} &= 0, \\
\dot{p_d} &= -\frac{d}{2} \sum_{k=1}^{2N-1} \frac{\left((-1)^k - 1\right)^2}{\left(4r^2\sin^2\left(\frac{k\pi}{2N}\right) + \left((-1)^k - 1\right)^2 d^2\right)^{3/2}}.
\end{aligned}
\tag{6}
$$

Since $\dot{p_\theta} = 0$, the angular momentum $p_\theta = \Theta$ is constant and can be calculated from the initial conditions. Then, once r is obtained, we will get θ from the second equation in (6)

3 Symmetric periodic solutions of the reduced problem

We call *reduced problem* the problem given by considering in (6) only the equations for the variables r and d, and *complete problem* the whole set of Eq. 6. Our aim is to find periodic solutions of this reduced problem. This solutions will be, in general, quasi–periodic solutions of the complete problem.

Consider the problem posed by the Hamiltonian (5) for a fixed value of the angular momentum $p_\theta = \Theta$. The equations of motion of the reduced problem are

$$
\begin{aligned}
\dot{r} &= p_r, \\
\dot{d} &= p_d, \\
\dot{p_r} &= \frac{\Theta^2}{r^3} - 2r \sum_{k=1}^{2N-1} \frac{\sin^2\left(\frac{k\pi}{2N}\right)}{\left(4r^2\sin^2\left(\frac{k\pi}{2N}\right) + \left((-1)^k - 1\right)^2 d^2\right)^{3/2}}, \\
\dot{p_d} &= -\frac{d}{2} \sum_{k=1}^{2N-1} \frac{\left((-1)^k - 1\right)^2}{\left(4r^2\sin^2\left(\frac{k\pi}{2N}\right) + \left((-1)^k - 1\right)^2 d^2\right)^{3/2}},
\end{aligned}
\tag{7}
$$

which has a unique equilibrium point $(r, d, p_r, p_d) = (a, 0, 0, 0)$, where $a = \Theta^2/K_N^2$ and

$$
K_N^2 = \frac{1}{4} \sum_{k=1}^{2N-1} \frac{1}{\sin\left(\frac{k\pi}{2N}\right)}. \tag{8}
$$

The matrix of the linearized equations around this equilibrium point is

$$
M = \begin{pmatrix} 0 & 0 & 1 & 0 \\ 0 & 0 & 0 & 1 \\ -\lambda_1^2 & 0 & 0 & 0 \\ 0 & -\lambda_2^2 & 0 & 0 \end{pmatrix},
$$

where

$$-\lambda_1^2 = \frac{-3\Theta^2}{a^4} + \frac{2K_N^2}{a^3} = -\frac{K_N^8}{\Theta^6},$$

$$-\lambda_2^2 = -\frac{1}{16a^3}\sum_{k=1}^{2N-1}\frac{((-1)^k-1)^2}{\sin^3\left(\frac{k\pi}{2N}\right)} = -\frac{S_N^2 K_N^6}{\Theta^6}$$

and

$$S_N^2 = \frac{1}{16}\sum_{k=1}^{2N-1}\frac{((-1)^k-1)^2}{\sin^3\left(\frac{k\pi}{2N}\right)}. \tag{9}$$

The matrix M has two pairs of imaginary eigenvalues $\pm i\lambda_1$ and $\pm i\lambda_2$. By Lyapunov's center theorem (Meyer and Hall, 1992), there exist two one–parameter families of periodic solutions, emanating from the equilibrium point provided that λ_2/λ_1 is not a rational number. That is, it suffices to ensure that

$$\left(\frac{\lambda_2}{\lambda_1}\right)^2 = \frac{1}{4}\frac{\sum_{k=1}^{2N-1}(((-1)^k-1)^2)/\sin^3\left(\frac{k\pi}{2N}\right)}{\sum_{k=1}^{2N-1}1/\sin\left(\frac{k\pi}{2N}\right)} \tag{10}$$

is not the square of a rational number. A number of values of this expression for different values of N are shown on Table 1.

The $d = p_d = 0$ plane is invariant and the r-mode solutions lie in this plane with periods approaching $2\pi/\lambda_1$. These are actually the homographic solutions near the relative equilibrium. The d-mode solutions are three-dimensional orbits whose periods tend to $2\pi/\lambda_2$ and this is the family we are interested in.

The equations of motion of the reduced problem (7) are invariant by the symmetries

$$\mathcal{S}_1\colon (t,r,d,p_r,p_d) \longrightarrow (-t,r,-d,-p_r,p_d),$$
$$\mathcal{S}_2\colon (t,r,d,p_r,p_d) \longrightarrow (-t,r,d,-p_r,-p_d)$$

and we have the following well known proposition.

Proposition 2 *Let* $\mathbf{q}(t) = (r(t),d(t),p_r(t),p_d(t))$ *be a solution of Eq. 7. If* $\mathbf{q}(t)$ *satisfies that* $(d(0),p_r(0)) = (0,0)$ *and* $(p_r(T),p_d(T)) = (0,0)$, *then* $\mathbf{q}(t)$ *is a doubly symmetric periodic solution of period* $4T$.

We will show the existence of doubly symmetric periodic orbits in system (7).

Let $\mathbf{q}_0 = (r_0,d_0,p_{r0},p_{d0}) = (r_0,0,p_{r0},0)$. The solution of system (7) with these initial conditions is given by $d(t) = 0$, for all t, together with any solution $r(t)$ of the

Table 1 Values of $(\lambda_2/\lambda_1)^2$ for different values of N

N	$(\lambda_2/\lambda_1)^2$
2	$4(4-\sqrt{2})/7$
3	$51/(15+4\sqrt{3})$
5	$(160+\sqrt{5})/(20+\sqrt{5}+4\sqrt{2+\sqrt{5}})$
6	$12(\sqrt{2}+12\sqrt{6})/(15+6\sqrt{2}+4\sqrt{3}+12\sqrt{6})$

Kepler problem

$$\ddot{r} = \frac{\Theta^2}{r^3} - \frac{K_N^2}{r^2}$$

with K_N^2 given by (8). As is well known, its solutions can be written as

$$r(t) = a(1 - e\cos E(t)),$$

where a is the semimajor axis, e the eccentricity and E the eccentric anomaly. The function $r(t)$ is periodic of period $T = 2\pi a^{3/2}/K_N = 2\pi/\lambda_1$ and $a(1-e^2)K_N^2 = \Theta^2$. These solutions will be called 'planar' as opposed to the 'spatial' or 'three-dimensional' solutions when $d(t)$ (the distance from the first body $\mathbf{r}_1$ to the $z = 0$ plane) is not identically zero.

In order to obtain periodic solutions of the reduced problem, we add perturbations to periodic planar orbits in the vertical direction. If the perturbation is small enough, the motion can be decoupled into a planar plus a vertical motion in a first approximation. Substituting d by ϵd on Eq. 7, and keeping terms in ϵ^2, we obtain

$$\begin{aligned}
\dot{r} &= p_r,\\
\dot{d} &= p_d,\\
\dot{p}_r &= \frac{\Theta^2}{r^3} - \frac{K_N^2}{r^2} + \frac{3S_N^2}{2r^4}\, d^2\epsilon^2 + O(\epsilon^4),\\
\dot{p}_d &= -\frac{S_N^2}{r^3}\, d + \frac{3W_N^2}{2r^5}\, d^3\epsilon^2 + O(\epsilon^4),
\end{aligned} \tag{11}$$

where

$$W_N^2 = \frac{1}{64}\sum_{k=1}^{2N-1} \frac{((-1)^k - 1)^4}{\sin^5\left(\frac{k\pi}{2N}\right)}. \tag{12}$$

System (11) can then be written as

$$\dot{q} = \mathcal{F}(\mathbf{q},\epsilon) = \mathcal{F}_0(\mathbf{q}) + \epsilon^2\mathcal{F}_1(\mathbf{q}) + \mathcal{O}(\epsilon^4), \tag{13}$$

where

$$\mathcal{F}_0(\mathbf{q}) = \left(p_r, p_d, \frac{\Theta^2}{r^3} - \frac{K_N^2}{r^2}, -\frac{S_N^2}{r^3}\, d\right), \qquad \mathcal{F}_1(\mathbf{q}) = \left(0, 0, \frac{3S_N^2}{2r^4}\, d^2, \frac{3W_N^2}{2r^5}\, d^3\right).$$

Let $\mathbf{q}_0$ be a vector of initial conditions. The solution of (13) with initial value $\mathbf{q}_0$ at $t = 0$ can be expanded as a power series in ϵ^2 as

$$\mathbf{q}(t, \mathbf{q}_0, \epsilon) = \mathbf{q}^{(0)}(t, \mathbf{q}_0) + \epsilon^2\mathbf{q}^{(1)}(t, \mathbf{q}_0) + \mathcal{O}(\epsilon^4),$$

where $\mathbf{q}^{(0)}(t, \mathbf{q}_0)$ is the solution of the unperturbed problem

$$\dot{\mathbf{q}}^{(0)}(t, \mathbf{q}_0) = \mathcal{F}_0(\mathbf{q}^{(0)}) \tag{14}$$

with initial conditions $\mathbf{q}_0$, and $\mathbf{q}^{(1)}(t, \mathbf{q}_0)$ is the solution of

$$\dot{\mathbf{q}}^{(1)}(t, \mathbf{q}_0) = \mathcal{F}_1(\mathbf{q}^{(0)}(t, \mathbf{q}_0)) + \mathcal{D}\mathcal{F}_0(\mathbf{q}^{(0)}(t, \mathbf{q}_0))\mathbf{q}^{(1)}(t, \mathbf{q}_0)$$

with initial conditions $\mathbf{q}^{(1)}(0,\mathbf{q}_0) = 0$. The entries of the matrix $\mathcal{DF}$ are the partial derivatives of $\mathcal{F}$ with respect to the $\mathbf{q}$ variable, and by the formula of Lagrange we have

$$\mathbf{q}^{(1)}(t,\mathbf{q}_0) = \mathcal{Q}(t,\mathbf{q}_0)\int_0^t \mathcal{Q}^{-1}(\tau,\mathbf{q}_0)\mathcal{F}_1(\mathbf{q}^{(0)}(\tau,\mathbf{q}_0))\mathrm{d}\tau, \tag{15}$$

where

$$\mathcal{Q}(t,\mathbf{q}_0) = \left.\frac{\partial\mathbf{q}^{(0)}(t,\xi)}{\partial\xi}\right|_{\xi=\mathbf{q}_0} \tag{16}$$

Theorem 1 *Let* $T_0^* = (2k+1)\pi/2\lambda_2$, $a = \frac{\Theta^2}{K_N^2}$ *and* $\mathbf{q}_0^* = (a,0,0,p_{d0})$. *Assume that* λ_2/λ_1 *given by* (10) *is not a rational number. Then there exist* $\Delta r_0, \Delta T$ *such that the solution* $\mathbf{q}(t,\mathbf{q}_0,\epsilon)$ *of system* (11) *with initial conditions* $\mathbf{q}_0 = (a+\Delta r_0,0,0,p_{d0})$ *is a doubly symmetric periodic solution of period* $4(T_0^*+\Delta T)$.

The functions Δr_0 *and* ΔT *are given by*

$$\Delta r_0 = -\epsilon^2\frac{3\,p_{d0}^2\,S_n^2}{a^4\,\lambda_1\,(\lambda_1^2-4\,\lambda_2^2)} + O(\epsilon^4),$$

$$\Delta T = \epsilon^2\,\frac{9\,p_{d0}^2}{a^5\,{\lambda_1}^3\,({\lambda_1}^2-4\,{\lambda_2}^2)^2}\left(B_1(k,N)+\frac{1}{32\,{\lambda_2}^5}B_2(k,N)\right)+O(\epsilon^4),$$

where

$$\begin{aligned} B_1(k,N) &= -\,{S_N}^2\,({\lambda_1}^2-2\,{\lambda_2}^2)\,\sin\left(\tfrac{(1+2\,k)\,\pi\,\lambda_1}{2\,\lambda_2}\right),\\ B_2(k,N) &= (1+2\,k)\,\pi\,{W_N}^2\,{\lambda_1}^3\,({\lambda_1}^2-4\,{\lambda_2}^2)^2+{S_N}^2\,{\lambda_2}^2\,B_{22} \end{aligned} \tag{17}$$

and

$$B_{22}(k,N) = \lambda_1\pi(1+2k)(\lambda_1^2-4\lambda_2^2)(3\lambda_1^2-8\lambda_2^2)+32\lambda_2^3(\lambda_1^2-2\lambda_2^2)\sin\left(\frac{(1+2k)\pi\,\lambda_1}{2\lambda_2}\right).$$

Note that for $\epsilon\to 0$ the periods of the solutions given by the theorem tend to $2\pi/\lambda_2$ and they belong to a symmetric Lyapunov family.

Proof Notice that the solution of the unperturbed problem (14) with initial condition $\mathbf{q}_0^*$ is

$$\mathbf{q}^{(0)}(t,\mathbf{q}_0^*) = \left(a, p_{d0}\frac{1}{\lambda_2}\sin(\lambda_2 t), 0, p_{d0}\cos(\lambda_2 t)\right). \tag{18}$$

This solution satisfies

$$\mathbf{q}^{(0)}(T_0^*,\mathbf{q}_0^*) = \left(a, p_{d0}\frac{(-1)^k}{\lambda_2}, 0, 0\right) \tag{19}$$

and $\mathbf{q}^{(0)}(t,\mathbf{q}_0^*)$ is a doubly symmetric periodic solution of the unperturbed system of period $4T_0^*$.

We must find initial conditions $\mathbf{q}_0 = (a+\Delta r_0,0,0,p_{d0})$ and $T = T_0^*+\Delta T$ such that the solution $\mathbf{q}(t,\mathbf{q}_0,\epsilon)$ of system (11) satisfies

$$\begin{aligned} p_r(T_0^*+\Delta T,\mathbf{q}_0,\epsilon) &= p_r^{(0)}(T_0^*+\Delta T,\mathbf{q}_0)+\epsilon^2 p_r^{(1)}(T_0^*+\Delta T,\mathbf{q}_0)+O(\epsilon^4)=0,\\ p_d(T_0^*+\Delta T,\mathbf{q}_0,\epsilon) &= p_d^{(0)}(T_0^*+\Delta T,\mathbf{q}_0)+\epsilon^2 p_d^{(1)}(T_0^*+\Delta T,\mathbf{q}_0)+O(\epsilon^4)=0. \end{aligned} \tag{20}$$

By Proposition 2, $\mathbf{q}(t, \mathbf{q}_0, \epsilon)$ will be a doubly symmetric periodic solution of period $4(T_0^* + \Delta T)$.

For a fixed value p_{d0}, and $k = 0, 1$, we consider $p_r^{(k)}(T_0^* + \Delta T, \mathbf{q}_0)$ and $p_d^{(k)}(T_0^* + \Delta T, \mathbf{q}_0)$ as functions of $(\Delta T, \Delta r_0)$. Expanding (20) as power series in the Δ's we get

$$\begin{pmatrix} 0 \\ 0 \end{pmatrix} = \begin{pmatrix} p_r^{(0)} & (T_0^*, \mathbf{q}_0^*) \\ p_d^{(0)} & (T_0^*, \mathbf{q}_0^*) \end{pmatrix} + \begin{pmatrix} \dfrac{\partial p_r^{(0)}}{\partial t} & \dfrac{\partial p_r^{(0)}}{\partial r} \\ \dfrac{\partial p_d^{(0)}}{\partial t} & \dfrac{\partial p_d^{(0)}}{\partial r} \end{pmatrix}_{(T_0^*, \mathbf{q}_0^*)} \begin{pmatrix} \Delta T \\ \Delta r_0 \end{pmatrix} + O_2(\Delta T, \Delta r_0)$$

$$+\epsilon^2 \left[\begin{pmatrix} p_r^{(1)} & (T_0^*, \mathbf{q}_0^*) \\ p_d^{(1)} & (T_0^*, \mathbf{q}_0^*) \end{pmatrix} + \begin{pmatrix} \dfrac{\partial p_r^{(1)}}{\partial t} & \dfrac{\partial p_r^{(1)}}{\partial r} \\ \dfrac{\partial p_d^{(1)}}{\partial t} & \dfrac{\partial p_d^{(1)}}{\partial r} \end{pmatrix}_{(T_0^*, \mathbf{q}_0^*)} \begin{pmatrix} \Delta T \\ \Delta r_0 \end{pmatrix} + O_2(\Delta T, \Delta r_0) \right]$$

$$+ O(\epsilon^4).$$

Now we have from (19) that

$$\begin{pmatrix} p_r^{(0)} & (T_0^*, \mathbf{q}_0^*) \\ p_d^{(0)} & (T_0^*, \mathbf{q}_0^*) \end{pmatrix} = \begin{pmatrix} 0 \\ 0 \end{pmatrix},$$

so that if

$$\begin{vmatrix} \dfrac{\partial p_r^{(0)}}{\partial t} & \dfrac{\partial p_r^{(0)}}{\partial r} \\ \dfrac{\partial p_d^{(0)}}{\partial t} & \dfrac{\partial p_d^{(0)}}{\partial r} \end{vmatrix}_{(T_0^*, \mathbf{q}_0^*)} \neq 0,$$

the system (20) can be solved for $(\Delta T, \Delta r_0)$ in a neigbourhood of $(0, 0)$ by means of the implicit function theorem. An approximation to $(\Delta T, \Delta r_0)$ can be easily computed from

$$\begin{pmatrix} \Delta T \\ \Delta r_0 \end{pmatrix} = -\epsilon^2 \left[\begin{pmatrix} \dfrac{\partial p_r^{(0)}}{\partial t} & \dfrac{\partial p_r^{(0)}}{\partial r} \\ \dfrac{\partial p_d^{(0)}}{\partial t} & \dfrac{\partial p_d^{(0)}}{\partial r} \end{pmatrix}_{(T_0^*, \mathbf{q}_0^*)}^{-1} \begin{pmatrix} p_r^{(1)} & (T_0^*, \mathbf{q}_0^*) \\ p_d^{(1)} & (T_0^*, \mathbf{q}_0^*) \end{pmatrix} + O_2(\Delta T, \Delta r_0) \right] + O(\epsilon^4). \tag{21}$$

The functions $\partial p_r^{(0)}/\partial t, \partial p_d^{(0)}/\partial t$ can be computed from (18). In order to get the terms $\partial p_r^{(0)}/\partial r, \partial p_d^{(0)}/\partial r$ we must compute (16). Then

$$\begin{pmatrix} \dfrac{\partial p_r^{(0)}}{\partial t} & \dfrac{\partial p_r^{(0)}}{\partial r} \\ \dfrac{\partial p_d^{(0)}}{\partial t} & \dfrac{\partial p_d^{(0)}}{\partial r} \end{pmatrix}_{(T_0^*, \mathbf{q}_0^*)} =$$

$$\begin{pmatrix} 0 & -\lambda_1 \sin\left(\dfrac{2k+1}{2}\dfrac{\lambda_1}{\lambda_2}\pi\right) \\ -p_{d0}\lambda_2(-1)^k & \dfrac{3\, p_{d0}\, \lambda_2\, (-1)^k\, \sin\left(((1+2k)\, \pi\, \lambda_1)/2\, \lambda_2\right)\, (\lambda_1^2 - 2\lambda_2^2)}{a\, \lambda_1\, \left({\lambda_1}^2 - 4\,{\lambda_2}^2\right)} \end{pmatrix} \tag{22}$$

Table 2 Numerical values of $(\Delta T, \Delta r_0)$ for $\Theta = 1, N = 3$ and $k = 0$

	$p_{d0} = 2$	$p_{d0} = 1$	$p_{d0} = 0.5$
$\epsilon = 0.03$	$(0.000660, 0.000496)$	$(0.000164, 0.000123)$	$(0.000041, 0.000030)$
$\epsilon = 0.06$	$(0.002661, 0.001997)$	$(0.000660, 0.000496)$	$(0.000164, 0.000123)$
$\epsilon = 0.09$	$(0.006061, 0.004534)$	$(0.001490, 0.001119)$	$(0.000371, 0.000279)$
$\epsilon = 0.12$	$(0.010963, 0.008166)$	$(0.002661, 0.001997)$	$(0.000660, 0.000496)$

and $p_r^{(1)}(T_0^*, \mathbf{q}_0^*), p_d^{(1)}(T_0^*, \mathbf{q}_0^*)$ are the last two components of

$$\mathbf{q}^{(1)}(T_0^*, \mathbf{q}_0^*) = \mathcal{Q}(T_0^*, \mathbf{q}_0^*) \int_0^{T_0^*} \mathcal{Q}^{-1}(\tau, \mathbf{q}_0) \mathcal{F}_1(\mathbf{q}^{(0)}(\tau, \mathbf{q}_0)) \mathrm{d}\tau,$$

which can be easily computed and are given by

$$\begin{aligned} p_r^{(1)}(z_0^*) &= \frac{-3\, p_{d0}^2 S_N^2}{a^4\, \lambda_1\, (\lambda_1^2 - 4\lambda_2^2)} \sin\left(\frac{(1+2k)\pi}{2} \frac{\lambda_1}{\lambda_2}\right), \\ p_d^{(1)}(z_0^*) &= \frac{9\,(-1)^k\, p_{d0}^3}{32\, a^5\, {\lambda_1}^3\, {\lambda_2}^4\, ({\lambda_1}^2 - 4\,{\lambda_2}^2)^2} B_2,(k,N), \end{aligned} \tag{23}$$

where $B_2(k,N)$ is given by (17). Finally we can substitute (23) and (22) in (21) and we obtain the approximation to $(\Delta T, \Delta r_0)$. □

Theorem 1 gives an approximation to initial conditions for doubly symmetric periodic orbits for ϵ sufficiently small. This results have been checked numerically and a good agreement has been obtained.

For fixed values of N, k, Θ, p_{d0} and ϵ, we compute numerically the values (T, r) near to (T_0^*, a) such that the orbit with initial conditions $q = (r, 0, 0, p_{d0})$ is a doubly symmetric periodic orbit of period $4T$. The integration of the differential equations has been done by means of a Runge–Kutta RK78 algorithm. Then we compute $\Delta T = T - T_0^*$ and $\Delta r_0 = r - a$. Table 2 shows the numerically computed values of $(\Delta T, \Delta r_0)$, for $\Theta = 1$ and different values of ϵ.

4 Hip-hop periodic orbits and choreographies

The question whether orbits which are periodic in the reduced system are periodic also in the inertial frame is of course only a question of commensurability between π and the angle rotated in the inertial system in a period. If this angle can be seen to change along the family of periodic solutions, then there will exist infinitely many periodic solutions in the inertial system (whenever its value is commensurable with π). It suffices then to see that its derivative with respect to ϵ^2 is different from zero for $\epsilon = 0$.

Now, a small variation on this simple argument shows that there exist infinitely many choreographies as well. Think of the orbit as having period $4T$, where T is the time spent from the initial planar position as a regular $2N$-gon to the maximum separation of the planes containing the N-gon configurations. After a time $2T$, the N bodies that at $t = 0$ where thrown upwards will hit the initial plane with a velocity symmetric to the initial one, which is exactly the initial velocity of the other N bodies, which were thrown downwards. If at $t = 2T$ the position of N bodies is the same as the position at $t = 0$ of the other N bodies, they will follow the same path, so we have the kind of motion that has been called a choreography.

We give an outline of the computation of the derivative. As we have seen in the previous Section, for small values of ϵ we can obtain periodic solutions $\mathbf{q}(t, \mathbf{q}_0, \epsilon)$ of the reduced problem for initial conditions $\mathbf{q}_0 = (a + \Delta r_0, 0, 0, p_{d0})$ and period $4T = 4T_0^* + 4\Delta T$. For a fixed value p_{d0}, the function $r(t)$ is given by

$$\begin{aligned} r(t, \mathbf{q}_0, \epsilon) &= \left[r^{(0)}(t,a) + \left(\frac{\partial r^{(0)}}{\partial r} \right)_{\mathbf{q}_0^*} \Delta r_0 + O_2(\Delta r_0) \right] \\ &\quad + \epsilon^2 \left[r^{(1)}(t,a) + \left(\frac{\partial r^{(1)}}{\partial r} \right)_{\mathbf{q}_0^*} \Delta r_0 + O_2(\Delta r_0) \right] + O(\epsilon^4) \\ &= a + \cos(\lambda_1 t)\Delta r_0 + \epsilon^2 r^{(1)}(a,t) + O(\epsilon^4), \end{aligned} \tag{24}$$

where Δr_0 is given in Theorem 1, and

$$r^{(1)}(a,t) = \frac{3 p_{d0}^2 S_N^2 \left(\sin(t\lambda_2)^2 \lambda_1{}^2 + 2\,(-1 + \cos(t\lambda_1))\,\lambda_2{}^2 \right)}{2\,a^4 \lambda_2{}^2 \left(\lambda_1{}^4 - 4\lambda_1{}^2 \lambda_2{}^2 \right)}$$

is the first component of the vector $\mathbf{q}^{(1)}(t)$ as given by (15).

Thus, in order to find periodic solutions in the inertial reference system, for a fixed Θ we must find solutions of the equation

$$\dot{\theta} = \frac{\Theta}{r^2} \tag{25}$$

with $\theta(0) = 0$ at such that

$$\theta(4qT) = 2\pi p \tag{26}$$

for some integers p and q.

Substituting (24) in (25), we get that

$$\dot{\theta} = \frac{\Theta}{a^2} \left(1 - \frac{2\cos(\lambda_1 t)\Delta r_0 + \epsilon^2 r^{(1)}(a,t)}{a} \right) + O(\epsilon^4).$$

Integrating and remembering that ΔT is $O(\epsilon^2)$, we have

$$\theta(4T) = \frac{\Theta}{a^2} 4T_0^* + \epsilon^2 \frac{3\,(1+2k)\pi p_{d0}^2 \Theta^{16}}{8\,a^7 K_N{}^{17} S_N{}^5 \left(K_N{}^2 - 4 S_N{}^2 \right)} H(N) + O(\epsilon^4)$$

where

$$H(N) = 8 S_N{}^6 + 3 K_N{}^4 W_N{}^2 + K_N{}^2 \left(S_N{}^4 - 12 S_N{}^2 W_N{}^2 \right) \tag{27}$$

and K_N, S_N, W_N are defined , respectively, in (8), (9) and (12). Thus, it is enough to see that the term $H(N)$ is different from zero to guarantee that there exist infinitely many values of the parameter ϵ^2 such that (26) holds. This is indeed the case for $2 \leq N \leq 10$, and probably for infinitely many values of N, although we do not have a formal proof of this fact.

Acknowledgements The first author is partially supported by DGES grant BFM2003-09504-C02-01. The second and fourth authors are partially supported by DGES grant number BFM2002-04236-C02-02 and by a DURSI grant number 2001SGR 00173. The third author is partially supported by CICYT grant number SEC2003-05112/ECO and by DURSI grant number SGR2001-164.

References

Chenciner, A.: Action minimizing solutions of the newtonian n-body problem: from homology to symmetry. In: Proceedings of the International Congress of Mathematicians, vol. III, Beijing, pp. 279–294 (2002) (Errata. *Proceedings of the International Congress of Mathematicians*) (Beijing, 2002) Vol. 1, 651–653

Chenciner, A., Féjoz, J.: L'équation aux variations verticales d'un équilibre relatif comme source de nouvelles solutions périodiques du problème des N corps. C. R. Math. Acad. Sci. Paris **340**(8), 593–598 (2005)

Chenciner, A., Gerver, J., Montgomery, R., Simó, C.: Simple Choreographic Motions of N Bodies: A Preliminary Study, Geometry, Mechanics, and Dynamics, pp. 287–308. Springer, New York (2002)

Chenciner, A., Montgomery, R.: A remarkable periodic solution of the three-body problem in the case of equal masses. Ann. Math. (2) **152**(3), 881–901 (2000)

Chenciner, A., Venturelli, A.: Minima de l'intégrale d'action du problème newtonien de 4 corps de masses égales dans $\mathbf{R}^3$: orbites "hip-hop". Celest. Mech. Dynam. Astron. **77**(2), 139–152 (2001)

Davies, I., Truman, A., Williams, D.: Classical periodic solution of the equal-mass $2n$-body problem, $2n$-ion problem and the n-electron atom problem. Phys. Lett. A **99**(1), 15–18 (1983)

Meyer, K.R., Hall, G.R.: Introduction to Hamiltonian Dynamical Systems and the N-body Problem, Vol. 90 of Applied Mathematical Sciences. Springer-Verlag, New York (1992)

Meyer, K.R., Schmidt, D.S.: Librations of central configurations and braided Saturn rings. Celest. Mech. Dynam. Astron. **55**(3), 289–303 (1993)

Simó, C.: New families of solutions in N-body problems. In: Proceedings of the third European Congress of Mathematics, in Prog. Math. 201, 101–115 (2001)

Terracini, S., Venturelli, A.: Symmetric trajectories for the $2n$-body problem with equal masses (2005) preprint

Celestial Mechanics and Dynamical Astronomy (2006) 95:67–80
DOI 10.1007/s10569-006-9030-0

ORIGINAL ARTICLE

Double choreographical solutions for *n*-body type problems

V. Barutello* · S. Terracini

Received: 15 November 2005 / Revised: 14 March 2006 /
Accepted: 16 May 2006 / Published online: 15 August 2006

Abstract In this work, we prove the existence of solutions for the double choreographical n-body problem as critical points which are not minimizers of the corresponding Lagrangian action functional. We use the topological features of the sublevels of the action when a strong force assumption is imposed. We conclude with a description of an operative method to determine these solutions in $\mathbb{R}^3$.

Keywords Critical ponts theory · n-body problem

1 Introduction

In the last few years many new outstanding results about the existence of periodic and symmetric solutions for the n-body problem have been proved (see Chen 2001; Chenciner and Montgomery 1999; Chenciner and Venturelli 2000; Ferrario and Terracini 2004; Marchal 2000). The variational techniques play a main role in these researches since classical solutions are determined as collisionless critical points for the Lagrangian action functional associated to the problem. In order to apply the direct methods in the calculus of variations the action is asked to be coercive on some suitable loops space. For this reason some authors restrict the domain of the Lagrangian action to spaces of symmetric loops: the coercivity of the functional is implied by a condition on the finite group which generates the symmetry (see Ferrario and Terracini 2004).

*This research has been supported by Istituto Nazionale di Alta Matematica.

V. Barutello (✉) · S. Terracini
Dipartimento di Matematica e Applicazioni,
Università di Milano–Bicocca, via Cozzi 53,
I-20125 Milan, Italy
e-mail: vivina.barutello@unimib.it

S. Terracini
e-mail: susanna.terracini@unimib.it

Among all possible symmetries, the simple choreographies have been widely studied (see Chenciner and Montgomery 1999; Chenciner et al. 2001; Barutello and Terracini 2004a; Arioli et al. 2006): under the simple choreography constraint all the bodies are forced to move on the same curve exchanging their positions after a fixed period of time. The Lagrangian action associated to this problem is coercive once we have fixed the center of mass. In this work we study *double choreographical* motions: the bodies separates in two groups and the bodies of each group form a simple choreography. This means that two bodies in the same group move on the same loop exchanging their position after a fixed period of time, τ that is the same for the two groups. Although this symmetry is seemingly slightly different from the simple choreographical one, it generates a completely different variational structure of the associated action functional.

As for the simple choreographical case, the Lagrangian action, afterwards denoted by f, only depends on the pair of loops $x = (x_1, x_2)$ on which the bodies move and can be written as the sum of two parts (Proposition 1): the first one depending uniquely on the simple choreographical motions that the bodies form on each curve while the second is due to the interaction among bodies lying on different curves. The presence of this last term makes the functional not coercive and the Palais Smale condition fails on the whole loops space. Hence, in this setting, the minimization techniques fail and our aim is to develop a suitable theory to determine critical points which are not necessarily minimizers for the action functional. To overcome the failure of the Palais Smale condition, in order to find some critical points, inspired by Majer and Terracini 1993, we prove a compactness property whenever we work in appropriate sublevels of the function $g(x) = |[x_1] - [x_2]|^2$ (see Proposition 2); indeed compactness fails whenever we consider sequences $(x_\nu)_\nu$ such that $g(x_\nu) \to +\infty$ and $[x_{1,\nu}] - [x_{2,\nu}] = 0$ for every ν. Hence, we look for a sufficient condition for the existence of critical points in the sublevels of the function g.

With this aim we define a descent flow for the action functional (see (17)) and we deform along the flow lines suitable subset of the loops space, called *admissible set* (Definition 1). To prove that every admissible set has at least one element whose ω-limit (with respect to the descent flow) is included in the critical points set of f we impose the *strong force* condition (see Gordon 1975) on the interaction between particles, i.e. we consider homogeneous potential of degree $-\alpha$ with $\alpha \geq 2$. This hypothesis imposes to the action functional to diverge to $+\infty$ whenever two bodies approach a collision and creates a sufficiently complex topology in the sublevels of f. More precisely the sublevels are not simply connected and, starting from some non-contractible sets, we guarantee the existence of critical points which are not minimizers as ω-limit of elements in such non contractible sets (for similar results on this subject we refer to Barutello and Terracini 2004b). We would like to emphasize the following fact: while the admissible sets which guarantee the existence of critical points has to verify some topological assumptions (they are not contractible), the critical points do not verify any. They just belong to some suitable sublevels of the function g.

We conclude this work with an operative section to determine numerical solutions in $\mathbb{R}^3$. As a first step we describe a simple method to construct an admissible set starting from a pair of orbits with non-trivial linking number. Since the strong force is assumed, the loop $A_x = \{x_\vartheta : \vartheta \in S^1\}$, where $x_\vartheta(t) = (x_1(t + T_1\vartheta), x_2(t))$ and T_1 the period of x_1, is not contractible, and then admissible, in some sublevels of the action functional; then in A_x there exists an element $x_{\bar{\vartheta}}$ that evolves with the descent flow in the critical points set of f. The last step consists in the description of an algorithm, based on a bisection method and on the excision property of the degree, that allows us to determine numerically the element $x_{\bar{\vartheta}}$.

2 The variational setting

Let $n_1, n_2 \in \mathbb{N} \setminus \{0\}$ and

$$L := \mathrm{LCM}(n_1, n_2) \qquad G := \mathrm{GCD}(n_1, n_2),$$

their Least Common Multiple and Great Common Divisor, respectively. We define the real numbers

$$\tau := \frac{2\pi}{L}, \qquad T_1 = n_1 \tau, \qquad T_2 = n_2 \tau \tag{1}$$

and the Hilbert space of 2π-periodic functions

$$H := \left\{ x = (x_1, x_2) : x_i \in H^1_{T_i}(\mathbb{R}, \mathbb{R}^d), i = 1, 2 \right\}, \tag{2}$$

where the dimension d is greater or equal 2. The space H is endowed with the norm

$$\|x\| = \left(\int_0^{2\pi} |\dot{x}(t)|^2 + |x(t)|^2 \mathrm{d}t \right)^{1/2},$$

whose corresponding scalar product is

$$x \cdot y = \int_0^{2\pi} \dot{x}(t)\dot{y}(t) + x(t)y(t)\mathrm{d}t, \quad x, y \in H.$$

$|\cdot|$ denotes the euclidean norm in $\mathbb{R}^d$ and xy is the scalar product between two vectors in $\mathbb{R}^d$.

We now consider $n_1 + n_2$ bodies with unitary masses and 2π-periodic trajectories q_i, $q_i(t + 2\pi) = q_i(t)$, $i = 1, \ldots, n_1 + n_2$, moving in $\mathbb{R}^d$ interacting according to the Newton's law

$$\ddot{q}_i = \sum_{j \neq i} \frac{\partial V_{ij}}{\partial q_i}(q_i - q_j), \quad i = 1, \ldots, n_1 + n_2, \tag{3}$$

where

$$V_{ij}(q_i - q_j) = \frac{1}{|q_i - q_j|^\alpha}, \quad \alpha > 2 \tag{4}$$

are the potential functions which generates an attractive force between each pair of bodies. The homogeneity of these functions is taken, for some topological reason that we will understand later, greater or equal 2. This assumption, termed *strong force*, has been introduced by Gordon (1975).

The *Lagrangian action functional* whose collisionless critical points are solutions for (3) is

$$\mathcal{A}(q) = \frac{1}{2} \sum_{i=1}^{n_1+n_2} \int_0^{2\pi} |\dot{q}_i|^2 + \sum_{i<j} \int_0^{2\pi} V_{ij}(q_i - q_j), \tag{5}$$

where $q = (q_1, \ldots, q_{n_1+n_2})$. The symmetry constraint that we impose is the following there exists $x = (x_1, x_2) \in H$ such that

$$q_i(t) = x_1(t + (i-1)\tau), \quad i \in N_1,$$
$$q_j(t) = x_2(t + (j - n_1 - 1)\tau), \quad j \in N_2, \tag{6}$$

where $N_1 = \{1, \ldots, n_1\}$ and $N_2 = \{n_1+1, \ldots, n_1+n_2\}$. When condition (6) holds, we say that the bodies form a *double choreography on the pair of loops* $x = (x_1, x_2)$.

When we consider the motion of n bodies interacting according to the Newton law (3) under a *simple choreography constraint*, that is the bodies are forced to move on the same T-periodic curve x exchanging their position after a fixed period of time (see Chenciner and Montgomery 1999; Chenciner et al. 2001; Barutello and Terracini 2004a; Arioli et al. 2006), the action functional in (5) only depends on the loop x (see Barutello and Terracini 2004a). More precisely in this situation the action is

$$\mathcal{A}_n(x) = \frac{1}{2}\int_0^T |\dot{x}|^2 + \frac{1}{2}\sum_{h=1}^{n-1}\int_0^T \frac{\mathrm{d}t}{|x(t) - x(t+h\tau)|^\alpha}. \tag{7}$$

In the following proposition we rewrite $\mathcal{A}$ in (5) when we impose the constraint (6). Similarly as in the simple choreography case, we have that when we impose a double choreography constraints the action functional only depends on the pair of loops (x_1, x_2) on which the bodies move.

Proposition 1 *Let Ω be the following open subset of H*

$$\Omega := \{x \in H : x_i(t + h_i\tau) \neq x_i(t) \quad \text{and} \quad x_1(t + h_1\tau) \neq x_2(t + h_2\tau)$$
$$\forall h_i = 1, \ldots, n_i - 1,\ i = 1, 2,\ \forall t \in \mathbb{R}\} \tag{8}$$

and let $f\colon \Omega \to \mathbb{R}$ be the C^2 functional

$$f(x) = \sum_{i=1}^{2} \mathcal{A}_{n_i}(x_i) + \int_0^{2\pi} V_{1,2}(x_1 - x_2), \tag{9}$$

where

$$V_{1,2}(x_1 - x_2) := \sum_{h=0}^{G-1} \frac{1}{2}\frac{1}{|x_1(t + h\tau) - x_2(t)|^\alpha} \tag{10}$$

and $\mathcal{A}_{n_i}(x_i)$ is defined in (7). *Then critical points of f in Ω are solutions of the dynamical system* (3) *under the double choreography constraint* (6).

Proof Being $x = (x_1, x_2)$ the pair of loops on which the bodies move under the double choreography constraint, we prove that

$$f(x) = \frac{1}{L}\mathcal{A}(q), \tag{11}$$

where $\mathcal{A}$ is defined in (5). The kinetic part of $\mathcal{A}$ can be written as

$$\sum_{i=1}^{n_1+n_2}\int_0^{2\pi} |\dot{q}_i|^2 = \sum_{i=1}^{2} n_i \int_0^{2\pi} |\dot{x}_i|^2$$
$$= \sum_{i=1}^{2}\frac{2\pi n_i}{T_i}\int_0^{T_i} |\dot{x}_i|^2 = L\sum_{i=1}^{2}\int_0^{T_i} |\dot{x}_i|^2, \tag{12}$$

while the potential one can be split in three parts: in the first one both $i,j \in N_1$, in the second $i,j \in N_2$ and in the third $i \in N_1$ and $j \in N_2$. More precisely we have

$$\begin{aligned}\sum_{i<j} \int_0^{2\pi} V_{ij}(q_i - q_j) \\ &= \sum_{\substack{h_i=1\\i=1,2}}^{n_i-1} \frac{n_i}{2} \int_0^{2\pi} \frac{dt}{|x_i(t) - x_i(t+h_i\tau)|^\alpha} + \sum_{\substack{i\in N_1, j\in N_2\\i<j}} \int_0^{2\pi} V_{ij}(q_i - q_j) \\ &= \sum_{\substack{h_i=1\\i=1,2}}^{n_i-1} \frac{L}{2} \int_0^{T_i} \frac{dt}{|x_i(t) - x_i(t+h_i\tau)|^\alpha} + \sum_{h=0}^{G-1} \frac{L}{2} \int_0^{2\pi} \frac{dt}{|x_1(t+h\tau) - x_2(t)|^\alpha}. \end{aligned} \quad (13)$$

We obtain (11) summing (12) and (13) and dividing by L. □

Under the strong force assumption the action functional in (9) diverges to $+\infty$ on every sequence $(x_\nu)_\nu \subset \Omega$ such that x_ν approaches the boundary of Ω as $\nu \to +\infty$. Being $f^{\leq c}$ the c-sublevel of f that is

$$f^{\leq c} = \{x \in \Omega : f(x) \leq c\}$$

from Gordon (1975) we deduce the following result

Lemma 1 *When $\alpha \geq 2$, for every $c > 0$ there exists $\delta = \delta(c)$ such that*

$$x \in f^{\leq c} \quad \Rightarrow \quad \min_{\substack{l,k\in N_1\cup N_2\\t\in\mathbb{R}}} |q_l(t) - q_k(t)| \geq \delta,$$

where q_i, $i = 1, \ldots, n_1 + n_2$, are defined in (6).

3 An abstract critical point theorem

We start this section with the following lemma that describes the elements $x = (x_1, x_2)$ in the sublevels of the functional f: each loop x_i, $i = 1, 2$ can be embedded in a ball centered in its mean value whose radius depends only on the sublevel and on the number of bodies moving on it.

Lemma 2 *For every $c > 0$ there exists $r = r(c)$ such that*

$$x \in f^{\leq c} \quad \Rightarrow \quad \|x_i - [x_i]\|_\infty = \max_{t\in[0,2\pi]} |x_i(t) - [x_i]| \leq r$$

for $i = 1, 2$, where $[x_i] = \dfrac{1}{2\pi}\displaystyle\int_0^{2\pi} x_i(t)\mathrm{d}t$.

Proof As a first step, remark that for every $x \in f^{\leq c}$ we have

$$\frac{1}{2}\int_0^{T_i} |\dot{x}_i(t)|^2 dt \leq c \quad \text{and} \quad \int_0^{2\pi} |\dot{x}_i(t)|^2 \mathrm{d}t \leq \frac{2cL}{n_i} \quad (14)$$

for $i = 1,2$. Moreover, the following chain of inequalities holds for $i = 1,2$ and for every $t \in \mathbb{R}$

$$\begin{aligned}
&|x_i(t) - [x_i]| \\
&= \frac{1}{2\pi}\left|\int_0^{2\pi}(x_i(t) - x_i(s))\mathrm{d}s\right| \leq \frac{1}{2\pi}\int_0^{2\pi}|x_i(t) - x_i(s)|\,\mathrm{d}s \\
&= \frac{1}{2\pi}\int_0^{2\pi}\left|\int_s^t \dot{x}_i(\tau)\mathrm{d}\tau\right|\mathrm{d}s \leq \frac{1}{2\pi}\int_0^{2\pi}\int_s^t|\dot{x}_i(\tau)|\,\mathrm{d}\tau\,\mathrm{d}s \\
&\leq \frac{1}{2\pi}\int_0^{2\pi}\int_0^{2\pi}|\dot{x}_i(\tau)|\,\mathrm{d}\tau\,\mathrm{d}s = \int_0^{2\pi}|\dot{x}_i(\tau)|\,\mathrm{d}\tau \leq \sqrt{2\pi}\sqrt{\int_0^{2\pi}|\dot{x}_i(t)|^2\mathrm{d}\tau}.
\end{aligned}$$

Hence from (14) we have $|x_i(t) - [x_i]| \leq \sqrt{\frac{4\pi cL}{n_i}}$ and we conclude taking

$$r(c) := \frac{\sqrt{4\pi cL}}{\min\left(\sqrt{n_1}, \sqrt{n_2}\right)}. \tag{15}$$

□

We define

$$H^c := \left\{x = (x_1, x_2) \in H : (x_1(t), x_2(t)) = (\bar{x}_1, \bar{x}_2) \in \mathbb{R}^{2d}, \forall t \in \mathbb{R}\right\}$$

the subset of pairs of constant loops in H, isomorphic to $\mathbb{R}^{2d}$. We term $\mathcal{H}$ the set of continuous deformations of $\Omega \subset H$ into the set H^c, that is

$$\mathcal{H} := \left\{h : [0,1] \times \Omega \to H \text{ continuous } : h(0,\cdot) = id_{|\Omega},\ h(1,\Omega) \subset H^c\right\}.$$

Definition 1 We say that a closed subset $A \subset \Omega$ is *admissible* if

$$\forall h \in \mathcal{H}, \exists x \in A, t \in \mathbb{R}, \lambda \in [0,1] \quad \text{such that } h_1(\lambda, x)(t) = h_2(\lambda, x)(t).$$

Let g be the real-valued function

$$g\colon H \to \mathbb{R}, \quad g(x) = |[x_1] - [x_2]|^2 = \left|\frac{1}{2\pi}\int_0^{2\pi}(x_1 - x_2)(t)\,\mathrm{d}t\right|^2$$

and for any positive value b we define its superlevel

$$g^{\geq b} = \{x \in H : g(x) \geq b\}.$$

Remark 1 Since g is defined on an Hilbert space, its gradient at a point $x \in H$ is an element of H that is isomorphic to its dual space. For this reason, after some computation, we can write

$$\nabla g(x) = \frac{1}{\pi}\left([x_1] - [x_2], [x_2] - [x_1]\right) \in H^c \subset H.$$

The following result asserts that, fixed a sublevel of the functional f, we can find a superlevel of g such that when $x = (x_1, x_2)$ lies in the intersection of these two sets then x_1 and x_2 are contained in two disjointed balls.

Lemma 3 *For every $c > 0$ there exists $b = b(c)$ such that the set $B = f^{\leq c} \cap g^{\geq b}$ is not admissible.*

Proof Take $b(c) > (2r(c))^2$, then Lemma 2 ensures that if $x = (x_1, x_2)$ lies in B, the trajectories $x_1(t), x_2(t), t \in [0, 2\pi]$ are contained in two disjointed balls centered, respectively, in $[x_1]$ and $[x_2]$. We consider the continuous deformation $h \in \mathcal{H}$ defined as

$$h(\lambda, x) = \lambda[x] + (1 - \lambda)x,$$

where $[x] = ([x_1], [x_2]) \in \mathbb{R}^{2d}$. Since $\lambda x_1(t) + (1-\lambda)[x_1] \neq \lambda x_2(t) + (1-\lambda)[x_2], \forall x \in B$, $\forall t \in \mathbb{R}, \forall \lambda \in [0, 1]$, we conclude that B is not an admissible subset of Ω. □

Lemma 4 *Fixed $c > 0$, let B be the not admissible set introduced in Lemma 3. Then*

$$\nabla f(x) \cdot \nabla g(x) < 0, \quad \forall x \in B.$$

Proof Since $\nabla g(x) = 1/\pi\ ([x_1] - [x_2], [x_2] - [x_1])$ (see Remark 1) is a constant vector, we compute that for every $x \in B$

$$\nabla f(x) \cdot \nabla g(x) = \frac{2}{\pi} \int_0^{2\pi} \nabla V_{1,2}(x_1 - x_2)([x_1] - [x_2]).$$

From (10) we then have

$$\begin{aligned}
&\nabla\ f(x) \cdot \nabla g(x) \\
&= -\frac{2\alpha}{\pi} \sum_{h=0}^{G-1} \int_0^{2\pi} \frac{1}{2} \frac{x_1(t + h\tau) - x_2(t)}{|x_1(t + h\tau) - x_2(t)|^{\alpha+2}} ([x_1] - [x_2])\mathrm{d}t \\
&= -\frac{\alpha}{\pi} \sum_{h=0}^{G-1} \int_0^{2\pi} \frac{|[x_1] - [x_2]|}{|x_1(t + h\tau) - x_2(t)|^{\alpha+1}} \cos\ \beta_h\ \mathrm{d}t,
\end{aligned}$$

where

$$\beta_h = \mathrm{ang}(x_1(t + h\tau) - x_2(t), [x_1] - [x_2]) \leq \arctan \frac{2r(c)}{\sqrt{b(c)}}.$$

Since $\sqrt{b(c)} > 2r(c)$ (see Lemma 3), $\beta_h \in \left(0, \frac{\pi}{4}\right)$ and $\cos\ \beta_h > 0$, for every $h = 0, \ldots, G - 1$. □

We term $\tilde{f}$ the terms of the Lagrangian action which do not depend on the interaction between bodies on different curves, that is

$$\tilde{f}(x) := \mathcal{A}_{n_1}(x_1) + \mathcal{A}_{n_2}(x_2),$$

since $\mathcal{A}_n$, the Lagrangian action associated to the choreographical n-body problem, is coercive on the space of zero-average periodic loops in H^1 (see Ferrario and Terracini 2004), similarly the sum $\tilde{f}$ is coercive on the pairs of zero-average periodic loops in H^1. This facts hold since the functionals $\mathcal{A}_n$ and $\tilde{f}$ are invariant under translations, indeed $\mathcal{A}_n(x+v) = \mathcal{A}_n(x), \forall x \in H^1_{2\pi}(\mathbb{R}, \mathbb{R}^d)$ and $v \in \mathbb{R}^d$, and $\tilde{f}(x) = \tilde{f}(x+w), \forall x = (x_1, x_2) \in H$ and $w = (w_1, w_2) \in \mathbb{R}^{2d}$.

The situation deeply changes when we consider the functional f. Indeed the presence of the interaction term $\int_0^{2\pi} V_{1,2}(x_1 - x_2)$ avoids the invariance of the functional under pairs of translations $w = (w_1, w_2) \in \mathbb{R}^{2d}$ with $w_1 \neq w_2$ and f is invariant under the action of the non-compact group of translations $\{(w, w) : w \in \mathbb{R}^d\}$. The functional f is then not coercive on the set of loops

$$H_0 = \{x = (x_1, x_2) \in H : [x_1] + [x_2] = 0\}.$$

Indeed, on every sequence $(x_\nu)_\nu$, $x_\nu = (x_{1,\nu}, x_{2,\nu}) \in H_0$, such that $x_{i,\nu}$ are critical points for $\mathcal{A}_{n_i}$ at some level c_i, $i = 1, 2$, and $g(x_\nu) \to +\infty$, as $\nu \to +\infty$, we have that $f(x_\nu) \to c_1 + c_2 < +\infty$.

Similarly $\tilde{f}$ verifies the Palais condition at every level c (see Proposition 2.2 (a) in Barutello 2004) but, concerning f, we prove in Proposition 2, that the Palais-Smale condition does not hold at the critical levels for the functional $\tilde{f}$, i.e. at every level c such that $c = c_1 + c_2$, where c_1 is a critical level for $\mathcal{A}_{n_1}$ and c_2 for $\mathcal{A}_{n_2}$. Anyway from Proposition 2 we will deduce that f verifies the Palais-Smale condition at every level c on the sublevels of the function g.

We define the *set of critical points for the functional* f as

$$\mathrm{Crit}(f) = \{x \in \Omega : \nabla f(x) = 0\}$$

and, for any positive δ, its δ-neighborhood

$$N_\delta(\mathrm{Crit}(f)) = \left\{x \in \mathrm{Crit}(f) : \inf_{y \in \mathrm{Crit}(f)} \|x - y\| < \delta\right\}.$$

Similarly we define the sets $\mathrm{Crit}(\tilde{f})$ and $N_\delta(\mathrm{Crit}(\tilde{f}))$.

Proposition 2 *For every* $b, c, \delta > 0$, *there exists* $\varepsilon > 0$ *such that*

$$\left\{x \in f^{\leq c} : \|\nabla f(x)\| < \varepsilon\right\} \subset N_\delta(\mathrm{Crit}(f)) \cup \left(g^{\geq b} \cap N_\delta(\mathrm{Crit}(\tilde{f}))\right).$$

Proof By the sake of contradiction, suppose there exist $\bar{b}, \bar{c}, \bar{\delta} > 0$, such that for every $\varepsilon_n = 1/n$, $n \in \mathbb{N}$, there exists $x_n = (x_{1,n}, x_{2,n})$ such that

$$f(x_n) \leq \bar{c}, \qquad \lim_{n \to +\infty} \|\nabla f(x_n)\| = 0$$

and

$$x_n \notin N_{\bar{\delta}}(\mathrm{Crit}(f)) \cup \left(g^{\geq \bar{b}} \cap N_{\bar{\delta}}(\mathrm{Crit}(\tilde{f}))\right). \tag{16}$$

Two different situations can occour: the sequence $g(x_n)$ can be bounded or it can diverge to $+\infty$ (up to subsequences). Suppose that $g(x_n) \to +\infty$ as $n \to +\infty$; since $x_n \in f^{\leq \bar{c}}$, we deduce from Lemma 2, that

$$\lim_{n \to +\infty} \|x_{1,n} - x_{2,n}\|_\infty = +\infty,$$

hence

$$\lim_{n \to +\infty} \|\nabla(f - \tilde{f})(x_n)\| = 0$$

and, since we have assumed that $\lim_{n \to +\infty} \|\nabla f(x_n)\| = 0$ we have

$$\lim_{n \to +\infty} \|\nabla \tilde{f}(x_n)\| = 0.$$

Since the functional $\tilde{f}$ verifies the Palais-Smale condition at every level, we conclude that there exists $N > 0$ such that, when $n > N$, $x_n \in N_{\bar{\delta}}(\mathrm{Crit}(\tilde{f}))$, which is in contradiction with (16).

When $g(x_n) < \bar{b}$, for every $n \in \mathbb{N}$, then there exists a converging subsequence $(x_{n_k})_{n_k} \subset (x_n)_n$, $x_{n_k} \to \bar{x}$. We can then conclude that

$$\lim_{n_k \to +\infty} \|\nabla f(x_{n_k})\| = \|\nabla f(\bar{x})\| = 0$$

and the existence of $\bar{N} \in \mathbb{N}$, such that, for every $n \geq \bar{N}$, $x_n \in N_{\bar{\delta}}(\mathrm{Crit}(f))$, that contradicts (16). □

Fixed $c > 0$ we take $b > 0$ such that $\sqrt{b} > 2r(c)$, where $r(c)$ is defined in (15). We consider a smooth function $\lambda\colon \mathbb{R} \to [0,1]$ such that $\lambda(t) = 0$ when $t \leq b$, $\lambda(t) = 1$ when $t \geq 2b$ and $\lambda(t) \in (0,1)$ when $t \in (b, 2b)$ and the flow $\eta\colon \mathbb{R}_+ \times f^{\leq c} \to \Omega$ defined as the solution of the Cauchy problem

$$\frac{\mathrm{d}}{\mathrm{d}t}\eta(t,x) = W\left(\eta(t,x)\right), \quad \eta(0,x) = x, \tag{17}$$

where

$$W(\eta) = -\left[1 - \lambda(g(\eta))\right]\frac{\nabla f(\eta)}{1 + \|\nabla f(\eta)\|} + \lambda(g(\eta))\frac{\nabla g(\eta)}{1 + \|\nabla g(\eta)\|}.$$

We say that a subset $\Omega_0 \subset \Omega$ is *positively invariant* for the flow η if $\{\eta(t,x_0), t \geq 0\} \subset \Omega_0$, for every $x_0 \in \Omega_0$. We term *ω-limit of $x \in \Omega$* for the flow η, the closed positively invariant set $\omega_x = \left\{\lim_{t_n \to +\infty} \eta(t_n, x) : (t_n)_n \subset \mathbb{R}_+\right\}$. In the following lemma (inspired by a similar result in Majer and Terracini 1993, while for a more general theory we refer to Bahri and Rabinowitz 1989) we prove that the functional f decreases on the flow η or, equivalently, that its sublevels are positively invariant.

Lemma 5 *If $\eta(t,x)$ is a solution of* (17), *then*

$$\frac{\mathrm{d}}{\mathrm{d}t}f(\eta(t,x)) < 0,$$

whenever $\eta(t,x) \in g^{\geq b}$.

Proof Since $\sqrt{b} > 2r(c)$, from Lemma 4 we have that $\nabla g(\eta) \cdot \nabla f(\eta) < 0$ whenever $\lambda(g(\eta)) > 0$, hence

$$\frac{\mathrm{d}}{\mathrm{d}t}f(\eta(t,x)) = -\left[1 - \lambda(g(\eta))\right]\frac{\|\nabla f(\eta)\|^2}{1 + \|\nabla f(\eta)\|} + \lambda(g(\eta))\nabla g(\eta) \cdot \nabla f(\eta) < 0.$$

□

Lemma 6 *If $\eta(t,x)$ is a solution of* (17), *then for every $t > 0$ such that $b < g(\eta(t,x)) < 2b$ there exists $\delta > 0$ such that*

$$\frac{\mathrm{d}}{\mathrm{d}t}f(\eta(t,x)) \leq -\delta.$$

Proof As in proof of Lemma 5 we compute

$$\begin{aligned}\frac{\mathrm{d}}{\mathrm{d}t}f(\eta(t,x)) &= -\left[1 - \lambda(g(\eta))\right]\frac{\|\nabla f(\eta)\|^2}{1 + \|\nabla f(\eta)\|} + \lambda(g(\eta))\nabla g(\eta) \cdot \nabla f(\eta)\\ &\leq \lambda(g(\eta))\nabla g(\eta) \cdot \nabla f(\eta)\\ &= -\lambda(g(\eta))\frac{\alpha}{\pi}\int_0^{2\pi}\sum_{h=0}^{G-1}\frac{|[x_1] - [x_2]|}{|x_1(t + h\tau) - x_2(t)|^{\alpha+1}}\cos\beta_h\,\mathrm{d}t.\end{aligned}$$

Since $\|[x_1] - [x_2]\| \geq \sqrt{b}$, $\cos\beta_h \geq \sqrt{2}/2$ (see Lemma 4) we have that

$$\frac{\mathrm{d}}{\mathrm{d}t}f(\eta(t,x)) \leq -\alpha\sqrt{2b}(G-1)\lambda(g(\eta))\min_{t \in S^1}\frac{1}{|x_1(t) - x_2(t)|^{\alpha+1}} = -\delta < 0.$$

□

Remark 2 Since the functional f is bounded from below, from Lemma 6 we deduce that every solution of (17) can stay only for a finite time in the set $\{x \in H : g(x) \in (b, 2b)\}$. Hence for every $x_0 \in H$ there exists t_0 such that

$$\{\eta(t, x_0) : t > t_0\} \cap \{x \in H : g(x) \in (b, 2b)\} = \emptyset.$$

The proof of the next lemma follows straightforwardly from the definition on the flow (17).

Lemma 7 *If $\eta(t,x)$ is a solution of* (17) *and $g(\eta(\bar{t},x)) \geq 2b$ for some time $\bar{t} > 0$ then for every $t \geq \bar{t}$ we have $g(\eta(t,x)) \geq 2b$.*

The flow η defined in (17) is then a descent flow for the functional f (see Lemma 6) such that the ω-limit of the points in $f^{\leq c}$ are subsets either of the set of critical points of f or of non-admissible subset of $g^{\geq 2b}$. The following theorem gives a sufficient condition on a subset of $f^{\leq c}$ to contain a point whose ω-limit is a subset of $\mathrm{Crit}(f)$.

Theorem 1 *Let A be an admissible compact subset of Ω such that $A \subset f^{\leq c}$ for some $c > 0$. Then for every $\delta > 0$ there exists $x_0 \in A$ and $t_0 > 0$ such that*

$$\{\eta(t, x_0), t \geq t_0\} \subset N_\delta(\mathrm{Crit}(f)). \tag{18}$$

Proof Since A is compact there exists c such that $A \subset f^{\leq c}$; moreover, from Proposition 2 and since f decreases along the flow η defined in (17), for every positive δ there exist $\varepsilon > 0$ and t_A such that

$$\begin{aligned}\{\eta(t,x) : t \geq t_A, x \in A\} &\subset \{x \in f^{\leq c} : \|\nabla f(x)\| < \varepsilon\} \\ &\subset N_\delta(\mathrm{Crit}(f)) \cup \left(g^{\geq b} \cap N_\delta(\mathrm{Crit}(\tilde{f}))\right).\end{aligned}$$

As we have observed in Remark 2 we can take t_A sufficiently large (t_A does not depend on x, but we can find a common value for every element in A since it is compact) to have

$$\{\eta(t,x) : t \geq t_A, x \in A\} \subset N_\delta(\mathrm{Crit}(f)) \cup \left(g^{\geq 2b} \cap N_\delta(\mathrm{Crit}(\tilde{f}))\right).$$

By the sake of contradiction suppose that

$$\{\eta(t,x) : t \geq t_A, x \in A\} \subset \left(g^{\geq 2b} \cap N_\delta(\mathrm{Crit}(\tilde{f}))\right). \tag{19}$$

Inclusion (19) means that we have deformed each element of the set A in a pair of not-linked loops and this contradicts the admissibility of the set A. Indeed the following continuous function

$$\begin{aligned} h\colon [0,1] \times A &\rightarrow H \\ (\lambda, x) &\mapsto h(\lambda, x) \end{aligned}$$

$$h(\lambda, x) = \begin{cases} \eta(2\lambda t_A, x), & \lambda \in [0, 1/2], \\ 2(1-\lambda)\eta(t_A, x) + (2\lambda - 1)[\eta(t_A, x)], & \lambda \in [1/2, 1] \end{cases} \tag{20}$$

is such that $h(0, \cdot) = \mathrm{id}|_A$, $h(1, A) \subset H^c$ and $h_1(\lambda, x)(t) \neq h_2(\lambda, x)(t)$ for every $(\lambda, x, t) \in [0,1] \times A \times [0, 2\pi]$. □

4 Applications

In this section, we describe an operative method to determine such critical points for the Lagrangian action functional whose existence is guaranteed by Theorem 1. This research consists in two steps: first we provide a sufficient condition on a pair of loops $x = (x_1, x_2) \in H$ to be the generator of an admissible set (according to Definition 1). Second, we propose an algorithm (see also Arioli et al. 2006; Barutello and Terracini 2004b) to find the element x_0 in the admissible set that verifies condition (18). This method is peculiar for the dimension $d = 3$ and it is based on the computation of the linking number between two curves and of the degree of a closed surface. Now on we then work in $\mathbb{R}^3$.

4.1 Construction of admissible sets

Given an element $x = (x_1, x_2) \in \Omega$, we consider the following loop in the space Ω

$$\gamma_x \colon S^1 \to \Omega, \qquad \vartheta \mapsto x_\vartheta := \gamma_x(\vartheta), \tag{21}$$

where $x_\vartheta(t) = (x_1(t + T_1\vartheta), x_2(t)), \forall t \in \mathbb{R}$. We define the compact subset of Ω, dependent on the element x

$$A_x := \gamma_x(S^1) - \{x_\vartheta : \vartheta \in S^1\} \tag{22}$$

Definition 2 When the set A_x is admissible, we will term it the admissible set associated to the element x.

Definition 3 Let $\gamma_i \colon S^1 \to \mathbb{R}^3$, $i = 1, 2$ be two C^1-loops such that $\gamma_1(t) \neq \gamma_2(t)$, for every $t \in S^1$, and such that $\gamma_1(S^1)$ is the boundary of a locally orientable surface. The *linking number* between γ_1 and γ_2 is the quantity

$$\{\gamma_1, \gamma_2\} = \frac{1}{4\pi} \int_0^{2\pi} \int_0^{2\pi} \left(\frac{\partial r_{12}}{\partial t_1} \wedge \frac{\partial r_{12}}{\partial t_2} \right) \frac{r_{12}}{|r_{12}|^3} \, dt_1 \, dt_2, \tag{23}$$

where $r_{12}(t_1, t_2) = \gamma_1(t_1) - \gamma_2(t_2)$.

Proposition 3 *Let $x = (x_1, x_2) \in \Omega$ such that we can define the linking number $\{x_1, x_2\}$. If $\{x_1, x_2\} \neq 0$ then the set A_x defined* (22) *is admissible.*

Proof Let $h \in \mathcal{H}$, our aim is to prove that there exists $\vartheta \in S^1$, $\lambda \in [0, 1]$ and $t \in \mathbb{R}$ such that $h_1(\lambda, x_\vartheta)(t) = h_2(\lambda, x_\vartheta)(t)$. We consider the continuous map on the torus

$$\begin{aligned} &F.\ \mathbb{T} = S^1 \times B^1 \to \mathbb{R}^3, \\ &F(\vartheta, \rho e^{Jt}) = h_2(1 - \rho, x_\vartheta)(t) - h_1(1 - \rho, x_\vartheta)(t). \end{aligned} \tag{24}$$

where $B^1 = \{\rho e^{Jt} : \rho \in [0, 1], t \in [0, 2\pi]\}$ is the unitary ball in $\mathbb{R}^2$. Our aim is then to show that the origin is in the image of the map F that is implied by the assertion

$$\deg(F, \mathbb{T}, 0) \neq 0.$$

We remark that $\deg(F, \mathbb{T}, 0)$ is well defined since $F(\vartheta, e^{Jt}) = x_{\vartheta,2}(t) - x_{\vartheta,1}(t) \neq 0$, for every $(\vartheta, t) \in S^1 \times S^1$ (indeed the linking number $\{x_1, x_2\}$ is defined) that is $0 \notin F(\partial \mathbb{T})$. To conclude we consider the map

$$G\colon \partial\mathbb{T} = S^1 \times S^1 \to \mathbb{R}^3,$$
$$G(\vartheta, e^{Jt}) = x_{\vartheta,2}(t) - x_{\vartheta,1}(t),$$

whose degree with respect to 0 is exactly the linking number $\{x_1, x_2\}$; being $\tilde{G}$ a continuous extension of G to the torus $\mathbb{T}$ we have that $\tilde{G}_{|\partial\mathbb{T}} \equiv F_{|\partial\mathbb{T}}$ and hence

$$\deg(F, \mathbb{T}, 0) = \deg(\tilde{G}, \mathbb{T}, 0) = \deg(G, \partial\mathbb{T}, 0) = \{x_1, x_2\} \neq 0.$$

□

From Theorem 1 we immediately deduce the following

Theorem 2 *Let c be such that the sublevel $f^{\leq c}$ contains an admissible set associated to an element x, A_x. Then there exists $x_{\bar{\vartheta}} \in A_x$ such that its ω-limit, $\omega_{x_{\bar{\vartheta}}}$, is a subset of* Crit(*f*).

4.2 A bisection algorithm for finding critical points

We now show an algorithm to determine the point $\bar{x} := x_{\bar{\vartheta}} \in A_x$ whose ω-limit is a subset of Crit(*f*) (see Theorem 2). Let h be the continuous deformation defined in (20) and F the parameterization of the three-dimensional torus (24). Since the set A_x is admissible, there exist $\bar{\rho} \in (0,1)$, $\bar{\vartheta} \in S^1$ and $\bar{t} \in [0, 2\pi]$ such that $F(\bar{\rho}e^{J\bar{t}}, \bar{\vartheta}) = 0$, or equivalently $0 \in F(\mathbb{T} \setminus \partial\mathbb{T})$.

To determine $x_{\bar{\vartheta}} \in A_x$ such that $\omega_{x_{\bar{\vartheta}}}$ is a critical point for $\mathcal{A}_{n_1,n_2}$ we use the following algorithm

Algorithm 1

Step 0. $\vartheta_1^0 := 0$, $\vartheta_2^0 := 2\pi$, $\vartheta_m^0 := \pi$
Step i+1. if $\omega_{x_{\vartheta_m^i}}$ is a critical point for f, STOP
$A^i := \{(\vartheta, \rho e^{Jt}) : \vartheta \in [\vartheta_1^i, \vartheta_m^i], \rho \in [0,1], t \in [0,1]\}$
if $\deg(F, A^i, 0) \neq 0$, $\vartheta_1^{i+1} := \vartheta_1^i$, $\vartheta_2^{i+1} := \vartheta_m^i$
else $\vartheta_1^{i+1} := \vartheta_m^i$, $\vartheta_2^{i+1} := \vartheta_2^i$

The control sentence if $\omega_{x_{\vartheta_m^i}}$ is a critical point for f, present at each step of Algorithm 1, needs some words of explanation. Before starting the implementation of the algorithm, we fix a "small" quantity $\epsilon > 0$ and a "suitable" time T such that $\omega_{x_{\vartheta_m^i}}$ surely verifies $\|\nabla f(\omega_{x_{\vartheta_m^i}})\| < \varepsilon$. We can always determine T sufficiently large to obtain this inequality since the flow is descent. The algorithm will then stop at the (i+1)th step if $g(\omega_{x_{\vartheta_m^i}}) < b$ where the quantity b depends on the sublevels in which the admissible set generated by a pair of linked orbits lies.

We conclude with the proof of the convergence of Algorithm 1. Let $(\vartheta_1^i)_i$ and $(\vartheta_2^i)_i$ be the sequences defined in Algorithm 1; when we consider the sequence of subset of the three-dimensional torus

$$B^i := \{(\vartheta, \rho e^{Jt}) : \vartheta \in [\vartheta_1^i, \vartheta_2^i], \rho \in [0,1], t \in [0,1]\}$$

the excision property of the degree ensures that for every index i

$$\deg(F, B^i, 0) \neq 0. \tag{25}$$

Since

$$0 = \vartheta_1^0 \le \vartheta_1^1 \le \cdots \le \vartheta_1^i < \vartheta_2^i \le \cdots \le \vartheta_2^1 \le \vartheta_2^0 = 2\pi,$$

then there exists $\tau \in [0, 2\pi]$ such that

$$\lim_{i\to+\infty} \vartheta_1^i = \lim_{i\to+\infty} \vartheta_2^i = \tau.$$

We now prove that τ necessarily coincides with $\tilde{\vartheta}$. By the sake of contradiction suppose $\tau \neq \tilde{\vartheta}$ and let

$$\delta := |\tau - \tilde{\vartheta}|$$

the construction of the set B^i ensures that $\tau \in [\vartheta_i^1, \vartheta_i^2]$, for every $i \in \mathbb{N}$. If we take $i > \log_2\left(\frac{2\pi}{\delta}\right)$, then $\deg(F, B^i, 0) = 0$, that is in contradiction with (25).

Figure 1 represents some numerical solutions found using Algorithm 1. One remark is due to explain the reason why the loops in the picture above on the right are not linked. A non-zero linking number is indeed necessary just to construct the admissible set and every element (pair of loops) belonging to an admissible set has this property. But the critical points that we find with this method are not elements of the admissible set but their deformations along the steepest descent flow. Is then absolutely natural that during this deformation their linking number change. The only condition that

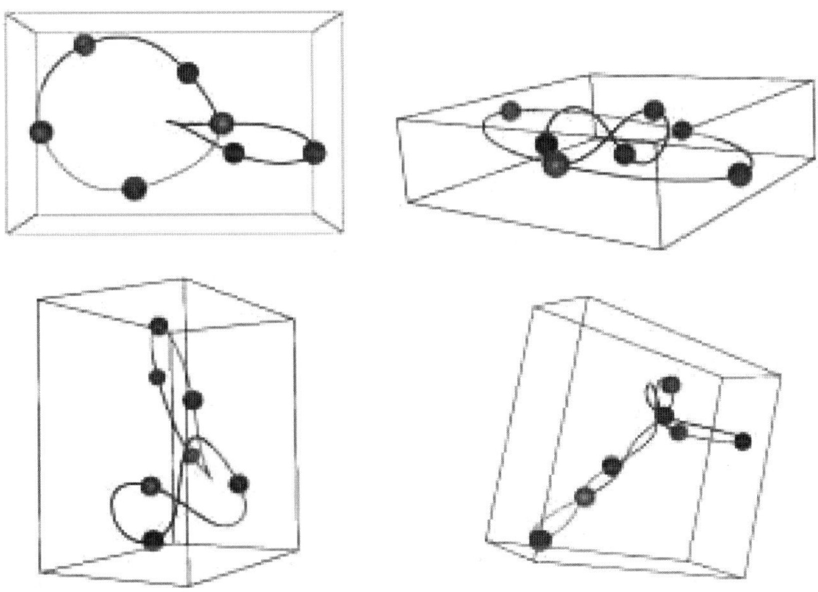

Fig. 1 Some example of double choreographies for n-body type problems with Strong Force. four bodies move on the first curve and three on the second. In all the pictures we can see that the distance between the mean-values of the two loops is small. This means, according with Proposition 2 and Theorem 1 that critical points belong to some sublevels of the function g. In the picture above on the right, the linking number between the two curves is zero; indeed this is not a necessary condition to be a critical point, but just to construct an admissible set. We moreover observed that the curves we have found are pairs of deformations of well known simple choreographic solutions: the eight-shape solution of Chenciner and Montgomery and the "super-eight" found by Gerver

such solutions have to satisfy is, according with Proposition 2, that they belong to a sublevel of the function $g = \|[x_1] - [x_2]\|^2$. The four pictures of Fig. 1 verify this condition.

References

Arioli, G., Barutello, V., Terracini, S.: A new branch of choreographic solutions for the n-body problem. Comm. Math. Phys. (2006) to appear

Bahri, A., Rabinowitz, P.H.: A minimax method for a class of Hamiltonian systems with singular potentials. J. Funct. Anal. **82**, 412–428 (1989)

Barutello, V.: On the n-body problem. Tesi di Dottorato, Università di Milano-Bicocca (2004)

Barutello, V., Terracini, S.: Action minimizing orbits in the n-body problem with choreography constraint. Nonlinearity **17**, 2015–2039 (2004a)

Barutello, V., Terracini, S.: A bisection algorithm for the numerical Mountain Pass. NoDEA (2004b) to appear

Chen, K.-C.: Action-minimizing orbits in the parallelogram four-body problem with equal masses. Arch. Ration. Mech. Anal. **158**, 293–318 (2001)

Chenciner, A., Gerver, J., Montgomery, R., Simó C.: Simple choreographies of N bodies: a preliminary study. Geometry, Mechanics and Dynamics, Springer, Berlin (2001)

Chenciner, A., Montgomery, R.: A remarkable periodic solution of the three body problem in the case of equal masses. Ann. of Math. **152**(3), 881–901 (1999)

Chenciner A., Venturelli A.: Minima de l'intégrale d'action du problème Newtonien de 4 corps de masses égales dans $\mathbb{R}^3$: orbites "hip-hop". Celest. Mech. Dynam. Astron. **77**, 139–152 (1999)

Ferrario, D., Terracini, S.: On the existence of collisionless equivariant minimizers for the classical n-body problem. Invent. Math. **155**(2), 305–362 (2004)

Gordon, W.: Conservative dynamical systems involving strong forces. Trans. Amer. Math. Soc. **204**, 113–135 (1975)

Marchal C.: The family P_{12} of the three-body problem—the simplest family of periodic orbits, with twelve symmetries per period. Celest. Mech. Dynam. Astron. **78**, 279–298 (2000)

Majer, P., Terracini, S.: Periodic solutions to some problems of n-body type. Arch. Ration. Mech. Anal. **124**, 381–404 (1993)

Celestial Mechanics and Dynamical Astronomy (2006) 95:81–99
DOI 10.1007/s10569-006-9023-z

ORIGINAL ARTICLE

From the circular to the spatial elliptic restricted three-body problem

J. F. Palacián · P. Yanguas

Received: 12 December 2005 / Revised: 6 March 2006 /
Accepted: 13 April 2006 / Published online: 15 August 2006

Abstract We deal with the study of the spatial restricted three-body problem in the case where the small particle is far from the primaries, that is, the so-called comet case. We consider the circular problem, apply double averaging and compute the relative equilibria of the reduced system. It appears that, in the circular problem, we find not only part of the equilibria existing in the elliptic case, but also new ones. These critical points are in correspondence with periodic and quasiperiodic orbits and invariant tori of the non-averaged Hamiltonian. We explain carefully the transition between the circular and the elliptic problems. Moreover, from the relative equilibria of elliptic type, we obtain invariant 3-tori of the original system.

Keywords Spatial restricted three-body problem · Averaging · Relative equilibria · Stability · Bifurcations · Invariant tori

1 Introduction

We study the three-dimensional motion of an infinitesimal particle under the gravitational action of two bodies (the primaries), which describe elliptic orbits in a plane around the centre of mass: the spatial (or three-dimensional) elliptic restricted three-body problem.

In a previous paper (Palacián et al. 2006), we consider the case in which the two primaries are moving in elliptic orbits with eccentricity $e_p \in (0, 1)$ and calculate periodic orbits and invariant manifolds. For that, we start by formulating the problem in the inertial frame and average over the time and over the mean anomaly. At certain

J. F. Palacián (✉) · P. Yanguas
Departamento de Matemática e Informática, Universidad Pública de Navarra,
Pamplona 31006, Spain
e-mail: palacian@unavarra.es

P. Yanguas
e-mail: yanguas@unavarra.es

step of the procedure, we get a generating function where e_{p} appears at the denominator in such a way that it is not defined for $e_{\mathrm{p}} = 0$. Consequently, the study performed in that paper is not valid for the circular case. Our purpose now is to extend the results obtained for $e_{\mathrm{p}} > 0$ to the case $e_{\mathrm{p}} = 0$ by applying similar techniques.

As pioneering works dealing with the three-body problem and the application of averaging procedures, implicit function theorems and KAM theory to prove the existence of some periodic and quasi-periodic motions, we have to quote the papers by Jefferys (1966) and Jefferys and Moser (1966) on the spatial three-body problem and on Moser (1970) on the spatial restricted problem. The paper (Jefferys and Moser 1966) also treats the restricted three-body problem as a limiting situation of the non-restricted case. A common feature of these works is that their study remains local.

We recall the main results and techniques appearing in Palacián et al. (2006). Through the combination of averaging and reduction techniques, we are able to simplify the original system and transform it into a Hamiltonian which, after truncating higher-order terms, defines an autonomous differential system of one degree of freedom. Then, the flow of this latter problem is readily discussed. We find six relative equilibria, two of them corresponding to orbits, which are coplanar with the primaries and the other four are related with orbits lying on a plane perpendicular to the one defined by the primaries. The orbital stability of these equilibria is characterised. Thereafter, we connect the flow of the reduced system with the one of the original problem, establishing the existence of six families of 3-tori and quasiperiodic orbits related with the six critical points. Additionally it is shown how some of the polar and equatorial families of quasiperiodic orbits are closed, by establishing rational relations among the characteristic frequencies of the orbits and taking also into account the discrete symmetries of the original Hamiltonian. In the special case of equal masses of the primaries, we can play with an extra independent symmetry to get doubly-symmetric closed orbits.

We also mention the work of Cors et al. (2005) where they find symmetric periodic solutions, which are perturbations of very large circular Keplerian orbits lying on a plane perpendicular to the one of the primaries. It is shown that they exist for a discrete sequence of values of the mean motion, and for all values of the eccentricity of the primaries' orbit. Their study is achieved by using an averaging method over the time combined with a version of the implicit function theorem valid for time-dependent systems which are periodic with a very high frequency together with the discrete symmetry of the problem.

The paper has been structured as follows. In Sect. 2, we formulate the spatial restricted three-body problem for the elliptic case both in an inertial and a rotating frame, giving explicitly the change of variables which brings the Hamiltonian from one chart to the other. We also define the comet case, introducing an appropriate small parameter so that the problem at hand could be expressed as the Keplerian two-body Hamiltonian plus a small perturbation.

Section 3 is devoted to simplify the Hamiltonian of the circular case in the inertial frame up to order four through two Lie transformations: an average over the time-dependent variable followed by a Delaunay normalisation removing the mean anomaly. With these two transformations, we complete the study of the elliptic problem performed in Palacián et al. (2006). Thus, the resulting Hamiltonian after truncation has two new integrals (the Delaunay action L associated with the mean anomaly and the third component of the angular momentum) and defines an autonomous system of one degree of freedom.

In Sect. 4, we express the transformed Hamiltonian in an adequate set of co-ordinates, resulting out of the reduction associated with the Lie transformations. Furthermore, we study the corresponding relative equilibria. We get up to six families of relative equilibria, which correspond with: (i) orbits whose eccentricity is very close to zero; (ii) orbits whose inclination with respect to the plane of the primaries is either close to zero or to π; (iii) and (iv) orbits whose inclination is close to the critical value arccos $(\pm 1/\sqrt{5})$ and their pericentre is near to 0 or π; (v) and (vi) orbits whose inclination is close to the critical value arccos $(\pm 1/\sqrt{5})$ and their pericentre is near to $\pi/2$ or $3\pi/2$.

The bifurcations arising when changing the number of equilibria and their stability is the aim of Sect. 5. We have characterised when the relative equilibria are either elliptic points (centres), saddles or degenerate in terms of the parameters L and H. Moreover, we have established the non-linear stability of the equilibria.

We have dealt with the persistence of true invariant tori surrounding the elliptic points of the doubly-reduced phase space in Sect. 6. This has been achieved using KAM theory.

2 Formulation: Rotating versus inertial frame

The spatial elliptic restricted three-body problem describes the motion of a body of infinitesimal mass m_3 under the gravitational attraction created by two bodies m_1 and m_2 called the primaries. We denote by μ the mass of the smallest primary (the mass of the biggest one is then $1-\mu$), with $\mu \in (0, 1/2]$. The two primaries are moving in elliptic orbits with eccentricity $e_{\rm p} \in [0, 1)$ and semimajor axes μ for m_1 and $1-\mu$ for m_2, around their centre of mass, which remains fixed at the origin. A representation of the problem is drawn in Fig. 1.

We may formulate the problem either in the rotating (also called synodical) frame or in the inertial one. In order to obtain the results appearing in the subsequent sections, we use the formulation in the inertial frame. In this frame, centred at the centre

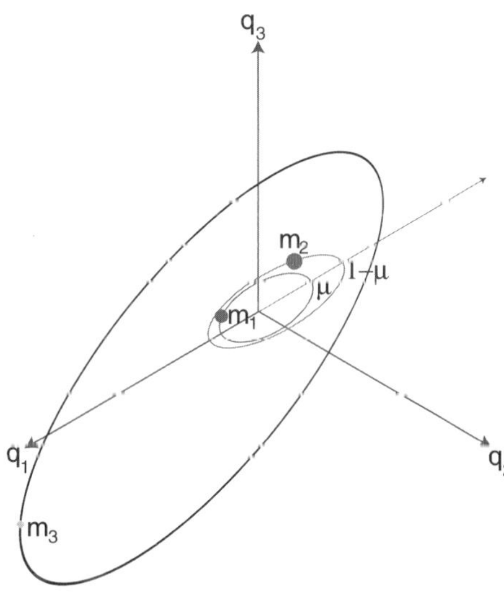

Fig. 1 The three-dimensional elliptic three-body problem in an inertial frame $q_1\, q_2\, q_3$

of mass of the two primaries, the Hamiltonian of the problem is given by:

$$\begin{aligned}\mathcal{H}(\boldsymbol{q},\boldsymbol{p},f_{\mathrm{p}};\mu,a_{\mathrm{p}},e_{\mathrm{p}}) &= \frac{1}{2}\,(p_1^2+p_2^2+p_3^2)\\ &-\frac{1-\mu}{\sqrt{[q_1+\mu\,\rho\,\cos(f_{\mathrm{p}})]^2+[q_2+\mu\,\rho\,\sin(f_{\mathrm{p}})]^2+q_3^2}}\\ &-\frac{\mu}{\sqrt{[q_1-(1-\mu)\,\rho\,\cos(f_{\mathrm{p}})]^2+[q_2-(1-\mu)\,\rho\,\sin(f_{\mathrm{p}})]^2+q_3^2}}.\end{aligned} \tag{1}$$

The axis q_1 is the line of apsides of the ellipses described by m_1 and m_2. Time is taken such that the gravitational constant is one. Co-ordinates $\boldsymbol{q} = (q_1,q_2,q_3)$ represent the position of the infinitesimal particle in the inertial frame and $\boldsymbol{p} = (p_1,p_2,p_3)$ are their conjugate momenta. The distance-like variable ρ stands for the radial distance between m_2 with respect to m_1 and it depends on time through the true anomaly according to the expression:

$$\rho = \frac{a_{\mathrm{p}}\,(1-e_{\mathrm{p}}^2)}{1+e_{\mathrm{p}}\,\cos(f_{\mathrm{p}})},$$

a_{p} being the semimajor axis of the orbit of m_2 with respect to m_1 and f_{p} its corresponding true anomaly (see for example, Szebehely 1967).

Hamiltonian (1) defines a differential system of three degrees of freedom which depends on time explicitly. Following the terminology appearing in Meyer and Hall (1992) when the infinitesimal particle is far from both primaries, the problem is called *comet*. Thus, we make a_{p} small by setting $a_{\mathrm{p}} = \varepsilon$, i.e. a_{p} represents the small parameter of the problem.

Although in this paper, we make use of the formulation in the inertial frame, the equations of motion are more commonly presented in a non-inertial co-ordinate system that rotates with the mean motion of the primaries. For this reason, we present in the following paragraphs the relation between both formulations.

In the rotating system the positions of both primaries are fixed. The passage from the inertial to the rotating system is done by a linear time-dependent change of variables and an appropriate rescaling of time. The planar version of this transformation can be read in Deprit and Rom (1970). First, we perform a linear change from $(\boldsymbol{q},\boldsymbol{p})$ to $(\boldsymbol{x},\boldsymbol{y})$, where $\boldsymbol{x} = (x_1,x_2,x_3)$ and $\boldsymbol{y} = (y_1,y_2,y_3)$ are, respectively, the new positions and momenta. The explicit change is defined through

$$\begin{aligned} q_1 &= \rho\,[x_1\,\cos(f_{\mathrm{p}}) - x_2\,\sin(f_{\mathrm{p}})], & p_1 &= \frac{1}{\rho}\,[y_1\,\cos(f_{\mathrm{p}}) - y_2\,\sin(f_{\mathrm{p}})],\\ q_2 &= \rho\,[x_1\,\sin(f_{\mathrm{p}}) + x_2\,\cos(f_{\mathrm{p}})], & p_2 &= \frac{1}{\rho}\,[y_1\,\sin(f_{\mathrm{p}}) + y_2\,\cos(f_{\mathrm{p}})],\\ q_3 &= \rho\,x_3, & p_3 &= \frac{y_3}{\rho}. \end{aligned}$$

This transformation carries out a change of independent variable, which is achieved through an adequate rescaling of time. In fact, we have to multiply $\mathcal{H}$ by ρ^2/p_{p}, where p_{p} stands for the semilatus rectum $a_{\mathrm{p}}\,(1-e_{\mathrm{p}}^2)$. Then, we add a remainder function to the Hamiltonian, as the change depends explicitly on time. The remainder function is given by

$$\mathcal{R} = \frac{1}{2\,\rho^2}\,(p_p - 1)\,(y_1^2 + y_2^2 + y_3^2) + \frac{1}{2\,\rho^2}\,(p_p - \rho)\,(x_1^2 + x_2^2 + x_3^2) + \frac{p_p}{\rho^2}\,(-y_2\,x_1 + y_1\,x_2).$$

Hence, the resulting Hamiltonian in the rotating frame yields: $\mathcal{H}_R = \rho^2\,(\mathcal{H}+\mathcal{R})/p_p$, and it gives:

$$\begin{aligned}\mathcal{H}_R(\boldsymbol{x},\boldsymbol{y},f_p;\mu,a_p,e_p) = {} & \frac{1}{2}\,(y_1^2 + y_2^2 + y_3^2) - y_2\,x_1 + y_1\,x_2 \\ & + \frac{1}{2}\left(1 - \frac{\rho}{p_p}\right)(x_1^2 + x_2^2 + x_3^2) \\ & - \frac{(1-\mu)}{\sqrt{(\mu + x_1)^2 + x_2^2 + x_3^2}}\,\frac{\rho}{p_p} \\ & - \frac{\mu}{\sqrt{(1-\mu - x_1)^2 + x_2^2 + x_3^2}}\,\frac{\rho}{p_p}.\end{aligned}$$

Thus, we recover the formulation of the three-dimensional elliptic restricted three-body problem as it appears in classical books, such as (Szebehely 1967). The independent variable in $\mathcal{H}_R$, say s, is related with the independent variable in $\mathcal{H}$ by means of $\mathrm{d}\,t/\mathrm{d}\,s = \rho^2/p_p$. Hence, the derivative of the true anomaly f_p with respect to s is given by $\mathrm{d}\,f_p/\mathrm{d}\,s = 1/\sqrt{p_p}$. When $e_p = 0, \mathcal{H}_R$ reduces to the spatial circular restricted three-body problem in the rotating frame.

For $e_p \in [0, 1)$, the Eulerian and Lagrangian equilibria in the rotating frame have co-ordinates:

$$(x_1, x_2, x_3, y_1, y_2, y_3) = (1/2 - \mu, \pm\sqrt{3}/2, 0, \mp\sqrt{3}/2, 1/2 - \mu, 0)$$

for $\mathcal{L}_4$ and $\mathcal{L}_5$, and the points $\mathcal{L}_1, \mathcal{L}_2$ and $\mathcal{L}_3$ appear on the axis x_1. In the three last cases x_1 is the solution of the same quintic equation as the one appearing in the circular problem. Expressed in the inertial frame, the equilibria are the solutions of Hamiltonian $\mathcal{H}$ which describe ellipses in the phase space. In Fig. 2 we have plotted

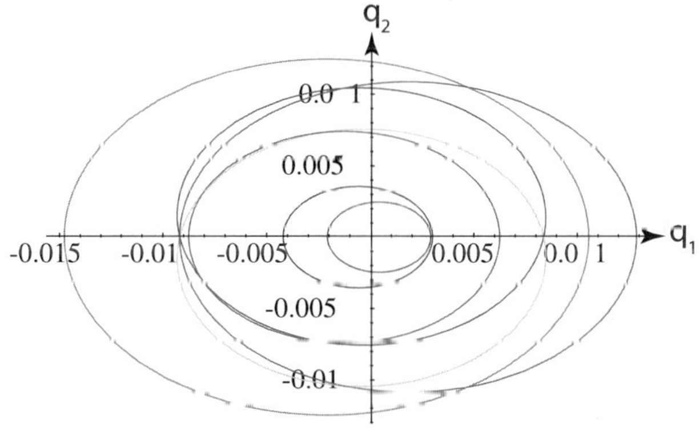

Fig. 2 Ellipses corresponding to the points $\mathcal{L}_1$ (*light blue*), $\mathcal{L}_2$ (*blue*), $\mathcal{L}_3$ (*green*), $\mathcal{L}_4$ (*yellow*) and $\mathcal{L}_5$ (*red*). The primary m_1 is describing the *brown ellipse*, whereas m_2 is moving along the *pink ellipse*. The value of e_p has been fixed to 1/6, whereas $\mu = 1/4$ and the semimajor axis $\varepsilon = 10^{-2}$

the projection of the ellipses corresponding to the equilibria in the plane $q_1 q_2$, as well as the ellipses traced by m_2 and m_1 around their centre of mass. Notice that the invariant objects, we look for are very far away from all the ellipses described by the Lagrangian and Eulerian equilibria.

For the circular case the time does not appear explicitly in the formulation in the rotating frame. However, it does appear explicitly if we express the Hamiltonian in the inertial frame. For $e_{\mathrm{p}} > 0$, the corresponding equations of motion are non-autonomous in both formulations.

3 Double averaging in the circular problem

3.1 Expansion of the Hamiltonian

We make $e_{\mathrm{p}} = 0$ and $a_{\mathrm{p}} = \varepsilon$ in (1) and expand it in powers of the small parameter ε. We arrive at:

$$\mathcal{H} = \mathcal{H}_0 + \varepsilon\,\mathcal{H}_1 + \frac{\varepsilon^2}{2}\,\mathcal{H}_2 + \frac{\varepsilon^3}{3!}\,\mathcal{H}_3 + \frac{\varepsilon^4}{4!}\,\mathcal{H}_4 + \frac{\varepsilon^5}{5!}\,\mathcal{H}_r, \tag{2}$$

where:

$$\begin{aligned}
\mathcal{H}_0 &= \frac{1}{2}\,(p_1^2 + p_2^2 + p_3^2) - \frac{1}{\sqrt{q_1^2 + q_2^2 + q_3^2}},\\
\mathcal{H}_1 &= 0,\\
\mathcal{H}_2 &= \frac{(-1+\mu)\,\mu}{2\,(q_1^2 + q_2^2 + q_3^2)^{5/2}}\\
&\quad \times \Big[q_1^2 + q_2^2 - 2\,q_3^2 + 3\,(q_1^2 - q_2^2)\,\cos(2 f_{\mathrm{p}}) + 6\,q_1\,q_2\,\sin(2 f_{\mathrm{p}})\Big],\\
\mathcal{H}_3 &= \frac{3\,(1-\mu)\,\mu\,(1-2\,\mu)\,[q_1\,\cos(f_{\mathrm{p}}) + q_2\,\sin(f_{\mathrm{p}})]}{2\,(q_1^2 + q_2^2 + q_3^2)^{7/2}}\\
&\quad \times \Big[q_1^2 + q_2^2 + 6\,q_3^2 - 5\,(q_1^2 - q_2^2)\,\cos(2 f_{\mathrm{p}}) - 10\,q_1\,q_2\,\sin(2 f_{\mathrm{p}})\Big],\\
\mathcal{H}_4 &= \frac{3\,(1-\mu)\,\mu\,[-1 + 3\,(1-\mu)\,\mu]}{8\,(q_1^2 + q_2^2 + q_3^2)^{9/2}}\\
&\quad \times \Big\{9\,(q_1^2 + q_2^2)^2 - 72\,(q_1^2 + q_2^2)\,q_3^2 + 24\,q_3^4\\
&\quad + 35\,(q_1^4 - 6\,q_1^2\,q_2^2 + q_2^4)\,\cos(4 f_{\mathrm{p}})\\
&\quad + 40\,q_1\,q_2\,(q_1^2 + q_2^2 - 6\,q_3^2)\,\sin(2 f_{\mathrm{p}})\\
&\quad + 20\,(q_1^2 - q_2^2)\,\cos(2 f_{\mathrm{p}})\Big[q_1^2 + q_2^2 - 6\,q_3^2 + 14\,q_1\,q_2\,\sin(2 f_{\mathrm{p}})\Big]\Big\},
\end{aligned}$$

and $\mathcal{H}_r$ contains higher-order terms in ε. Let us note that $\mathcal{H}_0$ corresponds to the Kepler problem centred at the origin of the inertial frame. This fact allows us to consider the motion of the infinitesimal particle with respect to the centre of mass of the primaries as a perturbation of the two-body problem when m_3 is far enough from the two primaries. We conclude that the expansion performed is convergent and no collision between the infinitesimal particle and the primaries occur provided that $\varepsilon/|\boldsymbol{q}| < 1/(1-\mu)$, which implies that the series expansion is valid for $\varepsilon < |\boldsymbol{q}|$.

Hamiltonian (3) defines a non-autonomous system of three degrees of freedom. In order to reduce the system, we perform a double averaging by means of Lie transformations of Deprit (1969). We follow the same steps as in Palacián et al. (2006), thus we will concentrate on the differences appearing in this case with respect to the one treated in that paper.

3.2 Averaging over the time

First, we average over the time. We begin by changing the time from t to $t' = t/\varepsilon$, hence $\mathcal{H}' = \varepsilon \sum_{i=0}^{4}(\varepsilon^i/i!)\,\mathcal{H}'_i$ with $\mathcal{H}'_0 = 0$ and $\mathcal{H}'_i = i\,\mathcal{H}_{i-1}$ for $i \geq 1$. This scaling can be applied because the relative motion between the primaries is much faster than that of the infinitesimal particle around them (see Cors et al. 2005). Next, we apply the Lie–Deprit method for time-dependent Hamiltonians (Deprit 1969) up to terms of order five, taking into account that at each step $\mathcal{K}'_i$ is selected as the average

$$\mathcal{K}'_i = \frac{1}{2\,\pi\,\varepsilon^{3/2}} \int_0^{2\pi} \tilde{\mathcal{H}}'_i \,\mathrm{d} f_{\mathrm{p}},$$

where $\tilde{\mathcal{H}}'_i$ is computed from the previous orders using the Lie–Deprit triangle. As $\mathcal{K}'_0 = 0$ the generating function at each order is calculated through the O.D.E.:

$$\frac{\partial\,\mathcal{W}_i}{\partial\,t'} = \tilde{\mathcal{H}}'_i - \mathcal{K}'_i$$

up to $i = 5$. We have obtained both $\mathcal{W}_i$ and $\mathcal{K}'_i$ for $i \in \{1, \ldots, 5\}$ in closed form. Instead of f_{p}, we could have used either the eccentric or the mean anomalies to perform the averaging as the three angles coincide because the eccentricity of the primaries is assumed to be zero.

Then, we change the time back and rescale the averaged Hamiltonian arriving at $\mathcal{K} = \mathcal{K}'/\varepsilon$ (i.e. $\mathcal{K}_{i-1} = \mathcal{K}'_{i/i}$ for $i \geq 1$), expressing also $\mathcal{W}$ in terms of t. We stress that $\mathcal{W}$ is periodic in t with period $2\,\pi\,\varepsilon^{3/2}$. The explicit expression of $\mathcal{K}$ yields:

$$\begin{aligned}
\mathcal{K}_0 &= \frac{1}{2}\,(p_1'^2 + p_2'^2 + p_3'^2) - \frac{1}{\sqrt{q_1'^2 + q_2'^2 + q_3'^2}},\\
\mathcal{K}_1 &= 0,\\
\mathcal{K}_2 &= \frac{(-1+\mu)\,\mu\,(q_1'^2 + q_2'^2 - 2\,q_3'^2)}{2\,(q_1'^2 + q_2'^2 + q_3'^2)^{5/2}},\\
\mathcal{K}_3 &= 0,\\
\mathcal{K}_4 &= 9\,(-1+\mu)\,\mu\,(1 - 3\,\mu + 3\,\mu^2)\\
&\quad\times \frac{\left[3\,(q_1'^2 + q_2'^2)^2 - 24\,(q_1'^2 + q_2'^2)\,q_3'^2 + 8\,q_3'^4\right]}{8\,(q_1'^2 + q_2'^2 + q_3'^2)^{9/2}}.
\end{aligned}$$

We do not print down the formula for $\mathcal{W}$ as its size is rather big. However, we can provide it (as well as the rest of Hamiltonians and generating functions) to the interested reader. We drop primes from now on.

We remark that the generating function corresponding to the averaging process over the time derived in Palacián et al. (2006), was not well defined for $e_{\mathrm{p}} = 0$. In particular, the denominator of $\mathcal{W}_3$ went to infinity when e_{p} approaches zero. Thus, we had to exclude the circular case in the analysis performed in that study and this fact is one of the motivations of the present paper.

3.3 Averaging over the mean anomaly

We eliminate the mean anomaly (see for example, Deprit 1982) up to order four. Polar-nodal and Delaunay variables (Brouwer and Clemence 1961; Deprit 1981) are better suited to normalise Keplerian systems because they have a geometric interpretation.

Polar-nodal variables are $(r, \vartheta, \nu, R, \Theta, N)$, where ϑ is the argument of the latitude. Its conjugated momentum, Θ, corresponds to the modulus of the angular-momentum vector $\boldsymbol{G} = \boldsymbol{q} \times \boldsymbol{p}$. The co-ordinate r is the modulus of $\boldsymbol{q}$ and its conjugate momentum R denotes the radial velocity. The action N, representing the projection of $\boldsymbol{G}$ onto the axis q_3, is the action conjugated to the argument of the node ν. The angular-momentum vector $\boldsymbol{G}$ is decomposed uniquely as $\boldsymbol{G} = \Theta\, \boldsymbol{n}$ with $\Theta > 0$ and $|\boldsymbol{n}| = 1$, where $\boldsymbol{n}$ indicates the normal direction and is orthogonal to the plane spanned by $\boldsymbol{q}$ and $\boldsymbol{p}$, i.e., the instantaneous orbital plane. Its inclination with respect to the plane $q_1\, q_2$ (the so-called equatorial plane) is given by the angle I in $(0, \pi)$ with $N = \Theta \cos I$. We define $c = \cos I$ and $s = \sin I$ for convenience of notation.

Delaunay variables (ℓ, g, h, L, G, H) represent a set of action-angle variables defined through Polar-nodal variables by means of a generating function (see Deprit 1981, 1982) for details. If $\mathcal{K}_0$ stands for the Hamiltonian of the two-body problem, the action L is related to the Keplerian energy by the identity $\mathcal{K}_0 = -1/(2\, L^2)$, where the gravitational constant is taken as the unity. The action G is defined as Θ, whereas the third component of $\boldsymbol{G}$ is H, so $H \equiv N$. The angle ℓ stands for the mean anomaly. The angle g is the argument of pericentre. It is reckoned from the pericentre of the orbit in the instantaneous orbital plane. Then, $g = \vartheta - f$ where f denotes the true anomaly. The angle h is the argument of the node, that is, $h \equiv \nu$. The eccentricity of the trajectory is designated by e and in terms of Delaunay actions it is expressed as $e = (1 - G^2/L^2)^{1/2}$. As $e \in [0, 1)$ then $G \in (0, L]$.

Hamiltonian $\mathcal{K}$ in mixed Polar-nodal and Delaunay co-ordinates yields:

$$
\begin{aligned}
\mathcal{K}_0 &= -\frac{1}{2\, L^2},\\
\mathcal{K}_1 &= 0,\\
\mathcal{K}_2 &= \frac{(1-\mu)\,\mu}{4\, r^3}\,[1 - 3\, c^2 - 3\, s^2 \cos(2\,\vartheta)],\\
\mathcal{K}_3 &= 0,\\
\mathcal{K}_4 &= \frac{9\,(1-\mu)\,\mu\,[-1 + 3\,(1-\mu)\,\mu]}{64\, r^5}\\
&\quad \times \Big[9 - 90\, c^2 + 105\, c^4 - 20\,(1 - 8\, c^2 + 7\, c^4)\cos(2\,\vartheta)\\
&\quad + 35\, s^4 \cos(4\,\vartheta)\Big].
\end{aligned}
$$

Notice that $\mathcal{K}$ defines an autonomous system of two degrees of freedom, either in Delaunay or Polar-nodal co-ordinates as it is independent of the argument of the node. This feature is a consequence of the elimination of time because the initial Hamiltonian $\mathcal{H}$ does indeed depend on h as it is a true non-autonomous Hamiltonian system with three degrees of freedom.

Now $\mathcal{K}$ is ready for the application of the Delaunay normalisation. Thus, we need to eliminate the mean anomaly including terms factored by ε^4. This time, the homological equation solved at each order i $(1 \le i \le 4)$ can be written in terms of Delaunay co-ordinates as:

$$\frac{1}{L^3}\frac{\partial \mathcal{V}_i}{\partial \ell} + \mathcal{L}_i = \tilde{\mathcal{K}}_i \tag{3}$$

for the unknowns $\mathcal{L}_i$, i.e. the transformed Hamiltonian, and $\mathcal{V}_i$ (the generating function). The terms $\tilde{\mathcal{K}}_i$ are known at each order as functions of $\tilde{\mathcal{K}}_j$ and of $\mathcal{V}_j$ $(1 \le j \le i-1)$. After taking $\mathcal{L}_i$ as the average of $\tilde{\mathcal{K}}_i$ with respect to ℓ, Eq. 3 is solved in $\mathcal{V}_i$ by calculating the corresponding primitive in ℓ.

The removal of ℓ avoiding Taylor and Fourier expansions is possible thanks to the routines for the Delaunay normalisation of functions depending on r, R, ϑ and f introduced in Osácar and Palacián (1994), thus preserving the closure of all the expressions involved in the process. The transformation induced through the Delaunay normalisation is valid for any type of elliptic motions. Moreover, $\mathcal{V}$ is a periodic function in the two angles ℓ and g. We do not write down the expression of $\mathcal{V}$ as it is too big, but it can be provided by the authors.

However, the Hamiltonian system free of the mean anomaly reads as $\mathcal{L} = \sum_{i=0}^{4}(\varepsilon^i/i!)\,\mathcal{L}_i$ where

$$\begin{aligned}
\mathcal{L}_0 &= -\frac{1}{2\,L'^2},\\
\mathcal{L}_1 &= 0,\\
\mathcal{L}_2 &= \frac{(1-\mu)\,\mu\,(G'^2 - 3\,H'^2)}{4\,L'^3\,G'^5},\\
\mathcal{L}_3 &= 0,\\
\mathcal{L}_4 &= \frac{3\,(1-\mu)\,\mu}{128\,L'^5\,G'^{11}}\\
&\quad\times\Big\{3\,(27 - 91\,\mu + 91\,\mu^2)\,G'^6 + 24\,(1-\mu)\,\mu\,L'\,G'^5\\
&\quad+\Big[-135\,(L'^2 + 6\,H'^2) + (1-\mu)\,\mu\,(451\,L'^2 + 2538\,H'^2)\Big]\,G'^4\\
&\quad-144\,(1-\mu)\,\mu\,L'\,H'^2\,G'^3\\
&\quad+3\,H'^2\Big[45\,(10\,L'^2 + 7\,H'^2) - \mu\,(1-\mu)\,(1402\,L'^2 + 955\,H'^2)\Big]\,G'^2\\
&\quad+216\,(1-\mu)\,\mu\,L'\,H'^4\,G' - 3\,(525 - 1553\,\mu + 1553\,\mu^2)\,L'^2\,H'^4\\
&\quad+6\,(L'^2 - G'^2)\,(G'^2 - H'^2)\\
&\quad\times\Big[15\,(G'^2 - 7\,H'^2) - (1-\mu)\,\mu\,(47\,G'^2 - 345\,H'^2)\Big]\cos(2\,g')\Big\}.
\end{aligned}$$

From now on, we drop primes, taking into account that in order to pass from the "non-prime" Delaunay variables to the "prime" ones, we have to use the change of variables defined through $\mathcal{V}$.

Four remarks are in order:

- We conclude, as expected, that $\mathcal{L}$ defines a Hamiltonian system of one degree of freedom and therefore, we can study its flow in a simpler way than the previous Hamiltonians.
- We could have taken as the initial Hamiltonian the one corresponding to the comet case of the spatial circular restricted three-body problem in the rotating frame (see Howison and Meyer 2000). Then we would have proceeded by eliminating first the argument of the node and then the mean anomaly, in a similar fashion as we did with the lunar problem (Palacián and Yanguas 2004) but

interchanging the order of the transformations, we had arrived at the same Hamiltonian $\mathcal{L}$, after dropping the Coriolis term $-H$ and scaling conveniently in the circular restricted three-body problem. Therefore, both formulations of the initial problem yield the same results. However, we have preferred to study the problem in the inertial frame as it may be understood as a subcase of the problem treated in Palacián et al. (2006) and we want to study the transition from the circular to the elliptic problem when the infinitesimal particle is very far from the primaries.

- In the case $e_p > 0$, at this step the doubly-averaged Hamiltonian does depend on the argument of the node, but not on the argument of pericentre. Moreover, in the purely elliptic situation the analysis of the reduced system can be performed by reaching only order two. Nevertheless, when $e_p = 0$ the reduced system up to order two in ε is not structurally stable. Thus, we have to go beyond and the degeneracy breaks down at order four. Hence, we need to reach this order to perform the analysis of the fully-reduced system, which will be the aim of the following section.
- We emphasize that the mean anomaly ℓ can be eliminated separately from time t without introducing small denominators as no resonance phenomenon can occur in the transformations. The reason is that the time t is $\varepsilon^{3/2}$—small compared to ℓ. More precisely, while ℓ is allowed to vary between 0 and 2π, t is bounded between 0 and $2\pi\,\varepsilon^{3/2}$. This fact has been pointed out previously in Palacián (2002) for the treatment of successive normal forms for perturbed Keplerian systems and by Féjoz (2002) for the averaging process of the planar three-body problem.

4 Reductions and analysis of the doubly-averaged problem

4.1 The Hamiltonian in invariants

The initial Hamilton function (1) defines a dynamical system whose flow lies on a seven-dimensional space. After averaging over the time, since the argument of the node is also removed in the same process and so H is a constant of motion up to terms of order ε^4, we may fix a value for it. Hence, the corresponding reduced space is the quotient space $\mathbf{R}^6/(S^1 \times S^1)$ (see for instance, Palacián, 2002; Palacián and Yanguas, 2005), which is a space of dimension four, the so-called first reduced phase space. Next, we average again, this time being over the mean anomaly and the resulting Hamiltonian is independent of the angle ℓ. Thus, L becomes a formal integral of the Hamiltonian up to the order of averaging, that is, order four. After fixing a value for L the resulting space is of dimension two.

The transformation is generated by three linearly-independent invariants with respect to both integrals L and H: τ_1, τ_2, τ_3 and are related by the equation, which defines the phase space:

$$\mathcal{T}_{L,H} = \{\boldsymbol{\tau} \in \mathbf{R}^3 \mid \tau_2^2 + \tau_3^2 = [(L+\tau_1)^2 - H^2]\,[(L-\tau_1)^2 - H^2]\}$$

for $0 \le |H| \le L$ and $L > 0$. Note that τ_2 and τ_3 always belong to the interval $[H^2 - L^2, L^2 - H^2]$, whereas τ_1 belongs to $[|H| - L, L - |H|]$. We note that τ_i's are called invariants as they remain invariant with respect to the flows generated by L and H. They are also called generators of the phase space $\mathcal{T}_{L,H}$ since this space can be parameterised in terms of them.

The two-dimensional phase spaces $\mathcal{T}_{L,H}$ and $\mathcal{T}_{L,0}$ (the phase space for $H = 0$) are plotted in Fig. 3 for fixed values of the parameters L and H. For more information on this reduction (see Cushman 1983; Palacián, 2002; Palacián and Yanguas 2004).

The relationship between g and G with τ_1, τ_2 and τ_3 is:

$$
\begin{aligned}
G^2 &= \tfrac{1}{2}\,(L^2 + H^2 - \tau_1^2 + \tau_3),\\
\cos g &= \frac{-\tau_2}{\sqrt{(L^2 - H^2)^2 - (\tau_1^2 - \tau_3)^2}},\\
\sin g &= \tau_1 \sqrt{\frac{2\,(L^2 + H^2 - \tau_1^2 + \tau_3)}{(L^2 - H^2)^2 - (\tau_1^2 - \tau_3)^2}}.
\end{aligned}
$$

Now, we want to write the Hamiltonian $\mathcal{L}$ in terms of the invariants τ_i's. After dropping the constant term $\mathcal{L}_0$ and dividing by ε^2, we get the following expression:

$$
\begin{aligned}
\mathcal{M}(\tau_1, \tau_2, \tau_3) = {} & \frac{(1-\mu)\,\mu\,(\tau_4^2 - 6\,H^2)}{\sqrt{2}\,L^3\,\tau_4^5} - \frac{\varepsilon^2\,(1-\mu)\,\mu}{64\,\sqrt{2}\,L^5\,\tau_4^{11}}\\
& \times \Big\{ 9\,(1-\mu+\mu^2)\,\tau_4^6 + 24\,\sqrt{2}\,L\,(1-\mu)\,\mu\,\tau_4^5\\
& + 2\Big[45\,(L^2 + 2\,H^2) - (1-\mu)\,\mu\,(169\,L^2 + 186\,H^2)\\
& + 12\,(15 - 47\,\mu + 47\,\mu^2)\,\tau_1^2 \Big]\,\tau_4^4 + 288\,\sqrt{2}\,(1-\mu)\,\mu\,L\,H^2\,\tau_4^3\\
& - 12\,H^2 \Big[105\,(2\,L^2 + H^2) - (1-\mu)\,\mu\,(618\,L^2 + 265\,H^2)\\
& + 60\,(7 - 23\,\mu + 23\,\mu^2)\,\tau_1^2 \Big]\,\tau_4^2 - 864\,\sqrt{2}\,(1-\mu)\,\mu\,L\,H^4\,\tau_4\\
& + 24\,(315 - 863\,\mu + 863\,\mu^2)\,L^2\,H^4 \Big\},
\end{aligned}
$$

where $\tau_4 = \sqrt{\tau_3 - \tau_1^2 + L^2 + H^2}$.

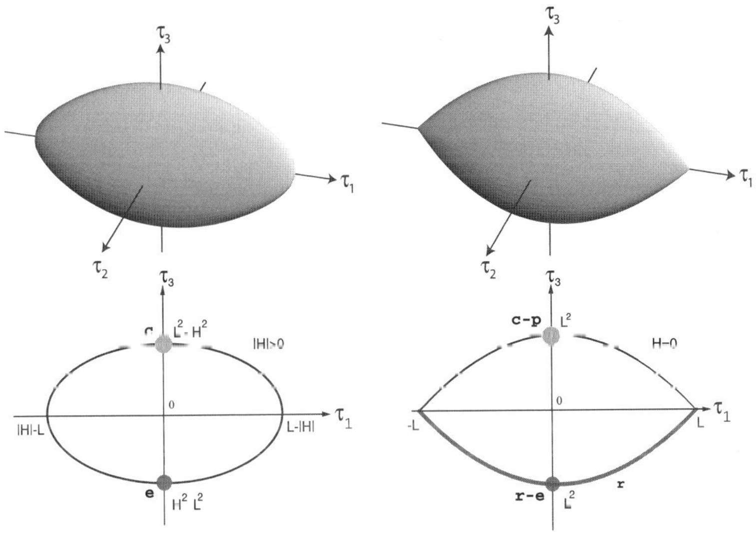

Fig. 3 Doubly-reduced phase spaces $\mathcal{T}_{L,H}$ and $\mathcal{T}_{L,0}$. Top: On the left $L = 10$ and $H = 2$, in this case $\mathcal{T}_{10,2}$ is diffeomorphic to a two-sphere. On the right, we have taken $L = 11$ and $H = 0$ hence $\mathcal{T}_{11,0}$ corresponds to a two-sphere with two singular points. Bottom: projections of $\mathcal{T}_{L,H}$ and $\mathcal{T}_{L,0}$ onto the plane $\tau_2 = 0$. The *green points* (`c` and `c-p`) are related with circular motions whereas the *red points* (`e` and `r-e`) correspond to equatorial trajectories. The *light blue arc* (`r`) represents rectilinear motions

At this point we stress the convenience of studying the circular case ($e_{\rm p} = 0$) apart from the elliptic one ($e_{\rm p} > 0$) as the corresponding doubly-reduced phase spaces are topologically different (the one related with the elliptic case is diffeomorphic to S^2).

4.2 Relative equilibria

Next, we discuss and calculate the relative equilibria associated with $\mathcal{M}$. To achieve this we need to construct the vector field related with the derivatives of τ_1, τ_2 and τ_3 with respect to time. We get:

$$\frac{d\,\tau_i}{\mathrm{d}t} = \sum_{1\le j\le 3} \{\tau_i\,,\,\tau_j\}\,\frac{\partial\,\mathcal{M}}{\partial\,\tau_j}, \quad i \in \{1,2,3\}, \tag{4}$$

where

$$\begin{aligned}
\{\tau_1\,,\,\tau_2\} &= -\{\tau_2\,,\,\tau_1\} = 2\,\tau_3,\\
\{\tau_1\,,\,\tau_3\} &= -\{\tau_3\,,\,\tau_1\} = -2\,\tau_2,\\
\{\tau_2\,,\,\tau_3\} &= -\{\tau_3\,,\,\tau_2\} = -4\,\tau_1\,(\tau_1^2 - L^2 - H^2).
\end{aligned}$$

Some points are obtained doing $\varepsilon = 0$ in the differential system (4), i.e., using the normal form $\mathcal{L}$ truncated at third-order. They are:

- Point $P_1 = (0, 0, L^2 - H^2)$ is an equilibrium for all values of the parameters L and H. This point corresponds to circular-type orbits.
- Point $P_2 = (0, 0, H^2 - L^2)$ is an equilibrium for $H \neq 0$ and all values of L. This point corresponds to equatorial-type orbits.
- Points

  $$P_{3,4} = (0, \mp 4\,|H|\,\sqrt{L^2 - 5\,H^2}, 9\,H^2 - L^2)$$

  are equilibria when $L \geq \sqrt{5}\,|H|$.

However, if $L > \sqrt{5}\,|H| > 0$ and $\tau_1 \neq 0$, the expression:

$$\left(\tau_1, \mp\sqrt{-20\,H^4 + 4\,L^2\,H^2 - 5\,H^2\,\tau_1^2}, 9\,H^2 - L^2 + \tau_1^2\right),$$

corresponds to an infinite set of relative equilibria for the truncation of $\mathcal{M}$ after second-order, hence the system is not structurally stable. This is the reason of pushing the normal form to order four. Then, provided that $L > \sqrt{5}\,|H| > 0$, we refine the points obtained above, arriving at:

$$P_{5,6} = \left(\mp 2\sqrt{\frac{L^2 - 5\,H^2}{5}}, \varepsilon^2 f_1, 5\,H^2 - \frac{L^2}{5} + \varepsilon^2 f_2\right)$$

with

$$\begin{aligned}
f_1 &= \mathcal{O}(\varepsilon^2),\\
f_2 &= \frac{15\,(123\,L^2 + 5\,H^2) + (1-\mu)\,\mu\,(5801\,L^2 - 17925\,H^2)}{1500\,L^2\,H^2} + \mathcal{O}(\varepsilon^2).
\end{aligned}$$

We remark that both f_1 and f_2 depend only on H, L, μ and ε. To determine the expressions of f_1 and f_2 factored by ε^2 one should push the Delaunay normalisation up to the next order. More precisely, f_2 might be obtained from $\mathcal{M}$ (that is, using only the fourth-order normal formal) although we have preferred to expand the cumbersome expression in powers of ε, thus its higher-order terms should be refined with the next order in the normalisation process. On the other hand, although f_1 cannot be completely determined at order four, it does not imply the appearance of an infinite set of equilibria.

- When $H = -L$, equilibria P_1 and P_2 coincide and there is no more equilibria.
- When $-L < H < -L/\sqrt{5}$ there are two equilibria: P_1 and P_2.
- When $H = -L/\sqrt{5}$ there are also two equilibria: P_1 and P_2, as $P_{3,4}$ coincide with P_1.
- When $-L/\sqrt{5} < H < 0$ there are six equilibria: P_1, P_2, $P_{3,4}$ and $P_{5,6}$. The four points $P_{3,4}$ and $P_{5,6}$ bifurcate from P_1.
- When $H = 0$ there is only one equilibria, namely P_1.
- When $0 < H < L/\sqrt{5}$ there are six equilibria: P_1, P_2, $P_{3,4}$ and $P_{5,6}$.
- When $H = L/\sqrt{5}$ there are two equilibria: P_1 and P_2, as $P_{3,4}$ are the same as P_1.
- When $L/\sqrt{5} < H < L$ there are two equilibria: P_1 and P_2.
- When $H = L$ there is only one equilibrium, P_1, which coincides with P_2.

We have depicted in Fig. 4 the bifurcation lines as well as the number of relative equilibria in each region. We have identified the line Γ_1 with the curve $H = -L, \Gamma_2$ with the curve $H = -L/\sqrt{5}, \Gamma_3$ with $H = 0, \Gamma_4$ with $H = L/\sqrt{5}$ and Γ_5 with the curve $H = L$.

Note that the picture of relative equilibria is symmetric with respect to $H = 0$.

Most of the discussion of the above paragraphs makes sense when the eccentricity of the primaries reaches the value 1 due to an argument given by Féjoz (2001). The corresponding phase–space can be compactified by adding degenerate eccentricity–one ellipses at infinity.

The stability of the equilibria is discussed in the next section.

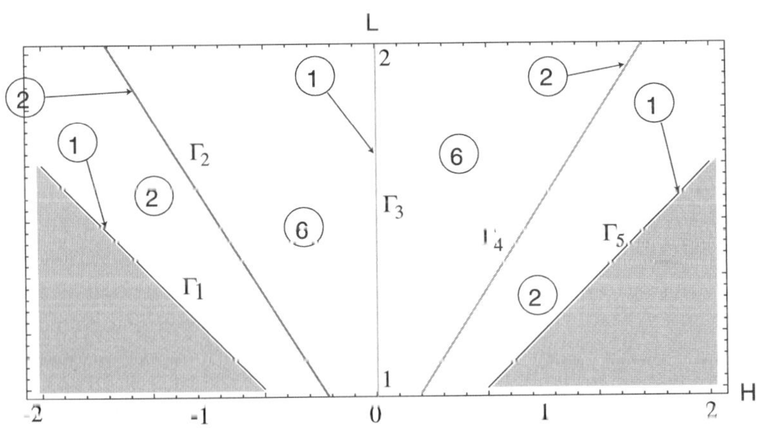

Fig. 4 Bifurcation diagram corresponding to $\mathcal{T}_{L,H}(0 \leq |H| \leq L)$, with the number of relative equilibria in each region encircled. We have drawn the action L varying between 1 and 2, which is consistent with the fact that $L^2\,(1+e) > \varepsilon$ so that we are in the comet case of the spatial restricted problem

5 Stability of the relative equilibria

5.1 Local analysis

In the next paragraphs our aim is to study the linear and non-linear stability of the relative equilibria. This is achieved by performing a local analysis at each equilibrium, calculating second-jets for the case of elliptic (centres) and hyperbolic points (saddles) or fourth-jets for the case of degenerations. So, we need to introduce appropriate local canonical co-ordinates in each case, say x and y, Taylor–expand around the equilibrium in order to bring these expansions to their local normal forms, thus making the type of stability of the equilibrium apparent.

More specifically, in the elliptic case we get

$$\sum_{k=0}^{2} b_k\,(x^2+y^2)^k + \text{higher order terms.}$$

The scalar b_k in this Birkhoff polynomial is the kth Birkhoff coefficient (cf. Arnold, 1978). The coefficient b_k is computed from the $2k$th-jet of the Hamilton function in the elliptic equilibrium. In our case it suffices to compute b_1, concluding that the centres persist as (non-linear) centres whenever $b_1 \neq 0$. On the other hand, we have calculated

$$b_0 + b_x\,x^2 + b_y\,y^2 + \text{higher order terms}$$

and deduced instability (hyperbolic points) when $b_x\,b_y < 0$.

We start by studying the relative equilibrium P_1. We define local co-ordinates as

$$x = \sqrt{2\,(L-G)}\,\cos g, \qquad y = \sqrt{2\,(L-G)}\,\sin g.$$

Using the Poisson brackets among the Delaunay variables, it is straight-forward to see that $\{x, y\} = 1$.

Now, after writing g and G in terms of x and y we substitute in $\mathcal{L}$ the corresponding expressions. Then, we expand the Hamiltonian in powers of x and y around the origin (which corresponds to circular motions) up to degree two. Dropping the constant terms we arrive at:

$$\mathcal{L}_C^{\text{2-jet}} = \frac{192\,(5\,H^2 - L^2)}{L^4}\,(x^2+y^2).$$

The quadratic part of $\mathcal{L}_C^{\text{2-jet}}$ gives the linear stability of the point P_1 and as the coefficient of x^2+y^2 only vanishes at $L = \sqrt{5}\,|H|$, we deduce that it is linearly stable (a centre) excepting in the bifurcation lines Γ_2 and Γ_4. To decide on the non-linear stability, we notice that $\mathcal{L}_C^{\text{2-jet}}$ corresponds to the quadratic terms of a Morse function with index 2 if $5\,H^2 > L^2$ or -2 when $5\,H^2 < L^2$. Thence, we conclude that whenever $L \neq \sqrt{5}\,|H|$, P_1 is a non-linear centre. The reader can look up about Morse theory in Milnor (1963) and Abraham and Marsden (1985).

For the bifurcations $L = \sqrt{5}\,|H|$, we compute the Taylor expansion of $\mathcal{L}$ to order four. After dropping the constant terms, the fourth-jet yields:

$$\mathcal{L}_{C,L=\sqrt{5}|H|}^{\text{4-jet}} = \frac{96}{L^3}\,(x^2+y^2)^2,$$

so it is a degenerate point. However, as this point corresponds to a local minimum it must be a stable point (see, for example, Meyer and Hall 1992).

We examine the stability of the point P_2. This time we define

$$x = \sqrt{2\,(G - |H|)}\ \sin g, \qquad y = \sqrt{2\,(G - |H|)}\ \cos g.$$

Proceeding similarly to the circular case, first, we put g and G in terms of x and y and replace conveniently in $\mathcal{L}$. Then we expand the Hamiltonian in terms of x and y around $(0, 0)$, which corresponds to the point P_2, up to polynomials of degree two. After dropping constant terms we end up with:

$$\mathcal{L}_E^{\text{2-jet}} = -\frac{768\,L^2}{H^4}\,(x^2 + y^2).$$

Note that, we have discarded previously the case $H = 0$ as then P_2 is not an equilibrium. It is readily deduced that P_2 corresponds to an elliptic point, i.e. a linear centre. Moreover, as $\mathcal{L}_E^{\text{2-jet}}$ represents the quadratic terms of a Morse function with index -2, we conclude that P_2 is a non-linear centre.

Now, we deal with the stability of the points $P_{3,4}$ when $0 < |H| < L/\sqrt{5}$. Written in terms of the Delaunay co-ordinates g and G, it is not hard to prove that P_3 is $(g_0, G_0) = (0, \sqrt{5}\,|H|)$ whereas P_4 is $(g_0, G_0) = (\pi, \sqrt{5}\,|H|)$ provided that we have dropped the normal form $\mathcal{L}$ at order three. Nevertheless, we need to refine these co-ordinates of the points in order to compute explicitly $\mathcal{L}^{\text{2-jet}}$. Thus, we consider $\mathcal{L}$ including terms of order four in ε. We get for P_3 that $(g_0, G_0) = (0, \sqrt{5}\,|H| + \varepsilon^2\, G_{3,4})$ and for P_4 the co-ordinates $(g_0, G_0) = (\pi, \sqrt{5}\,|H| + \varepsilon^2\, G_{3,4})$ where

$$G_{3,4} = \frac{-45\,(7\,L^2 - 5\,H^2) + (1 - \mu)\,\mu\,(793\,L^2 - 75\,H^2)}{3000\,\sqrt{5}\,L^2\,H^2\,|H|}.$$

For P_3 we define the symplectic change:

$$g = x, \qquad G = y + \sqrt{5}\,|H| + \varepsilon^2\, G_{3,4}$$

whereas for P_4 we take:

$$g = x + \pi, \qquad G = y + \sqrt{5}\,|H| + \varepsilon^2\, G_{3,4}.$$

Next, we replace these changes in $\mathcal{L}$ and expand the resulting expressions in terms of x and y around $(0, 0)$ up to polynomials of degree two. In both situations, after dropping constant terms we arrive at:

$$\begin{aligned}\mathcal{L}_{3,4}^{\text{2-jet}} = &-\frac{96\,\varepsilon^2\,(3 - 11\,\mu + 11\,\mu^2)\,(L^2 - 5\,H^2)}{625\,\sqrt{5}\,H^6\,|H|}\,x^2\\ &+ \left\{\frac{384\,L^2}{25\,\sqrt{5}\,H^4\,|H|} + \frac{72\,\varepsilon^2\,(21\,L^2 - 85\,H^2)}{3125\,\sqrt{5}\,H^8\,|H|}\right.\\ &\left. - \frac{72\,\varepsilon^2\,(1 - \mu)\,\mu\,[\sqrt{5}\,(161\,L^2 - 625\,H^2) - 100\,L\,|H|]}{78125\,H^8\,|H|}\right\} y^2.\end{aligned}$$

As $3 - 11\,\mu + 11\,\mu^2$ is positive when $\mu \in (0, 1/2]$ and $L^2 - 5\,H^2$ is strictly positive when $P_{3,4}$ exist, we deduce that the sign of the coefficient of x^2 is negative. On the other hand, we know that the term of y^2 independent of ε is much bigger than the other two terms, provided that ε is sufficiently small, so the sign of the coefficient of

y^2 is the sign of $384\,L^2/(25\sqrt{5}\,H^4\,|H|)$ which is positive. That means that both P_3 and P_4 are saddles, thence unstable linear and non-linear points and we do not need to calculate their fourth-jets. We stress that P_3 and P_4 share the same value of energy. It is easy to deduce after replacing the co-ordinates of P_3 and P_4 in $\mathcal{M}$.

Finally, we tackle the analysis of the points $P_{5,6}$ when $0 < |H| < L/\sqrt{5}$. This time we know a priori that we need to consider the fourth-order normal form $\mathcal{L}$ in order to decide on the stability of the relative equilibria. Expressed in terms of g and G, we conclude that P_5 is $(g_0, G_0) = (\pi/2, \sqrt{5}\,|H| + \varepsilon^2\, G_{5,6})$ whereas P_6 results in $(g_0, G_0) = (3\,\pi/2, \sqrt{5}\,|H| + \varepsilon^2\, G_{5,6})$. In both cases one gets

$$G_{5,6} = \frac{-45\,(41\,L^2 - 125\,H^2) + (1-\mu)\,\mu\,(5801\,L^2 - 17925\,H^2)}{6000\sqrt{5}\,L^2\,H^2\,|H|}.$$

We need to move the origin either to P_5 or to P_6 through symplectic changes. Thus, for P_5 we define:

$$g = x + \pi/2, \qquad G = y + \sqrt{5}\,|H| + \varepsilon^2\, G_{5,6}$$

and for P_6 we make:

$$g = x + 3\,\pi/2, \qquad G = y + \sqrt{5}\,|H| + \varepsilon^2\, G_{5,6}.$$

The next step consists in replacing these changes in $\mathcal{L}$ (including $\mathcal{L}_4$), expanding the resulting expressions in terms of x and y around $(0, 0)$ up to polynomials of degree two. In both situations, after dropping constant terms, we end up with:

$$\begin{aligned}\mathcal{L}_{5,6}^{\text{2-jet}} &= \frac{96\,\varepsilon^2\,(3 - 11\,\mu + 11\,\mu^2)\,(L^2 - 5\,H^2)}{625\sqrt{5}\,H^6\,|H|}\,x^2\\ &\quad + \left\{\frac{384\,L^2}{25\sqrt{5}\,H^4\,|H|} - \frac{72\,\varepsilon^2\,(L^2 + 275\,H^2)}{3125\sqrt{5}\,H^8\,|H|}\right.\\ &\quad \left. + \frac{24\,\varepsilon^2\,(1-\mu)\,\mu}{78125\,H^8\,|H|}\left[\sqrt{5}\,(327\,L^2 + 12725\,H^2) + 300\,L\,|H|\right]\right\} y^2.\end{aligned}$$

The sign of the coefficient of x^2 is positive while the one of y^2 is the same as the sign of $384\,L^2/(25\sqrt{5}\,H^4|H|)$ provided that ε keeps small enough. Thus, we conclude that it is also positive, which implies that $P_{5,6}$ are both centres. Next, as $\mathcal{L}_{5,6}^{\text{2-jet}}$ corresponds to the main terms of a Morse function with index 2, the points $P_{5,6}$ are non-linear centres.

5.2 Bifurcations

When $L = |H|$, i.e., in the lines Γ_1 and Γ_5 the phase space $\mathcal{T}_{L,H}$ gets reduced to a point, the one corresponding to circular equatorial orbits. So, the points P_1 and P_2 collapse into a stable point. This case could be studied in the product of spheres $S^2 \times S^2$ (see for instance, Palacián 2002) but we do not deal with this particular situation in this paper.

In the bifurcation lines Γ_2 and Γ_4 the point P_1 becomes degenerate with zero linearisation. This makes these bifurcations rather special. Indeed, when $L < \sqrt{5}\,|H|$ there are two centres, P_1 and P_2, which give rise to four centres (P_1, P_2 and $P_{5,6}$) and two saddles $P_{3,4}$ having the same energy when $L > \sqrt{5}\,|H| > 0$. In the lines $L = \sqrt{5}\,|H|$ there are only two points: P_2, which remains a centre and P_1, which is degenerate but non-linearly stable. When $L > \sqrt{5}\,|H|$, the two separatrices forming the saddles $P_{3,4}$ are the same lines and surround four elliptic points: P_1, P_2 and $P_{5,6}$.

When $H = 0$ (i.e., the line Γ_3), we have to remove the line of $\mathcal{T}_{L,0}$ corresponding to rectilinear orbits from the valid reduced phase space, that is, the arc $\tau_3 = \tau_1^2 - L^2 - H^2$. Thus, the point P_2 cannot be studied whereas $P_{3,4}$ collapse into P_2 and therefore they do not remain as critical points. The equilibria $P_{5,6}$ disappear as they are not present when H vanishes.

Let us note that when $e_{\rm p} = 0$ the dynamics of the problem is richer than in the case $e_{\rm p} > 0$. On the one hand, we recover part of the equilibria appearing for $e_{\rm p} > 0$ (in the elliptic situation the polar-type of orbits were equilibria whereas in the case $e_{\rm p} = 0$, the polar orbits are equilibria only if they are also circular). On the other hand, we get new equilibria and bifurcation lines. In particular, we obtain circular-type trajectories having any inclination and orbits whose inclination is closed to $63^{\rm o}.4349\ldots$ or to $116^{\rm o}.565\ldots$ but having any eccentricity. Recall that for $e_{\rm p} \neq 0$ no bifurcations appear while in the circular case the five bifurcation lines determine the evolution of the flow.

Approximate invariant tori and quasiperiodic orbits for Hamiltonian (1) may be obtained from the relative equilibria using the changes of co-ordinates provided by the generating functions responsible of the transformations. In fact these objects may be computed explicitly using the direct change of the Delaunay normalisation followed by the direct change of the time-averaging transformation, and inserting thereafter the co-ordinates of the selected equilibrium point. Using the same arguments as in Palacián and Yanguas (2004) and Palacián et al. (2006), we can also determine approximate periodic orbits of the initial system, some of them symmetric.

We notice that $\mathcal{T}_{L,H}(0 \leq |H| < L)$ is compact and its Euler-Poincaré characteristic is 2. That means that the sum of the indexes of all relative equilibria must be $+2$ in all cases. Furthermore, we recall that the index of a centre is $+1$, the index of a saddle is -1 whereas a degenerate point has index 0. This argument has been used to check the number of critical points and their stability at any point of the parametric plane defined by L and H (excepting the lines $L = |H|$ and $H = 0$), yielding the right results in all situations. Hence, when the phase space has only two equilibria, they must be centres (even if there is an equilibrium whose linearisation has null eigenvalues) whereas when the doubly-reduced phase space contains six critical points, as no degeneration occurs, there must be four centres and two saddles.

6 Existence of KAM tori

We may normalise Hamiltonian $\mathcal{L}$ with respect to the argument of the pericentre in order to obtain a Hamiltonian of zero degrees of freedom. We achieve this step and arrive at the Hamilton function $\mathcal{Z} = \sum_{i=0}^{4}(\varepsilon^i/i!)\,\mathcal{Z}_i$, where

$$
\begin{aligned}
\mathcal{Z}_0 &= -\frac{1}{2\,L^2},\\
\mathcal{Z}_1 &= 0,\\
\mathcal{Z}_2 &= \frac{(1-\mu)\,\mu\,(G^2 - 3\,H^2)}{4\,L^3\,G^5},\\
\mathcal{Z}_3 &= 0,
\end{aligned}
$$

$$\mathcal{Z}_4 = \frac{3\,(1-\mu)\,\mu}{128\,L^5\,G^{11}}$$
$$\times \Big\{3\,(27 - 91\,\mu + 91\,\mu^2)\,G^6 + 24\,(1-\mu)\,\mu\,L\,G^5$$
$$+ \Big[-135\,(L^2 + 6\,H^2) + (1-\mu)\,\mu\,(451\,L^2 + 2538\,H^2)\Big]\,G^4$$
$$-144\,(1-\mu)\,\mu\,L\,H^2\,G^3$$
$$+3\,H^2\Big[45\,(10\,L^2 + 7\,H^2) - \mu\,(1-\mu)\,(1402\,L^2 + 955\,H^2)\Big]\,G^2$$
$$+216\,(1-\mu)\,\mu\,L\,H^4\,G - 3\,(525 - 1553\,\mu + 1553\,\mu^2)\,L^2\,H^4\Big\},$$

after dropping the double primes in L'', G'' and H''. The accompanying generating function is $\mathcal{N} = \varepsilon^4\,\mathcal{N}_4/4!$ where

$$\mathcal{N}_4 = \frac{(L^2 - G^2)\,(G^2 - H^2)}{128\,L^2\,G^5\,(G^2 - 5\,H^2)}$$
$$\times \Big[-15\,(G^2 - 7\,H^2) + (1-\mu)\,\mu\,(47\,G^2 - 345\,H^2)\Big]\,\sin(2\,g).$$

We notice that $\mathcal{N}$ vanishes at circular and equatorial orbits which makes the expression to be valid for these types of trajectories. Indeed, the generating function is singular when $G = 0$ and $G = \sqrt{5}\,|H|$.

However, we are interested in using the normal form $\mathcal{Z}$ around the elliptic points, that is, for the equilibrium P_1 in the parametric plane formed by L and H excepting for the lines $L = \sqrt{5}\,|H|$, for P_2 excepting for the cases $H = 0$ and $L = |H|$ (note that if $L = |H|$ then P_2 and P_1 coincide) and for $P_{5,6}$, which are equilibria only if $L > \sqrt{5}\,|H| > 0$. For $P_{5,6}$, $\mathcal{N}$ is also well defined as one should replace G by $\sqrt{5}\,|H| + \varepsilon^2\,G_{5,6}$ in the generating function but $G_{5,6}$ never vanishes for the allowed values of μ, L and H, hence the denominator of $\mathcal{Z}_4$ is bounded from below by a positive quantity. So, we may conclude that this last transformation is valid for all the elliptic points of the problem.

Now, we can apply KAM theory to the Hamiltonian $\mathcal{Z}$ (see Arnold et al. 1997). Jorba and Villanueva (1997) develop a theory for proving the persistence of KAM tori for time-dependent periodic Hamiltonians which can be applied to our case. First, we observe that $\mathcal{Z}$ is isoenergetically non-degenerate. In fact, $\mathcal{Z}_0$ depends only on L and $\partial^2\mathcal{Z}_0/\partial L^2 = -3/L^4 \neq 0$. Its next non-null term, $\mathcal{Z}_2$, satisfies

$$\det\begin{pmatrix} \dfrac{\partial^2\mathcal{Z}_2}{\partial G^2} & \dfrac{\partial^2\mathcal{Z}_2}{\partial G\,\partial H} \\[2ex] \dfrac{\partial^2\mathcal{Z}_2}{\partial H\,\partial G} & \dfrac{\partial^2\mathcal{Z}_2}{\partial H^2} \end{pmatrix} = -\frac{9\,(1-\mu)^2\,\mu^2\,(G^2 + 5\,H^2)}{2\,L^6\,G^{12}},$$

which never vanishes. This ensures the persistence of true invariant 3-tori for the time-dependent Hamiltonian (1) around the elliptic points. In a similar way, we could determine the existence of KAM 3-tori for the elliptic problem (i.e., $e_{\mathrm{p}} > 0$) but there we should apply a local treatment around each elliptic point, in order to eliminate the last angle, that is, the argument of the node, checking the isoenergetic hypothesis later on. However, we do not push these details here.

Acknowledgements We appreciate the judicious remarks and suggestions of two anonymous referees. This work has been partially supported by Project # BFM2002-03157 of Ministerio de Ciencia y Tecnología (Spain), Project # MTM2005-08595 of Ministerio de Educación y Ciencia (Spain) and Project Resolución 18/2005 of Departamento de Educación y Cultura, Gobierno de Navarra (Spain).

References

Abraham, R., Marsden, J.E.: Foundations of Mechanics. Dover, Redwood City, CA (1985)
Arnold, V.I.: Mathematical Methods of Classical Mechanics. Springer-Verlag, Berlin (1978)
Arnold, V.I., Kozlov, V.V., Neishtadt, A.I.: Mathematical Aspects of Classical and Celestial Mechanics. Dynamical Systems. III, Encyclopaedia Mathematics Science, 3. Springer-Verlag, Berlin (1997)
Brouwer, D., Clemence, G.M.: Methods of Celestial Mechanics. Academic Press, New York, London (1961)
Cors, J.M., Pinyol, C., Soler, J.: Analytic continuation in the case of non-regular dependency on a small parameter with an application to Celestial Mechanics. J. Diff. Eq. **219**, 1–19 (2005)
Cushman, R.H.: Reduction, Brouwer's Hamiltonian, and the critical inclination. Celest. Mech. **31**, 401–429 (1983)
Deprit, A.: Canonical transformations depending on a small parameter. Celest. Mech. **1**, 12–30 (1969)
Deprit, A.: The elimination of the parallax in satellite theory. Celest. Mech. **24**, 111–153 (1981)
Deprit, A.: Delaunay normalisations. Celest. Mech. **26**, 9–21 (1982)
Deprit, A., Rom, A.: Characteristic exponents at L_4 in the elliptic restricted problem. Astronom. & Astrophys. **5**, 416–425 (1970)
Féjoz, J.: Averaging the planar three-body problem in the neighborhood of double inner collisions. J. Diff. Eq. **175**, 175–187 (2001)
Féjoz, J.: Quasiperiodic motions in the planar three-body problem. J. Diff. Eq. **183**, 303–341 (2002)
Howison, C., Meyer, K.R.: Doubly-symmetric periodic solutions of the spatial restricted three-body problem. J. Diff. Eq. **163**, 174–197 (2000)
Jefferys, W.H.: Periodic orbits in the three-dimensional three-body problem. Astronom. J. **71**, 566–567 (1966)
Jefferys, W.H., Moser, J.: Quasi-periodic solutions for the three-body problem. Astronom. J. **71**, 568–578 (1966)
Meyer, K.R., Hall, G.R.: Introduction to Hamiltonian Dynamical Systems and the N-Body Problem. Applied Mathematical Sciences 90. Springer-Verlag, New York (1992)
Milnor, J.: Morse Theory. Annals of Mathematical Studies 51. Princeton University Press, Princeton, NJ (1963)
Moser, J.: Regularization of Kepler's problem and the averaging method on a manifold. Comm. Pure Appl. Math. **23**, 609–636 (1970)
Osácar, C., Palacián, J.F.: Decomposition of functions for elliptic orbits. Celes. Mech. Dynam. Astronom. **60**, 207–223 (1994)
Palacián, J.F.: Normal forms for perturbed Keplerian systems. J. Diff. Eq. **180**, 471–519 (2002)
Palacián, J.F., Yanguas, P.: Invariant manifolds of spatial restricted three-body problems: the lunar case. In: Delgado, J. Lacomba, E.A., Llibre J., Pérez–Chavela E. (eds.) New Advances in Celestial Mechanics and Hamiltonian Systems, pp. 199–221. Kluwer Academic/Plenum Publishers, Dordrecht (2004)
Palacián, J.F., Yanguas, P.: Asymptotic invariant tori of perturbed two-body problems. J. Symbolic Comput. **40**, 1256–1268 (2005)
Palacián, J.F., Yanguas, P., Fernández, S., Nicotra, M.A.: Searching for periodic orbits of the spatial elliptic restricted three-body problem by double averaging. Phys. D **213** 15–24 (2006)
Szebehely, V.: Theory of Orbits. Academic Press, New York (1967)

Celestial Mechanics and Dynamical Astronomy (2006) 95:101–116
DOI 10.1007/s10569-006-9015-z

ORIGINAL ARTICLE

Stability of axial orbits in galactic potentials

Cinzia Belmonte · Dino Boccaletti · Giuseppe Pucacco

Received: 10 November 2005 / Revised: 8 February 2006 /
Accepted: 25 March 2006 / Published online: 15 August 2006

Abstract We investigate the dynamics in a galactic potential with two reflection symmetries. The phase-space structure of the real system is approximated with a resonant detuned normal form constructed with the method based on the Lie transform. Attention is focused on the stability properties of the axial periodic orbits that play an important role in galactic models. Using energy and ellipticity as parameters, we find analytical expressions of bifurcations and compare them with numerical results available in the literature.

Keywords Normal forms of Hamiltonian systems · Stability of periodic orbits · Galactic potentials

1 Introduction

To determine salient features of the orbital structure of non-integrable potentials is an important topic in dynamical astronomy. Techniques based on the various versions of perturbation theory have been applied to several examples and with various degrees of approximation (for a review, see, e.g., Contopoulos 2002). Of particular interest is to understand motion in potentials which seem to be suitable to describe elliptical galaxies. Among other features, the knowledge of the stability properties of the main periodic orbits is of paramount importance, since the bulk of density distribution

C. Belmonte
Dipartimento di Fisica, Università di Roma "la Sapienza", Rome, Italy

D. Boccaletti
Dipartimento di Matematica, Università di Roma "la Sapienza", Rome, Italy

G. Pucacco (✉)
Dipartimento di Fisica, Università di Roma "Tor Vergata" and INFN—Sezione Tor Vergata,
Via della Ricerca Scientifica, 1, Rome,
I-00133, Italy
e-mail: pucacco@roma2.infn.it

is shaped by the stars in regular phase-space regions around stable periodic orbits (Binney and Tremaine 1987). In particular, for triaxial ellipsoids, periodic orbits along symmetry axes play a special role. An enormous effort has therefore been devoted to investigate families and bifurcations of periodic orbits, starting with the study of models based on perturbed oscillators (again, for a review, Contopoulos 2002) and gradually exploring more realistic galactic potentials with numerical (Miralda-Escudé and Schwarzschild 1989; Fridman and Merritt 1997) and semi-analytical (de Zeeuw and Merritt 1983; Scuflaire 1995) approaches.

One of the most powerful analytic tools is the normal form approximation of a non integrable system. Although the normal form approach is quite widespread in galactic dynamics, its use in studying stability of periodic orbits has not been as systematic as the theory could allow (Sanders and Verhulst 1985). Aim of the present paper is to apply the Lie transform normalization method (Dragt and Finn 1976; Finn 1984; Koseleff 1994) to approximate the dynamics of a Binney logarithmic potential (Binney and Tremaine 1987). We compare the findings to that of Miralda-Escudé and Schwarzschild (1989), who employ purely numerical techniques to implement the Floquet method and to that of Scuflaire (1995), who studies the stability of axial orbits by solving the Hill-like perturbation equation with the Lindstedt–Poincaré approach. Another example that is briefly treated is provided by the galactic Schwarzschild (1979) potential with a comparison to the results of de Zeeuw and Merritt (1983). These authors based their approach on the averaging procedure of normalization: it is therefore interesting a comparison with that method also. We remark that, in careful numerical computations, the accuracy of predictions is usually much higher than in approximate analytical approaches. However, a reliable analytic tool is of invaluable help to gather a global overview of the behavior of the system.

To study the linear orbital stability of the main periodic orbits with a truncated normal form one can proceed in essentially two ways: the most general and exhaustive is that of determining the explicit form of the normal modes and solve the equation of their perturbation. A less general but easier approach is that of determining the nature of the fixed points on a surface of section. This is constructed with the aid of the approximate integral of motion provided by the normalization. The first method is in general quite cumbersome and can be applied when the procedure of reduction to a single degree of freedom Hamiltonian system and the use of action-angle variables lead to a reasonably simple system of equations. The second one is clearly less general but relies on simple geometric arguments related to the Hessian of a polynomial in its critical points and is, at least in principle, quite easy to implement. In this work we are going to apply both methods to perform the comparison mentioned in the paragraph above.

In galactic dynamics, the periodic orbits along the axes of symmetry (axial orbits) play a particularly important role; moreover, they are easily identified both as normal modes of the reduced system and as "central" fixed points on the surfaces of section. Therefore, we will limit the detailed evaluation of the stability characteristics in the parameter space to these axial orbits. However, both procedures we have followed are quite general and can be directly applied to all periodic orbits of sufficiently low commensurability.

From the results obtained, we can state that the predictive power of the normal form ranges well outside the neighborhood in which the expansion of the original Hamiltonian is performed. It is rather related to the extent of the asymptotic convergence radius of the approximate integrals of motion. However, in concrete applications, the validity of the prediction has to be corroborated with an independent evaluation of the best suited resonant normal form of the problem at hand. A criterion for the choice is illustrated in the last section devoted to the applications and is connected to the ratio between the frequency of the periodic orbit and that of a normal perturbation to it.

The plan of the paper is as follows: in Sect. 2 we recall the procedure of normalization as applied to reflection symmetry potentials; in Sects. 3 and 4 we study the 1:1 and 1:2 resonances, respectively; in Sect. 5 we reconstruct approximate integrals of motion; in Sect. 6 we compare our analytical results with those available in the literature.

2 The procedure

2.1 General

Suppose the original system is given by a Hamiltonian

$$H(\mathbf{p},\mathbf{q}) = \frac{1}{2}(p_x^2 + p_y^2) + V(x^2, y^2), \tag{1}$$

with V a smooth potential with an absolute minimum and reflection symmetry with respect to both axes. In our case, we will use the Schwarzschild and logarithmic potentials described below. We expand the potential up to some given degree so that

$$V(x, y; \varepsilon) = \sum_{n=0}^{\infty} \varepsilon^n V_n(x, y) \tag{2}$$

and look for a new Hamiltonian given by (Boccaletti and Pucacco 1999)

$$K(\mathbf{P},\mathbf{Q};\varepsilon) = \sum_{n=0}^{\infty} \varepsilon^n K_n(\mathbf{P},\mathbf{Q};\varepsilon) = M_g^{-1} H(\mathbf{p},\mathbf{q};\varepsilon), \tag{3}$$

where $\mathbf{P}, \mathbf{Q}$ result from the canonical transformation

$$(\mathbf{P},\mathbf{Q}) = M_g(\mathbf{p},\mathbf{q}). \tag{4}$$

By expanding (3) in power series of ε and equating the coefficients of the same order, one has

$$\begin{aligned} K_0 &= H_0\,, \\ &\vdots \\ K_n &= H_n + \sum_{m=0}^{n-1} M_{n-m} H_m = H_n + M_n H_0 + \sum_{m=1}^{n-1} M_{n-m} H_m\,, \\ &\vdots \end{aligned} \tag{5}$$

The linear differential operator M_g is defined by

$$M_g \equiv e^{-\varepsilon L_{g_1}} e^{-\varepsilon^2 L_{g_2}} \cdots e^{-\varepsilon^n L_{g_n}} \cdots, \tag{6}$$

where the functions g_n are the coefficients in the expansion of the generating function of the canonical transformation and the linear differential operator L_S is defined through the Poisson bracket

$$L_S f \equiv \{S, f\} \equiv \sum_{l=1}^{2} \left(\frac{\partial S}{\partial q_l} \frac{\partial f}{\partial p_l} - \frac{\partial S}{\partial p_l} \frac{\partial f}{\partial q_l} \right). \tag{7}$$

The exponentials in the definition of M_g are intended as the formal sum of a power series so that it gives rise to a near identity coordinate transformation known as Lie series.

The unperturbed part of the Hamiltonian, H_0, determines the form of the transformation. In fact, the new Hamiltonian K is said to be *in normal form* if

$$\{H_0, K\} = 0. \tag{8}$$

This condition is used at each step of the procedure to determine each function g_n in order to eliminate as much as possible terms in the new Hamiltonian. The only terms of which K is made of are those staying in the kernel of the operator L_{H_0} associated to H_0 through the definition above. The procedure is stopped at some "optimal" order and therefore in all ensuing discussion we refer to a "truncated" normal form. H_0 must be considered a function of the new coordinates at each step in the process: it is therefore an integral of the motion for the new Hamiltonian K. The function

$$I = K - H_0 \tag{9}$$

can be therefore used as a second integral of the motion conveying approximate informations on the dynamics of the original system. For practical applications (for example to compare results with numerical computations) it is useful to express approximating functions in the original physical coordinates. Inverting the coordinate transformation, the new integral of motion can be expressed in terms of the original variables. Denoting it as the power series

$$I = \sum_{n=0}^{\infty} \varepsilon^n I_n, \tag{10}$$

its terms can be recovered by means of

$$I_n = H_n - K_n + \sum_{m=1}^{n-1} M_{n-m}\big[H_m - I_m\big], \quad n \geq 1\,. \tag{11}$$

We remark that in all subsequent applications involving series expansions, the role of the perturbation parameter can also be played by the size of the neighbourhood of the origin where the Hamiltonian is considered. Therefore, the powers of the parameter ε are left in all expansion formulas just to indicate their order and are treated as unity in the computations.

2.2 Galactic potentials

The model potentials we will consider are the Binney logarithmic potential (Binney and Tremaine 1987) and the Schwarzschild (1979) potential. In both cases, we will actually need an expansion of the form (2) and we will assume that each term can be written as a homogeneous polynomial of degree $k+2$ of the form

$$V_k(x,y) = \sum_{j=0}^{k+2} \frac{1}{k+2} a_{(j,k+2-j)} x^j y^{k+2-j}. \tag{12}$$

The logarithmic potential is

$$V = \frac{1}{2} \log(1 + x^2 + y^2/q^2) \tag{13}$$

and plays a very important role in galactic dynamics because, despite its very simple form, it has realistic features like a density distribution compatible at large radii with flat rotation curves. The form written here is simplified by the choice of fixing the length scale (the "core radius" R_c) equal to one, but this is not a limitation due to the invariance in both the length scale and the energy scale. With these units, the energy E may take any non negative value

$$0 \le E < \infty. \tag{14}$$

The parameter giving the "ellipticity" of the figure ranges in the interval

$$0.6 \le q \le 1. \tag{15}$$

Lower values of q can in principle be considered but correspond to a non physical density distribution. Values greater than unity are included in the treatment by reversing the role of the coordinate axes. The series expansion of the logarithmic potential is

$$V = \sum_{j=1}^{\infty} \frac{(-1)^{j+1}}{2j} (x^2 + y^2/q^2)^j, \tag{16}$$

so that the lowest order coefficients are

$$a_{20} = \omega_1^2 = 1, \quad a_{02} = \omega_2^2 = 1/q^2, \tag{17}$$

$$a_{40} = -1, \quad a_{22} = -1/q^2, \quad a_{04} = -1/q^4, \tag{18}$$

$$a_{60} = 1, \quad a_{42} = 3/q^2, \quad a_{24} = 3/q^4, \quad a_{06} = 1/q^6. \tag{19}$$

The Schwarzschild (1979) potential is to be considered more for its historical role rather than for its practical usefulness. However, it has been deeply investigated and is therefore a good benchmark for comparison. It can be written as

$$V = u(r) + \frac{x^2 - y^2}{r^2} w(r) + 1, \tag{20}$$

where

$$u = -\frac{1}{r} \log\left(\sqrt{1+r^2} + r\right) - c_1 \frac{r^2}{2(1+c_2 r^2)^{3/2}}, \tag{21}$$

$$w = -c_3 \frac{r^2}{(1+c_4 r^2)^{3/2}}, \tag{22}$$

are two functions of $r = \sqrt{x^2 + y^2}$ such that $0 < u, w < 1$ and the c's are fixed constants. With the choice of de Zeeuw and Merritt (1983)[1]

$$c_1 = 0.064, \quad c_2 = 0.655, \quad c_3 = 0.015, \quad c_4 = 0.481, \tag{23}$$

the lowest order coefficients are

$$\omega_1 = 0.421, \quad \omega_2 = 0.601, \tag{24}$$

$$a_{40} = -0.042, \quad a_{22} = -0.174, \quad a_{04} = -0.307, \tag{25}$$

$$a_{60} = -0.006, \quad a_{42} = 0.221, \quad a_{24} = 0.460, \quad a_{06} = 0.233. \tag{26}$$

The energy range in the Schwarzschild potential is

$$0 \leq E \leq 1. \tag{27}$$

2.3 Detuning the normal form

The natural setting in which one can perform a low order normalization is therefore that of a perturbed quadratic Hamiltonian with a potential starting with a harmonic term. In the general case in which the frequencies are rationally independent, the kernel of the operator associated to

$$H_0 = \frac{1}{2}(p_x^2 + p_y^2 + \omega_1^2 x^2 + \omega_2^2 y^2) \tag{28}$$

is trivial, consisting only of functions of the partial energies in the harmonic potential: it is customary to refer to the normal form constructed in this case as a "Birkhoff" normal form (Birkhoff 1927). The presence of terms with small denominators in the expansion, forbids in general its convergence. It is therefore more effective to work since the start with a "resonant" normal form, with a "richer" kernel that allows to reconstruct the main natural resonances shaping the phase-space of the system. To catch the main features of the orbital structure, we therefore approximate the frequencies with an integer ratio plus a small "detuning" that we assume $O(\varepsilon^2)$

$$\frac{\omega_1}{\omega_2} = \frac{m}{n} + \varepsilon^2 \delta \tag{29}$$

and we speak of a *detuned* (*m*:*n*) *resonance*, with $m + n$ the *order* of the resonance.

We have to put the system in a form suitable to apply the normalization procedure: we rescale variables in order to put the Hamiltonian in the form

$$H = \frac{1}{2}[(m + n\delta)(p_x^2 + x^2) + n(p_y^2 + y^2)] + \sum_{k=2}^{\infty} \sum_{j=0}^{k+2} \frac{b_{(j,k+2-j)}}{k+2} x^j y^{k+2-j} \tag{30}$$

where we have used the same notation for the rescaled variables and

$$b_{(j,k+2-j)} = \frac{n a_{(j,k+2-j)}}{\omega_1{}^{j/2} \omega_2^{(k+4-j)/2}}. \tag{31}$$

The procedure is now that of an ordinary resonant "Birkhoff–Gustavson" normalization (Gustavson 1966; Moser 1968) with two variants: the coordinate transformations

[1] Note the correction in c_4 with respect to the value reported in the Appendix of de Zeeuw and Merritt (1983), necessary to comply with the other reported constants.

are performed through the Lie series and the detuning quadratic term is treated as a term of higher order and put in the perturbation.

2.4 Choice of the resonance

Given an arbitrary pair of unperturbed frequencies, it could seem better to approximate their ratio as close as possible with a suitable pair of integers. However, beside possible computational problems, there are arguments on which a more effective choice can be based. Actually, the resonance should be of the lowest possible commensurability giving rise to a frequency ratio compatible with the dynamics of the actual system. The reason for this is that, the lower the order of the resonance, the richer the family of terms compatible with it that are available to construct the normal form.

Moreover, another argument in favor of low order resonances comes from their role in the stability properties of periodic orbits. A typical situation is that in which a family of periodic orbits becomes unstable when a low order resonance occurs between its fundamental frequency and that of a normal perturbation: the simplest case is given by an axial orbit that, depending on the specific form of the potential, can be unstable through bifurcation of loop orbits (1:1 resonance), "banana" orbits (1:2 resonance), "fish" orbits (2:3 resonance), etcetera. Therefore, a detuned low-order resonant normal form can be quite accurate in describing the corresponding bifurcations.

Finally, it must be emphasized that the structure of a resonant normal form is also affected by the symmetries of the original system. The normal form must preserve these symmetries and this in general also leads to a criterion for truncation. In the present instance of a double reflection symmetry, given a resonance ratio m/n, the normal form must contain at least terms of degree $2(m+n)$ (see, e.g. Tuwankotta and Verhulst 2000). Therefore, the criterion we have adopted in this paper has been that of working with the lowest order truncated normal form incorporating the symmetries of a typical galactic potential: the 1:1 symmetric resonance which allows to truncate the normal form to degree 4 and the 1:2 symmetric resonance which requires to truncate the normal form to degree 6. A systematic investigation of the optimal order of truncation has recently been performed by Contopoulos et al. (2003) and Efthymiopoulos et al. (2004). Their results confirm the rapid decrease of the optimal order with the radius of the phase-space domain in which expansions are computed: we may conjecture that if we are interested in the global dynamics and accept a moderate level of accuracy, with this very conservative approach we can get reliable information up to the breakdown of the regular dynamics.

3 1:1 Symmetric resonance and first order normalization

A Lie transform normalization truncated to the second order gives the following expression of the first-order normal form (cfr. Belmonte et al. 2006)

$$\begin{aligned} K_2^{(1:1)} &= \frac{1}{2}\delta(P_1^2+Q_1^2)+\frac{3}{32}\left[b_{40}(P_1^2+Q_1^2)^2+b_{04}(P_2^2+Q_2^2)^2\right]\\ &\quad+\frac{b_{22}}{32}\left(P_2^2(3P_1^2+Q_1^2)+Q_2^2(P_1^2+3Q_1^2)+4P_1P_2Q_1Q_2\right),\\ K_2^{(m:n)} &= \frac{n}{2}\delta(P_1^2+Q_1^2)+k_1\left[b_{40}(P_1^2+Q_1^2)^2+b_{04}(P_2^2+Q_2^2)^2\right]\\ &\quad+k_2\left[b_{22}(P_1^2+Q_1^2)(P_2^2+Q_2^2)\right], \end{aligned}$$

where k_1, k_2 are rational numbers dependent on m and n, Eq. 29 has been used, that in the present instance reads

$$\omega_1 = (1 + \varepsilon^2\delta)\omega_2 \tag{32}$$

and the canonical variables $\mathbf{P}, \mathbf{Q}$ are as in Eq. 3. We see that in this case we have the same situation as in the first order averaging approach (Verhulst 1996): the 1:1 resonance *or all other resonances*.

To show more clearly how the symmetries influence the structure of the normal form, we have that it can be written as

$$K = J_1 + J_2 + \varepsilon^2 \left[\delta J_1 + \frac{3}{8} \left(b_{40} J_1^2 + b_{04} J_2^2 + \frac{2}{3} b_{22} J_1 J_2 (2 + \cos(2\theta_1 - 2\theta_2)) \right) \right] \tag{33}$$

where an overall rescaling by a factor ω_2 has been performed and the action-angle variables are introduced according to

$$Q_1 = \sqrt{2J_1} \sin\theta_1, \tag{34}$$
$$P_1 = \sqrt{2J_1} \cos\theta_1, \tag{35}$$
$$Q_2 = \sqrt{2J_2} \sin\theta_2, \tag{36}$$
$$P_2 = \sqrt{2J_2} \cos\theta_2. \tag{37}$$

In fact, inverting these expressions and putting them into (33) we get

$$\omega_2 K = H_0 + \omega_2 \varepsilon^2 K_2^{(1:1)}. \tag{38}$$

The structure of (33) displays the effect of the symmetries on the resonant part: angles appear only through the combination $2\theta_1 - 2\theta_2$ and this shows why the symmetric 1:1 resonance can also be dubbed a "2:2" resonance.

We can use (33) to identify the main periodic orbits. The procedure is the following (Sanders and Verhulst 1985, sect. 7.4). We perform the following canonical transformation to "adapted resonance coordinates"

$$\psi = 2(\theta_1 - \theta_2), \tag{39}$$
$$\chi = 2(\theta_1 + \theta_2), \tag{40}$$
$$J_1 = (\mathcal{E} + \mathcal{R})/2, \tag{41}$$
$$J_2 = (\mathcal{E} - \mathcal{R})/2. \tag{42}$$

In this way we get

$$\widetilde{K} = \frac{1}{2}\delta(\mathcal{E} + \mathcal{R}) + A(\mathcal{E}^2 + \mathcal{R}^2) + B\mathcal{E}\mathcal{R} + C(\mathcal{E}^2 - \mathcal{R}^2)(2 + \cos\psi) \tag{43}$$

where the new action $\mathcal{E}$ is the additional integral of motion and has therefore been subtracted to get the effective Hamiltonian

$$\widetilde{K} = \frac{K - \mathcal{E}}{\varepsilon^2}. \tag{44}$$

The coefficients

$$A = \frac{3}{32}(b_{40} + b_{04}), \tag{45}$$

$$B = \frac{3}{16}(b_{40} - b_{04}), \tag{46}$$

$$C = \frac{1}{32}b_{22}, \tag{47}$$

appearing in $\widetilde{K}$, give it the simplest form.

Considering the dynamics at a fixed value of $\mathcal{E}$, we have that $\widetilde{K}$ defines a one-degree of freedom $(\psi, \mathcal{R})$ system. We get the following equations of motion

$$\dot{\psi} = \widetilde{K}_{\mathcal{R}} = \frac{1}{2}\delta + B\mathcal{E} + 2\left(A - C(2 + \cos\psi)\right)\mathcal{R}, \tag{48}$$

$$\dot{\mathcal{R}} = -\widetilde{K}_{\psi} = C\left(\mathcal{E}^2 - \mathcal{R}^2\right)\sin\psi. \tag{49}$$

Let us determine the fixed points of this system: these in turn give the periodic orbits of the original system. The right hand of (49) vanishes either for $\mathcal{R} = \pm\mathcal{E}$ or for $\psi = 0, \pm\pi$. In the first case, the right hand of (48) vanishes when

$$\delta + 2\left[B \pm 2\left(A - C(2 + \cos\psi)\right)\right]\mathcal{E} = 0 \tag{50}$$

and the two periodic orbits

$$\mathcal{R} = \mathcal{E}, \quad J_2 = 0, \quad \text{(Type Ia)}, \tag{51}$$

$$\mathcal{R} = -\mathcal{E}, \quad J_1 = 0, \quad \text{(Type Ib)}, \tag{52}$$

ensue. In the second case, the right hand of (48) vanishes either when

$$\mathcal{R} = \frac{\delta + 2B\mathcal{E}}{4(3C - A)}, \quad (\psi = 0), \tag{53}$$

or when

$$\mathcal{R} = \frac{\delta + 2B\mathcal{E}}{4(C - A)}, \quad (\psi = \pi). \tag{54}$$

The fixed point in (53) determines the "inclined" orbit

$$J_1 = \frac{\delta + 2(B + 2(3C - A))\mathcal{E}}{8(3C - A)}, \quad \text{(Type II)}. \tag{55}$$

Note that

$$0 \le J_1 \le \mathcal{E} \tag{56}$$

and this range determines the condition for existence of the orbit of Type II. The fixed point in (54) determines the elliptic orbit

$$J_2 = \frac{\delta + 2(B + 2(C - A))\mathcal{E}}{8(C - A)}, \quad \text{(Type III)}. \tag{57}$$

The range (56) still determines the condition for existence of the orbit of Type III.

Let us now consider the question of the stability of the periodic orbits. In particular, we are interested in what happens in the case of axial orbits of Type I: unfortunately, action-angle variables have singularities on these orbits and these affect also the adapted resonance coordinates. However, the remedy is quite straightforward: to use a mixed combination of action-angle variables on the normal mode and Cartesian variables for the other degree of freedom. The ensuing procedure is then first to determine the condition for the normal mode to be a critical curve of the Hamiltonian in these coordinates. Second, to assess its nature (Kummer 1977; Contopoulos 1978; Sanders and Verhulst 1985, sect.7.4.4): the condition is found by considering the function

$$K^{(\mu)} = K + \mu H_0, \tag{58}$$

where μ has to be considered as a *Lagrange multiplier* to take into account that there is the constraint $H_0 = \mathcal{E}$. The Lagrange multiplier is found by imposing

$$\mathrm{d}\, K^{(\mu)} = 0 \tag{59}$$

on the normal mode. Its nature is assessed by computing the matrix of the second derivatives of $K^{(\mu)}$: if the Hessian determinant of the second variation is positive definite the mode is elliptic stable; if it is negative definite the mode is hyperbolic unstable.

In the case of the y-axis orbit of Eq. 52, good coordinates are

$$Q_1 = X, \tag{60}$$

$$P_1 = U, \tag{61}$$

$$Q_2 = \sqrt{2J}\sin\theta, \tag{62}$$

$$P_2 = \sqrt{2J}\cos\theta, \tag{63}$$

so that the periodic orbit is given by

$$X = U = 0, \;\; J = \mathcal{E}. \tag{64}$$

The terms in the normal form are then

$$H_0 = \frac{1}{2}(X^2 + U^2) + J \tag{65}$$

and

$$\begin{aligned} K_2 = {} & \frac{1}{2}\delta(X^2+U^2) + \frac{3}{32}\left[b_{40}(X^2+U^2)^2 + 4b_{04}J^2\right] \\ & + \frac{1}{16}b_{22}J\left[2(X^2+U^2) + (X^2-U^2)\cos 2\theta + 2XU\sin 2\theta\right]. \end{aligned} \tag{66}$$

It is straightforward to check that, in this case, imposing Eq. 59 on the periodic orbit defined by Eq. 64, we get the equation

$$\mu + 1 + \frac{3}{4}b_{04}\mathcal{E} = 0, \tag{67}$$

which allows to find the required value of the Lagrange multiplier. With this result, the matrix of the second derivatives of $K^{(\mu)}$ on the normal bundle to the periodic orbit is

$$\frac{1}{8}\begin{pmatrix} 8\delta + \mathcal{E}[b_{22}(2+\cos 2\theta) - 6b_{04}] & \mathcal{E}b_{22}\sin 2\theta \\ \mathcal{E}b_{22}\sin 2\theta & 8\delta + \mathcal{E}[b_{22}(2-\cos 2\theta) - 6b_{04}] \end{pmatrix}. \tag{68}$$

The equation $\det K^{(\mu)}(\mathcal{E}) = 0$ gives

$$(36b_{04}^2 - 24b_{04}b_{22} + 3b_{22}^2)\mathcal{E}^2 - 32(3b_{04} - b_{22})\delta\mathcal{E} + 64\delta^2 = 0 \tag{69}$$

with roots

$$\mathcal{E}_1 = \frac{8\delta}{6b_{04} - b_{22}}, \quad \mathcal{E}_2 = \frac{8\delta}{3(2b_{04} - b_{22})}. \tag{70}$$

Recalling the rescaling in (33), the physical energy is given by

$$E = \omega_2 \mathcal{E}. \tag{71}$$

If, as in the application cases that will be examined later, the first coefficient in Eq. 69 is positive, the range of instability of the y-axis orbit is

$$\frac{8\omega_2\delta}{6b_{04} - b_{22}} < E < \frac{8\omega_2\delta}{3(2b_{04} - b_{22})}. \tag{72}$$

Proceeding in the same way in the case of the x-axis orbit of Eq. 51, the analogous expressions

$$\frac{8}{3}\frac{\omega_2\delta}{b_{22} - 2b_{40}} < E < \frac{8\omega_2\delta}{b_{22} - 6b_{40}} \tag{73}$$

can be written for the other conditions.

4 1:2 Symmetric resonance and second order normalization

In the case of the 1:2 resonance in presence of reflection symmetries about both axes, we know that the normal form must be pushed at least to degree $2\times(1+2) = 6$. We therefore have to perform a further step of normalization and include K_4 in the normal form. Its expression is quite involved (cfr. Belmonte et al. 2006), but we can exploit the change of variables to action-angle coordinates to see that the normal form has the structure

$$K = J_1 + 2J_2 + \varepsilon^2(2\delta J_1 + P_2(J_1,J_2)) + \varepsilon^4(P_3(J_1,J_2) + kJ_1^2J_2\cos(4\theta_1 - 2\theta_2)), \tag{74}$$

where Eq. 29 has been used, that in the present instance reads

$$\omega_1 = \left(\frac{1}{2} + \varepsilon^2\delta\right)\omega_2, \tag{75}$$

the polynomials P_2 and P_3 are homogeneous of degree 2 and 3, respectively

$$P_2 = \frac{3}{8}b_{40}J_1^2 + \frac{1}{4}b_{22}J_1J_2 + \frac{3}{8}b_{04}J_2^2$$

$$P_3 = -\frac{1}{384}\Big((102b_{40}^2 - 160b_{60})J_1^3 + (10b_{22}^2 + 72b_{22}b_{40} - 96b_{42})J_1^2J_2$$
$$+(36b_{04}b_{22} + 10b_{22}^2 - 96b_{24})J_1J_2^2 + (51b_{04}^2 - 160b_{06})J_2^3\Big)$$

and

$$k = -\frac{1}{192}\left(3b_{22}^2 - 3b_{22}b_{40} - 8b_{42}\right). \tag{76}$$

The study of existence and stability of periodic orbits proceeds in the same way as in the previous section: we limit the presentation to the axial orbits. In the present case they are given by

$$\mathcal{R} = 2\mathcal{E}, \quad J_2 = 0, \quad \text{(Type Ia)}, \tag{77}$$
$$\mathcal{R} = -\mathcal{E}/2, \quad J_1 = 0, \quad \text{(Type Ib)}. \tag{78}$$

The procedure to determine the condition of stability of axial orbits lead to analyze critical curves of the function (58) where K is given by (74). Considering the x-axis (type Ia, Eq. 77) orbit, good coordinates are

$$Q_1 = \sqrt{2J}\sin\theta, \tag{79}$$
$$P_1 = \sqrt{2J}\cos\theta, \tag{80}$$
$$Q_2 = Y, \tag{81}$$
$$P_2 = V, \tag{82}$$

so that the periodic orbit is given by

$$Y = V = 0, \quad J = \mathcal{E}, \tag{83}$$

and

$$H_0 = J + Y^2 + V^2. \tag{84}$$

The condition Eq. 59 allows to find the Lagrange multiplier

$$\mu = -1 - 2\delta - \frac{\mathcal{E}}{4}\left[3b_{40}\left(1 - \frac{17}{16}b_{40}\mathcal{E}\right) + 5b_{60}\mathcal{E}\right]. \tag{85}$$

The equation $\det K^{(\mu)}(\mathcal{E}) = 0$, obtained by computing the matrix of the second derivatives of $K^{(\mu)}$ on the normal bundle to the periodic orbit (83), is

$$(A_2\mathcal{E}^2 + A_1\mathcal{E} + 768\delta)(A_2'\mathcal{E}^2 + A_1\mathcal{E} + 768\delta) = 0, \tag{86}$$

where

$$A_1 = 48(6b_{40} - b_{22}), \tag{87}$$
$$A_2 = 2b_{22}^2 + 33b_{22}b_{40} - 306b_{40}^2 - 56b_{42} + 480b_{60}, \tag{88}$$
$$A_2' = 2b_{22}^2 + 39b_{22}b_{40} - 306b_{40}^2 - 40b_{42} + 480b_{60}. \tag{89}$$

We see that the inequalities to be satisfied in this case are quadratic in the energy, contrary to the linear case of (72) and (73). Below we discuss the specific example of the x-axis orbit in the logarithmic potential.

5 Approximate integral in the original variables

The general procedure illustrated so far can be applied to study the dynamics associated to every reduced normal form. Specifically, it can be extended to forms truncated to an arbitrary high order and to all normal modes and periodic orbits associated to the given resonance. However, if the 1:1 and 1:2 cases treated here are rather simple, computations become quite cumbersome for higher orders. It can be useful to exploit an alternative, less general, but easier procedure. Recalling the generic expression of the second invariant (11), if we truncate at first order, we have

$$I_2 = H_0 + \varepsilon^2(V_2 - K_2). \tag{90}$$

This is the best approximate integral of motion of a symmetric perturbed 1:1 oscillator to order ε^4 in the perturbation parameter: in fact, due to the symmetries of the problem, odd-degree terms are absent both in the normal form and in the approximate integral. To assess the stability of, e.g., the y-axis orbit, we may proceed in this way: to use I_2 and the conserved energy

$$E = \frac{1}{2}(p_x^2 + p_y^2) + V_0(x^2, y^2) + \varepsilon^2 V_2(x^2, y^2) \tag{91}$$

to construct an $x - p_x$ Poincaré section by means of the intersection of the function $I_2(x, y, p_x; E)$ with the $y = 0$ hyperplane and the level curves of the function

$$F = I_2(x, 0, p_x; E); \tag{92}$$

to study the nature of the origin as a critical point determining if it is either an extremum (elliptic fixed point = stable periodic orbit) or a saddle (hyperbolic fixed point = unstable periodic orbit) by using the Hessian determinant

$$F_{p_x p_x} F_{xx} - F_{xp_x}^2. \tag{93}$$

Clearly, for resonant periodic orbits even the location of the critical points can be already quite difficult and this limits the generality of the approach. However, in the case of axial orbits, the approach is straightforward: in the case of the y-axis orbit, we have that the second derivatives in the origin are

$$F_{p_x p_x} = 4\omega_2 \delta - 3E\left((1+\delta)b_{04} - \frac{1}{2}b_{22}\right), \tag{94}$$

$$F_{xx} = 4\omega_2 \delta - E\left(3(1+\delta)b_{04} - \frac{1}{2}b_{22}\right). \tag{95}$$

The mixed second derivative F_{xp_x} is identically zero in the origin. The range of instability (namely, the range where the two second derivatives have different sign) is

$$\frac{8\omega_2\delta}{6b_{04}(1+\delta) - b_{22}} < E < \frac{8\omega_2\delta}{3(2b_{04}(1+\delta) - b_{22})}. \tag{96}$$

Comparing with (72) we have an order δ^2 discrepancy whose origin is connected with the construction of the function (92) where the term with the detuning is not considered as a perturbation. Analogous inequalities can be written for the x-axis orbit by constructing a $y - p_y$ Poincaré section and studying the level curves of the function

$$F = I_2(0, y, p_y; E) \tag{97}$$

obtained by means of the intersection of the function $I_2(x, y, p_y; E)$ with the $x = 0$ hyperplane.

On the other side, we can develop this parallel approach of determining the nature of fixed points on the surface of section for higher order approximate integrals too. Using (11), if we truncate at second order, we have

$$I_4 = H_0 + \varepsilon^2(V_2 - K_2) + \varepsilon^4(V_4 - \{g_2, K_2\} - K_4) \tag{98}$$

where g_2 is the first order generating function determined in the first step of the normalization. This is the best approximate integral of motion of a symmetric perturbed 1:2 oscillator to order ε^6 in the perturbation parameter. Let us consider the x-axis orbit: the construction of the $y - p_y$ Poincaré section and the study of the critical points of the function

$$F = I_4(0, y, p_y; E), \tag{99}$$

obtained by means of the intersection of the function $I_4(x, y, p_y; E)$ with the $x = 0$ hyperplane, proceeds as above. Concerning the nature of the fixed point in the origin (linked to the stability of the x-axis orbit), we get quadratic inequalities in the energy analogous to those arising from (86): the discrepancy between the two methods is still of order δ^2 and is discussed in the subsequent section about a specific example.

6 Comparison of analytical and numerical results

We have applied the theory discussed in the previous sections to the logarithmic and the Schwarzschild potentials. We can compare these analytical predictions concerning the stability energy ranges of axial orbits with other results, mainly numerical, from the literature. In particular, we have chosen the work by Miralda–Escudé and Schwarzschild (1989, MES in what follows) who made an accurate numerical exploration of the phase-space structure of the logarithmic potential: they give the existence parameter ranges of the main families of periodic orbits and the bifurcation ensuing from instability thresholds in two models (corresponding to ellipticity values $q = 0.7$ and $q = 0.9$) determined by solving the perturbation equations with the Floquet method. The authors use the core radius R_c to parametrize the sequence of periodic orbits because they were interested in comparing the results with the case of the *singular* scale-free case $R_c = 0$: in fact they fix the energy $E = 0$ in all their computations and vary R_c and q. For our purposes, it is more natural to use the energy as a parameter and therefore we have made the conversion

$$E = -\log R_c \tag{100}$$

to compare the results. Another approach to the study of the stability of axial orbits in the logarithmic potential has been followed by Scuflaire (1995, SC in what follows) who solved the Hill-like equation of the perturbation by means of a Poincaré–Lindstedt series expansion up to order 20. SC uses a, the amplitude of the axial orbit, as a parameter: the conversion to the energy is given by

$$E = \frac{1}{2}\log(1 + a^2). \tag{101}$$

In Tables 1 and 2 we display the results in the case of the logarithmic potential: in the first column are the two values of the parameter q for which reliable numerical estimates are available; in the second are the analytical predictions according to the procedures of Sect. 3 for Table 1 and 4 for Table 2; in the third are the analytical predictions according to the procedures of Sect. 5; in the fourth the numerical prediction of MES; in the last the semi-analytical prediction of SC.

The values shown in Table 1 represent the energy thresholds to instability of the short y-axis orbit: this orbit becomes unstable when its frequency falls to that of a normal perturbation and as soon as this happens a loop orbit bifurcates. The values of the second and third columns are therefore computed using the results of the 1:1 resonance treatment of Sects. 3 and 5, namely the first inequality in (72) and the first in (96). The computations are performed by using the coefficients given in (18)

Table 1 Energy threshold of instability for the short y-axis orbit in the logarithmic potential

q	Normal-form	Invariant	MES	SC
0.7	0.52	0.86	0.72	0.80
0.9	0.19	0.22	0.21	0.25

Table 2 Energy intervals of instability for the long x-axis orbit in the logarithmic potential

q	Normal-form	Invariant	MES	SC
0.7	$1.32 \div 4.82$	$1.71 \div 4.86$	$1.52 \div 4.29$	$1.42 \div -$
0.9	0.85	0.80	3.64	–

recalling definition (31). Clearly the accuracy of both predictions is better for $q = 0.9$ corresponding to a smaller detuning of the 1:1 resonance. Overall, we see that the prediction using the invariant curves performs significantly better than that obtained directly working with the normal form.

The long x-axis orbit can suffer instability, but not through a 1:1 resonance. This orbit becomes unstable when its frequency falls to $1/2$ of that of a normal perturbation and, as soon as this happens, a banana orbit bifurcates. Therefore, we turn the attention to Table 2. where the values of the second and third columns are computed using the results of the 1:2 resonance treatment of Sects. 4 and 5: namely, the condition in order to the quartic in (86) is negative, using the coefficients computed from (18) and (19) and the nature of the fixed point in the origin as determined through function (99). An important prediction of the numerical approaches is that, if the ellipticity is less than a given value, there is an instability *interval* in energy: actually in the figure of SC it is not possible to determine the upper extremum of it, but this interval is very well determined in the computations of MES. We see that the analytical estimates of the instability interval are not very far from them, with an acceptable prediction of the upper boundary of the interval too. Our treatment also provides a critical value of $q_c = 0.78$ of the ellipticity, beyond which the interval disappears. It would be interesting to numerically test this prediction. For higher values of q the analytical estimates of the threshold of instability are not so good and the reason is that the detuning of the 1:2 resonance becomes excessively large.

Finally, in Table 3. we report the comparisons concerning the short y-axis orbit in the Schwarzschild potential, with the coefficients listed in (24) and (25): in the third column we report the prediction given by de Zeeuw and Merritt (1983, DZM) using the first order averaging of the equations of motion and in the fourth the corresponding prediction of the same authors using a numerical integration. We see that, as usual, the analytical prediction based on the invariant is quite accurate.

Table 3 Energy threshold of instability for the short y axis orbit in the Schwarzschild potential

Normal-form	Invariant	DZM averaging	DZM numerical
0.19	0.28	0.21	0.27

We conclude observing that applications like those presented here confirm that this very conservative strategy provides sufficient qualitative and quantitative agreement with other more accurate but less general approaches. An important issue to clarify is if, in order to improve the degree of accuracy, it is enough to simply truncate the normal form to higher orders or if it is rather necessary to incorporate in a more careful way the detuning in higher order terms of the perturbation.

Acknowledgements Our thanks to Giuseppe Gaeta and Ferdinand Verhulst for very fruitful discussions.

References

Belmonte, C., Boccaletti, D., Pucacco, G.: Approximate first integrals for a model of galactic potential with the method of Lie transform normalization. In: Perez-Chavela, E., Xia, J. (eds.) Submitted to the Proceedings of the Saarifest (2005)

Birkhoff, G.D.: Dynamical systems, Amer. Math. Soc. Coll. Publ., vol. 9, New York, USA (1927)

Binney, J., Tremaine, S.: Galactic Dynamics, Princeton University Press (1987)

Boccaletti, D., Pucacco, G.: Theory of Orbits, vol. 2, Springer-Verlag, Berlin (1999)

Contopoulos, G.: Higher order resonances in dynamical systems. Cel. Mech. **18**, 195–204 (1978)

Contopoulos, G.: Order and Chaos in Dynamical Astronomy. Springer-Verlag, Berlin (2002)

Contopoulos, G., Efthymiopoulos, C., Giorgilli, A.: Nonconvergence of formal integrals of motion. J. Phys. A: Math. Gen. **36**, 8639–8660 (2003)

de Zeeuw, T., Merritt, D.: Stellar orbits in a triaxial galaxy. I. Orbits in the plane of rotation. Astrophys. J. **267**, 571–595 (1983)

Dragt, A., Finn, J.M.: Lie series and invariant functions for analytic symplectic maps. J. Mat. Phys. **17**, 2215–2227 (1976)

Efthymiopoulos, C., Giorgilli, A., Contopoulos, G.: Nonconvergence of formal integrals: II. Improved estimates for the optimal order of truncation. J. Phys. A: Math. Gen. **37**, 10831–10858 (2004)

Finn, J.M.: Lie series: a perspective. Local and global methods of nonlinear dynamics. Lecture Notes in Physics, vol. 252, pp. 63–86 (1984)

Fridman, T., Merritt, D.: Periodic orbits in triaxial galaxies with weak cusps. Astron. J. **114**, 1479–1487 (1997)

Gustavson, F.: On constructing formal integrals of a Hamiltonian system near an equilibrium point. Astron. J. **71**, 670–686 (1966)

Koseleff, P.V.: Comparison between Deprit and Dragt-Finn perturbation methods. Cel. Mech. & Dynam. Astron. **58**, 17–36 (1994)

Kummer, M.: On resonant Hamiltonians with two degrees of freedom near an equilibrium point. Lecture Notes in Physics, vol. 93, pp. 57–75 (1977)

Miralda-Escudé, J., Schwarzschild, M.: On the orbit structure of the logarithmic potential. Astrophys. J. **339**, 752–762 (1989)

Moser, J.: Lectures on Hamiltonian systems. Mem. Am. Math. Soc. **81**, 1–60 (1968)

Sanders, J.A., Verhulst, F.: Averaging Methods in Nonlinear Dynamical Systems. Springer-Verlag, New York (1985)

Schwarzschild, M.: A numerical model for a triaxial stellar system in dynamical equilibrium. Astrophys. J. **232**, 236–247 (1979)

Scuflaire, R.: Stability of axial orbits in analytic galactic potentials. Cel. Mech. & Dynam. Astron. **61**, 261–285 (1995)

Tuwankotta, J.M., Verhulst, F.: Symmetry and resonance in Hamiltonian systems. SIAM J. Appl. Math. **61**, 1369–1385 (2000)

Verhulst, F.: Nonlinear Differential Equations and Dynamical Systems. Springer-Verlag, Berlin (1996)

Celestial Mechanics and Dynamical Astronomy (2006) 95:117–139
DOI 10.1007/s10569-005-6215-x

REVIEW ARTICLE

KAM tori for N-body problems: a brief history

A. Celletti · L. Chierchia

Received: 28 November 2005 / Accepted: 24 December 2005
Published online: 17 August 2006

Abstract We review analytical (rigorous) results about the existence of invariant tori for planetary many-body problems.

Keywords Computer-assisted proofs · Invariant tori · KAM theory · N-body problem · Small divisor problems

1 Introduction

In this paper, we review analytical results concerning the existence of KAM tori (smooth invariant tori, for a nearly integrable Hamiltonian system, on which the flow is quasi-periodic with Diophantine frequencies) in the context of the planetary many-body problem.

The main body of the paper is divided in two sections and two appendices.

In Sect. 2, general existence theorems for the planetary $(1 + n)$-body problem are discussed. In particular, after a brief reminder about the Hamiltonian setting for the many-body problem (Sect. 2.1) and about classical KAM theory (Sect. 2.2), it is shown how Kolmogorov's 1954 theorem yields easily the existence of KAM tori in the special non-degenerate case of the restricted, planar, circular three-body problem (Sect. 2.2.2). Kolmogorov's theorem, on the other hand, does not apply to the general case because of the proper degeneracy of the $(1 + n)$-body problem, when $n \geqslant 2$. In this context, Arnold (1963), stated a general result, which he proved only in the

A. Celletti (✉)
Dipartimento di Matematica, Università di Roma Tor Vergata,
Via della Ricerca Scientifica 1, I-00133 Roma Italy
e-mail: celletti@mat.uniroma2.it

L. Chierchia
Dipartimento di Matematica, Università "Roma Tre",
Largo S. L. Murialdo 1, I-00146 Roma, Italy
e-mail: luigi@mat.uniroma3.it

planar three-body case; Arnold's theorem was proven in Fejóz (2004), who completed Herman's work on the matter (Sect. 2.3).

In Sect. 3 rigorous "computer-assisted" results about the existence of KAM tori for Hamiltonian models of solar subsystems are reviewed. In particular, in Sect. 3.2, the KAM stability of the subsystem Sun–Jupiter–Victoria, modelled by a truncated restricted, planar, circular three-body problem, obtained recently by the authors, is discussed.

In Sect. 4, several sets of symplectic variables relevant for analytical investigations of the many-body problem are reviewed.

In Sect. 5, a numerical comparison between the dynamics of the truncated model considered in Sect. 3.2 and the non-truncated model is discussed.

2 KAM tori for general many-body problems

2.1 Hamiltonian models for planetary many-body problems

2.1.1 Newton's equations

Newton's equations for $n+1$ bodies (point masses), interacting only through gravitational attraction, are given by

$$\ddot{u}^{(i)} = \sum_{\substack{0\leqslant j\leqslant n \\ j\neq i}} m_j \frac{u^{(j)} - u^{(i)}}{|u^{(i)} - u^{(j)}|^3}, \qquad i = 0, 1, \ldots, n, \tag{2.1}$$

where $u^{(i)} = \left(u_1^{(i)}, u_2^{(i)}, u_3^{(i)}\right) \in \mathbb{R}^3$ are Cartesian coordinates of the ith body of mass m_i, $|u| = \sqrt{u \cdot u} = \sqrt{\sum_i u_i^2}$ is the standard Euclidean norm, "dot" denotes time derivative, and the gravitational constant has been normalized to one (by rescaling time t). Equation (2.1) is invariant by change of inertial frames, i.e., by change of variables of the form $u^{(i)} \to u^{(i)} - (a + ct)$ with fixed $a, c \in \mathbb{R}^3$. This allows to restrict the attention to the manifold of initial data given by [1]

$$\sum_{i=0}^{n} m_i u^{(i)}(0) = 0, \quad \sum_{i=0}^{n} m_i \dot{u}^{(i)}(0) = 0\,. \tag{2.2}$$

The total linear momentum $M_{\text{tot}} := \sum_{i=0}^{n} m_i \dot{u}^{(i)}$ does not change along the flow of (2.1), i.e., $\dot{M}_{\text{tot}} = 0$ along trajectories; therefore, by (2.2), $M_{\text{tot}}(t)$ vanishes for all t. But, then, also the position of the barycenter $B(t) := \sum_{i=0}^{n} m_i u^{(i)}(t)$ is constant ($\dot{B} = 0$) and, again by (2.2), $B(t) \equiv 0$. In other words, the manifold of initial data (2.2) is invariant under the flow (2.1).

[1] Replace the coordinates $u^{(i)}$ by $u^{(i)} - (a + ct)$ with

$$a := \mathrm{m}_{\text{tot}}^{-1} \sum_{i=0}^{n} m_i u^{(i)}(0) \quad \text{and} \quad c := \mathrm{m}_{\text{tot}}^{-1} \sum_{i=0}^{n} m_i \dot{u}^{(i)}(0), \quad \mathrm{m}_{\text{tot}} := \sum_{i=0}^{n} m_i.$$

2.1.2 Hamiltonian point of view

Equations (2.1) are the Hamiltonian equations generated by the Hamiltonian function

$$\widehat{\mathcal{H}}_{\text{New}} := \sum_{i=0}^{n} \frac{|U^{(i)}|^2}{2m_i} - \sum_{0\leqslant i<j\leqslant n} \frac{m_i m_j}{|u^{(i)} - u^{(j)}|}, \tag{2.3}$$

where $(U^{(i)}, u^{(i)})$ are standard symplectic variables ($U^{(i)} = m_i \dot{u}^{(i)}$ is the momentum conjugated to $u^{(i)}$) and the phase space is the "collisionless" open domain in $\mathbb{R}^{6(n+1)}$ given by

$$\widehat{\mathcal{M}} := \{U^{(i)}, u^{(i)} \in \mathbb{R}^3 : u^{(i)} \neq u^{(j)}, \quad 0 \leqslant i \neq j \leqslant n\}$$

endowed with the standard symplectic form

$$\sum_{i=0}^{n} \mathrm{d}U^{(i)} \wedge \mathrm{d}u^{(i)} := \sum_{\substack{0\leqslant i\leqslant n\\ 1\leqslant k\leqslant 3}} \mathrm{d}U_k^{(i)} \wedge \mathrm{d}u_k^{(i)}. \tag{2.4}$$

As explained above, the physically relevant motions governed by (2.3) lie on

$$\widehat{\mathcal{M}}_0 := \left\{ (U,u) \in \widehat{\mathcal{M}} : \sum_{i=0}^{n} m_i u^{(i)} = 0 = \sum_{i=0}^{n} U^{(i)} \right\}$$

(which corresponds to the manifold described in (2.2)). The submanifold $\widehat{\mathcal{M}}_0$ is symplectic (i.e., the restriction of the form (2.4) to $\widehat{\mathcal{M}}_0$ is again a symplectic form) and the $\widehat{\mathcal{H}}_{\text{New}}$–flow on it is best described in terms of heliocentric coordinates. Let ϕ_{hel}: $(R,r) \to (U,u)$ be the linear symplectic transformation given by

$$\phi_{\text{hel}} : \begin{cases} u^{(0)} = r^{(0)}, & u^{(i)} = r^{(0)} + r^{(i)}, \quad (i = 1, \ldots, n), \\ U^{(0)} = R^{(0)} - \sum\limits_{i=1}^{n} R^{(i)}, & U^{(i)} = R^{(i)}, \quad (i = 1, \ldots, n). \end{cases} \tag{2.5}$$

In such variables $\widehat{\mathcal{M}}_0$ reads

$$\left\{ (R,r) \in \mathbb{R}^{6(n+1)} : R^{(0)} = 0, r^{(0)} = -\mathrm{m}_{\text{tot}}^{-1} \sum_{i=1}^{n} m_i r^{(i)} \text{and } 0 \neq r^{(i)} \neq r^{(j)}\ \forall 1 \leqslant i \neq j \leqslant n \right\};$$

the restriction of the 2-form (2.4) on $\widehat{\mathcal{M}}_0$ is simply $\sum_{i=1}^{n} \mathrm{d}R^{(i)} \wedge \mathrm{d}r^{(i)}$ and

$$\begin{aligned}(\widehat{\mathcal{H}}_{\text{New}} \circ \phi_{\text{hel}})|_{\widehat{\mathcal{M}}_0} &= \sum_{i=1}^{n} \left(\frac{|R^{(i)}|^2}{2\frac{m_0 m_i}{m_0+m_i}} - \frac{m_0 m_i}{|r^{(i)}|} \right) \\ &\quad + \sum_{1\leqslant i<j\leqslant n} \left(\frac{R^{(i)} \cdot R^{(j)}}{m_0} - \frac{m_i m_j}{|r^{(i)} - r^{(j)}|} \right) =: \mathcal{H}_{\text{New}}.\end{aligned}$$

Thus, the dynamics generated by $\widehat{\mathcal{H}}_{\text{New}}$ on $\widehat{\mathcal{M}}_0$ is equivalent to the dynamics generated by the Hamiltonian $(R,r) \in \mathbb{R}^{6n} \to \mathcal{H}_{\text{New}}(R,r)$ on

$$\mathcal{M}_0 := \left\{ (R,r) = (R^{(1)}, \ldots, R^{(n)}, r^{(1)}, \ldots, r^{(n)}) \in \mathbb{R}^{6n} : 0 \neq r^{(i)} \neq r^{(j)}\ \forall\, 1 \leqslant i \neq j \leqslant n \right\}$$

with respect to the standard symplectic form $\sum_{i=1}^n \mathrm{d}R^{(i)} \wedge \mathrm{d}r^{(i)}$; to recover the full dynamics on $\widehat{\mathcal{M}}_0$ from the dynamics on $\mathcal{M}_0$ one will simply set $R^{(0)}(t) \equiv 0$ and $r^{(0)}(t) := -\mathrm{m}_{\mathrm{tot}}^{-1} \sum_{i=1}^n m_i r^{(i)}(t)$.

Motivated by the planetary case, let us perform the trivial rescaling by a small positive parameter ε:

$$\bar{m}_0 := m_0,\ m_i = \varepsilon \bar{m}_i\ (i \geqslant 1), \quad X^{(i)} := \frac{R^{(i)}}{\varepsilon},\ x^{(i)} := r^{(i)}, \quad \mathcal{H}_{\mathrm{plt}}(X,x) := \frac{1}{\varepsilon} \mathcal{H}_{\mathrm{New}}(\varepsilon X, x),$$

which leaves unchanged Hamilton's equations. Explicitly, if $\mu_i := \frac{\bar{m}_0 \bar{m}_i}{\bar{m}_0 + \varepsilon \bar{m}_i}$ and $M_i := \bar{m}_0 + \varepsilon \bar{m}_i$, then

$$\begin{aligned} \mathcal{H}_{\mathrm{plt}}(X,x) &:= \sum_{i=1}^n \left(\frac{|X^{(i)}|^2}{2\mu_i} - \frac{\mu_i M_i}{|x^{(i)}|} \right) + \varepsilon \sum_{1 \leqslant i < j \leqslant n} \left(\frac{X^{(i)} \cdot X^{(j)}}{\bar{m}_0} - \frac{\bar{m}_i \bar{m}_j}{|x^{(i)} - x^{(j)}|} \right) \\ &=: \mathcal{H}_{\mathrm{plt}}^{(0)}(X,x) + \varepsilon \mathcal{H}_{\mathrm{plt}}^{(1)}(X,x) \end{aligned} \tag{2.6}$$

the phase space being

$$\mathcal{M} := \left\{ (X,x) = (X^{(1)}, \ldots, X^{(n)}, x^{(1)}, \ldots, x^{(n)}) \in \mathbb{R}^{6n} :\ 0 \neq x^{(i)} \neq x^{(j)}\ \forall\, 1 \leqslant i \neq j \leqslant n \right\}$$

endowed with the standard symplectic form $\sum_{i=1}^n \mathrm{d}X^{(i)} \wedge \mathrm{d}x^{(i)}$.

Remarks

(i) The reduction of the general dynamics to the $\mathcal{H}_{\mathrm{plt}}$-dynamics on $\mathcal{M}$ is sometimes referred to as the "reduction of the linear momentum". Notice that a reflection of such reduction is that there is no more "conservation of the total linear momentum", as $\sum_{i=1}^n X^{(i)}$ is obviously not an integral[2] for $\mathcal{H}_{\mathrm{plt}}$. On the other hand, the transformation (2.5) does preserve the total angular momentum $\sum_{i=0}^n U^{(i)} \times u^{(i)}$, where "$\times$" denotes the standard vector product in $\mathbb{R}^3$. Thus, the Hamiltonian $\mathcal{H}_{\mathrm{plt}}$ admits, besides the energy, three more integrals, which are the three components of the total angular momentum

$$C = (C_1, C_2, C_3) := \sum_{i=1}^n X^{(i)} \times x^{(i)}. \tag{2.7}$$

Such integrals do not commute (i.e., their Poisson brackets do not vanish):

$$\{C_1, C_2\} = C_3, \quad \{C_2, C_3\} = C_1, \quad \{C_3, C_1\} = C_2$$

but, for example, $|C|^2$ and C_3 are two commuting, independent integrals.

(ii) The two-body case (corresponding to $n = 1$ and no $\mathcal{H}_{\mathrm{plt}}^{(1)}$ term) is integrable for any $\varepsilon > 0$ (Kepler). Therefore, also the term $\mathcal{H}_{\mathrm{plt}}^{(0)}$ in the planetary Hamiltonian (2.6) is integrable, being the sum of n decoupled two-body problems. In Delaunay

[2] We recall that $F(X,x)$ is an integral for $\mathcal{H}(X,x)$ if $\{F, \mathcal{H}\} = 0$, where $\{F, G\} = F_X \cdot G_x - F_x \cdot G_X$ denotes the (standard) Poisson bracket.

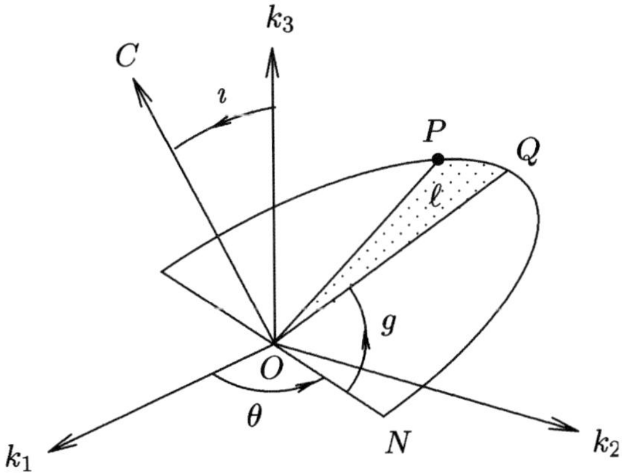

Fig. 1 Spatial Delaunay angle variables

action-angle variables $((L,G,\Theta),(\ell,g,\theta))$ defined on the phase space[3]

$$\mathcal{M}_{\rm plt} := \left\{ (L,G,\Theta) \in \mathbb{R}^{3n} \,:\, L_i > G_i > \Theta_i > 0\,,\ \frac{L_i}{\mu_i\sqrt{M_i}} \neq \frac{L_j}{\mu_j\sqrt{M_j}}\,,\ \forall\, i \neq j \right\} \times \mathbb{T}^{3n} \tag{2.8}$$

the Hamiltonian $\mathcal{H}^{(0)}_{\rm plt}$ takes the form

$$\mathcal{H}^{(0)}_{\rm plt} = -\sum_{i=1}^{n} \frac{\mu_i^3 M_i^2}{2L_i^2}; \tag{2.9}$$

the phase space $\mathcal{M}_{\rm plt}$, which corresponds to an open subset of $\mathcal{M}$ in (2.6), is endowed with the standard symplectic form

$$\sum_{i=1}^{n} \mathrm{d}L_i \wedge \mathrm{d}\ell_i + \mathrm{d}G_i \wedge \mathrm{d}g_i + \mathrm{d}\Theta_i \wedge \mathrm{d}\theta_i = \sum_{i=1}^{n}\sum_{j=1}^{3} \mathrm{d}X_j^{(i)} \wedge \mathrm{d}x_j^{(i)}$$

(for more information on Delaunay variables, see Sect. 4.1.) (Fig. 1).

Notice that the $6n$-dimensional phase space $\mathcal{M}_{\rm plt}$ is foliated by $3n$-dimensional $\mathcal{H}^{(0)}_{\rm plt}$-invariant tori $\{L,G,\Theta\} \times \mathbb{T}^3$, which, in turn, are foliated by n-dimensional tori $\{L\} \times \mathbb{T}^n$, expressing geometrically the degeneracy of the integrable Keplerian limit of the $(1+n)$-body problem.

2.2 KAM theory

The perturbative approach to the many-body problem is based on the modern theory of conservative dynamical systems as developed, mainly, by Poincaré, Birkhoff, Siegel, Kolmogorov, Arnold, Moser, and Herman. We recall here, briefly, some classical results.

[3] $\mathbb{T}^n$ denotes the standard flat torus $\mathbb{T}^n := \mathbb{R}^n/(2\pi\mathbb{Z}^n)$.

2.2.1 Quasi-periodic motions and KAM tori

Consider a smooth Hamiltonian $(p,q) \in \mathcal{M} \to H(p,q)$ on a $2d$-dimensional phase space $\mathcal{M}$ endowed with standard symplectic coordinates (p,q). A (maximal) *KAM torus* for H is a d-dimensional H-invariant torus, on which the H-flow is conjugated to $\theta \in \mathbb{T}^d \to \theta + \omega t$ with $\omega \in \mathbb{R}^d$ Diophantine. We recall that ω is Diophantine if there exist positive constants γ and τ such that

$$|\omega \cdot k| \geqslant \frac{\gamma}{|k|^\tau}, \quad \forall\, k \in \mathbb{Z}^d \backslash \{0\}. \tag{2.10}$$

In particular, the motion on a KAM torus is quasi-periodic with frequencies $\omega_1, \ldots, \omega_d$.

2.2.2 Kolmogorov's 1954 Theorem

In Kolmogorov (1954), Kolmogorov stated (and gave a beautiful albeit sketchy proof) of his famous theorem on the persistence of invariant tori, which may be formulated as follows:

Consider a "nearly-integrable" Hamiltonian system with phase space $\mathcal{M} := V \times \mathbb{T}^d$, V being an open bounded region in $\mathbb{R}^d$, and with Hamiltonian function given by

$$H_\varepsilon(I,\varphi) := h(I) + \varepsilon f(I,\varphi) \tag{2.11}$$

with real-analytic functions h, f, and ε a small real parameter. The variables (I,φ) are standard symplectic "action-angle" variables, the symplectic form being $\mathrm{d}I \wedge \mathrm{d}\varphi := \sum_{i=1}^d \mathrm{d}I_i \wedge \mathrm{d}\varphi_i$.

Theorem 2.1 (Kolmogorov 1954) *In any neighborhood of any torus $\{I_0\} \times \mathbb{T}^d \subset \mathcal{M}$ such that*

$$\det h''(I_0) := \det \left(\frac{\partial^2 h}{\partial I_i \partial I_j}(I_0) \right)_{i,j=1,\ldots,d} \neq 0, \tag{2.12}$$

there exists a positive measure set of phase points belonging to analytic KAM tori for H_ε, provided ε is small enough.

A simple variation of the proof of Kolmogorov's theorem leads to the "iso–energetic" version of Theorem 2.1, namely:

Theorem 2.2 *Let I_0 be such that*[4]

$$\det \begin{pmatrix} h''(I_0) & h'(I_0) \\ h'(I_0) & 0 \end{pmatrix} \neq 0; \tag{2.13}$$

let $\mathcal{M}_0 := \{(I,\varphi) \in \mathcal{M} : H_\varepsilon(I,\varphi) = h(I_0)\}$ be the energy level corresponding to the "unperturbed" energy $h(I_0)$. Then, there exists on $\mathcal{M}_0$ a positive measure set of phase points belonging to analytic KAM tori for H_ε, provided ε is small enough.

Clearly, the measure referred to in Theorem 2.1 is the $2d$-dimensional Liouville measure in phase space, while the measure referred to in Theorem 2.2 is the restriction of the Liouville measure on the energy level $\mathcal{M}_0$.

[4] The matrix in (2.13) is a $(d+1) \times (d+1)$-matrix and the gradient $h'(I_0) := (\partial_{I_1} h(I_0), \ldots, \partial_{I_d} h(I_0))$ has to be thought of as a column in the upper right corner and as a row in lower left corner.

2.2.3 Proper degeneracies

A nearly-integrable system with Hamiltonian (2.11) for which h does not depend upon all the actions $I_1, \dots, I_d$ is called *properly degenerated*. This is the case of the many-body problem since $\mathcal{H}^{(0)}_{\text{plt}}$ in (2.9) depends only on the actions L's.

For properly degenerate systems neither condition (2.12) nor (2.13) holds and KAM tori may not exist at all.[5] To establish the existence of KAM tori in properly degenerate systems it is necessary to have more information on the perturbation f. In Arnold (1963), Arnold proved the following theorem, which he intended (and partially succeeded) to apply to the planetary many-body problem.

Let $\mathcal{M}$ denote the phase space

$$\mathcal{M} := \Big\{(I,\varphi,p,q) : \ (I,\varphi) \in V \times \mathbb{T}^d \text{ and } (p,q) \in B\Big\},$$

where V is an open bounded region in $\mathbb{R}^d$ and B is a ball around the origin in $\mathbb{R}^{2m}$; $\mathcal{M}$ is equipped with the standard symplectic form

$$\mathrm{d}I \wedge \mathrm{d}\varphi + \mathrm{d}p \wedge \mathrm{d}q = \sum_{i=1}^{d} \mathrm{d}I_i \wedge \mathrm{d}\varphi_i + \sum_{i=1}^{m} \mathrm{d}p_i \wedge \mathrm{d}q_i.$$

Let, also, H_ε be a real analytic Hamiltonian on $\mathcal{M}$ of the form

$$H_\varepsilon(I,\varphi,p,q) := h(I) + \varepsilon f(I,\varphi,p,q) \tag{2.14}$$

and denote by $\bar{f}$ the average of f over the "fast angles" φ:

$$\bar{f}(I,p,q) := \int_{\mathbb{T}^d} f(I,\varphi,p,q)\frac{\mathrm{d}\varphi}{(2\pi)^d}. \tag{2.15}$$

Theorem 2.3 (Arnold 1963) *Assume that $\bar{f}$ is of the form*

$$\bar{f} = f_0(I) + \sum_{j=1}^{m} \Omega_j(I) J_j + \frac{1}{2} A(I) J \cdot J + o_4, \quad J_j := \frac{p_j^2 + q_j^2}{2}, \tag{2.16}$$

where A is a symmetric $(m \times m)$-matrix and $\lim_{(p,q)\to 0} |o_4|/|(p,q)|^4 = 0$. Assume, also, that $I_0 \in V$ is such that

$$\det h''(I_0) \neq 0, \tag{2.17}$$

$$\sum_{j=1}^{m} \Omega_j(I_0) k_j \neq 0, \qquad \forall\, k \in \mathbb{Z}^m \text{ with } 0 < \sum_{j=1}^{m} |k_j| \leqslant 6, \tag{2.18}$$

$$\det A(I_0) \neq 0. \tag{2.19}$$

Then, in any neighborhood of $\{I_0\} \times \mathbb{T}^d \times \{(0,0)\} \subset \mathcal{M}$ there exists a positive measure set of phase points belonging to analytic KAM tori for H_ε, provided ε is small enough.

This theorem has been generalized by Herman (1998), as we shall, now, briefly explain. To formulate the non-degeneracy assumption of Herman's theorem, we need the notion of *non-planar map* introduced by Pyartli (1969). A smooth curve $u \in U \subset$

[5] Trivially, any unperturbed properly degenerate system on a $2d$-dimensional phase space with $d \geqslant 2$ will have motions with frequencies not rationally independent over $\mathbb{Z}^d$.

$\mathbb{R} \to \omega(u) \in \mathbb{R}^n$, U open non-empty interval, is called non-planar at $u_0 \in U$ if all the u-derivatives up to order $(n-1)$ at $u_0, \omega(u_0), \omega'(u_0), \ldots, \omega^{(n-1)}(u_0)$ are linearly independent over $\mathbb{R}^n$; a smooth map $u \in U \subset \mathbb{R}^d \to \omega(u) \in \mathbb{R}^n, d \leqslant n$, is called non-planar at $u_0 \in U$ if there exists a smooth curve $\alpha : \hat{U} \subset \mathbb{R} \to U$ such that $\omega \circ \alpha$ is non-planar at $t_0 \in \hat{U}$ with $\alpha(t_0) = u_0$.

Theorem 2.4 (Herman 1998) *Let H_ε and $\bar{f}$ be C^∞ functions as in* (2.14) *and* (2.15). *Assume that $\bar{f}$ is of the form*

$$\bar{f} = f_0(I) + \sum_{j=1}^{m} \Omega_j(I) J_j + o_2, \quad J_j := \frac{p_j^2 + q_j^2}{2},$$

where $\lim_{(p,q)\to 0} |o_2|/|(p,q)|^2 = 0$. *Assume, also, that $I_0 \in V$ is such that the "frequency map"*

$$I \in V \to \big(h'(I), \Omega_1(I), \ldots, \Omega_m(I)\big) \in \mathbb{R}^{d+m} \tag{2.20}$$

is non-planar at I_0. Then, in any neighborhood of $\{I_0\} \times \mathbb{T}^d \times \{(0,0)\} \subset \mathcal{M}$ there exists a positive measure set of phase points belonging to C^∞ KAM tori for H_ε, provided ε is small enough.

This theorem is based on a C^∞ local inversion theorem on "tame" Frechet spaces due to F. Sergeraert and R. Hamilton (which, in turn, is related to the Nash–Moser implicit function theorem; see Bost 1986). A non-properly-degenerate version of Theorem 2.4 was established by Rüssmann (2001). A proof of Herman's Theorem 2.4 can be found in Féjoz (2004).

2.3 Arnold's theorem on planetary motions

The main question, longly studied by astronomers and mathematicians, which Arnold addressed in his 1963 paper is the following (Arnold 1963, Ch III, p. 125):

> *"Do there exist, in the n-body problem, a set of initial conditions having positive measure such that, if the initial position and velocities of the bodies belong to this set, then the distances of the bodies from each other will remain perpetually bounded?"*

Indeed, in a (very) special case, Kolmogorov's theorem yields immediately a positive answer to such a question: it is the case of the restricted, planar, circular three-body problem (RPC3BP, for short).

The RPC3BP, largely investigated by Poincaré, consists in studying the motion of a "zero mass" asteroid moving on the plane containing the trajectories of two unperturbed major bodies (say, Sun and Jupiter) revolving on a Keplerian circle. The mathematical model for the restricted three-body problem is obtained by taking $n = 2$ and setting $m_2 = 0$ in (2.1): the equations for the two major bodies $(i = 0, 1)$ decouple from the equation for the asteroid $(i = 2)$ and form an integrable two-body-system; the problem consists, then, in studying the evolution of the asteroid $u^{(2)}(t)$. In the circular, planar case the motion of the two primaries is assumed to be circular and the motion of the asteroid is assumed to take place on the plane containing the motion of the two primaries; in fact (to avoid collisions) one considers either inner or outer (with respect to the circle described by the relative motion of the primaries) asteroid motions. Using "rotating" planar Delaunay variables (see Sect. 4.2)

$$\big((L,G),(\ell,g)\big) \in \{(L,G) \subset \mathbb{R}^2 : L > G > 0\} \times \mathbb{T}^2$$

the Hamiltonian $\mathcal{H}_{\rm rcp}$ governing the motion of the RCP3BP problem, in suitably normalized units, is given by

$$\mathcal{H}_{\rm rcp}(L,G,\ell,g;\varepsilon) := -\frac{1}{2L^2} - G + \varepsilon\mathcal{H}_1(L,G,\ell,g;\varepsilon), \tag{2.21}$$

where the perturbation is given by

$$\mathcal{H}_1 := x^{(2)} \cdot x^{(1)} - \frac{1}{|x^{(2)} - x^{(1)}|} \tag{2.22}$$

expressed in the above Delaunay variables, $x^{(2)}$ being the heliocentric coordinate of the asteroid and $x^{(1)}$ that of the planet (Jupiter); the parameter ε represents essentially the mass ratio of the two main bodies (see Appendix 4.2 for more information).

The integrable limit Hamiltonian $\mathcal{H}_{\rm rcp}|_{\varepsilon=0} = -\frac{1}{2L^2} - G$ satisfies (2.13) in a neighborhood of any point of the phase space (the determinant in (2.13) being equal, in the present case, to $3/L^4$) and, therefore, Theorem 2.2 yields the existence of a positive measure set of initial data, in each energy level $\mathcal{M}_0 := \{\mathcal{H}_{\rm rcp} = -\frac{1}{2L_0^2} - G_0\}$, that belong to KAM tori for $H_{\rm rcp}$, provided ε is small enough. In particular, the distance between the asteroid and the Keplerian circle described by the major bodies remains forever bounded.

Remarks 2.1 Indeed, in this very special case, much more is true: since two-dimensional KAM tori separate the three-dimensional energy levels, also all trajectories starting between two KAM tori remain forever trapped in the region bounded by such two tori; compare Fig. 4 below.

As for the general planetary many body problem, Arnold (1963) stated the following:

Theorem 2.5 (Arnold's theorem on planetary motions) *Let $n \geqslant 2$. Then if ε is small enough, the Hamiltonian $\mathcal{H}_{\rm plt}$ in* (2.6) *admits a positive measure set of phase points, in a neighborhood of circular and coplanar Keplerian motions, leading to quasi-periodic motions with $3n-1$ frequencies.*

This statement is taken from Féjoz (2004), where a proof of Arnold's Theorem, in this generality, appeared for the first time. Actually, in Arnold (1963) a somewhat stronger result was announced,[6] but the proof was given only for the planar three-body case.[7] A brief history of the proof of Arnold's Theorem is the following.

1. In Arnold (1963), Arnold gave a complete proof for the case of three coplanar bodies: $n = 2$ and $(X,x) \in \mathbb{R}^2 \times \mathbb{R}^2$ in (2.6). In such a case, the word "coplanar" in Theorem 2.5, is redundant and $3n-1$ has to be replaced by 4. Arnold's proof

[6] "*If the masses, eccentricities and inclinations of the planets are sufficiently small, then for the majority of initial conditions the true motion is conditionally periodic and differs little from Lagrangian motion with suitable initial conditions throughout an infinite interval of time* $-\infty < t < \infty$" (Arnold 1963, Ch. III, p. 127).

[7] In fact, Arnold gave indications on how to generalize his approach to the general case, but, apparently, nobody has succeeded in implementing Arnold's indications.

is based upon his KAM Theorem 2.3: first, by means of planar Poincaré variables (see Sect. 4.5 with $n = 2$), the Hamiltonian $\mathcal{H}_{\text{plt}}$ is put in the form (2.14), (2.16) (with $d = m = 2$); then conditions (2.18) and (2.19) ((2.17) is trivial) are checked by means of Leverrier's tables in the asymptotic regime $a_1/a_2 \to 0$ (a_i being the semimajor axis of the osculating Keplerian ellipse of the ith planet).

2. The spatial three-body case was proven in Laskar and Robutel (1995) and Robutel (1995). The strategy is similar to that of Arnold and, in particular, it is again based upon Theorem 2.5: first, by means of spatial "osculating" Poincaré variables, Jacobi's "reduction of the nodes" (see, e.g., Sect. 4.4) and Birkhoff theory of normal form (see, e.g., Siegel and Moser 1971), the Hamiltonian $\mathcal{H}_{\text{plt}}$ is put in the form (2.14), (2.16) (again, $d = m = 2$); then, the non-degeneracy conditions (2.18) and (2.19) are numerically checked, with the aid of computers, in a relatively large region of semiaxes.
3. The full proof of Theorem 2.5, as mentioned, was published in 2004 by Féjoz (2004), where Herman's work[8] on the subject was presented for the first time in a complete manner. The first step is to introduce Poincaré variables (see Sect. 4.3) and, in view of the conservation of the total angular momentum (2.7), to restrict the attention to the symplectic manifold of vertical total angular momentum, $\mathcal{M}_{\text{vert}} := \{C_1 = 0 = C_2\}$. The idea is then to use the KAM Theorem 2.4 and hence to check the non-planarity of the frequency map (2.20). However, this strategy fails for $\mathcal{H}_{\text{plt}}$ (expressed in Poincaré variables and restricted to $\mathcal{M}_{\text{vert}}$); the reason being the presence of an extra resonance ("Herman's resonance"). To overcome this problem, following Poincaré, Féjoz considers the modified Hamiltonian $\mathcal{H}^{\delta}_{\text{plt}} := \mathcal{H}_{\text{plt}} + \delta C_3^2$. For such Hamiltonian the non-planarity condition of the frequency map is satisfied; but since the Hamiltonians $\mathcal{H}^{\delta}_{\text{plt}}$ and $\mathcal{H}_{\text{plt}}$ commute they have the same Lagrangian tori and hence the result is established also for $\mathcal{H}_{\text{plt}}$.

3 KAM tori in solar subsystems

3.1 Results

Certainly the main motivation for KAM theory was the existence of regular (relatively bounded) motions in the Solar System. In fact, as soon as the first KAM theorems were established, astronomers tried to apply them to astronomical models. However, such direct applications lead to very poor "practical" results, the restriction on ε (i.e., the size of the mass ratios) being far too strong to allow for applications to the Solar System (or solar subsystems). At this regard, in a 1966 paper (Hénon 1966), Hénon concludes: "Ainsi, ces théorèmes, bien que d'un très grand intérêt théorique, ne semblent pas pouvoir en leur état actuel être appliqués à des problèmes pratiques."[9]

[8] Herman worked for long time on the planetary problem and gave several lectures and seminars on it in the mid 1990s but his untimely death (November 2, 2000) did not allow him to publish the complete results of his researches. Herman's work on the planetary problem was, then, taken up by friends and colleagues in Paris and completed in Féjoz (2004).

[9] (Hénon 1966, p. 64): "Les théorèmes d'Arnold et Moser ne s'appliquent qu'à des problèmes qui diffèrent d'un problème intégrable par une perturbation extrêmement petite. [$\cdots$] Par exemple, dans la démonstration d'Arnold (1963, *Russian math. Surveys*, 18, 9, p. 16) on a: [$\cdots$] Dans le cas du problème restreint, on a: $n = 2$. D'autre part, le cas intégrable est représenté par $\mu = 0$; on retrouve

A major breakthrough towards applications of KAM theory to physical models came from the interaction between KAM theory and techniques for computer-assisted proofs. Such techniques, which are based upon the so-called interval arithmetic,[10] allow to perform long computations on computers keeping rigorously track of the rounding errors introduced by the machine.

For more information about computer-aided proofs and computer-assisted KAM theory applied to model problems (such as the standard map or a simple forced pendulum), see, e.g., Celletti and Chierchia 1987, 1988, 1995; Celletti et al. 1987, 2000; Rana 1987; Celletti and Giorgilli, 1988; Llave and Rana 1990, and references therein.

Computer-aided existence of KAM tori for three-body problems with mass ratios within at most three orders of magnitude of the observed values have been (rigorously) established in the following three papers.

1. In Celletti and Chierchia (1997) the Sun–Jupiter–Ceres problem has been investigated in the context of the RPC3BP using rotating planar Delaunay variables. The observed average frequency of Ceres is about $\Omega_C \simeq 2.577107$, while $e_C \simeq 0.0766$ is the observed eccentricity. The perturbing function has been expanded in Fourier–Taylor series, retaining only the terms whose size is bigger than the gravitational influence due to Saturn and the Jupiter/Sun mass ratio (which is about 10^{-3}) has been replaced by ε. Implementing computer-assisted KAM estimates, existence of quasi-periodic tori with Diophantine frequencies close to Ω_C has been established for any mass-ratio $\varepsilon \leqslant 10^{-6}$.
2. In Locatelli and Giorgilli (2000) the planetary problem formed by the Sun, Jupiter, and Saturn has been considered. After Jacobi's reduction of the nodes (see, e.g., Sect. 4.4), one obtains a Hamiltonian function with four degrees of freedom. Such Hamiltonian is expanded up to the second-order in the masses and averaged over the fast angles $(\lambda_1^*, \lambda_2^*)$ (in the notation of Sect. 4.4). In this way, a two degree–of-freedom Hamiltonian is obtained, which nearly gives the slow motion of the parameters characterizing the Keplerian approximation (e.g., the eccentricities). Looking for invariant tori in the proximity of an equilibrium elliptic point, the perturbation, written in Poincaré variables, is expanded up to the order 6 in the eccentricities. Then, a Birkhoff normal form, combined with a computer-assisted implementation of a KAM theorem, provides the existence of two invariant tori

Footnote continued

alors le problème des deux corps. Pour $\mu \neq 0$, la perturbation est proportionelle à la masse μ du second corps. M et μ sont donc du même ordre de grandeur. Des ínegalités ci–dessus, on tire:

$$M < 10^{-333}. \tag{14}$$

Une estimation du même genre peut être faite dans la démonstration de Moser (1962, *Nach. Akad. Wiss. Göttingen*, Math. Phys. Kl., 1); on aboutit à:

$$M < 10^{-48}. \tag{15}$$

Ainsi, ces théorèmes, bien que d'un très grand intérêt théorique, ne semblent pas pouvoir en leur état actuel être appliqués á des problèmes pratiques, où les perturbations sont toujours beaucoup plus grandes que les limites (14) ou (15) ."

[10] Roughly speaking, computers work with special classes of rational numbers ("representable numbers"). In general, an elementary operation $(+, -, *, \div)$ between two representable numbers is no more a representable number, since the result is affected by rounding–off and propagation errors. Therefore, one needs to provide the result as an interval, whose endpoints are representable numbers and which yield lower and upper bounds on the result of the elementary operation.

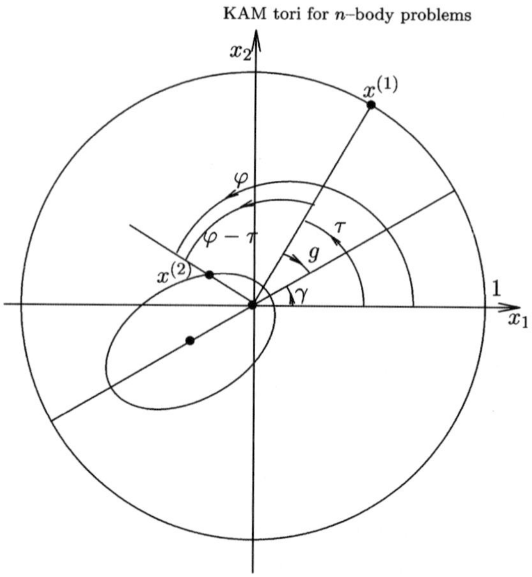

Fig. 2 Planar Delaunay angle variables

bounding the *secular* motions of Jupiter and Saturn for the observed values of the parameters.[11]

3. In Celletti and Chierchia (2005), which is extensively reviewed in Sect. 3.2, a truncated RPC3BP model for Sun, Jupiter, and Asteroid 12 Victoria is investigated. On a fixed energy level,[12] invariant KAM tori trapping the motion of Victoria have been established for the astronomical value of the Jupiter–Sun mass-ratio.

For other computer-aided KAM results of interest for Celestial Mechanics (see Celletti 1990a,b Celletti and Falcolini 1992) (spin-orbit problem) and (Celletti 1993) (librational tori).

3.2 KAM stability of the Sun–Jupiter–Victoria system modelled by a truncated RPC3BP

Here, we describe with some details the results in Celletti and Chierchia (2005) mentioned in item 3 above. Let us begin by describing precisely the mathematical model. The framework is that of RPC3BP as described in Sect. 2.3 and 4.2; see in particular (2.21), (2.22), and Fig. 2. As main bodies we take the Sun (P_0) and Jupiter (P_1), which are therefore assumed to revolve on a circle of radius one. In such a case the perturbative parameter ε is the Jupiter/Sun mass ratio, which amounts to

$$\varepsilon = \varepsilon_J := 0.954 \times 10^{-3} \tag{3.23}$$

(the normalizations are described in Sect. 4.2; see, in particular, Eq. (4.30) and (4.31)). We, then, proceed to select a minor ("zero mass") body, P_2, within the asteroidal belt;

[11] For interesting numerical results related to Locatelli and Giorgilli (2000), see Locatelli and Giorgilli (2005a, b).

[12] In comparing this result with Celletti and Chierchia (1997), keep in mind that there the energy level is not a priori fixed as it is done here.

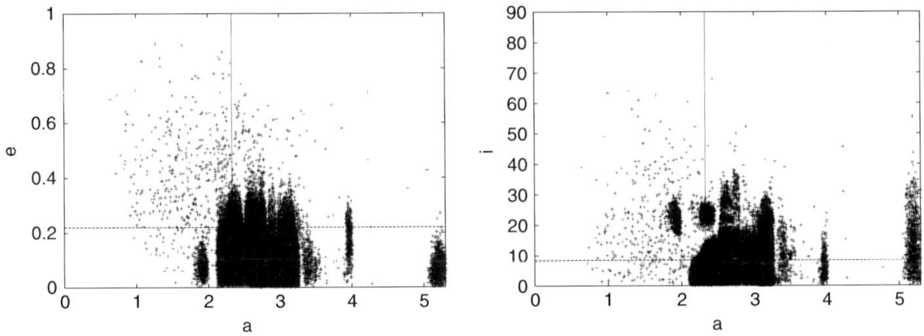

Fig. 3 Orbital elements of the numbered asteroids: semimajor axis versus eccentricity (left panel), semimajor axis versus inclination (right panel). The internal lines locate the position of the asteroid 12 Victoria

in order to avoid the introduction of another small parameter, we privileged those asteroids whose eccentricity is not too small (which also happen to be quite common in the asteroidal belt, as we shall shortly explain). We pick Asteroid 12 Victoria, whose orbital elements are:

$$\begin{aligned} a_V &\simeq 2.334\ \mathrm{AU} \quad e_V \simeq 0.220 \quad \iota_V \simeq 8.363^\circ, \\ \hat{g} &\simeq 69.717^\circ \quad \Omega \simeq 235.548^\circ \quad M \simeq 135.908^\circ, \end{aligned}$$

where ι_V is the inclination with respect to the ecliptic, $\hat{g}$ the argument of perihelion, Ω the longitude of the ascending node, and M is the mean anomaly referred to the epoch MJD 53400.

In order to explore the peculiarity of this choice, we report in Fig. 3 the elements of the numbered asteroids.[13] The majority of the asteroids lie within the region $1.8 \leqslant a \leqslant 3.5$, while the eccentricity is typically confined to $0 \leqslant e \leqslant 0.35$ and, as Fig. 3 shows, the orbital elements of Victoria (which are located by the internal lines) appear to be rather typical in the nearly planar, non-too eccentric region of the orbital elements of the numbered asteroids.

In our model we disregarded the eccentricity of Jupiter, the mutual inclinations, the gravitational effects of the other bodies (notably those of Mars and Saturn), any dissipative phenomena like tides, solar winds, Yarkovsky effect, etc. As empirical criterion, we decide to expand the perturbation in the eccentricity and semimajor axes ratio, disregarding the contributions smaller than the most important term we have neglected in our model, which is actually due to the eccentricity of the orbit of Jupiter. Moreover, in order to balance the fact that lower harmonics are physically more relevant than higher ones, we reintroduce in the lowest order harmonics the first discarded term. We are thus led to consider the one-parameter family of Hamiltonians

$$\begin{aligned} H_{\mathrm{SJV}}(\ell, g, L, G; \varepsilon) &:= -\frac{1}{2L^2} - G - \varepsilon P_{\mathrm{SJV}}(\ell, g, L, G) \\ &=: H_0(L, G) + \varepsilon H_1(\ell, g, L, G), \end{aligned} \tag{3.24}$$

[13] The elements of the numbered asteroids are provided by the JPL's DASTCOM database at http://ssd.jpl.nasa.gov/?sb_elem

where $0 < G < L$ (being $e = \sqrt{1 - G^2/L^2}$; see Eq. (4.29), Sect. 4.1) and, setting $a_0 := L^2$, the perturbing function is given by

$$\begin{aligned} P_{\rm SJV}(\ell, g, L, G) := \; & 1 + \frac{{a_0}^2}{4} + \frac{9}{64}{a_0}^4 + \frac{3}{8}{a_0}^2 e^2 - \left(\frac{1}{2} + \frac{9}{16}{a_0}^2\right){a_0}^2 e \, \cos\ell \\ & + \left(\frac{3}{8}{a_0}^3 + \frac{15}{64}{a_0}^5\right)\cos(\ell + g) - \left(\frac{9}{4} + \frac{5}{4}{a_0}^2\right){a_0}^2 e \, \cos(\ell + 2g) \\ & + \left(\frac{3}{4}{a_0}^2 + \frac{5}{16}{a_0}^4\right)\cos(2\,\ell + 2\,g) + \frac{3}{4}{a_0}^2 e \, \cos(3\,\ell + 2\,g) \\ & + \left(\frac{5}{8}{a_0}^3 + \frac{35}{128}{a_0}^5\right)\cos(3\,\ell + 3\,g) + \frac{35}{64}{a_0}^4 \cos(4\,\ell + 4\,g) \\ & + \frac{63}{128}{a_0}^5 \cos(5\ell + 5g). \end{aligned} \tag{3.25}$$

Fixing the perturbing parameter $\varepsilon = \varepsilon_J$ as in (3.23), we obtain the Sun–Jupiter–Victoria Hamiltonian:

$$\begin{aligned} H^*_{\rm SJV}(\ell, g, L, G) &:= -\frac{1}{2L^2} - G - \varepsilon_J P_{\rm SJV}(\ell, g, L, G) \\ &= H_0(L, G) + \varepsilon_J H_1(\ell, g, L, G). \end{aligned}$$

We next fix the energy level. To this end, we remark that the observed values of the Delaunay's action variables are $\sqrt{a_{\rm V}} \simeq 0.670 =: L_{\rm V}$ and $L_{\rm V}\sqrt{1 - e_{\rm V}^2} \simeq 0.654 =: G_{\rm V}$. Let

$$\begin{aligned} E^{(0)}_{\rm V} &:= -\frac{1}{2L_{\rm V}^2} - G_{\rm V} \simeq -1.768, \quad E^{(1)}_{\rm V} := \langle H_1(\cdot, L_{\rm V}, G_{\rm V})\rangle \simeq -1.060, \\ E_{\rm V}(\varepsilon) &:= E^{(0)}_{\rm V} + \varepsilon E^{(1)}_{\rm V}. \end{aligned}$$

We define the osculating energy level of the Sun–Jupiter–Victoria model as

$$E^*_{\rm V} := E_{\rm V}(\varepsilon_J) = E^{(0)}_{\rm V} + \varepsilon_J E^{(1)}_{\rm V} \simeq -1.769. \tag{3.26}$$

On $\mathcal{S}^*_{\rm SJV} := (H^*_{\rm SJV})^{-1}(E^*_{\rm V})$ we want to prove the existence of two invariant tori, bounding from above and below the observed values $L_{\rm V}$ and $G_{\rm V}$. More precisely, if $\tilde{L}_\pm = L_{\rm V} \pm 0.001$ we consider the frequencies

$$\tilde{\omega}_\pm := \left(\frac{\partial H_0}{\partial L}, \frac{\partial H_0}{\partial G}\right) = \left(\frac{1}{\tilde{L}^3_\pm}, -1\right) =: (\tilde{\alpha}_\pm, -1).$$

In order to obtain two bounding *Diophantine* frequencies we compute the continued fraction expansion up to the order 5 of $\tilde{\alpha}_\pm$ and we add a tail of one's to obtain the following Diophantine numbers:

$$\begin{aligned} \alpha_- &:= [3; 3, 4, 2, 1^\infty] = 3.30976937631389\ldots, \\ \alpha_+ &:= [3; 2, 1, 17, 5, 1^\infty] = 3.33955990647860\ldots \end{aligned}$$

Finally, we define

$$\omega_\pm := (\alpha_\pm, -1),$$

which satisfy the Diophantine condition (2.10) with constants

$$\tau_\pm := \tau = 1, \quad \gamma_- := 7.224496 \times 10^{-3}, \quad \gamma_+ := 3.324329 \times 10^{-2}.$$

The stability of the asteroid Victoria is an immediate consequence of the following theorem, which yields the existence of the KAM continuations of the unperturbed tori $\mathcal{T}_0^\pm := \{(L_\pm, G_\pm)\} \times \mathbb{T}^2$.

Theorem 3.1 *For $|\varepsilon| \leqslant \varepsilon_* := 10^{-3}$ the unperturbed tori $\mathcal{T}_0^\pm$ can be analytically continued into invariant KAM tori $\mathcal{T}_\varepsilon^\pm$ on the energy level $\mathcal{S}_\varepsilon := H_{\rm SJV}^{-1}(E_{\rm V}(\varepsilon))$ keeping fixed the ratio of the frequencies.*

As a consequence (recall Remark 2.1), the orbital elements corresponding to the semimajor axis and to the eccentricity (which are simply related to the Delaunay's variables L and G) stay forever ε-close to their unperturbed values.

The idea of the proof relies on the combination of a new KAM iso-energetic theorem with accurate computer-assisted construction of approximate solutions. First, one observes that the parametric representation $\theta \in \mathbb{T}^2 \to (x, y) = (u(\theta), v(\theta))$ of a KAM torus lying with Diophantine frequencies (ω_1, ω_2), on the energy level E satisfies the following semilinear PDE

$$\begin{aligned} &Du = \frac{\partial H}{\partial y}(u, v), \quad Dv = -\frac{\partial H}{\partial x}(u, v), \\ &H(u(0), v(0)) = E, \end{aligned} \tag{3.27}$$

where D denotes the vector field $\left(\omega_1 \frac{\partial}{\partial \theta_1} + \omega_2 \frac{\partial}{\partial \theta_2}\right)$. Then, the system (3.27) is solved by a "hard implicit function theorem" à la Nash–Moser (compensating the effect of the small divisors with a quadratic scheme). To apply effectively this implicit function theorem, we first compute explicitly an "approximate solution", say $z^{(1)}$, and, then, we prove that close to it there exists a much better approximate solution, $z^{(2)}$, to which the stringent smallness condition dictated by the KAM implicit function theorem applies. In fact, $z^{(1)}$ is a Fourier–Taylor polynomial function (depicted in Figs. 4 and 5), while $z^{(2)}$ is obtained via iteration of a certain non-linear operator and can only be controlled by estimating suitable norms. The construction of $z^{(1)}$ is based on an algorithm for computing iso-energetic Lindstedt series.[14]

Remark 3.1 From the mathematical point of view, the Fourier-truncation introduced in this model is rather unsatisfactory. However, we believe that a similar strategy to that leading to Theorem 3.1, could be applied to the full RPC3BP. From the physical point of view, instead, the truncation does not seem to affect much the dynamics. In fact, numerical studies suggest that for the frequencies and parameter values considered in Theorem 3.1, the truncated Hamiltonian (3.24) and (3.25) provides results very close to those obtained using the complete perturbing function (see Celletti et al. 2004, briefly reviewed in Sect. 5).

14 Lindstedt series—already known at the times of Poincaré—are formal Fourier–Taylor series expansions of the solution of system (3.27).

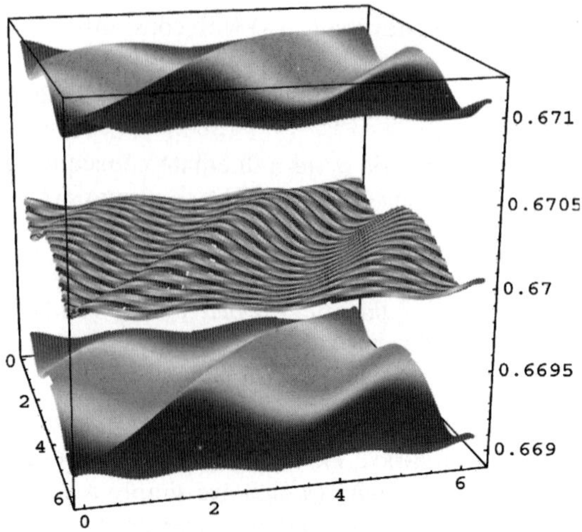

Fig. 4 The upper and lower surfaces are the graphs on the three-dimensional energy level $(H^*_{\rm SJV})^{-1}(E^*_V)$ of the approximate solutions $z^{(1)}$ described in the text for the two frequency vectors $\omega_- = (3.30976937631389\ldots,-1)$ and $\omega_+ = (3.33955990647860\ldots,-1)$; the intermediate surface is obtained integrating numerically a Sun–Jupiter–Victoria sample motion on the same energy level. The coordinates used are the (rotating) Delaunay angles $(\ell,g) \in \mathbb{T}^2$ in abscissa and the action $L > 0$ in ordinates; the perturbing parameter is set equal to the actual Jupiter–Sun mass ratio $\varepsilon_{\rm J} = 0.954 \times 10^{-3}$

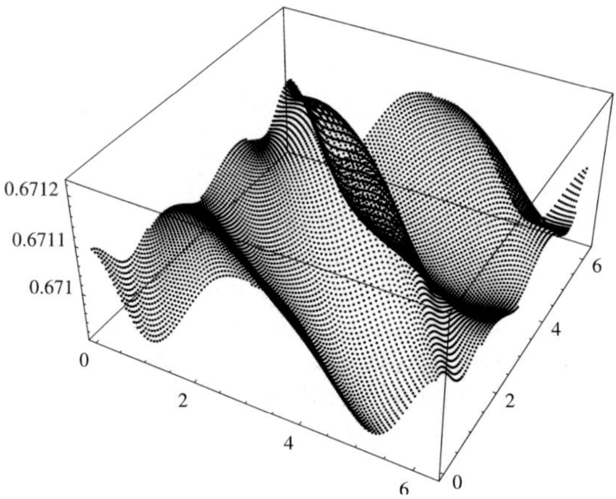

Fig. 5 The upper bounding surface, on a different scale, showing the oscillatory structure of the KAM trapping tori

4 Appendix: symplectic variables for many-body problems

4.1 Delaunay variables

We begin by briefly describing the Delaunay variables for the Keplerian two-body problem.

Let $\mathcal{H}_{\text{Kep}} = \frac{|X|^2}{2\mu} - \frac{\mu M}{|x|}$ denote the (reduced) two-body Hamiltonian with $(X, x) \in \mathbb{R}^3 \times \mathbb{R}^3 \backslash \{0\}$, where M denotes the total mass of the two bodies and μ is a free rescaling parameter, and consider negative energies $\mathcal{H}_{\text{Kep}} < 0$. In such a case, if $(X(t), x(t))$ denotes the $\mathcal{H}_{\text{Kep}}$-flow, then $x(t)$ describes an ellipse lying in the plane π_C orthogonal to $C := X \times x$, with focus in the origin and fixed symmetry axes. Assume that the angular momentum C is not vertical and that the ellipse is not a circle. Introduce the following notations:

a is the semimajor axis of the ellipse spanned by x;
ι (the inclination) is the angle between the x_3-axis and C;
$G = |C| = \sqrt{C_1^2 + C_2^2 + C_3^2}$;
$\Theta = G \cos \iota = C_3$;
$L = \mu\sqrt{Ma}$;
ℓ is the mean anomaly of x (:=2π times the normalized area spanned by x measured from the perihelion Q, which is the point of the ellipse closest to the origin);
θ is the angle between the x_1-axis and the node line N (i.e. the intersection of the (x_1, x_2)-plane with π_C);
g is the argument of the perihelion (:= the angle between N and (O, Q)).

Then

$$((L, G, \Theta), (\ell, g, \theta)) \in \mathcal{M}_{\text{Kep}} := \{L > G > \Theta > 0\} \times \mathbb{T}^3 \tag{4.28}$$

are conjugated symplectic coordinates (i.e., $\mathrm{d}L \wedge \mathrm{d}\ell + \mathrm{d}G \wedge \mathrm{d}g + \mathrm{d}\Theta \wedge \mathrm{d}\theta = \sum_{i=1}^{3} \mathrm{d}X_i \wedge \mathrm{d}x_i$) and if ϕ_{Del} is the corresponding symplectic map, then

$$\mathcal{H}_{\text{Kep}} \circ \phi_{\text{Del}} = -\frac{\mu^3 M^2}{2L^2}.$$

The eccentricity e of the Keplerian ellipse with energy $-\mu^3 M^2/2L^2$ and absolute value of angular momentum G is, then, given by

$$e = \sqrt{1 - \frac{G^2}{L^2}}. \tag{4.29}$$

Thus, the inequalities in (4.28) are seen to correspond to regions in phase space of non-degenerate elliptical motions (i.e., ellipses with $0 < e < 1$) taking place on the plane transversal with and not perpendicular to the (x_1, x_2)-plane.

In expressing the planetary $(1 + n)$-problem in Delaunay action-angle variables one considers Delaunay variables $(L_i, G_i, \Theta_i), (\ell_i, g_i, \theta_i)$ associated to the limiting two-body problem formed by the Sun ($i = 0$) and the ith planet ($1 \leqslant i \leqslant n$). The (clearly sympletic) variables $(L_i, G_i, \Theta_i), (\ell_i, g_i, \theta_i)$ are well defined in the Cartesian product of the Keplerian phase spaces

$$\prod_{1 \leqslant i \leqslant n} \{L_i > G_i > \Theta_i > 0\} \times \mathbb{T}^{3n}$$

and the relations

$$a_i = \frac{L_i}{\mu_i \sqrt{M_i}} \neq a_j = \frac{L_j}{\mu_j \sqrt{M_j}}, \quad \forall\, 1 \leqslant i \neq j \leqslant n$$

avoid collisions; this accounts for the definition of $\mathcal{M}_{\text{plt}}$ given in (2.8).

Complete details may be found, e.g., in Biasco et al. (2003, Sect. C.1, pp. 117–119) and Celletti and Chierchia (2005, Sect. 3.2).

4.2 Planar delaunay variables and the RPC3BP Hamiltonian

We start by describing planar Delaunay variables $\big((L,G),(\ell,\hat{g})\big)$ and then describe the "rotating" planar Delaunay variables $((L,G),(\ell,g))$. Consider a planar two-body problem with $(X,x)\in\mathbb{R}^2\times\mathbb{R}^2\backslash\{0\}$ and $\mathcal{H}_{\text{Kep,pl}}=\frac{|X|^2}{2\mu}-\frac{\mu M}{|x|}$, M being the total mass of the two body and μ a free rescaling parameter. Introduce the following notations:

a is the semi-major axis of the ellipse spanned by x;
$L=\mu\sqrt{Ma}$;
e is the eccentricity of the ellipse spanned by x and $G=L\sqrt{1-e^2}$;
ℓ is the mean anomaly of x;
$\hat{g}$ is the argument of the perihelion.

Then

$$\big((L,G),(\ell,\hat{g})\big)\in\mathcal{M}_{\text{Kep,pl}}:=\{L>G>0\}\times\mathbb{T}^2$$

are conjugated symplectic coordinates (i.e., $\mathrm{d}L\wedge\mathrm{d}\ell+\mathrm{d}G\wedge\mathrm{d}\hat{g}=\sum_{i=1}^2\mathrm{d}X_i\wedge\mathrm{d}x_i$) and if $\phi_{\text{Del,pl}}$ is the corresponding symplectic map, then

$$\mathcal{H}_{\text{Kep,pl}}\circ\phi_{\text{Del,pl}}=-\frac{\mu^3M^2}{2L^2}.$$

The rotating planar Delaunay variables for the RPC3BP for P_0 (main body), P_1 (planet), and P_2 (zero-mass asteroid) are, then, given by

$$((L,G),(\ell,g))\in\mathcal{M}_{\text{Kep,pl}}:=\{L>G>0\}\times\mathbb{T}^2,\quad g:=\hat{g}-\tau,$$

τ being the longitude of P_1 (i.e., the angle between the x_1-axis and $x^{(1)}(\tau)$, which denotes the relative position P_1–P_0). The units are choosen so that:

$$m_0+m_1=1,\quad |x^{(1)}(\tau)|=1, \tag{4.30}$$

where m_i denote the masses of P_i. With such normalization the period of the P_0–P_1 motion is 2π (so that $\tau\in\mathbb{T}$).
Now, if we also set

$$\mu:=\frac{1}{m_0^{2/3}},\quad \varepsilon:=\frac{m_1}{m_0^{2/3}}=\frac{m_1}{(1-m_1)^{2/3}}, \tag{4.31}$$

then the Hamiltonian of the RCP3BP, in rotating planar Delaunay variables, takes the form (2.21) with

$$\mathcal{H}_1(L,G,\ell,g;\varepsilon):=x^{(2)}\cdot x^{(1)}_{\text{circ}}(\tau)-\frac{1}{|x^{(2)}-x^{(1)}_{\text{circ}}(\tau)|},\quad x^{(1)}_{\text{circ}}(\tau):=(\cos\tau,\sin\tau),$$

where, of course, $x^{(2)}$ (the heliocentric position of the asteroid) has to be expressed in term of the rotating planar Delaunay variables.

Complete details may be found, e.g., in Celletti and Chierchia (2005, Sect. 3.2 and 3.3).

4.3 Poincaré variables

The spatial Poincaré variables for the planetary $(1+n)$- body-problem is a set of symplectic variables for an open (physically relevant) subset of the phase space $\mathcal{M}_{\rm plt}$; in particular such variables are well defined (and analytic) in a neighborhood of circular and co-planar motions. For $1 \leqslant i \leqslant n$, let $((L_i, G_i, \Theta_i), (\ell_i, g_i, \theta_i))$ denote the Delaunay variables associated to the two–body system Sun-ith planet. The (spatial) Poincaré variables are given by $((\Lambda_i, \lambda_i), (\eta_i, \xi_i), (p_i, q_i))$ where

$$\Lambda_i = L_i, \quad \lambda_i = \ell_i + g_i + \theta_i,$$

and

$$\eta_i = \sqrt{2(L_i - G_i)}\ \cos(g_i + \theta_i), \qquad p_i = \sqrt{2(G_i - \Theta_i)}\ \cos\theta_i,$$
$$\xi_i = -\sqrt{2(L_i - G_i)}\ \sin(g_i + \theta_i), \qquad q_i = -\sqrt{2(G_i - \Theta_i)}\ \sin\theta_i.$$

Then, for any $\Lambda_+ > \Lambda_- > 0$ there exists $r > 0$ such that the Poincaré variables are symplectic and analytic on the domain

$$\Lambda_- < \Lambda_i < \Lambda_+ \quad \text{for } 0 \leqslant i \leqslant n\,, \quad (\lambda_1, \ldots, \lambda_n) \in \mathbb{T}^n,$$
$$\eta_i^2 + \xi_i^2 < r^2 \quad \text{and} \qquad p_i^2 + q_i^2 < r^2 \quad \text{for } 0 \leqslant i \leqslant n.$$

If e_i, $C^{(i)}$ and ι_i denote, respectively, the eccentricity, angular momentum and inclination of the (instantaneous or osculating) two–body system Sun-ith planet, then the following relations hold

$$\frac{\eta_i^2 + \xi_i^2}{2} = \Lambda_i \left(1 - \sqrt{1 - e_i^2}\right),$$
$$|C^{(i)}| = \Lambda_i \sqrt{1 - e_i^2}\,,$$
$$\frac{p_i^2 + q_i^2}{2} = |C^{(i)}|(1 - \cos\iota_i)$$

(for details, see, e.g., Biasco et al. 2003, Sect. C.1).

4.4 Osculating Poincaré Variables and Jacobi's reduction of the nodes

Poincaré introduced another set of symplectic variables, particularly suited to describe the classical Jacobi's reduction of the nodes, which allows to give a representation of the spatial three-body in terms of a four-degree-of-freedom Hamiltonian system.[15]

Let, for $i = 1, 2$, $\big((L_i, G_i, \Theta_i), (\ell_i, g_i, \theta_i)\big)$ denote the Delaunay variables introduced in Sect. 4.1. Then the variables

$$\big((\Lambda_i^*, \lambda_i^*), (\eta_i^*, \xi_i^*), (\Theta_i, \theta_i)\big) \tag{4.32}$$

defined by

$$\Lambda_i^* = L_i, \qquad \eta_i^* = \sqrt{2(L_i - G_i)}\cos g_i,$$
$$\lambda_i^* = \ell_i + g_i, \quad \xi_i^* = -\sqrt{2(L_i - G_i)}\sin g_i$$

15 This description is borrowed from Chierchia (2005)

are symplectic and analytic near circular, non-co-planar motions (for details, see, e.g., Biasco et al. 2003). Denote by

$$\mathcal{H}_{3\text{bp}} := \mathcal{H}^{(0)}(\Lambda^*) + \varepsilon \mathcal{H}^{(1)}(\Lambda^*, \lambda^*, \eta^*, \xi^*, \Theta, \theta)$$

the Hamiltonian (2.6) (with $n = 2$) expressed in terms of the symplectic variables (4.32), $\Lambda^* = (\Lambda_1^*, \Lambda_2^*)$, etc. Then, $\Theta_1 + \Theta_2$ is the vertical component, $C_3 = C \cdot k_3$, of the total argument $C = C^{(1)} + C^{(2)}$. Introduce, now, the symplectic variables

$$(\Lambda^*, \lambda^*, \eta^*, \xi^*, \Psi, \psi) = \phi(\Lambda^*, \lambda^*, \eta^*, \xi^*, \Theta, \theta),$$

where $(\Psi_1, \Psi_2, \psi_1, \psi_2) := (\Theta_1, \Theta_1 + \Theta_2, \theta_1 - \theta_2, \theta_2)$ and let $\mathcal{H}_{3\text{bp}}^* := \mathcal{H}_{3\text{bp}} \circ \phi^{-1}$ denote the Hamiltonian of the spatial three-body problem in these symplectic variables. Since the Poisson bracket of $\Psi_2 = \Theta_1 + \Theta_2$ and $\mathcal{H}_{3\text{bp}}^*$ vanishes (C_3 being an integral for the $\mathcal{H}_{3\text{bp}}$–flow), the conjugate angle ψ_2 is cyclic for $\mathcal{H}_{3\text{bp}}^*$, i.e.,

$$\mathcal{H}_{3\text{bp}}^* = \mathcal{H}_{3\text{bp}}^*(\Lambda^*, \lambda^*, \eta^*, \xi^*, \Psi_1, \Psi_2, \psi_1).$$

Because the total angular momentum C is preserved, one can restrict the attention to the 10-dimensional invariant (and symplectic) submanifold $\mathcal{M}_{\text{ver}}$ defined by fixing the total angular momentum to be vertical. Such submanifold, in terms of Delaunay variables, is given by

$$\theta_1 - \theta_2 = \pi \quad \text{and} \quad G_1^2 - \Theta_1^2 = G_2^2 - \Theta_2^2,$$

so that $\mathcal{M}_{\text{ver}}^* := \phi(\mathcal{M}_{\text{ver}}) = \{\psi_1 = \pi\ , \quad \Psi_1 = \widehat{\Psi}_1(\Lambda^*, \eta^*, \xi^*; \Psi_2)\}$ with

$$\widehat{\Psi}_1 := \frac{\Psi_2}{2} + \frac{(\Lambda_1^* - H_1^*)^2 - (\Lambda_2^* - H_2^*)^2}{2\Psi_2}, \qquad H_i^* := \frac{\eta_i^{*2} + \xi_i^{*2}}{2}.$$

Since $\mathcal{M}_{\text{ver}}^*$ is invariant for the flow ϕ_*^t of $\mathcal{H}_{3\text{bp}}^*$, $\psi_1(t) := \pi$ and $\dot{\psi}_1 := 0$ for motions starting on $\mathcal{M}_{\text{ver}}^*$, which implies that $(\partial_{\Psi_1} \mathcal{H}_{3\text{bp}}^*)|_{\mathcal{M}_{\text{ver}}^*} = 0$. This fact allows to introduce, for fixed values of the vertical angular momentum $\Psi_2 = c \neq 0$, the following reduced Hamiltonian:

$$\mathcal{H}_{\text{red}}^c(\Lambda^*, \lambda^*, \eta^*, \xi^*) := \mathcal{H}_{3\text{bp}}^*(\Lambda^*, \lambda^*, \eta^*, \xi^*, \widehat{\Psi}_1(\Lambda^*, \eta^*, \xi^*; c), c, \pi)$$

on the eight-dimensional phase space $\mathcal{M}_{\text{red}} := \{\Lambda_i^* > 0, \lambda \in \mathbb{T}^2, (\eta^*, \xi^*) \in B^4\}$ endowed with the standard symplectic form $d\Lambda^* \wedge d\lambda^* + d\eta^* \wedge d\xi^*$ (B^4 being a ball around the origin in $\mathbb{R}^4$). In fact, the (standard) Hamilton's equations for $\mathcal{H}_{\text{red}}^c$ are immediately recognized to be a subsystem of the full (standard) Hamilton's equations for $\mathcal{H}_{3\text{bp}}$ when the initial data are restricted on $\mathcal{M}_{\text{ver}}^*$ and the constant value of Ψ_2 is chosen to be c.

4.5 Planar Poincaré Variables and the planar $(1 + n)$-body problem

The planar Poincaré variables for $(1 + n)$ co-planar bodies are defined as follows. For $0 \leqslant i \leqslant n$, let $(L_i, G_i), (\ell_i, g_i)$ be the planar Delaunay variables (as defined in Sect. 4.2) associated to the two–body system Sun-ith planet and let

$$\Lambda_i = L_i, \qquad \eta_i = \sqrt{2(L_i - G_i)}\ \cos \hat{g}_i,$$

$$\lambda_i = \ell_i + \hat{g}_i, \quad \xi_i = -\sqrt{2(L_i - G_i)}\ \sin \hat{g}_i.$$

Table 1

	Truncated	Complete
$\varepsilon_c \in$	$[0.07, 0.09]$	0.08
Intermediate value	$[0.08, 0.1]$	0.09

Then, for any $\Lambda_+ > \Lambda_- > 0$ there exists $r > 0$ such that the planar Poincaré variables are symplectic and analytic on the domain

$$\Lambda_- < \Lambda_i < \Lambda_+ \quad \text{for } 0 \leqslant i \leqslant n, \;\; (\lambda_1, \ldots, \lambda_n) \in \mathbb{T}^n,$$
$$\eta_i^2 + \xi_i^2 < r^2 \;\; \text{for } 0 \leqslant i \leqslant n.$$

If e_i denotes the eccentricity of the (instantaneous or osculating) two-body system Sun-ith planet then

$$\frac{\eta_i^2 + \xi_i^2}{2} = \Lambda_i \left(1 - \sqrt{1 - e_i^2}\right)$$

(for complete details, see, e.g., Biasco et al. 2005, Appendix A).

5 Appendix: Numerical investigation of the RPC3BP

A complementary numerical study of the stability of the asteroid Victoria has been performed in Celletti et al. (2004) using frequency analysis as introduced in Laskar et al. (1992) and Laskar (1993). The dynamical system described by (3.24) and (3.25) has been compared to the system where no truncation of the perturbing function has been performed.[16] If (ω_L, ω_G) are the fundamental frequencies, we denote by $\gamma := |\frac{\omega_L}{\omega_G}|$ the frequency ratio.

In practice one can proceed as follows. Fix $E = E_0$ and $\varepsilon = \varepsilon_0$; set the initial data as $L = L_0$, $\ell = 0$, $g = g_0$, where L_0, g_0 vary over a grid (which corresponds to consider a slice projection by fixing $\ell = 0$). Find G_0 by solving the relation

$$E_0 = -\frac{1}{2L_0^2} - G + \varepsilon_0 R(L_0, G, 0, g_0).$$

Using the solution of the equations of motion, frequency analysis is implemented to compute (ω_L, ω_G). We remark that according to a standard criterion (see Laskar et al. 1992), the dynamics is discriminated on the basis of the graph of γ versus the initial conditions L_0, g_0. More precisely: a region of invariant tori is characterized by a regular (i.e., monotonically increasing or decreasing) behavior of the frequency-map; no variation of the frequency ratio corresponds to a resonant regime; a chaotic region is characterized by consecutive sudden jumps of the frequency map.

Having fixed the energy level according to (3.26), let ε_c be the critical value of the perturbing parameter at which the transition from stability to instability occurs. The results are shown in Table 1, where we provide an interval, say $\varepsilon_c \in [\varepsilon_-, \varepsilon_+]$ such that if $\varepsilon_c \leqslant \varepsilon_-$, then both lower and upper bounding tori (with frequencies ω_+) exist;

[16] In Celletti et al. (2004), also more realistic models, like those in which Jupiter moves on an eccentric orbit or where the relative inclination of Jupiter and of the asteroid is not neglected, have been considered.

whenever $\varepsilon_c \geqslant \varepsilon_+$ we have numerical evidence of the disappearance of both tori. Due to the topology of the model (compare Remark 2.1), for $\varepsilon_c \leqslant \varepsilon_-$ the motion of the asteroid is confined on the given energy level between the two bounding tori. We also provide an intermediate value at which one of the two tori still survives. The results provided in Table 1 suggest that the truncated model provides a good approximation of the complete model, at least as far as the above energy level and frequencies are considered.

References

Arnold, V.I.: Small denominators and problems of stability of motion in classical and celestial mechanics. Uspehi Mat. Nauk. **18**(6) (114), 91–192 (1963)

Biasco, L., Chierchia, L., Valdinoci, E.: Elliptic two-dimensional invariant tori for the planetary three-body problem. Arch. Rational Mech. Anal. **170**, 91–135 (2003)

Biasco, L., Chierchia, L., Valdinoci, E.: N-dimensional invariant tori for the planar $(N+1)$-body problem. SIAM J. Math. Anal. **37**, 1560–1580 (Preprint downloadable in http://www.mat.uniroma3.it/users/chierchia/WWW/english_version.html). (2006)

Bost, J.B.: Tores invariants des systèmes dynamiques hamiltoniens', (d'aprés Kolmogorov, Arnol'd, Moser, Rüssmann, Zehnder, Herman, Pöschel,...) Sémin. Bourbaki, **1984/85**, (Astérisque 133–134), 113–157 (1986)

Celletti, A.: Analysis of resonances in the spin-orbit problem in celestial mechanics: the synchronous resonance (Part I). J. Appl. Math. Phys. (ZAMP). **41**, 174–204 (1990a)

Celletti, A.: Analysis of resonances in the spin-orbit problem in celestial mechanics: higher order resonances and some numerical experiments (Part II). J. Appl. Math. Phys (ZAMP). **41**, 453–479 (1990b)

Celletti, A.: Construction of librational invariant tori in the spin-orbit problem. J. Appl. Math. Phys. (ZAMP). **45**, 61–80 (1993)

Celletti, A., Chierchia, L.: Rigorous estimates for a computer-assisted KAM theory. J. Math. Phys. **28** 2078–2086 (1987)

Celletti, A., Chierchia, L.: Construction of analytic KAM surfaces and effective stability bounds. Commun. Math. Phys. **118**, 119–161 (1988)

Celletti, A., Chierchia, L.: A constructive theory of Lagrangian tori and computer-assisted applications. In: Jones, C.K.R.T., Kirchgraber, U., Walther, H.O. (eds.), Dynamics Reported. vol. 4 Springer-Verlag, Berlin, (New Series), pp. 60–129 (1995)

Celletti, A., Chierchia, L.: On the stability of realistic three-body problems. Commun. Math. Phys. **186**, 413–449 (1997)

Celletti, A., Chierchia, L.: KAM stability and celestial mechanics. Memoirs AMS, to appear. Preprint downloadable in http://www.mat.uniroma3.it/users/chierchia/WWW/english_version.html. (2005)

Celletti, A., Falcolini, C.: Construction of invariant tori for the spin-orbit problem in the Mercury–Sun system. Celestial Mech. Dyn. Astronom. **53**, 113–127 (1992)

Celletti, A., Falcolini, C., Porzio, A.: Rigorous numerical stability estimates for the existence of KAM tori in a forced pendulum. Ann. Inst. H. Poincaré. **47**, 85–111 (1987)

Celletti, A., Froeschlé, C., Lega, E.: Frequency analysis of the stability of asteroids in the framework of the restricted, three-body problem. Celest. Mech. Dyn. Astr. **90**(3–4), 245–266 (2004)

Celletti, A., Giorgilli, A.: On the numerical optimization of KAM estimates by classical perturbation theory. J. Appl. Math. Phys (ZAMP). **39**, 743–747 (1988)

Celletti, A., Giorgilli, A., Locatelli, U.: Improved estimates on the existence of invariant tori for Hamiltonian systems. Nonlinearity. **13**(2), 397–412 (2000)

Chierchia, L.: KAM theory and celestial mechanics. Encyclopedia Math Phys (to appear). Preprint downloadable in http://www.mat.uniroma3.it/users/chierchia/WWW/english_version.html (2005)

Féjoz, J.: Démonstration du théorème d'Arnold sur la stabilité du système planétaire (d'après Michael Herman). Ergod. Th. Dyn. Sys. **24**, 1–62 (2004)

Herman, M.R.: Démonstration d'un théorème de V.I. Arnold Various seminars at University of Paris. Jussieu and manuscripts. (1998)

Hénon, M.: Explorationes numérique du problème restreint IV: masses egales, orbites non periodique. Bull. Astronom. **3**(1, fasc. 2), 49–66 (1966)
Kolmogorov, A.N.: On the conservation of conditionally periodic motions under small perturbation of the Hamiltonian. Dokl. Akad. Nauk. SSR. **98**, 527–530 (1954)
Laskar, J.: Frequency analysis for multi–dimensional systems. Global dynamics and diffusion. Physica D. **67**, 257. (1993)
Laskar, J., Froeschlé, C., Celletti, A.: The measure of chaos by the numerical analysis of the fundamental frequencies. Application to the standard mapping. Physica D. **56**, 253–269 (1992)
Laskar, J., Robutel, P.: Stability of the planetary three–body problem. I. Expansion of the planetary Hamiltonian. Celest. Mech. Dyn. Astr. **62**(3), 193–217 (1995)
de la Llave, R., Rana, D.: Accurate strategies for small divisor problems. Bull. Am. Math. Soc. (N. S.) **22**(1), 85–90 (1990)
Locatelli, U., Giorgilli, A.: Invariant tori in the secular motions of the three-body planetary systems. Celest. Mech. Dyn. Astr. **78**, 47–74 (2000)
Locatelli, U., Giorgilli, A.: Construction of the Kolmogorov's normal form for a planetary system. Regular Chaotic Dyn. **10**(2), 153–171 (2005a)
Locatelli, U., Giorgilli, A.: Invariant tori in the Sun–Jupiter–Saturn system, Preprint (2005b)
Pyartli, A.S.: Diophantine approximations on Euclidean submanifolds; (in Russian) Funckcional. Anal. i Prilozen. **3**, 59–62; (in English) Funct. Anal. Appl. **3**, 303–306 (1969)
Rana, D.: Proof of accurate upper and lower bounds to stability domains in small denominator problems. Ph.D. thesis, Princeton University Press, Princeton, NJ (1987)
Robutel, P.: Stability of the planetary three–body problem. II. KAM theory and existence of quasi–periodic motions. Celest. Mech. Dyn. Astr. **62**(3), 219–261 (1995)
Rüssmann, H.: Invariant tori in non–degenerate nearly–integrable Hamiltonian systems, Regul. Chaotic Dyn. **6**, 119–204 (2001)
Siegel, C.L., Moser, J.K.: Lectures on Celestial Mechanics. Springer-Verlag, Berlin (1971)

Celestial Mechanics and Dynamical Astronomy (2006) 95:141–153
DOI 10.1007/s10569-006-9004-2

ORIGINAL ARTICLE

Analysis of the chaotic behaviour of orbits diffusing along the Arnold web

Claude Froeschlé · Elena Lega · Massimiliano Guzzo

Received: 14 November 2005/Accepted: 31 January 2006 /
Published online: 18 March 2006

Abstract In a previous work [Guzzo et al. DCDS B 5, 687–698 (2005)] we have provided numerical evidence of global diffusion occurring in slightly perturbed integrable Hamiltonian systems and symplectic maps. We have shown that even if a system is sufficiently close to be integrable, global diffusion occurs on a set with peculiar topology, the so-called Arnold web, and is qualitatively different from Chirikov diffusion, occurring in more perturbed systems. In the present work we study in more detail the chaotic behaviour of a set of 90 orbits which diffuse on the Arnold web. We find that the largest Lyapunov exponent does not seem to converge for the individual orbits while the mean Lyapunov exponent on the set of 90 orbits does converge. In other words, a kind of average mixing characterizes the diffusion. Moreover, the Local Lyapunov Characteristic Numbers (LLCNs), on individual orbits appear to reflect the different zones of the Arnold web revealed by the Fast Lyapunov Indicator. Finally, using the LLCNs we study the ergodicity of the chaotic part of the Arnold web.

Keywords KAM and Nekhoroshev theorem · Arnold's diffusion · Chaos detection

1. Introduction

The characterization of mechanisms for diffusion of orbits in quasi-integrable Hamiltonian systems and symplectic maps is a relevant topic for many fields of physics, such as celestial mechanics, dynamical astronomy (Vergassola 1998, Contopoulos 2002; Morbidelli 2002), statistical physics, plasma physics and particle accelerators. Any dynamical state of an integrable system can be characterized with a set of action-angle conjugate variables where the actions are constants of motion, while the angles simply change linearly with time. The properties of the system which are relevant for the stability are determined by the actions.

C. Froeschlé · E. Lega (✉)
Observatoire de la Côte d'Azur, B.P. 4229, 06304 Nice Cedex 4, France
e-mails: claude@obs-nice.fr; elena@obs-nice.fr

M. Guzzo
Dipartimento di Matematica pura e Applicata Università di Padova, Via Belzoni 7, 35131 Padova, Italy
e-mail: guzzo@math.unipd.it

Small perturbations of integrable systems can produce a slow drift of the initial value of the actions, and after certain time these drifts can cumulate in such a way to drive the orbit far away its initial value.

In 1979 Chirikov described (Chirikov 1979) a possible model for global drift valid when the perturbation is greater than some critical value. One of the reasons of the broad detection of the Chirikov diffusion is that its typical times fall within the simulation abilities of modern computers as far back as the seventies.

For smaller perturbations the systems fall within the range of celebrated perturbation theories such as KAM (Kolmogorov 1954; Moser 1958; Arnold 1963) and Nekhoroshev (1977) theorems which leave the possibility for global drift only on a subset of the possible dynamical states with peculiar topology, the so-called Arnold web, and force diffusion times to be at least exponentially long with an inverse power of the norm of the perturbation. Following Nekhoroshev (1977), we call Arnold diffusion a diffusion occurring on the Arnold web of a system which satisfies the hypotheses of both KAM and Nekhoroshev theorems.

The theoretical possibility of Arnold diffusion has been first shown by Arnold (1964) for a specific system.

Being interested to applications to specific systems, and in particular to systems of interest for physics, we have used a numerical approach (Lega et al. 2003; Guzzo et al. 2005) which, avoiding theoretical difficulties, measures directly the quantitative features of long-term diffusion. More precisely, we consider systems satisfying the hypothesis of both KAM and Nekhoroshev theorems and small values of the perturbation parameter so that both theorems apply. Therefore, beyond estimates of diffusion and instability we provide a direct numerical investigation of the geometry of the resonances of the system (Froeschlé et al. 2000) which clearly allows to identify the dynamical regime in which the diffusion is detected.

Following this strategy, we detected (Lega et al. 2003) a very slow local diffusion confined to the Arnold web of a system which satisfies both the hypotheses of KAM and Nekhoroshev theorems. In that work, we estimated the diffusion coefficient to decrease faster than a power law of the perturbing parameter, consistently with the exponential estimates provided by the Nekhoroshev theorem. More recently, we have provided evidence that indeed those orbits diffuse in macroscopic domains of the action space (Froeschlé et al. 2005; Guzzo et al. 2005).

In the present paper we analyse the chaotic behaviour of such orbits. We show that the largest Lyapunov exponent does not seem to converge for the individual orbits while the mean Lyapunov exponent on a set of 90 orbits does converge. We have computed the Local Lyapunov Characteristic Numbers (Froeschlé et al. 1993, LLCNs hereafter)[1] on individual orbits showing that their values reflect the different zones of the Arnold web.

Considering an ergodic approach, we test the equivalence between spatial and temporal averages in the chaotic part of the Arnold web. Of course, the ergodic hypothesis is trivially not valid if we consider generic open sets in the phase space, because of the existence of the invariant tori. Instead the ergodic hypothesis could be potentially true if we restrict it to the subset of the phase space with Fast Lyapunov Indicator (FLI hereafter,introduced in Froeschlé et al. 1997, defined here in Section 2, Eq. 3) higher that the FLI which characterizes the KAM tori. Even in this case, what we find is that the equivalence between spatial and temporal averages is not so evident even for large values of the perturbation parameter.

In Section 2 we describe the geometry of the Arnold web of symplectic maps and we recall the numerical method used to detect the Arnold web and the global diffusion on it. We report in Section 3 the computation of the Largest Lyapunov Exponent of orbits diffusing on

[1] Sometimes called stretching numbers (Voglis and Contopoulos 1994).

the Arnold web and we recall the definition of the Local Lyapunov Characteristic Numbers showing how they are related to the geography of the resonances.

In Section 4 we look for the application of the ergodic hypothesis for different values of ϵ. Conclusion is provided in Section 5.

2. Numerical detection of the Arnold web

We consider the following quasi-integrable symplectic map:

$$\begin{cases} \phi_1' = \phi_1 + I_1, & \phi_2' = \phi_2 + I_2 \\ I_1' = I_1 + \epsilon \dfrac{\partial f}{\partial \phi_1}(\phi_1 + I_1, \phi_2 + I_2), & I_2' = I_2 + \epsilon \dfrac{\partial f}{\partial \phi_2}(\phi_1 + I_1, \phi_2 + I_2) \end{cases} \tag{1}$$

where $f = 1/(\cos(\phi_1) + \cos(\phi_2) + 4)$. At small ϵ, the KAM and Nekhoroshev theorems (Kuksin 1993; Kuksin and Pöschel 1994; Guzzo 2004) apply to this kind of maps. The resonances of this system are defined by the straight lines $k_1 I_1 + k_2 I_2 + 2\pi k_3 = 0$, with $k_1, k_2, k_3 \in \mathbb{Z}\backslash 0$. To describe the topology of the Arnold web, it is convenient to refer to the subset of the phase space determined by the section $S = \{(I_1, I_2) \in \mathbb{R}^2, \phi_i = 0, i = 1, 2\}$. From the KAM theorem, it follows that any invariant torus cuts the section S in only one point (I_1, I_2). Moreover, the Diophantine condition on the frequencies implies that these points are outside a neighbourhood of lines $k_1 I_1 + k_2 I_2 + 2\pi k_3 = 0$ proportional to $\gamma/|k|^\tau$, for any $k = (k_1, k_2, k_3) \in \mathbb{Z}^3\backslash 0$. Therefore, the intersection between the Arnold web and the section S consists of all these straight lines with a neighbourhood which decreases as $|k|$ increases. Any of these straight lines is called resonance, and the integer $|k|$ is called resonance order. The Arnold web is open, dense, and has a small relative measure (proportional to γ).

A precise numerical detection of the Arnold web is possible with the so-called FLI method (Froeschlé et al. 1997, 2000).

The FLI is related to the computation of the tangent map for a given choice of an initial tangent vector and was first introduced to detect weak chaos (Froeschlé et al. 1997).

We recall that given a set of differential equations:

$$\frac{d}{dt}\vec{X} = \vec{F}(\vec{X}), \quad \vec{X} = (x_1, x_2, \ldots, x_n) \tag{2}$$

under some suitable regularity conditions, the evolution $\vec{v}(t)$ of any vector $\vec{v}(0) \in R^n$ is obtained by integrating the variational equations:

$$\frac{d\vec{v}}{dt} = \left(\frac{\partial \vec{F}}{\partial X}\right)\vec{v}.$$

The largest Lyapunov exponent is defined in such a way that, unless $\vec{v}(0)$ belongs to some lower dimensional linear spaces, the quantity $\ln ||\vec{v}(t)||/t$ tends to it as t goes to infinity. Of course numerically we work on finite times and we call Lyapunov Characteristic Indicator (LCI hereafter) the quantity $\ln ||\vec{v}(t)||/t$ at a given time t. If Eq. (2) is Hamiltonian and if the motion is regular (except for some peculiar hyperbolic structures, such as whiskered tori) then the largest Lyapunov exponent is zero, otherwise it is positive. This property has been largely used to discriminate between chaotic and ordered motion. However, among regular motions the Lyapunov exponent does not distinguish between KAM tori and regular resonant motions.

We recall that the FLI (Froeschlé et al. 2000; Guzzo et al. 2002) is defined as:

$$\mathrm{FLI}(t) = \sup_{0<k<t} \log \|v(k)\| \tag{3}$$

As detailed in (Guzzo et al. 2002) the FLI distinguishes between resonant and non-resonant regular motions.

In (Guzzo et al. 2005) we showed that the Arnold web of the above mapping indeed corresponds to the above theoretical description. Moreover, we have shown how to choose initial conditions which are good candidates to diffuse thanks to the accurate detection of the web. In fact, within resonances, both chaotic and regular motions are observed. Of course, regular motions do not diffuse, and therefore diffusive orbits must be chosen in the small subset of the Arnold web made of chaotic motions. We are aware that the set of chaotic orbits selected (having FLI larger than the reference value for the tori) could include periodic orbits with very large periods which formally do not diffuse but numerically seem to diffuse. Anyway the measure of such orbits is close to zero.

Moreover, even within chaotic motions, to have the chance to observe the phenomenon one has to restrict the choice of initial conditions to the single order resonances, i.e. to the portion of resonant lines which are far from the main crossing between resonances. The reason is that stability times are expected to be much longer up to a distance of order $\sqrt{\epsilon}$ from these crossings (see, for example, Pöschel 1993). With this selection of initial conditions, we have been able to show, in Lega et al. (2003), that indeed after times that increase faster than power law, initial conditions diffuse along the chaotic border of resonant lines. Though the computational time used in Lega et al. (2003) was already quite long, due to the slowness of Arnold diffusion, it was not sufficient to evidence a global instability and diffusion, i.e. the wandering of motions from one resonance to the others. Thanks to a much longer computational time, and the use of a suitable symplectic map, we have displayed in (Froeschlé et al. 2005; Guzzo et al. 2005) for the first time a global diffusion phenomenon.

Figure 1 shows the Arnold web detected computing the FLI of the mapping 1 for $\epsilon = 0.6$ on a set of 500×500 initial conditions taken on section S and regularly spaced in I_1, I_2.

In the plot, we have associated to each point of the action plane (I_1, I_2) a colour corresponding to the FLI value according to the colour scale reported below the figure.[2] The intermediate value of the FLI (grey in the colour scale) corresponds to KAM tori. Darker regions correspond to initial conditions (I_1, I_2) on the section S which produce resonant regular motions, while light grey regions correspond to initial conditions which produce chaotic motions. Each resonance appears as a darker grey region with a light grey border, or as a single light grey line (depending on the resonance and on the value of the angles chosen to define the section S; see Froeschlé et al. 2000 for more details). Therefore, the light grey region on the FLI map corresponds to the chaotic subset of the Arnold web, where an eventual Arnold diffusion should be confined.

Following the explained criterion of choice of the initial conditions we have repeated the computation of Guzzo et al. (2005) on a larger set (90 instead of 20) of initial conditions near $(I_1, I_2) = (2.35, 1.07)$. Then, we computed numerically the map up to 2×10^{10} iterations. The results are reported in the four panels of Fig. 1: on the FLI map of the action plane (I_1, I_2) we plotted as black dots all points of the orbits which have returned after some time on the section S. Of course, because computed orbits are discrete we represented points on the double section $|\phi_1| \le 0.005$, $|\phi_2| \le 0.005$; reducing the tolerance 0.005 reduces only the number of points on the section, but does not change their diffusion properties. Figure 1a

[2] Colour pictures can be found at: http://www.obs-nice.fr/elena/Global_diffu/images.html

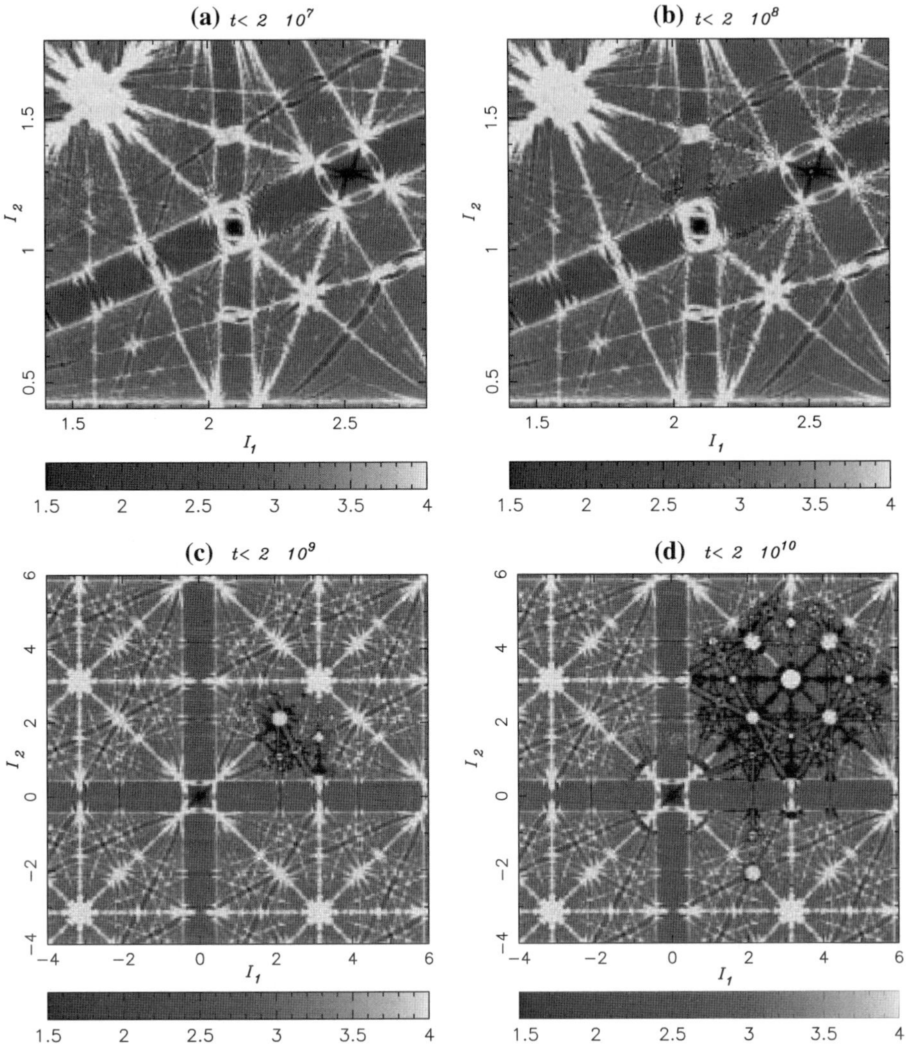

Fig. 1 The four panels correspond to the FLI map of the action plane (I_1, I_2) for the map (1), with initial condition on the section S, with different magnifications. The light grey region corresponds to the chaotic part of the Arnold web. On panel (**a, b, c, d**) we mark with a black dot all points of the 90 orbits which have returned after some time on the section S. We consider 2×10^7 iterations for panel (**a**); 2×10^8 iterations for panel (**b**), 2×10^9 iterations for panel (**c**) and 2×10^{10} iterations for panel (**d**)

shows the local diffusion along the resonance $I_1 = 2I_2$, while Fig. 1b shows the result after the intermediate time 2×10^8. We observe that the orbits diffuse also along resonant lines different from the initial one.

Figure 1c, d show the result after much longer times. To properly display such long-term evolutions we used a zoomed out map of the action plane.

We observe that the orbits filled a macroscopic region of the action plane whose structure is clearly that of the Arnold web. The orbits have moved along the single resonances, and avoided the centre of the main resonance crossings, in agreement with the theoretical results which predict longer stability times for motions in these regions. The larger resonances

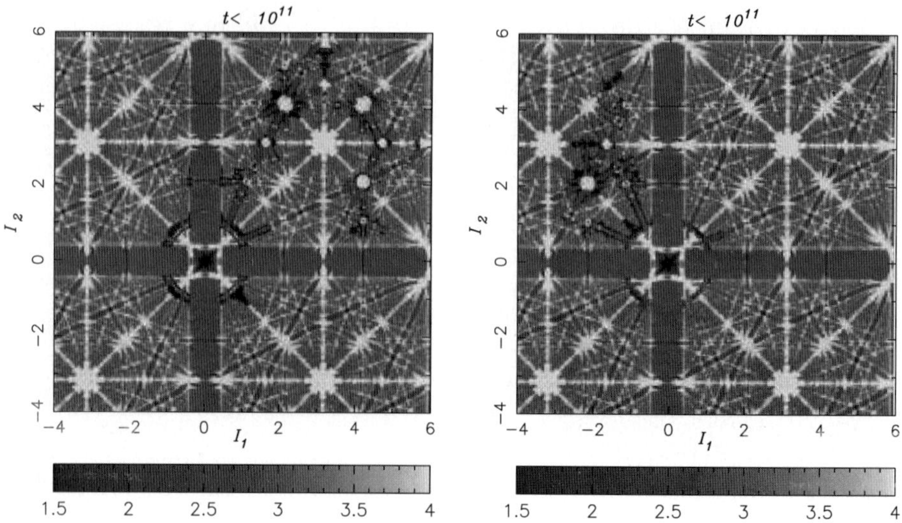

Fig. 2 The two panels correspond to the FLI map of the action plane (I_1, I_2) for the map (2), with initial condition on the section S. The light grey region corresponds to the chaotic part of the Arnold web. The different panels correspond to the intersections with section S up to a time $t = 10^{11}$ of two orbits with initial conditions on the resonant line $I_1 = 2I_2$. More precisely, left panel represent the evolution of one orbit with initial conditions $I_1(0) = 1.71640$ and $I_2(0) = 0.8160$, while the right panel shows the evolution of one orbit with initial conditions $I_1(0) = 1.71643$, $I_2(0) = 0.8159$. Let us remark that although the initial conditions are very close the two orbits undergo completely different paths

(which correspond to the smallest orders $|k|$) are practically all visited, while this is not the case for the thinnest ones (which correspond to the highest orders $|k|$). This is in agreement with the theoretical results of Morbidelli and Giorgilli (1995b), which predict that the speed of diffusion on each resonance becomes smaller for resonances of high order. Therefore, the possibility of visiting all possible resonances is necessarily limited by finite computational time.

In Fig. 2 we plotted the orbits corresponding to two very close initial conditions. The figure allows to appreciate that, although the differences in the initial actions is only 10^{-4}, the two orbits undergo completely different paths. In order to estimate the precision of the calculation we have integrated over 10^{11} an initial condition corresponding to a KAM torus finding that the supremum of the variation of the actions, on the double section S, with respect to their initial value is of the order of 10^{-4}. Moreover, the integration of chaotic orbits of a 2 degree of freedom Hamiltonian did not show a diffusion (E. Lega 2002, Thèse d'Habilitation à diriger des recherches) violating the conservation of energy.

3. The largest Lyapunov exponent

A classical way to quantify the strength of chaos is to compute the largest Lyapunov characteristic exponent (LCE hereafter). It is well known that the LCE is defined as:

$$\lambda = \lim_{N\to\infty} \frac{1}{Nk} \sum_{n=1}^{N} \log\left(\frac{||v(nk)||}{||v((n-1)k)||}\right). \tag{4}$$

In the practical computation of λ, the vector v is renormalized every k iterations.

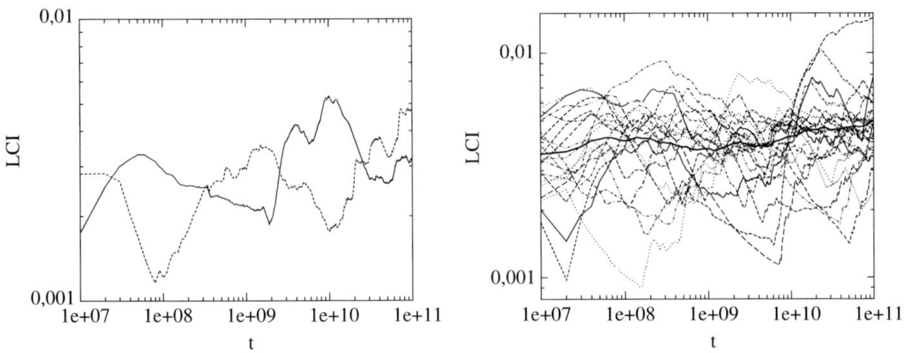

Fig. 3 Evolution with time of the LCI of the two orbits of Fig. 2 (left panel). Evolution with time of the LCI of 20 of the 90 orbits of Fig. 1 (right panel). The more regular curve (thick line) is the evolution with time of mean value of the 90 LCI

Let us also recall the definition (Froeschlé et al. 1993) of the LLCNs as the set of numbers

$$\alpha(n) = \frac{1}{k} \log \left(\frac{||v(nk)||}{||v((n-1)k)||} \right)$$

For a given interval of time the mean of the LLCNs is nothing but the LCI whose limit when the time goes to infinity is the LCE.

Figure 3, (left) shows the evolution with time of the LCI of the 2 orbits of Fig. 2. We clearly observe that not only the two LCI are different but also they do not show clear convergence up to $t = 10^{11}$ iterations. However, Fig. 3 (right) showing the evolution of the LCI of 20 orbits within the set of 90 orbits of Fig. 1 displays excursions of the LCIs within a constant interval $[10^{-3}, 10^{-2}]$ for t going from 10^7 to 10^{11}. This is corroborated by the fact that the mean value of the 90 LCIs shows convergence (Fig. 3, right thick line). Moreover, let us remark that the quantity $\log ||v(t)||/t$, when computed on a single stochastic region, converges to the largest Lyapunov Exponent in a time called the Lyapunov time; while it does not converge (even on a very large interval of time: $T = 10^{11}$ iterations, Fig. 3) when the diffusion occurs on the Arnold web. However, Fig. 3 (right) shows that the mean on a set of N orbits converges on a time shorter than $T/N \simeq 10^9$. Therefore, in order to compute the LCI one has to do it on a set of orbits instead that on a single one.

3.1. On the relationship between LLCNs and phase space

When computing the LCI only the mean value of the tangent vectors is considered while the distribution of such vectors can give hints about the history of the orbit. In order to explore the complex chaotic evolution of an orbit we consider first the distribution of the LLCNs of the orbit of Fig. 2 (left).

Figure 4 shows the distribution of the LLCNs for this orbit. Let us remark that, for a KAM torus at the same value of the perturbation parameter $\epsilon = 0.6$ the LLCNs range in the interval $[-5 \times 10^{-5}, 5 \times 10^{-5}]$. For the chaotic orbit the majority of the LLCNs range from $[2 \times 10^{-5}, 10^{-3}]$ and a queue of larger but rare values goes up to 1.75×10^{-2}. In order to see if there is a relation between the LLCNs and the motion of the orbit we have associated to the points of the orbit intersecting S the corresponding LLCNs and plotted in Fig. 5, (left) the points having LLCNs < 0.005 and in Fig. 5 (right) the points having LLCNs ≥ 0.005. It appears clearly that the larger values of the LLCNs are associated to the neighbourhood of

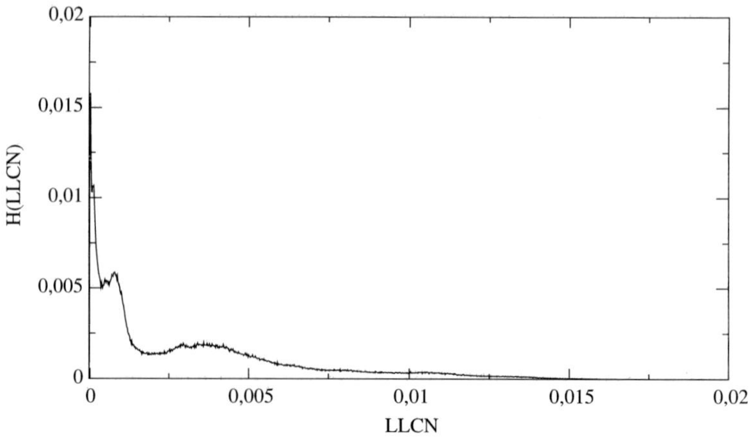

Fig. 4 Distribution of the LLCNS of the orbit of Fig. 2(left)

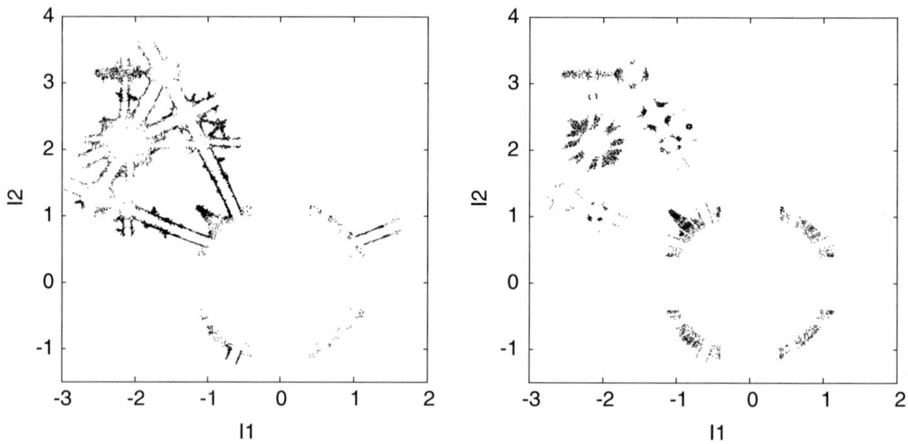

Fig. 5 In the two panel the intersections of the point of the orbit of Fig. 2 (left) with the section S are plotted. More precisely the points intersecting S and having LLCN < 0.005 are plotted in left panel the others in the right panel

resonance crossing while the lower values correspond to resonant lines. The great percentage of low LLCNs values is significative of the long time of residence on the resonant lines. The distribution gives informations about the different regions of the phase space visited by the orbits.

4. An ergodic approach

The aim of this section is to revisit the old idea that in a chaotic zone all orbits have the same LCE (Contopoulos 2002, p. 255) and that it can be computed using a set of different orbits. At this purpose, we have considered a set of 100 orbits with initial conditions in the chaotic domain taken at random in the interval $0 < I_1 < \pi$, $0 < I_2 < \pi$ ($\phi_1 = 0, \phi_2 = 0$) and we have integrated them for 10^8 iterations computing the LLCNs every 10^4 iterations.

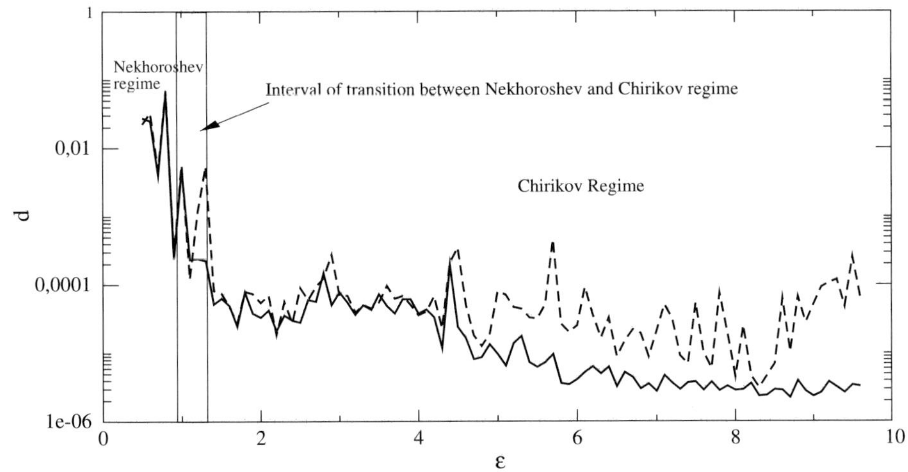

Fig. 6 Variation of the distance d as a function of ϵ. All points of data set 1 and data set 2 are considered (dashed line). Only 90 orbits of data set 1 and 90 sets of data set 2 having similar mean values of the LLCNs have been analysed (continuous line). The interval $0.9 < \epsilon < 1.3$ correspond to the transition between the Nekhoroshev and the Chirikov regime

Considering an ergodic approach, i.e. the well-known hypothesis which states the equivalence between spatial and temporal means, we have integrated one orbit of the previous set for 10^{10} iterations computing the LLCNs every 10^4 iterations.

Since the distribution of the LLCNs are closely related to the detailed chaotic history of an orbit (Fig. 5) we consider the distance between the distributions of the LLCNs as a good indicator of ergodicity. More precisely, we introduce the chi-squared distance:

$$d = \sum_{h=1}^{M} (P_h - \tilde{P}_h)^2, \tag{5}$$

where P_h is the percentage of LLCN's in the bin h, $h = 1, \ldots, 1000$ of the distribution of the LLCNs of the 100 orbits (Data set 1 in the following) and $\tilde{P}_h$ is the same as P_h for the single orbit (Data set 2 in the following).

Figure 6 (dashed line) shows the variation of d as a function of the perturbation parameter ϵ.

The distance, although noisy, drops from 10^{-2} to about 10^{-4} for ϵ going from 0.5 to 1.3. For greater values of ϵ, when the system is in Chirikov regime, the distance shows noisy variations around 5×10^{-5} up to $\epsilon = 9.8$. The result for large value of the perturbation parameter is puzzling. For large values of ϵ we have large chaotic domains and we expected that the orbits of the two data sets had the same chaotic properties. To clarify the situation we have focused the attention on one case, let's say $\epsilon = 9.5$. When looking at the 2 distribution for such value of ϵ, Fig. 7 shows that the difference takes place at very low values of the LLCNs.

Following the same approach than in Fig. 5 we plot the points of section S having LLCNs larger than 0.01 in Fig. 8 (left); and those having LLCNs lower than 0.01 in Fig. 8 (right) for the orbit of data set 2.

At first glance, in opposition to the previous case, no difference appears between the two sets of points. More precisely, there is a difference in the density distribution of the points but no structures appear neither in the left nor in the right picture, mainly because of the large domain on which the fast Chirikov diffusion takes place.

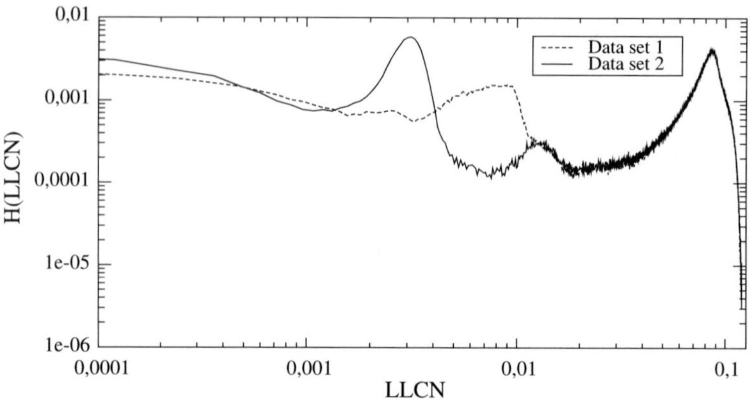

Fig. 7 Histogram of the distribution of the LLCNs for data set 1 (100 orbits integrated for 10^8 iterations) and data set 2 (1 single orbit integrated for 10^{10} iterations)

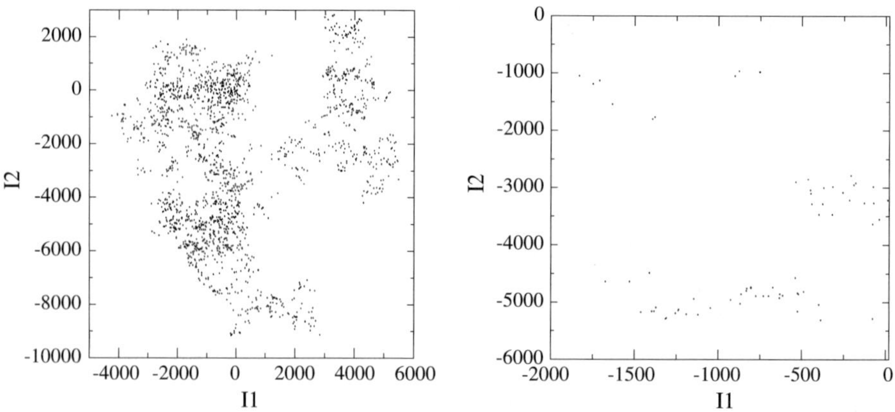

Fig. 8 In the two panel the intersections of the point of the orbit of data set 2 with the section S are plotted. More precisely the points intersecting S and having LLCN < 0.01 are plotted in left panel the others in the right panel

The computation of the FLI in a small portion of the phase space (indicated by a square in Fig. 8 (left), allows to show the presence of some regular orbits. We observe that the points having low values of the LLCNs (superimposed in black in Fig. 9) lie in a close neighbourhood of regular orbits.

This phenomenon, called sticking, has been extensively studied (Contopoulos 2002 and references therein). In addition, we observe here an orbit which does not start in a sticking region but is trapped as time goes on. For two-dimensional maps this is more rare. According to the theory (Morbidelli and Giorgilli 1995a) the time of capture in these regions may be very long (we have found $\simeq 1/10$ of the total integration time for the analysed orbit). One way of selecting the sticked orbits is to consider the mean values of the LLCNs of the 100 orbits (data set 1) and of the 100 sets of 10^8 iterations of the single orbit (data set 2) as shown in Fig. 10 for $\epsilon = 9.5$.

It appears that the 90% of the mean values of the LLCNs are very similar while the remaining 10% are up to an order of magnitude lower.

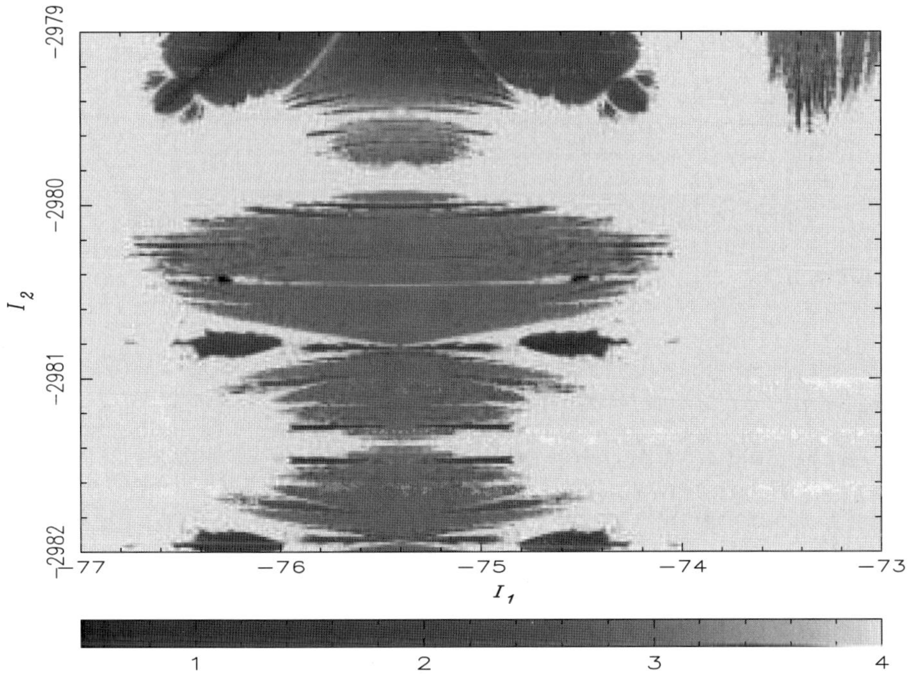

Fig. 9 FLI chart of the zone of the phase space within the box of Fig. 8. The points of the orbit of data set 2 intersecting the section S and having LLCN < 0.01 are plotted in black. They appear as two spots at about $I_2 \simeq -2980.5$ lying in a close neighbourhood of regular orbits

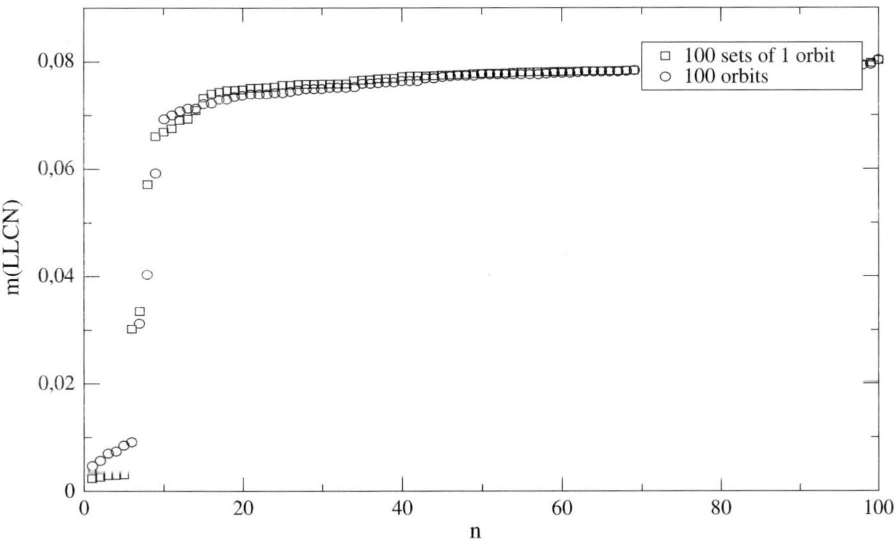

Fig. 10 Mean values, ordered from minimum to maximum value, of the distributions of the LLCNs of the 100 orbits (data set 1) and of the 100 sets of 10^8 iterations of the single orbit (data set 2) for $\epsilon = 9.5$

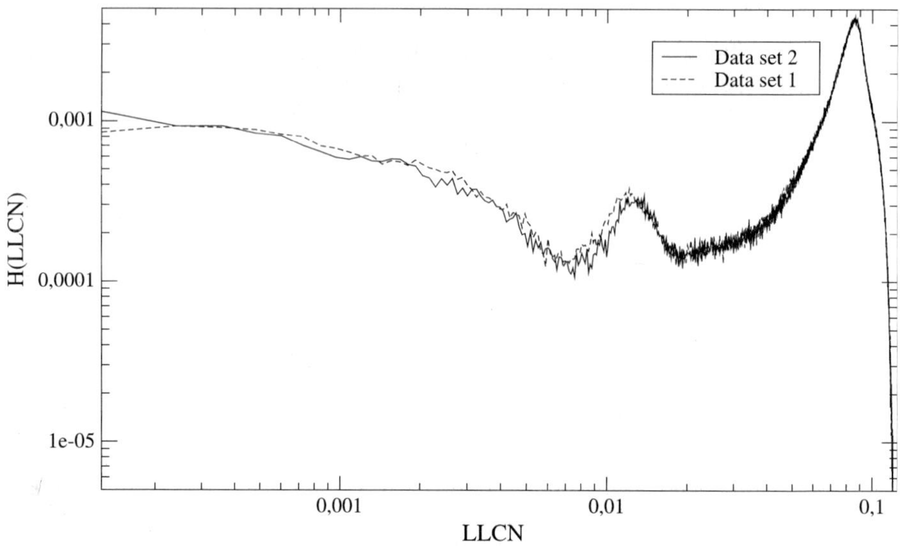

Fig. 11 Histogram of the distribution of the LLCNs of data set 1 (90 orbits selected as explained in the text) and of data set 2 (90 sets of the single orbit selected as explained in the text)

Let us recall that the mean of the LLCNs is nothing but the LCI and we obtain very similar values when considering respectively data set 1 and 2. However, if we consider the distribution of the LLCNs the discrepancies can be removed getting rid of the sticking parts.

The distribution of the 2 data sets containing the 90 orbits though selected appear much more similar (Fig. 11).

The same procedure applied to the whole set of values of ϵ shows that, at least for large values of ϵ the noise is reduced and the values of d have decreased (Fig. 6, continuous line). Finally, for low values of ϵ (Nekhoroshev regime and local Chirikov regime) we observe that even when removing the orbits, as explained above, the distance between the two sets does not change.

This is a consequence of the existence, for such low values of ϵ of the Arnold web structure. More precisely, some initial conditions in the crossing of resonances do not diffuse.

According to the theory (Pöschel 1993) such regions have stability times longer than in the single resonance.

Of course such results are time depending, some orbits which show intermittent phenomena with long time sticking may finally appear ergodic for a much larger time interval. We have shown that for this mapping two phenomena of complete different nature give the same macroscopic results as far as diffusion is concerned. On one side, we have the well-known sticking phenomena extensively studied for two-dimensional mappings and on the other side the trapping into effective stable chaotic crossing. The first phenomena is characterized by very low values of the LLCNs while the second one exhibits relatively large LLCNs. For orbits, i.e. region of the phase space which do not have such peculiar geometry, we have found an ergodic-like behaviour.

5. Conclusion

Using the FLI method, we have recalled some previous results showing evidence of global diffusion occurring in slightly perturbed systems.

In the present paper we have studied in more detail the chaotic behaviour of orbits diffusing on the Arnold web. We have shown that the LCI of such orbits does not converge on a quite long interval of time (10^{11}) iterations. However, the mean value of the LCI of the 90 explored orbits does converge showing a kind of average mixing.

Thanks to the computation of the LLCNs we were able to associate the LLCNs to geometrically well-defined regions of the phase space. Moreover, we have tested the ergodicity of the system for increasing values of the perturbation parameter. We have shown that the existence of the Arnold web structure prevents ergodicity on the large but finite time $t = 10^{10}$ iterations. Conversely, for large values of the perturbation parameter, temporal and spatial means are more close, although the existence of some sticking phenomena.

References

Arnold, V.I.: Proof of a theorem by A.N. Kolmogorov on the invariance of quasi-periodic motions under small perturbations of the Hamiltonian. Russ. Math. Surveys **18**, 9–36 (1963)

Arnold, V.I.: Instability of dynamical systems with several degrees of freedom. Sov. Math. Dokl **6**, 581–585 (1964)

Contopoulos, G.: Order and Chaos in Dynamical Astronomy. Springer-Verlag, Berlin, Heidelberg (2002)

Chirikov, B.V.: An universal instability of many dimensional oscillator system. Phys. Rep. **52**, 263–379 (1979)

Froeschlé, C. Froeschlé, Ch., Lohinger, E.: Generalized Lyapunov characteristics indicators and corresponding like entropy of the standard mapping, Celest. Mech. Dynam. Astron. **56**, 307–315 (1993)

Froeschlé, C., Guzzo, M., Lega, E.: Graphical evolution of the Arnold web: from order to chaos. Science **289**, 2108–2110 (2000)

Froeschlé, C., Lega, E., Gonczi, R.: Fast Lyapunov indicators. Application to asteroidal motion. Celest. Mech. Dynam. Astron. **67**, 41–62 (1997)

Froeschlé, C., Guzzo, M., Lega, E.: Local and global diffusion along resonant lines in discrete quasi-integrable dynamical systems. Celest. Mech. Dynam. Astron. **92**, 243–255 (2005)

Guzzo, M.: A direct proof of the Nekhoroshev theorem for nearly integrable symplectic maps. Ann. Henry Poincaré **5**, 1013–1039 (2004)

Guzzo, M., Lega, E., Froeschlé, C.: On the numerical detection of the effective stability of chaotic motions in quasi-integrable systems. Physica D **163**, 1–25 (2002)

Guzzo, M., Lega, E., Froeschlé, C.: First numerical evidence of global Arnold diffusion in quasi-integrable systems. DCDS-B, **5**, 687–698 (2005)

Kolmogorov, A.N.: On the conservation of conditionally periodic motions under small perturbation of the Hamiltonian. Dokl. Akad. Nauk SSSR **98**, 527 (1954)

Kuksin, S. B.: On the inclusion of an almost integrable analytic symplectomorphism into a Hamiltonian flow. Russ. J. Math. Phys. **1**, 191–207 (1993)

Kuksin S.B., Pöschel, J.: On the inclusion of analytic symplectic maps in analytic Hamiltonian flows and its applications. Nonlinear Diff. Eq. Appl. **12**, 96–116 (1994)

Lega, E., Guzzo, M., Froeschlé, C.: Detection of Arnold diffusion in Hamiltonian systems. Physica D **182**:179–187 (2003)

Morbidelli, A., Giorgilli, A.: Super-exponential stability of KAM tori. J. Stat. Phys. **78**, 1607 (1995a)

Morbidelli, A.: Modern Celestial Mechanics. Aspects of Solar System Dynamics. Taylor and Francis (2002)

Morbidelli, A., Giorgilli, A.: On a connection between KAM and Nekhoroshev's theorems. Physica D **86**, 514–516 (1995b)

Moser, J.: On invariant curves of area-preserving maps of an annulus. Comm. Pure Appl. Math. **11**, 81–114 (1958)

Nekhoroshev, N.N.: Exponential estimates of the stability time of near-integrable Hamiltonian systems. Russ. Math. Surveys **32**, 1–65 (1977)

Poschel, J.: Nekhoroshev estimates for quasi-convex Hamiltonian systems. Math. Z. **213**, 187 (1993)

Vergassola, M.: Standard and anomalous diffusion in dynamical systems. In: Benest, D., Froeschlé, C. (eds.), Analysis and Modeling of Discrete Dynamical Systems. Gordon and Breach Science Publishers (1998)

Voglis, N., Contopoulos, G.: Invariant spectra of orbits in dynamical systems. J. Phys. A: Math. Gen. **27**, 4899–4909 (1994)

Celestial Mechanics and Dynamical Astronomy (2006) 95:155–171
DOI 10.1007/s10569-006-9010-4

ORIGINAL ARTICLE

The scattering map in the planar restricted three body problem

E. Canalias · A. Delshams · J. J. Masdemont · P. Roldán

Received: 15 November 2005 /Accepted: 31 January 2006 /
Published online: 15 August 2006

Abstract We study homoclinic transport to Lyapunov orbits around a collinear libration point in the planar restricted three body problem. A method to compute homoclinic orbits is first described. Then we introduce the scattering map for this problem (defined on a suitable normally hyperbolic invariant manifold) and we show how to compute it using the information already obtained for the homoclinic orbits. An example application to Astrodynamics is also proposed.

Keywords Restricted three body problem · Homoclinic orbits · Normally hyperbolic invariant manifolds · Scattering map

1 Introduction

Interest in the dynamics around the Lagrangian points has been increasing in the last decades, as L_1 and L_2 of the Sun–Earth and Earth–Moon systems have been selected for a wide variety of spatial missions (see Farquhar 1972; Breakwell et al. 1974; Gómez et al. 1991; Canalias and Masdement 2005). The introduction of invariant manifolds as a means to describe the phase space around these points yields not only a more efficient determination of the desired transfers but also an adaptable procedure for mission analysis. Invariant manifolds dominate the dynamics and mass transport in the studied system and in the whole Solar System. In this context, homoclinic trajectories provide zero cost transfers from a libration orbit around L_i $(i = 1, 2)$ to another one. In the planar case, there is only one periodic motion around L_1. In particular, homoclinic trajectories are not a means to go from one orbit to another one, but to leave the periodic orbit at a certain point and approach it again in a different point. This feature has possible astrodynamical applications, such as the placement of different satellites in previously chosen points of the same orbit.

E. Canalias · A. Delshams · J. J. Masdemont · P. Roldán (✉)
Departament de Matemàtica Aplicada I,
ETSEIB-UPC, Avda. Diagonal 647, 08028 Barcelona, Spain
e-mail: Pablo.Roldan@upc.edu

This paper includes a detailed explanation of the numerical method used for finding homoclinic connections by intersecting the stable and unstable manifold of planar Lyapunov orbits around L_1, which was presented for the first time in Canalias and Masdemont (2006).

Further, we look at the set of Lyapunov orbits around a collinear libration point as a normally hyperbolic invariant manifold Λ, with associated stable and unstable invariant manifolds W^s_Λ, W^u_Λ. These manifolds are important because they organize the dynamics of the system. The hyperbolic invariant manifolds W^s_Λ, W^u_Λ intersect transversally along families of homoclinic orbits $H^{j,i}$.

Following Delshams et al. (2000b, 2006), we define the scattering map $S^{j,i}\colon \Lambda \mapsto \Lambda$ associated to a family of homoclinic orbits. The scattering map relates two trajectories in Λ through a homoclinic orbit in the family $H^{j,i}$.

Exploiting the fact that there is only one Lyapunov orbit around the libration point for each value of the Jacobi constant, we show that the scattering map for the planar restricted three body problem (PRTBP) has a simple form. Specifically, it is an integrable twist map.

Using only three points from each homoclinic orbit in the family $H^{j,i}$, we are able to compute the associated scattering map. For illustration purposes, in this paper we compute the scattering map for several families of homoclinic orbits to the L_1 libration point in the Earth–Moon model.

Our future goal is to give a complete description of homoclinic and heteroclinic phenomena in the spatial restricted three body problem (RTBP). This paper is a first step toward that goal, where we have used the PRTBP as a test case for the concepts and methods involved.

2 Methodology

The RTBP is a model of the motion of an infinitesimal particle, affected by the gravitational attraction of two massive bodies (primaries), which describe circular orbits around their common center of masses. When the movement of the massless particle is only studied in the plane of motion of the primaries, the problem becomes planar (PRTBP).

Taking adequate units of mass, length and time we can simplify the equations of motion. Furthermore, the following synodical system of reference is taken: the origin set at the center of masses of the two primaries, the *X*-axis given by the line that goes from the smallest to the biggest primary (with this orientation) and the Y-axis taken so that *XY* is a planar positively oriented coordinate system. In our synodical system the primary of mass μ (small) is always located at $(\mu - 1, 0)$ and the primary of mass $1-\mu$ at $(\mu,0)$ (Szebehely 1967). The equations of motion are:

$$\begin{aligned} \ddot{x} - 2\dot{y} &= \frac{\partial \Omega}{\partial x}, \\ \ddot{y} + 2\dot{x} &= \frac{\partial \Omega}{\partial y}, \end{aligned} \tag{1}$$

where,

$$\begin{aligned} \Omega(x,y) &= \frac{1}{2}(x^2+y^2) + \frac{1-\mu}{r_1} + \frac{\mu}{r_2} + \frac{1}{2}\mu(1-\mu), \\ r_1^2 &= (x-\mu)^2 + y^2, \qquad r_2^2 = (x+1-\mu)^2 + y^2. \end{aligned}$$

The PRTBP in synodical coordinates has five equilibrium points. Three of them are located on the *X*-axis, and are called *Eulerian* or *collinear* points ($L_{1,2,3}$). L_4 and L_5 (*Lagrangian*

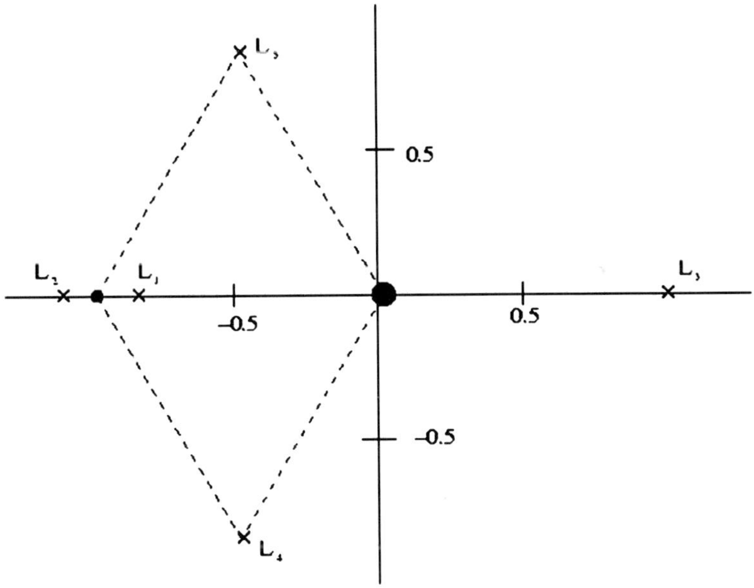

Fig. 1 The five equilibrium points of the RTBP

points) are found as the third vertex of the two equilateral triangles that can be formed using the primaries as vertices (Fig. 1).

Introducing momenta $p_x = \dot{x} - y$ and $p_y = \dot{y} + x$, the PRTBP equations of motion cast into a Hamiltonian system. The Hamiltonian function is

$$\mathcal{H}(x, y, p_x, P_y) = \frac{1}{2}(p_x^2 + p_y^2) + yp_x - xp_y - \frac{1-\mu}{r_1} - \frac{\mu}{r_2} - \frac{1}{2}\mu(1-\mu)$$

We define the Jacobi constant, $\mathcal{C}$,

$$\mathcal{C}(x, y, \dot{x}, \dot{y}) = -(\dot{x}^2 + \dot{y}^2) + 2\Omega(x, y) \tag{2}$$

with Ω as in (1). It is easily proved that $\mathcal{C} = -2\mathcal{H}$. Consequently, the Jacobi constant is a first integral of the RTBP.

In this work, we will focus on the dynamics around the equilibrium point L_1. The linear behavior of this point is of the type center $\times$ saddle. The hyperbolicity introduced by the saddle component is inherited by all the orbits in a large vicinity of L_1. Moreover, it is responsible for the existence of unstable and stable manifolds arising from the aforementioned orbits.

2.1 Planar Lyapunov orbits

For each value of the Jacobi constant, there is a unique planar Lyapunov orbit homeomorphic to S^1 around L_1. This is the unique periodic motion around the L_1 for the planar case, and the one we use in this work.

Planar Lyapunov orbits and their hyperbolic manifolds can be computed using Lindstedt–Poincaré procedures. In this way one obtains their expansions in convenient RTBP coordinates.

To this end, we set the origin of coordinates at the libration point and scale the variables in such a way that the distance from the equilibrium point to the small primary is equal to one. The expansion of the equations of motion (1) in these variables $(\bar{x}, \bar{y})$ takes the form,

$$\begin{aligned} \ddot{\bar{x}} - 2\dot{\bar{y}} - (1+2c_2)\bar{x} &= \frac{\partial}{\partial \bar{x}} \sum_{n\geq 3} c_n \rho^n P_n\left(\frac{\bar{x}}{\rho}\right), \\ \ddot{\bar{y}} + 2\dot{\bar{x}} + (c_2-1)\bar{y} &= \frac{\partial}{\partial \bar{y}} \sum_{n\geq 3} c_n \rho^n P_n\left(\frac{\bar{y}}{\rho}\right), \end{aligned} \tag{3}$$

where $\rho^2 = \bar{x}^2 + \bar{y}^2$, P_n is the Legendre polynomial of degree n, and c_n are constants which depend only on μ and the selected equilibrium point (see Masdemont 2005). We note that in (3) the linear terms appear in the left hand side part of the equations and the nonlinear ones on the right-hand side. The solution of the linear part of Eq. (3) is:

$$\begin{aligned} \bar{x}_1(t) &= \alpha_1 \exp(\lambda_0 t) + \alpha_2 \exp(-\lambda_0 t) + \alpha_3 \cos(\omega_p t + \phi), \\ \bar{y}_1(t) &= \kappa_2\alpha_1 \exp(\lambda_0 t) - \kappa_2\alpha_2 \exp(-\lambda_0 t) + \kappa_1\alpha_3 \sin(\omega_p t + \phi) \end{aligned} \tag{4}$$

where κ_1, κ_2, ω_p, and $\lambda_0 > 0$ are constants for a given model.

The αs are free amplitudes. α_1 and α_2 are the ones associated with the hyperbolic manifolds. If $\alpha_1 = \alpha_2 = 0$, we have the linear part of the Lyapunov orbit with amplitude α_3. When $\alpha_1 = 0$ and $\alpha_2 \neq 0$ we have orbits tending exponentially fast to the Lyapunov orbit of amplitude α_3 when time tends to infinity (stable manifold). On the contrary when $\alpha_2 = 0$ and $\alpha_1 \neq 0$, orbits leave the vicinity of the Lyapunov exponentially fast in forward time (unstable manifold).

When we consider also the nonlinear terms of (3), solutions close to L_1 are obtained by means of formal series in powers of the amplitudes of the form:

$$\begin{aligned} \bar{x}(t) &= \sum e^{(i-j)\theta_2}[x_{ijk}^p \cos(p\theta_1) + \bar{x}_{ijk}^p \sin(p\theta_1)]\alpha_1^i\alpha_2^j\alpha_3^k, \\ \bar{y}(t) &= \sum e^{(i-j)\theta_2}[y_{ijk}^p \cos(p\theta_1) + \bar{y}_{ijk}^p \sin(p\theta_1)]\alpha_1^i\alpha_2^j\alpha_3^k, \end{aligned} \tag{5}$$

where $\theta_1 = \omega t + \phi$, $\theta_2 = \lambda t$, and $\omega = \sum \omega_{ijk}\alpha_1^i\alpha_2^j\alpha_3^k$, $\lambda = \sum \lambda_{ijk}\alpha_1^i\alpha_2^j\alpha_3^k$. Summation is extended over all $i, j, k \in \mathbb{N}$, and $p \in \mathbb{Z}$ (note: p in x_{ijk}^p is an index, not an exponent).

However, due to symmetries, many of the coefficients x_{ijk}^p, $\bar{x}_{ijk}^p$, y_{ijk}^p, $\bar{y}_{ijk}^p$, ω_{ijk}, λ_{ijk} are zero. It is important to note that for the frequency series, the coefficient $\omega_{ijk} \neq 0$ only if $i = j$. Thus, we can write,

$$\omega = \sum \omega_{lk}(\alpha_1\alpha_2)^l\alpha_3^k, \qquad \lambda = \sum \lambda_{lk}(\alpha_1\alpha_2)^l\alpha_3^k. \tag{6}$$

In the computations, the series are truncated at a certain (high) order (Masdemont 2005). Nevertheless, we note that the meaning of the amplitudes in the nonlinear expansions (5) is the same one as in the linear solutions (4).

Therefore, in these coordinates, we can write the Lyapunov orbit of amplitude α as,

$$\lambda_\alpha = \{X = (\alpha_1, \alpha_2, \alpha_3, \phi) \in \mathbb{R}^3 \times \mathbb{T}\colon \alpha_1 = \alpha_2 = 0, \alpha_3 = \alpha, \phi \in \mathbb{T}\}.$$

Notice that in the PRTBP there is a bijective correspondence between values of the Jacobi constant and values of the amplitude α_3. Thus, we can also consider coordinates $X = (\alpha_1, \alpha_2, C, \phi) \in \mathbb{R}^3 \times \mathbb{T}$ and write the above Lyapunov orbit as

$$\lambda_C = \{X = (\alpha_1, \alpha_2, C, \phi) \in \mathbb{R}^3 \times \mathbb{T}\colon \alpha_1 = \alpha_2 = 0, C = C, \phi \in \mathbb{T}\},$$

where C is the Jacobi constant associated to α.

We can now define the local unstable and stable manifolds associated to λ_C:

$$W^s_{\lambda_C,\mathrm{loc}} = \{(0, \alpha_2, C, \phi)\colon\ \alpha_2 \in I_2 \subset \mathbb{R}, \phi \in \mathbb{T}\},$$
$$W^u_{\lambda_C,\mathrm{loc}} = \{(\alpha_1, 0, C, \phi)\colon\ \alpha_1 \in I_1 \subset \mathbb{R}, \phi \in \mathbb{T}\},$$

where $I_1 \times I_2 \subset \mathbb{R}^2$ is an open rectangle containing the origin.

2.2 Hyperbolic manifolds

In the previous section, we saw how the Lyapunov orbits and their stable and unstable manifolds can be described semi-analytically up to a certain order. Nevertheless, as far as manifolds are concerned, the analytical description is used only in a small vicinity of the Lyapunov orbit. For the rest of the tube, the manifold is obtained by numerically integrating this semi-analytical expression.

That is, there are two different parts in the description of each manifold. By setting $\alpha_1 = \pm\epsilon$ and $\alpha_2 = 0$ (respectively, $\alpha_2 = \pm\epsilon$ and $\alpha_1 = 0$), and the rest of the amplitudes and phases fixed to the chosen Lyapunov orbit values, a point on the local unstable (respectively, stable) invariant manifold is obtained in the series (5). Set $t = 0$, so we can refer to these points as *initial conditions*. Furthermore, the aforementioned initial conditions are at a distance of order ϵ from the Lyapunov orbit. The part of the local manifold which asymptotically joins them with the Lyapunov orbit is described analytically by the same series for values of time $t \neq 0$. In the case $\alpha_1 \neq 0$, negative values of time tending to $-\infty$ result in points belonging to the local unstable manifold and approaching the Lyapunov orbit exponentially fast (respectively, positive times tending to $+\infty$ for $\alpha_1 \neq 0$). This part is represented, schematically, in Fig. 2 as the small gap between the 'Lyapunov orbit' and the '*initial conditions*'.

For other values of time (positive for the unstable case and negative for the stable), the manifold tubes can be continued by integrating the initial conditions: in forward time for the case $\alpha_1 \neq 0$ and backward in time for $\alpha_2 \neq 0$ (part labelled as 'integrated manifold' in Fig. 2). Note that this numerical integration is necessary, as it provides us with an accurate description of the manifolds in regions where the Lindstedt–Poincaré expansions are no longer valid due to the large distance to the equilibrium point.

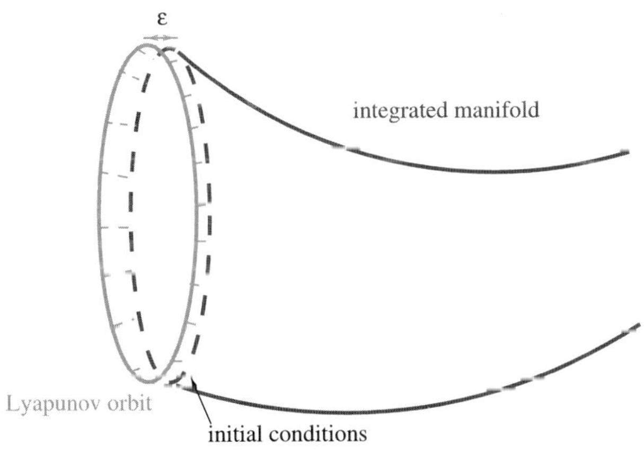

Fig. 2 Two different parts in the description of the hyperbolic manifolds: analytical series and numerical integration

2.3 Numerical computation of homoclinic trajectories

In order to look for homoclinic orbits of a Lyapunov orbit emerging from L_1, we will use a Poincaré section. This is a common technique in the study of dynamical systems, and consists of observing the flow of the system only in the points that cross a given section. In this way, the flow is studied in a lower dimensional space, but relevant information can still be derived. The stable and unstable manifolds of Lyapunov planar orbits are two-dimensional objects (see Sect. 2.2). If we study them only in their cut with a Poincaré section (i.e. $\Sigma = \{(x, y, \dot{x}, \dot{y})\colon x = K\}$, where K is a constant), the resulting objects are closed curves, and therefore, much easier to study and represent. These two closed curves may intersect transversally in several points. Each intersection point belongs to a different homoclinic connection to the Lyapunov orbit.

One can consider the first, second, and so on, cut of the invariant manifolds with the Poincaré section, and obtain in this way different closed curves which may have a different number of intersections and therefore, a different number of homoclinic connections (which perform several loops around the primary).

So, take Σ as the plane in the state space of the points with $x = -1 + \mu$. Now, the process of finding homoclinic connections, γ_C, to a Lyapunov orbit, λ_C, consists of three parts:

1. Integration of the manifolds ($W^s_{\lambda_C}$, $W^u_{\lambda_C}$) arising from the Lyapunov orbits (λ_C) until they cut the Poincaré section Σ (see Fig. 3).
2. Detection of intersections between the cuts of the asymptotic manifolds ($W^s_{\lambda_C} \cap \Sigma$ and $W^u_{\lambda_C} \cap \Sigma$) in the Poincaré section (see Fig. 3).
3. Refinement of the intersections up to a certain precision.

2.3.1 Integration of the manifolds

As it has been said, when the value of C is chosen, α_3 is fixed. In order to take initial conditions on the unstable (respectively, stable) local manifolds in (5) we set $\alpha_1 = \pm\epsilon$ (respectively, $\alpha_2 = \pm\epsilon$). ϵ is a small parameter ($\approx 10^{-3}$), and the sign indicates toward which branch of the manifold we will integrate: In L_1, $+\epsilon$ means toward the small primary, while $-\epsilon$ means toward the interior region.

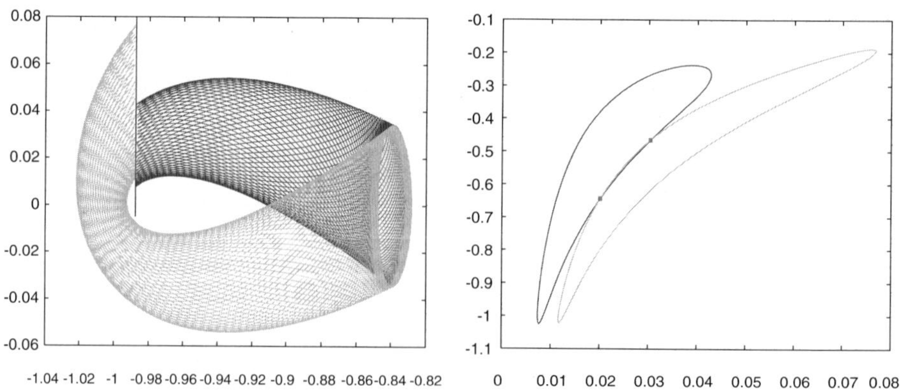

Fig. 3 (*left*) Hyperbolic manifolds of a planar Lyapunov orbit around L_1 (xy projection). The stable one is integrated until the second cut with the Poincaré section, while the unstable one until the first cut. (*right*) $y\dot{y}$ representation of the cut of the manifolds on the Poincaré section. The *blue dots* represent the intersecting points between both manifolds, that is the homoclinic trajectories

The discretization on the initial conditions arises from taking n phases around the Lyapunov orbit λ_C, $\theta_1{}^i(0) = \phi^i = 2\pi i/n$, $i = 0, \ldots, n-1$.

The bunch of orbits launched by these kind of initial conditions, with the series in (5) truncated at order 15, form manifold tubes. A Runge–Kutta–Fehlberg method of orders 7–8 is then used for the integration. Note that order 15 provides a precise approximation to the manifolds which is adequate to fulfil very stringent numerical requirements (in particular, enough for the purposes of this work).

As the equation of the Poincaré section is $x = \mu - 1$, the number of cuts (N) with the section can be easily controlled by studying the sign of $x(t) + 1 - \mu$. When N matches the desired number of cuts, a one-dimensional Newton procedure is used in order to obtain the point on the section with an accuracy of at least 10^{-12} (i.e. $|x(t) + 1 - \mu| < 10^{-12}$).

2.4 Detection of intersections between the cuts of asymptotic manifolds in the Poincaré section

Once the integration is completed, we have a set of points on the Poincaré section belonging to a discrete bunch of trajectories lying on the manifolds (n trajectories if n phases where used). So, the result of joining these points is a polygon made of n segments. The integration is performed for both the unstable manifold and the stable one, so two curves are obtained on the Poincaré section (see the right picture in Fig. 3).

The next step that has to be taken is to check for intersections on the section Σ between each segment of one of the curves and all the segments of the other curve. All in all, $(n_1) \times (n_2) \approx O(n^2)$ (n_i number of θ_1 in manifold i) comparisons between segments. This is why it is not worth taking large n values, because it does not enhance the precision of the final cut, but increases the computational effort. Nevertheless, n has to be large enough for the cuts on the Poincaré section to be well determined with an approximated S^1 structure when the segments are joined together (for instance, $n_1 = n_2 = 100$).

There is a correspondence between each segment (a, b) on the section and a segment $(\phi^a, \phi^b) \in \mathbb{T}$, such that a (respectively b) is the integrated point corresponding to the initial condition on the manifold with $\theta_1 = \phi^a$ (respectively ϕ^b) Therefore, at the end of this process we store four phase values for each detected cut. Two of them for the unstable manifold, (ϕ_u^a, ϕ_u^b) and the other two (ϕ_s^a, ϕ_s^b) for the stable manifold.

2.4.1 Refinement of the intersections

We start with the intervals found in the previous step, and we look for $\phi_u \in [\phi_u^a, \phi_u^b]$ and $\phi_s \in [\phi_s^a, \phi_s^b]$, which define the exact initial conditions of the connecting trajectory.

We use a secant-like method to find the intersection between the segments $s_u^0 = (\phi_u^{a_0}, \phi_u^{b_0})$ and $s_s^0 = (\phi_s^{a_0}, \phi_s^{b_0})$, which we know have a nonempty intersection if $a_0 = a$ and $b_0 = b$. The pair of intersecting phases divides each of the original segments in two, so four new segments are defined. Now, we look for intersections between the segments associated with $W^u_{\lambda_C}$ and the ones associated with $W^s_{\lambda_C}$. Only one of the four checks should be positive. So, we store only the four phases which are in the border of the new intersecting segments s_u^1 and s_s^1. Thus, we have again $[\phi_u^{a_1}, \phi_u^{b_1}]$, $[\phi_s^{a_1}, \phi_s^{b_1}]$, but the longitude of the intervals has been reduced.

We iterate the process k times until $|s_u^k| + |s_s^k| = |\phi_u^{a_k} - \phi_u^{b_k}| + |\phi_s^{a_k} - \phi_u^{b_k}| < 10^{-12}$.

When the three steps have been successfully carried out for an appropriate $\mathcal{C}$, we have (ϕ_u, ϕ_s), the phases on the Lyapunov orbit defining the connection. We have the intersecting point on the Poincaré section Σ, and the corresponding integration times as well.

So, the connecting trajectory can be reconstructed by taking the following initial conditions:

- For the unstable manifold: $(\alpha_1, \alpha_2, \alpha_3, \theta_1) = (10^{-3}, 0, \alpha_3(\mathcal{C}), \phi_u)$.
- For the stable manifold: $(\alpha_1, \alpha_2, \alpha_3, \theta_1) = (0, -10^{-3}, \alpha_3(\mathcal{C}), \phi_s)$

The integration of these initial conditions yields the connection. Obviously, the one on the unstable manifold has to be integrated forward in time, while the other one backward.

Once a homoclinic connection to a Lyapunov orbit $\lambda_{\mathcal{C}}$ has been found, some other connecting trajectories can be found to nearby Lyapunov orbits $\lambda_{\mathcal{C}'}$, by slightly varying the initial conditions (due to the analytic continuous dependence of the solutions of the dynamical equations w.r.t. these initial conditions). A group of trajectories which are coincident in some of the initial parameters giving rise to one-parameter family of homoclinic trajectories will be called a family of homoclinic trajectories and denoted by H. We can classify the families of homoclinic trajectories to L_1 planar Lyapunov orbits depending on the number of cuts with the Poincaré section or equivalently, the number of loops, j, around the small primary that the trajectories do. In addition, for the same number of cuts with the section Σ, more than one family is obtained. In particular, the number of families is equal to the number of intersections between the manifolds on the section.

To construct these continuous one-parameter families $H^{j,i}$ of homoclinic trajectories, we consider only those intervals $(\mathcal{C}_-^{j,i}, \mathcal{C}_+^{j,i})$ of Jacobi constants for which the intersections between the manifolds are transversal. The specific values $\mathcal{C}_\pm^{j,i}$ are candidates to bifurcation homoclinic intersections.

Thus, we will write,

$$H^{j,i} = \bigcup_{\mathcal{C}\in(\mathcal{C}_-^{j,i}, \mathcal{C}_+^{j,i})} \gamma_{\mathcal{C}}^{j,i}, \tag{7}$$

where $(\mathcal{C}_-^{j,i}, \mathcal{C}_+^{j,i})$ is the interval of Jacobi constants for which the family is obtained, j is the number of loops and i is the integer identifying the family depending on the y coordinate on the Poincaré section ($i = 1$ corresponding to the families yielding the intersection with smallest y).

Notice that, by construction, for any j and i, and any $\mathcal{C} \in (\mathcal{C}_-^{j,i}, \mathcal{C}_+^{j,i})$ there exists a unique homoclinic orbit $\gamma_{\mathcal{C}}^{j,i} \subset H^{j,i}$ to the Lyapunov orbit $\lambda_{\mathcal{C}}$.

Figure 4 is a representation of the families that have been obtained using the procedure explained above for the Earth–Moon mass parameter μ. The notation is simplified in the figure to H^j. Each one of the branches corresponds to $H^{j,i}$ with i increasing upwards. The x-axis in the figures corresponds to the value of the Jacobi constant, and the y-axis to the y-coordinate of the trajectory in one of the cuts with the Poincaré section in adimensional RTBP coordinates.

Notice that this figure can be thought of as a bifurcation diagram of the homoclinic orbits as a function of the parameter $\mathcal{C}$. As the Jacobi constant decreases, new families of homoclinic orbits $H^{j,i}$ appear (typically in pairs). From this interpretation it is clear that the bifurcations occuring in the diagram do not depend on the position of the section. Thus our definition of homoclinic families $H^{j,i}$ is independent of Σ.

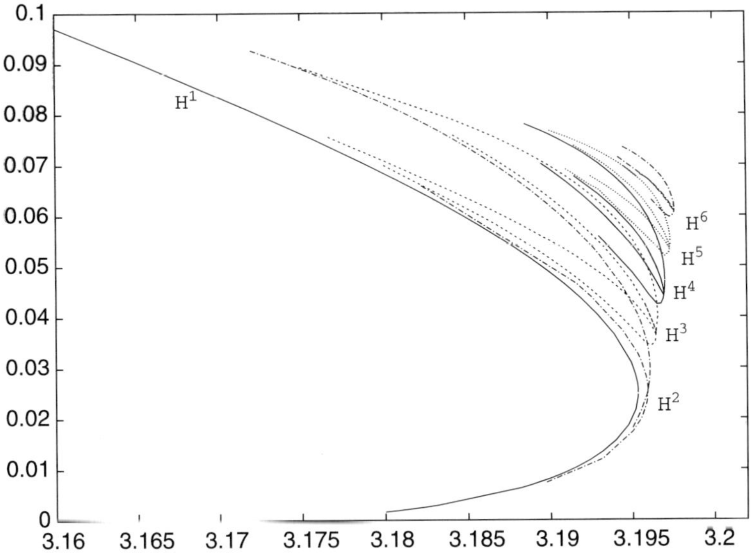

Fig. 4 Families of L_1 homoclinic orbits in the Earth–Moon case. On the *horizontal axis*, value of the Jacobi constant, and on the *vertical axis*, y coordinate of the connecting orbit on the Poincaré section

3 Scattering map

3.1 The NHIM Λ and associated hyperbolic manifolds

In Sect. 2.1, Lindstedt–Poincaré variables for the restricted three body problem around the fixed point have been introduced. A point in phase space close to L_1 is written in these variables as $X = (\alpha_1, \alpha_2, \alpha_3, \phi) \in \mathbb{B} \times \mathbb{R} \times \mathbb{T}$, where $\mathbb{B} = I_1 \times I_2 \subset \mathbb{R}^2$ is an open rectangle containing the origin in $\mathbb{R}^2$.

These variables are convenient because they parametrize the invariant manifolds present in the system in a natural way.

Let $[\mathcal{C}_1, \mathcal{C}_2]$ be an interval of energy values such that $\forall \mathcal{C} \in [\mathcal{C}_1, \mathcal{C}_2]$ there exists (at least) one orbit homoclinic to $\lambda_{\mathcal{C}}$. Following Delshams et al. (2006), we introduce the set

$$\Lambda = \{X \in \mathbb{B} \times \mathbb{R} \times \mathbb{T} : \alpha_1 = \alpha_2 = 0, \mathcal{C} \in [\mathcal{C}_1, \mathcal{C}_2]\} \tag{8}$$

$$= \bigcup_{\mathcal{C} \in [\mathcal{C}_1, \mathcal{C}_2]} \lambda_{\mathcal{C}}. \tag{9}$$

This set is a two-dimensional manifold with boundary. It is invariant for the flow Φ_t of the system. The manifold Λ is topologically $[\mathcal{C}_1, \mathcal{C}_2] \in \mathbb{T}$. A point in the manifold is parametrized by $(\mathcal{C}, \phi)$. Λ is a subset of the so-called center manifold (Fig. 5).

Notice that Λ is actually a normally hyperbolic invariant manifold for the flow (see Delshams et al. 2006 for the definition and properties of a normally hyperbolic invariant manifold).

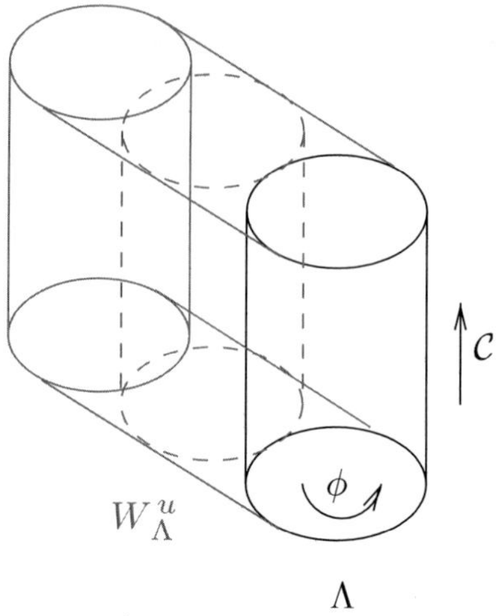

Fig. 5 Schematic representation of the normally hyperbolic invariant manifold Λ and one of its asymptotic manifolds W^u_Λ

The sets

$$W^s_\Lambda = \bigcup_{C\in[C_1,C_2]} W^s_{\lambda_C},$$
$$W^u_\Lambda = \bigcup_{C\in[C_1,C_2]} W^u_{\lambda_C}$$

are called stable and unstable manifolds of Λ in normally hyperbolic theory. They are three-dimensional manifolds. A local parametrization is given by

$$W^s_{\Lambda,\text{loc}} = \{(0, \alpha_2, C, \phi)\colon\ \alpha_2 \in I_2 \subset \mathbb{R}, C \in [C_1, C_2], \phi \in \mathbb{T}\},$$
$$W^u_{\Lambda,\text{loc}} = \{(\alpha_1, 0, C, \phi)\colon\ \alpha_1 \in I_1 \subset \mathbb{R}, C \in [C_1, C_2], \phi \in \mathbb{T}\}.$$

Recall the definition of the stable manifold of a point X in a hyperbolic set Λ:

$$W^s_X = \{Y \in \mathbb{R}^3 \times \mathbb{T}\colon\ \text{dist}(\Phi_t(X), \Phi_t(Y)) \le K(X)\exp^{-\lambda t},\quad t \to \infty\}. \tag{10}$$

Notice the same time t in $\Phi_t(X)$ and $\Phi_t(Y)$ and an exponent of convergence $\lambda > 0$. In general, this manifold is *not* invariant for the flow of the system. (It is the nonlinear analog of the contracting subspace $\mathbb{E}^s_x$ in the hyperbolic splitting of a hyperbolic invariant set. For an introduction to hyperbolic invariant sets, see Robinson 1999). However, the stable manifold of a point satisfies the property

$$\Phi_t(W^s_X) = W^s_{\Phi_t(X)}. \tag{11}$$

In the case of the PRTBP, the local stable manifold of a point $X_+ = (0, 0, C_+, \phi_+) \in \Lambda$ is given by

$$W^s_{X_+,\text{loc}} = \{(0, \alpha_2, C_+, \phi_+)\colon\ \alpha_2 \in I_2\} \tag{12}$$

as one can easily check from Eqs. (5) and (6). Analogously, given $X_- = (0, 0, C_-, \phi_-) \in \Lambda$, its local unstable manifold is given by

$$W^u_{X_-,\mathrm{loc}} = \{(\alpha_1, 0, \mathcal{C}_-, \phi_-) \colon \alpha_1 \in I_1\}. \tag{13}$$

It is thus evident that

$$W^s_{\Lambda,\mathrm{loc}} = \bigcup_{X\in\Lambda} W^s_{X,\mathrm{loc}} \quad \text{and} \quad W^s_{X,\mathrm{loc}} \cap W^s_{Y,\mathrm{loc}} = \emptyset, \quad \text{if } X \neq Y$$

and by extension,

$$W^s_\Lambda = \bigcup_{X\in\Lambda} W^s_X \quad \text{and} \quad W^s_X \cap W^s_Y = \emptyset, \quad \text{if } X \neq Y. \tag{14}$$

There is a similar foliation for the unstable manifold.

3.2 The dynamics in Λ

In the case of the PRTBP, the dynamics restricted to Λ is very simple. The normally hyperbolic invariant manifold is foliated by invariant tori corresponding to periodic Lyapunov orbits. The flow on Λ is given by

$$\Phi_t(0, 0, \mathcal{C}_0, \phi_0) = (0, 0, \mathcal{C}_0, \phi_0 + \omega t), \tag{15}$$

where ω is given in (6), and we will denote by

$$\lambda_t(\mathcal{C}_0, \phi_0) = (0, 0, \mathcal{C}_0, \phi_0 + \omega t) \tag{16}$$

the orbits in the torus $\lambda_{\mathcal{C}_0}$.

3.3 Definition of the scattering map

In this section, we are going to define a scattering map $S^{j,i}$ for every family $H^{j,i}$ of homoclinic orbits to Λ. For ease of presentation, through this section we will drop the indices that identify a homoclinic family.

Consider a homoclinic family H defined on a given interval $(\mathcal{C}_-, \mathcal{C}_+)$. Take a point $Z^* \in H$ and let $\mathcal{C} \in (\mathcal{C}_-, \mathcal{C}_+)$ be its Jacobi constant, so that $Z^* \in \gamma_{\mathcal{C}}$. By definition of the stable and unstable manifolds of Λ (see (10)), there exist $X^*_-, X^*_+ \in \Lambda$ such that

$$\mathrm{dist}(\Phi_t(Z^*), \Phi_t(X^*_\pm)) \leq K(Z^*) \exp^{-\lambda|t|}, \quad t \to \pm\infty,$$

which are unique by (14). Indeed, by preservation of the Jacobi constant, $X^*_-, X^*_+ \in \lambda_{\mathcal{C}}$.

Consider now an arbitrary point X_- in the Lyapunov periodic orbit $\lambda_{\mathcal{C}}$, so that $X_- = \Phi_\tau(X^*_-)$, with $\tau \in [0, T_{\mathcal{C}})$, where $T_{\mathcal{C}}$ is the period of $\lambda_{\mathcal{C}}$. We define $X_+ = \Phi_\tau(X^*_+) \in \lambda_{\mathcal{C}} \subset \Lambda$, $Z = \Phi_\tau(Z^*) \in \gamma_{\mathcal{C}}$ and notice that

$$\mathrm{dist}(\Phi_t(Z), \Phi_t(X_\pm)) \leq K(Z) \exp^{-\lambda|t|}, \quad t \to \pm\infty. \tag{17}$$

The map

$$X_- \in \mathcal{D} \subset \Lambda \overset{S}{\longmapsto} X_+ \in \mathcal{D} \subset \Lambda, \tag{18}$$

where $\mathcal{D} = \cup_{\mathcal{C}\in(\mathcal{C}_-,\mathcal{C}_+)} \lambda_{\mathcal{C}}$, is called the scattering map and is characterized by the property that

$$X_+ = S(X_-) \text{ if and only if there exists } Z \in H \text{ such that (17) holds.}$$

Notice that the scattering map $S = S^{j,i}$ and its domain of definition $\mathcal{D} = \mathcal{D}^{j,i}$ depend on the homoclinic family $H^{j,i}$.

Next we check this characterization. Assume that there exist $Z_1, Z_2 \in H$ and $X_{+,1}, X_{+,2} \in \Lambda$ with

$$\mathrm{dist}(\Phi_t(Z_k), \Phi_t(X_-)) \leq K(Z_k) \exp^{-\lambda|t|}, \quad t \to -\infty \tag{19}$$

but

$$\mathrm{dist}(\Phi_t(Z_k), \Phi_t(X_{+,k})) \leq K(Z_k) \exp^{-\lambda|t|}, \quad t \to +\infty. \tag{20}$$

Since Z_1, Z_2 belong to γ_C, $Z_2 = \Phi_{t_1}(Z_1)$, and imposing (19) and the fact that λ_C is periodic, one easily gets that $t_1 = nT_C$ for some integer n. Imposing now (20), one finds that $X_{+,1} = X_{+,2}$. This proves that the scattering map S is well defined and that it does not depend on the choice of the point Z in (17). Analogously, one can check that $S^{-1} : X_+ \mapsto X_-$ is also well defined.

An interesting property of the scattering map in the case of the planar restricted three body problem is that it commutes with the flow: for any $t \in \mathbb{R}$

$$S(\Phi_t(X_-)) = \Phi_t(X_+), \tag{21}$$

which follows readily from the characterization (17) and the fact that if (17) holds for two points Z_1 and Z_2 then $Z_2 = \Phi_{t_1}(Z_1)$ with $t_1 = nT_C$.

We now look for the expression of the scattering map in the coordinates $(\mathcal{C}, \phi)$ of Λ introduced in (8).

Due to the low dimensional nature of the PRTBP, one can easily describe the scattering map associated to a homoclinic manifold. Let $X_-, X_+ \in \Lambda$ have Lindstedt–Poincaré variables

$$X_- = (\mathcal{C}, \phi_-),$$
$$X_+ = (\mathcal{C}, \phi_+),$$

so that $S(\mathcal{C}, \phi_- + \omega t) = (\mathcal{C}, \phi_+ + \omega t)$ by property (21). Introduce the phase shift

$$\Delta = \Delta(\mathcal{C}) = \phi_+ - \phi_-, \tag{22}$$

which depends on the torus. Then we can express the scattering map as

$$S(\mathcal{C}, \phi) = (\mathcal{C}, \phi + \Delta(\mathcal{C})) \tag{23}$$

for any $(\mathcal{C}, \phi) \in (\mathcal{C}_-, \mathcal{C}_+) \times \mathbb{T}$. Hence, the scattering map has the form of an integrable twist map. As a matter of fact, it is a nontrivial twist map, that is $\Delta'(\mathcal{C}) \not\equiv 0$, as will be shown in the next section.

3.4 Computation of the scattering map

As explained in the previous section, the particular geometrical features of the PRTBP impose a strong structure for the scattering map associated to a homoclinic manifold. For every j, i,

- $S = S^{j,i}$ can be computed independently on each torus $\lambda_C \in \mathcal{D}^{j,i}$.
- On a given torus λ_C, the scattering map is just a translation $\phi \mapsto \phi + \Delta$ by a constant shift $\Delta = \Delta^{j,i}(\mathcal{C})$.

Therefore, the problem of computing S reduces to computing $\Delta(\mathcal{C})$ for any torus in the range $\mathcal{C} \in (\mathcal{C}_-^{j,i}, \mathcal{C}_+^{j,i})$, where $\Delta(\mathcal{C})$ is given in (22).

Remark 1 *The characterization of the scattering map in terms of a scalar function $\Delta(\mathcal{C})$ is very specific of this problem, because $\mathcal{C}$ is a constant of motion and there is only one torus $\lambda_{\mathcal{C}}$ for every $\mathcal{C}$. Obviously, this will not happen in the spatial RTBP, where the $\lambda_{\mathcal{C}}$ have to be replaced by two-dimensional KAM invariant tori.* □

Let us fix $\mathcal{C} \in (\mathcal{C}_-^{j,i}, \mathcal{C}_+^{j,i})$ and consider the corresponding 1-torus $\lambda_{\mathcal{C}}$ and homoclinic orbit $\gamma_{\mathcal{C}} = \gamma_{\mathcal{C}}^{j,i}$. We seek $X_- = (\mathcal{C}, \phi_-)$ and $X_+ = (\mathcal{C}, \phi_+)$ in $\lambda_{\mathcal{C}}$ such that

$$Z \in W^u_{X_-}$$

and

$$Z \in W^s_{X_+}$$

for some $Z \in \gamma_{\mathcal{C}}$.

Note that X_- and X_+ are related to Z *asymptotically*, so these two points cannot be found from Z by direct iteration. The algorithm used in Sect. 2.3 to find the homoclinic connection $\gamma_{\mathcal{C}}$ provides us with $X_u \in W^u_{\lambda_{\mathcal{C}}}$, $X_s \in W^s_{\lambda_{\mathcal{C}}}$ and $Z \in \gamma_{\mathcal{C}}$ related in the following way:

$$\begin{aligned} \Phi_{t_u}(X_u) &= Z, \\ \Phi_{t_s}(X_s) &= Z. \end{aligned}$$

The algorithm already gives X_u and X_s in Lindstedt–Poincaré coordinates

$$\begin{aligned} X_u &= (\alpha_1, \alpha_2, \mathcal{C}, \phi) = (\epsilon, 0, \mathcal{C}, \phi_u), \\ X_s &= (\alpha_1, \alpha_2, \mathcal{C}, \phi) = (0, \epsilon, \mathcal{C}, \phi_s). \end{aligned}$$

From (13), X_u belongs to the unstable manifold of the point $\tilde{X}_- = (0, 0, \mathcal{C}, \phi_u)$. Similarly, X_s belongs to the stable manifold of the point $\tilde{X}_+ = (0, 0, \mathcal{C}, \phi_s)$. By assertion (11),

$$\begin{aligned} Z &= \Phi_{t_u}(X_u) \in W^u_{\Phi_{t_u}(\tilde{X}_-)}, \\ Z &= \Phi_{t_s}(X_s) \in W^s_{\Phi_{t_s}(\tilde{X}_+)}. \end{aligned}$$

Thus the points we seek are

$$\begin{aligned} X_- &= \Phi_{t_u}(\tilde{X}_-) = \lambda_{t_u}(\tilde{X}_-), \\ X_+ &= \Phi_{t_s}(\tilde{X}_+) = \lambda_{t_s}(\tilde{X}_+). \end{aligned}$$

In coordinates, they are given by

$$\begin{aligned} X_- &= (0, 0, \mathcal{C}, \phi_-) = (0, 0, \mathcal{C}, \phi_u + t_u\omega), \\ X_+ &= (0, 0, \mathcal{C}, \phi_+) = (0, 0, \mathcal{C}, \phi_s + t_s\omega). \end{aligned}$$

Therefore,

$$\begin{aligned} \phi_- &= \phi_u + t_u\omega, \\ \phi_+ &= \phi_s + t_s\omega, \end{aligned}$$

so that $\Delta = \phi_+ - \phi_- = \phi_s - \phi_u + (t_s - t_u)\omega \pmod{2\pi}$.

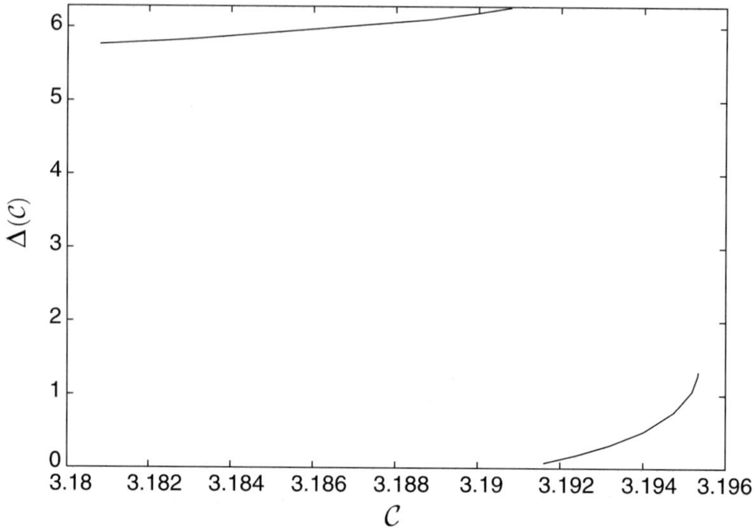

Fig. 6 Rotation angle associated to $S^{1,1}$

3.5 Some numerical results

Using the method described in the previous section, we have computed the scattering maps associated to several homoclinic families in the PRTBP. The results presented here are for Lyapunov orbits around L_1 in the Earth–Moon model.

Recall from Sect. 3.3 that the scattering map $S^{j,i}$ associated to a given homoclinic family $H^{j,i}$ is characterized by a scalar function $\Delta^{j,i}(\mathcal{C})$ defined on a domain $\mathcal{C} \in (\mathcal{C}_-^{j,i}, \mathcal{C}_+^{j,i})$. The function $\Delta^{j,i}(\mathcal{C})$ is the rotation angle that results from applying $S^{j,i}$ to any point in $\lambda_{\mathcal{C}}$.

Consider the simplest homoclinic families $H^{1,1}$ and $H^{1,2}$, which loop once around the small primary. Figures 6 and 7 show $\Delta^{1,1}$, respectively, $\Delta^{1,2}$ in the same domains where the homoclinic families were computed (compare to Fig. 4). We also plot both functions together in Fig. 8 to show how they are a natural continuation of each other. In this way it is also easier to compare the intervals of Jacobi constants where $\Delta^{j,i}$ are defined. The functions are represented in the universal cover, instead of $[0, 2\pi)$, to emphasize that each one is continuous.

Note that the slope of $\Delta(\mathcal{C})$ goes to infinity at the bifurcation point. Note also that $\Delta^{1,2}(\mathcal{C})$ is a twist map for all $\mathcal{C} \in (\mathcal{C}_-^{1,2}, \mathcal{C}_+^{1,2})$ except for a point where $\Delta^{1,2}$ attains a maximum, giving rise there to a nontwist map (see Delshams and De la Llave 2000a).

3.6 Example application to astrodynamics

This section describes an example application of the scattering map to the design of spacecraft trajectories in the restricted three body problem. The goal of this example is to illustrate the convenience and usefulness of the scattering map as a tool to study homoclinic and heteroclinic phenomena in dynamical systems. Other applications can probably be found in the field of mission design or in a different field.

As a simplified example using our planar restricted three body problem, let us consider two spacecraft orbiting in a given Lyapunov orbit λ about L_1 in the Earth–Moon system. For

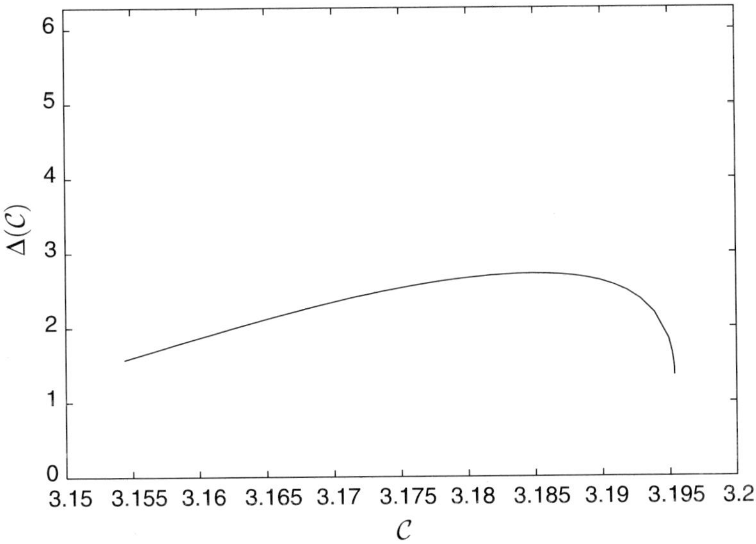

Fig. 7 Rotation angle associated to $S^{1,2}$

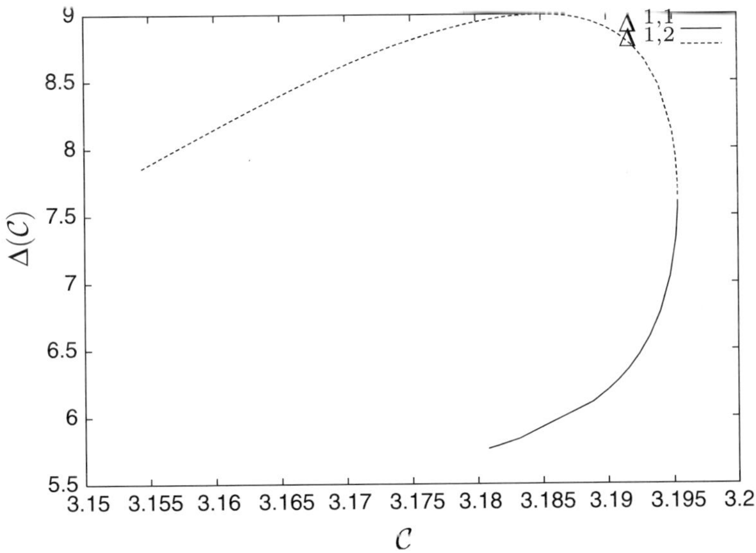

Fig. 8 Rotation functions associated to $S^{1,1}$ (*solid line*) and $S^{1,2}$ (*dotted line*)

any reason (for instance due to a contingency plan or because of launch dispersion) the two spacecraft ended up in a circular like choreography at very different phases in the Lyapunov orbit, while the actual purpose would be to have them close to each other (for instance because of a formation flight requirement). In this situation, performing the rendezvous and overcoming "the gap" between the satellites using classical transfer maneuvers is expensive in terms of delta-v but, using homoclinic connections and with the help of the scattering map, it can be performed at almost zero cost.

Table 1 Rotation angle for several homoclinic families for $\mathcal{C} = 3.19$

$\Delta^{1,1}$	$\Delta^{1,2}$	$\Delta^{2,3}$	$\Delta^{2,4}$	$\Delta^{3,1}$	$\Delta^{3,2}$	$\Delta^{3,3}$	$\Delta^{3,4}$	$\Delta^{4,1}$	$\Delta^{4,2}$	$\Delta^{4,4}$
6.197	2.615	2.529	5.767	2.437	5.611	5.662	2.702	5.594	2.578	5.876

The main drawback of this method is that the transfer times are longer than when using classical maneuvers.

In our example, let us fix some Lyapunov orbit for definiteness, say $\lambda = \lambda_{\mathcal{C}}$ with $\mathcal{C} = 3.19$. There are several scattering maps S^H defined on $\lambda \subset \Lambda$, depending on the choice of homoclinic family H. Recall from Sect. 3.3 that λ is parametrized by an angle ϕ, and the scattering map restricted to λ is just a rigid rotation

$$\phi \mapsto \phi + \Delta.$$

Having computed several scattering maps numerically in Sect. 3.4, we know the values of Δ for several homoclinic manifolds. Table 1 shows these values (in radians).

Each application of the scattering map S^H corresponds to performing a homoclinic excursion through a trajectory $\gamma \subset H$. In the example application, a homoclinic excursion consists in injecting a satellite into the unstable manifold of the Lyapunov orbit using station keeping like maintenance maneuvers, and then following the selected homoclinic trajectory, which naturally comes back to the Lyapunov orbit following its stable manifold. The result is that the satellite is shifted by an angle Δ with respect to its initial phase.

Therefore, the example problem can be formulated as follows: let $\mathcal{I}$ be an index set that will be used to index the finite set of scattering maps (or rotation angles) shown in Table 1. Equivalently, $\mathcal{I}$ also indexes the homoclinic manifolds considered. Suppose that $x_A = (\mathcal{C}, \phi_A)$ and $x_B = (\mathcal{C}, \phi_B) \in \Lambda$ are the initial conditions of the two spaceships on the Lyapunov orbit. Find a sequence $(i_1, i_2, \ldots, i_n) \in \mathcal{I}^n$ such that

$$\text{dist}\,(S^{i_n} \circ \cdots \circ S^{i_2} \circ S^{i_1}(x_A), x_B) < d$$

or equivalently

$$\text{dist}\,(\phi_A + \Delta^{i_1} + \Delta^{i_2} + \cdots + \Delta^{i_n}, \phi_B) < d,$$

where d is a prescribed distance.

This would correspond to n homoclinic excursions for spaceship A. Obviously, one could apply the scattering map to both ϕ_A and ϕ_B, i.e., applying maneuvers to both spaceships.

Acknowledgements This research has been supported by the MCyT-FEDER grant BFM2003-09504, and the Catalan grant 2003XT-00021. E.C. acknowledges the fellowship AP2002–1409 of the Spanish Government. The authors thank R. de la Llave for his insightful comments on the scattering map.

References

Breakwell, J.V., Kamel, A.A., Ratner, M.J.: Station-keeping for a translunar communication station. Celestial Mech. **10**, 357–373 (1974)

Canalias, E., Masdemont, J.J.: Lunar space station for providing services to solar libration point missions. In 56th International Astronautical Federation Congress, Fukuoka, Japan (2005)

Canalias, E., Masdemont, J.J.: Homoclinic and heteroclinic transfer trajectories between Lyapunov orbits in the Sun–Earth and Earth–Moon Systems. Discrete Contin. Dyn. Syst. **14**, 261–279 (2006)

Delshams, A., de la Llave, R.: KAM theory and a partial justification of Greene's criterion for nontwist maps. SIAM J. Math. Anal. **31**(6), 1235–1269 (2000a)

Delshams, A., de la Llave, R., Seara, T.M.: A geometric approach to the existence of orbits with unbounded energy in generic periodic perturbations by a potential of generic geodesic flows of $\mathbb{T}^2$. Comm. Math. Phys. **209**(2), 353–392 (2000b)
Delshams, A., de la Llave, R., Seara, T.M.: A geometric mechanism for diffusion in Hamiltonian systems overcoming the large gap problem: heuristics and rigorous verification on a model. Mem. Amer. Math. Soc. **179**(844), 1–141 (2006)
Farquhar, R.W.: A halo orbit lunar station. Astronautics and Aeronautics **10**(6), 52–63 (1972)
Gómez, G., Jorba, A., Masdemont, J.J., Simó, C.: A dynamical systems approach for the analysis of the SOHO mission. Third International Symposium on Spacecraft Flight Dynamics, ESA/ESOC, pp. 449–454, (1991)
Masdemont, J.J.: High-order expansions of invariant manifolds of libration point orbits with applications to mission design. Dyn. Syst. **20**(1), 59–113 (2005)
Robinson, C.: Dynamical systems. Studies in Advanced Mathematics 2nd edn. Stability, symbolic dynamics and chaos, CRS Press, Boca Raton (1999)
Szebehely, V.: Theory of orbits. Academic Press, New York (1967)

Celestial Mechanics and Dynamical Astronomy (2006) 95:173–200
DOI 10.1007/s10569-006-9018-9

ORIGINAL ARTICLE

On final evolutions in the restricted planar parabolic three-body problem

Martha Alvarez · Josep Maria Cors · Joaquin Delgado

Received: 15 November 2005 / Revised: 13 March 2006 /
Accepted: 28 March 2006 / Published online: 15 August 2006

Abstract In this paper, we prove the existence of special type of motions in the restricted planar parabolic three-body problem, of the type exchange, emission–capture, and emission–escape with close passages to collinear and equilateral triangle configuration, among others. The proof is based on a gradient-like property of the Jacobian function when equations of motion are written in a rotating–pulsating reference frame, and the extended phase space is compactified in the time direction. Thus a phase space diffeomorphic to $[-\pi/2, \pi/2] \times \mathbb{C}\backslash\{-\mu_1, \mu_2\} \times \mathbb{C}$ -coordinates (θ, ζ, ζ') is obtained with the boundary manifolds $\theta = \pm\pi/2$ corresponding to escapes of the binaries when time tends to $\pm\infty$. It is shown there exists exactly five critical points on each boundary, corresponding to classic homographic solutions. The connections of the invariant manifolds associated to the collinear configurations, and stable/unstable sets associated to binary collision on the boundary manifolds, are obtained for arbitrary masses of the primaries. For equal masses extra connections are obtained, which include equilateral configurations. Based on the gradient-like property, a geometric criterion for capture is proposed and is compared with a criterion introduced by Merman (1953b) in the fifties, and an example studied numerically by Kocina (1954).

Keywords Restricted three-body problem · Final evolutions · Heteroclinic orbits

M. Alvarez · J. Delgado
Departamento de Matemáticas, Universidad Autónoma Metropolitana, Iztapalapa,
A.P. 55-534, Iztapalapa 09340, México, D.F., México
e-mails: mar@xanum.uam.mx; jdf@xanum.uam.mx

J. M. Cors (✉)
Departament de Matemàtica Aplicada III, Universitat Politècnica de Catalunya,
08240 Manresa, Spain
e-mail: cors@eupm.upc.es

1 Introduction

The investigation of the solution of the three-body problem has been in the direction of studying the growth of the system for long intervals of time. Chazy (1922) gave the first general classification of the motion as time $t \to \pm\infty$.

He enumerated seven types of motion for positive masses $m_i, i = 0, 1, 2$ in terms of the order of magnitude (as $t \to \infty$, and similarly as $t \to -\infty$) of the three mutual distances $r_{ij}, i \neq j = 0, 1, 2$.

In what follows, we will say that $f(t) \sim t$ (resp. $f(t) \sim t^{2/3}$) if $f(t) = Ct + o(t)$ (resp. $f(t) = Ct^{2/3} + o(t)$), for some positive constant C.

Motion is called *parabolic* (P) if $r_{ij} \sim t^{2/3}$ for all i, j, while if $r_{ij} \sim t$ for all i, j, motion is called *hyperbolic* (H). If motion is parabolic (resp. hyperbolic) as $t \to -\infty$, and there exists $R > 0$ such that $r_{ij} < R$ for some i, j as $t \to \infty$, then it is called parabolic-elliptic (resp. hyperbolic-elliptic).

A *capture* of m_i by m_j occurs (it is a symmetric relationship) if $r_{ij} \to \infty$ as $t \to -\infty$ and there exists $R > 0$ such that $r_{ij} < R$ as $t \to \infty$. Similarly, particle m_i is *emitted* by m_j if there exists $R > 0$ such that $r_{ij} < R$ as $t \to -\infty$ and $r_{ij} \to \infty$ as $t \to \infty$.

An exchange of the k-th particle occurs if there exists $R_1, R_2 > 0$ such that $r_{ik} < R_1$ as $t \to -\infty$ and $r_{jk} < R_2$ for some $i \neq j$ as $t \to \infty$.

In the restricted parabolic problem, one of the masses is zero, say $m_0 = 0$, while the primaries perform a two-body parabolic motion, so only the distances of the infinitesimal to the primaries and its velocity as $t \to \pm\infty$ is of concern here.

Chazy (1922) showed that in the general three-body problem for zero energy, hyperbolic–elliptic motions or parabolic motions are possible; parabolic–hyperbolic and parabolic–elliptic motions are impossible due to conservation of energy. Some motions which are impossible in the non-restricted case can occur in the restricted one. Merman (1954b) showed that in the restricted parabolic problem the capture motion is possible if certain inequalities are satisfied. N.G. Kocina (1954) showed by means of a numerical example the possibility of a capture in the restricted parabolic three-body problem.

The main results of the paper are summarized in Propositions 10 and 13, which we now briefly explain: Let $W^{u,s}(L_i)$ denote the unstable (stable) manifold associated to critical points $L_i, i = 1, 2, \ldots, 5$, of the system (25), see Sect. 4. Naturally extend this definition to the stable (unstable) sets of binary collision $W^{u,s}(B_{\mu_i})$, $i = 1, 2$, and infinity $W^{u,s}(\infty)$ (see Sect. 7, Definition 4). Then the diagram in Proposition 10 shows the possible connections among the stable/unstable sets for any value of the mass parameter $\mu_1 (\mu_2 = 1 - \mu_1)$, while that of Proposition 13 shows the connections we can prove to exist for equal masses. In the proof of both Propositions the existence of a gradient-like (or Lyapunov) function plays a crucial role. In our case, the analogous to the Jacobian function in the restricted circular problem C, see (20), plays this role. The critical levels in those diagrams are $C_i = C(L_i)$.

From Propositions 10 and 13 the existence of several types of motions as described in Propositions 14–18 follows as a corollary. For example, in Proposition 15 the existence of a kind of motion called exchange is proved. This is shown numerically in Fig. 10.

The paper is organized as follows: in Sect. 2, the equations of motion in a rotating-pulsating coordinate system are obtained; in Sect. 3, a compactification of phase space in the time direction is achieved and the main autonomous system (16) is obtained. In Sect. 3.1, it is shown how the homographic solutions appear associated to critical points or homothetic orbits. The gradient-like property is proved in Sect. 3.2. Section 4 continues with the linear analysis of the critical points and the dimensions of the various invariant manifolds are obtained. In Sect. 5, it is shown that parabolic orbits are unions of immersed manifolds of dimensions 4 and 3, a classical result which is extended to the restricted case. One of several Merman's criteria is recalled in Sect. 6 and Kocina's example is revisited. In Sect. 6.1, a geometric criterion based on the gradient-like property is introduced and validated in Kocina's example. The rest of this paper is devoted to study some heteroclinic connections of the various invariant manifolds on the boundary manifolds corresponding to escape of the primaries. This is done in Sect. 7. Combining the connecting diagrams on the boundaries and together with the homothetic orbits, some dynamical consequences are obtained and the existence of some type of motion is proved. The description is not exhaustive for two reasons: the first is that the study of all possible connections is extremely difficult due to the high dimensions of the manifolds involved and we give only the "easiest ones," for arbitrary masses and then specialize to equal masses. A complete study would require new analytical tools, which are not at our disposal, or at least a serious numerical study. In any case this will be the purpose of a future paper; the second reason is mainly of brevity of the paper and a description of the type of motions is mainly intended to be of some interest of the astrophysical community.

1.1 Related works to the present paper

The techniques used in the present research are closely related to previous investigations on the planar three-body problem and other particular cases. This section pretends to point out similarities as well as differences with previous research. In a seminal paper on the collinear three-body problem, McGehee (1974) introduced the technique of blow up of total collision. This amounts to glue a boundary manifold to phase space, substituting the singularity locus due to total collision. A proper change of variables and a scaling of time extends the flow to this boundary manifold in an invariant manner. The collision manifold is topologically equivalent to two sphere minus four points (two symmetric pants glued at the wrist) having two symmetric saddle points associated to the collinear central configurations. They differ by the sign of the scaled radial velocity coordinate (herein denoted by v). For one of these critical points two unstable branches escape through distinct legs while the behavior of the stable branches depend on the mass ratio (the other critical point behaves symmetrically). He showed that for a proper choice of masses the saddles do not connect and thus proves the existence of particular motions coming close to total collision and escaping with arbitrarily high radial velocity with successive binary collisions. Devaney (1980) applies this technique to the isosceles three-body problem getting for the collision manifold the same topological type as in the collinear case (two symmetric pants glued at the wrist with a closed neck). The flow has six critical points on the collision manifold. Let us denote by $W^{u,s}(L_i^+)$ the unstable (stable) manifold associated to the central configuration L_i, $i = 1, 2, 3, 4, 5$ with positive v. Devaney proved the existence of the connection $W^u(L_4^+) \to W^s(L_2^+)$ and also $W^u(L_5^+) \to W^s(L_2^+)$ forced by the gradient-like property.

In 1983, Moeckel studied the planar three-body problem (three degrees of freedom) and proves for the full three-body problem the existence of the connection of the type shown by Devaney in the isosceles case. He uses a topological argument and a reduction of dimension to show this. In general the connections among critical points with different signs of v are much more difficult to prove. Moeckel showed that under the hypothesis of existence of a transversal topological connection among certain critical points with opposite signs of v, forces further connections.

More recently, Simó and Susín (1991) performed a thorough numerical study of connections among critical points with different signs of v. They used Murnaghan-Lemâitre variables in order to regularize previously double collisions and then a blow up.

For zero energy we mention the following relevant papers. Chazy (1922), Saari (1971), Hulkower (1978) and Lacomba and Bryant (1983), Llibre and Waldwogel (1996). The contributions of the first three authors are discussed in Sect. 5. Lacomba and Bryant showed that the equations of motion for the n-body problem on the collision manifold can be viewed as a contact vector field on a quotient space of the usual phase space. This space is obtained by projecting along a "homothetical vector field" chosen with the homogeneity of the Hamiltonian. They use the term "projectable" when the flow on the fixed energy level projects to the flow on the collision manifold and this happens in the zero energy case, as shown previously by Saari et al. (1981) using different techniques. Llibre and Waldwogel (1996) studied the global behavior of the solutions of the flow on the zero energy level for a class of mechanical systems including the n-body problem.

McGhee's coordinates are suggested by the homogeneity of the potential, which is scaled by the total moment of inertia r: Let $T(p)$ be the kinetic energy, homogeneous of degree 2 in the momenta, and $U(q)$ the potential, homogeneous of degree -1 in the positions. Then, from the energy level $T(p) - U(q) = h$ and taking r as the total moment of inertia, $q = rQ$, transforms the energy relationship into $rT(p) - U(Q) = rh$, which suggests a scaling of momenta $P = r^{-1/2}p$. A further decomposition into radial and tangential velocities yields the desired change of variables.

McGehee's blow up does not carry *vis-a-vis* when one or several masses are zero. The main reason is that the mass matrix defining becomes degenerate and the energy relationship does not hold. Nevertheless so much of the ideas can be continued. In this vein, Martínez and Simó (1987/88) studied the limiting case of one mass going to zero in the isosceles three-body problem by introducing ad hoc coordinates. ElBialy (1989) studied the isosceles three-body problem with an infinitesimal mass (positive or zero) by adapting McGehee's blow-up using an equivalent to the euclidean norm instead of the mass matrix. Following similar ideas, Alvarez and Delgado (2003) obtained the full equations of the planar restricted three-body problem recovering ElBialy's for the isosceles case. The gradient-like property is lost as shown in detail by Alvarez and Delgado.

Rotating pulsating coordinates have been used by several authors. Broucke (1972) deduces the equations of motion in these coordinates for the elliptic restricted isosceles three-body problem and computes periodic orbits. In a less known paper of the 40s, Monroe H. Martin (1942) applies these coordinates to the restricted hyperbolic, elliptic and parabolic three-body problem. He studies in particular the case where the primaries move parabolically along a line while the massless particle moves perpendicularly. He uses systematically the concept of acquisitive and dissipative system, a great resemble to the gradient-like property used in the present paper.

On the other hand, rotating–pulsating coordinates uses strongly the homogeneity of the acceleration on the massless particle—as the blow up uses the homogeneity of the potential—but combines a scaling of the position of the massless particle with the motion of the primaries—as the blow up uses the moment of inertia. Nevertheless so much of the techniques used in previous examples for studying the total collision manifold are conveniently used in the present paper.

In a rotating–pulsating coordinate system the particles with positive mass remain fixed so total collision is eliminated from phase space, of course it remains present in the reparametrization of time of the form $dt/ds = \text{const.}r^{3/2}$ (see (10)).

2 Equations in pulsating coordinates

Let two finite masses m_1, m_2 move in parabolic orbits. The relative position vector from m_2 to m_1 in complex is

$$z = q(\sigma^2 - 1) + 2iq\sigma \tag{1}$$

where $(-q, 0)$ is the vertex of the parabola opening towards the x-axis, and σ is the true anomaly. Here and in what follows we use indistinctly, complex or vector notation to denote positions in the plane.

For a parabolic motion the following relations hold (Kocina 1954):

$$r = |z| = q(\sigma^2 + 1), \tag{2}$$

$$n(t - T) = \sigma + \frac{\sigma 3}{3}, \tag{3}$$

$$2n^2q^3 = m_1 + m_2 \equiv m \tag{4}$$

where T is the time of perihelion passage, n the mean motion, q the semi-latus rectum and m the total mass.

The following derivatives will be used

$$\frac{d\sigma}{dt} = \frac{nq}{r}, \quad \frac{dr}{d\sigma} = 2q\sigma. \tag{5}$$

The positions of the primaries relative to the center of mass at the origin are

$$z_1 = \mu_2 z, \; z_2 = -\mu_1 z, \quad \text{where } \mu_i = \frac{m_i}{m}.$$

The equations of motion of a massless particle acted by the gravitational forces of the primaries are

$$\ddot{z}_0 = -\frac{m_2(z_0 - z_2)}{|z_0 - z_2|^3} - \frac{m_1(z_0 - z_1)}{|z_0 - z_1|^3}. \tag{6}$$

We will now introduce a rotating and pulsating coordinate where the primaries remain fixed along the new x-axis. Let

$$z_0 = \zeta z$$

then

$$\dot{z}_0 = \zeta\dot{z} + \dot{\zeta}z, \quad \ddot{z}_0 = \ddot{\zeta}z + 2\dot{\zeta}\dot{z} + \zeta\ddot{z}.$$

Substituting these expressions in (6) yields

$$\ddot{\zeta}z + 2\dot{\zeta}\dot{z} + \zeta\ddot{z} = -\frac{mz}{|z|^3}\left(\frac{\mu_2(\zeta + \mu_1)}{|\zeta + \mu_1|^3} + \frac{\mu_1(\zeta - \mu_2)}{|\zeta - \mu_2|^3}\right). \tag{7}$$

Substituting the equation for Kepler's problem

$$\ddot{z} = -\frac{mz}{|z|^3},$$

in the last expression we get

$$\ddot{\zeta} + 2\frac{\dot{\zeta}\dot{z}}{z} = -\frac{m}{|z|^3}\left(\frac{\mu_2(\zeta+\mu_1)}{|\zeta+\mu_1|^3} + \frac{\mu_1(\zeta-\mu_2)}{|\zeta-\mu_2|^3 - \zeta}\right). \tag{8}$$

In the second term straightforward computation yields

$$\frac{\dot{z}}{z} = \frac{\dot{z}\bar{z}}{r^2} = \frac{2nq2(\sigma - i)}{r^2}.$$

Substitution of this expression in (8) yields

$$\ddot{\zeta} + \frac{4nq^2}{r^2}(\sigma - i)\dot{\zeta} = -\frac{m}{r^3}\left(\frac{\mu_2(\zeta+\mu_1)}{|\zeta+\mu_1|^3} + \frac{\mu_1(\zeta-\mu_2)}{|\zeta-\mu_2|^3} - \zeta\right). \tag{9}$$

Let us perform the reparametrization of time

$$\frac{dt}{ds} = \frac{1}{n}\left(\frac{r}{q}\right)^{3/2} \tag{10}$$

then using (5),

$$\begin{aligned}
\frac{d\zeta}{dt} &= \frac{nq^{3/2}}{r^{3/2}}\frac{d\zeta}{ds},\\
\frac{d^2\zeta}{dt^2} &= \frac{nq^2}{r^3}\frac{d^2\zeta}{ds^2} - \frac{3}{2}\frac{nq^{3/2}}{r^{5/2}}\frac{dr}{d\sigma}\frac{d\sigma}{dt}\frac{d\zeta}{ds}\\
&= \frac{nq^2}{r^3}\frac{d^2\zeta}{ds^2} - \frac{3n^2q^{7/2}}{r^{5/2}}\sigma\frac{d\zeta}{ds}.
\end{aligned}$$

Upon substitution in (9) yields

$$\begin{aligned}
\frac{nq^3}{r^3}\frac{d^2\zeta}{ds^2} + \frac{n^2q^{7/2}}{r^{7/2}}(\sigma - 4i)\frac{d\zeta}{ds} &= \frac{nq^3}{r^3}\left(\frac{d^2\zeta}{ds^2} + \left(\frac{q}{r}\right)^{1/2}(\sigma - 4i)\frac{d\zeta}{ds}\right)\\
&= -\frac{m}{r^3}\left(\frac{\mu_2(\zeta+\mu_1)}{|\zeta+\mu_1|^3} + \frac{\mu_1(\zeta-\mu_2)}{|\zeta-\mu_2|^3} - \zeta\right)
\end{aligned}$$

Using (4) and (2) simplifies to

$$\frac{d^2\zeta}{ds^2} + \left(\frac{\sigma - 4i}{\sigma^2+1}\right)\frac{d\zeta}{ds} = \zeta - \frac{\mu_2(\zeta+\mu_1)}{|\zeta+\mu_1|^3} - \frac{\mu_1(\zeta-\mu_2)}{|\zeta-\mu_2|^3}. \tag{11}$$

To express the anomaly σ in terms of the new independent variable s, use (5), (10), (2) and integrate to get

$$\begin{aligned}
\frac{d\sigma}{ds} &= \frac{d\sigma}{dt}\frac{dt}{ds} = \left(\frac{r}{q}\right)^{1/2} = \sqrt{\sigma^2+1}\\
\sigma &= \sinh(s).
\end{aligned}$$

Substitution of σ in (11) yields finally,

$$\frac{d^2\zeta}{ds^2} + (\tanh(s) - 4i\,\mathrm{sech}(s))\frac{d\zeta}{ds} - 2\zeta = -\frac{2\mu_2(\zeta+\mu_1)}{|\zeta+\mu_1|^3} - \frac{2\mu_1(\zeta-\mu_2)}{|\zeta-\mu_2|^3}. \tag{12}$$

Or in terms of the real and imaginary parts $\zeta = x + iy$,

$$x'' + x' \tanh(s) + 4y' \operatorname{sech}(s) = 2x - \frac{2\mu_2(x+\mu_1)}{((x+\mu_1)^2+y^2)^{3/2}} - \frac{2\mu_1(x-\mu_2)}{((x-\mu_2)^2+y^2)^{3/2}},$$
$$y'' - 4x' \operatorname{sech}(s) + y' \tanh(s) = 2y - \frac{2\mu_2 y}{((x+\mu_1)^2+y^2)^{3/2}} - \frac{2\mu_1 y}{((x-\mu_2)^2+y^2)^{3/2}},$$

where the primes (′) now denote derivatives with respect to s.

3 Compactification of phase space

System (12) can be written as

$$\zeta'' + (\tanh(s) - 4i \operatorname{sech}(s))\zeta' = \nabla\Omega \tag{13}$$

or as a non-autonomous system

$$\begin{aligned} \zeta' &= \beta, \\ \beta' &= -(\tanh(s) - 4i \operatorname{sech}(s))\beta + \nabla\Omega \end{aligned} \tag{14}$$

where

$$\Omega = |\zeta|^2 + \frac{2\mu_2}{|\zeta+\mu_1|} + \frac{2\mu_1}{|\zeta-\mu_2|},$$

System (14) has the disadvantage of being non-autonomous. In order to get an autonomous system the time variable s has to be included as a dynamical variable. Instead of performing the naive change $s' = 1$, let us introduce the variable θ by the requirement

$$\sin\theta = \tanh(s) \tag{15}$$

then (14) transforms into

$$\begin{aligned} \frac{d\theta}{ds} &= \cos\theta, \\ \frac{d\zeta}{ds} &= \beta, \\ \frac{d\beta}{ds} &= -(\sin\theta - 4i\cos\theta)\beta + \nabla\Omega \end{aligned} \tag{16}$$

Observe that the original system (14) is defined only for $\theta \in (-\pi/2, \pi/2)$, but (16) extends the system invariantly to the boundaries $\theta = \pm\pi/2$, therefore the extended phase space is

$$\mathcal{H} = [-\pi/2, \pi/2] \times (\mathbb{C}\backslash\{-\mu_1, \mu_2\}) \times \mathbb{C}. \tag{17}$$

In the following subsection we study critical elements of (16).

3.1 Critical points and homothetic orbits

The critical points of system (16) are given by

$$\theta = \pm\frac{\pi}{2}, \quad \beta = 0, \quad \nabla\Omega(\zeta^*) = 0, \tag{18}$$

and are therefore associated to central configurations[1] ζ^*. In fact for one of these configurations

$$\theta = \arcsin(\tanh(s)), \quad \beta = 0, \quad \zeta(s) = \zeta^*$$

is an homothetic orbit in the pulsating coordinate system connecting the critical points associated to the same configuration at the bound-aries $\theta = \pm\pi/2$. (In the original coordinate system the homothetic orbit is a homographic solution since $z_0(t) = z(t)\zeta^*$.)

Remark 1 The condition $\nabla\Omega(\zeta^*) = 0$ can be written as $\nabla\bar{\Omega}(\zeta) = 0$ where

$$\bar{\Omega} = \frac{1}{2}|\zeta|^2 + \frac{\mu_2}{|\zeta + \mu_1|} + \frac{\mu_1}{|\zeta - \mu_2|}$$

Comparing this one with (47) of Szebehely (1967), we obtain that the critical points coincide with those of the restricted circular three-body problem in rotating coordinates. We use the classical notation L_1 to L_3 for collinear configurations and L_4, L_5 for equilateral.

For $\theta = \pm\pi/2$ we obtain the reduced systems

$$\begin{aligned} \frac{d\zeta}{ds} &= \beta, \\ \frac{d\beta}{ds} &= \mp\beta + \nabla\Omega, \end{aligned} \tag{19}$$

respectively, which will be studied in the next section.

3.2 The gradient-like property

We derive some general properties of system (16).

Proposition 1 *Let*

$$C(\zeta, \beta) = 2\Omega(\zeta) - |\zeta'|^2 \tag{20}$$

then along solutions of system (13),

$$C' = 2\sin\theta|\zeta'|^2. \tag{21}$$

Proof Let $\langle v, \zeta\rangle = \mathrm{Re}(v\bar{\zeta})$ denote the scalar product in $\mathbb{R}^2$, then

$$\begin{aligned} \frac{C'}{2} &= \langle\nabla\Omega(\zeta), \zeta'\rangle - \langle\zeta', \zeta''\rangle \\ &= \langle\nabla\Omega(\zeta), \zeta'\rangle - \langle\zeta', -(\sin\theta - 4i\cos\theta)\zeta' + \nabla\Omega(\zeta)\rangle \\ &= \langle\nabla\Omega(\zeta), \zeta'\rangle + \langle\zeta', (\sin\theta - 4i\cos\theta)\zeta'\rangle - \langle\zeta', \nabla\Omega(\zeta)\rangle \\ &= \langle\zeta', (\sin\theta - 4i\cos\theta)\zeta'\rangle \\ &= \mathrm{Re}\left(\zeta'(\sin\theta + 4i\cos\theta)\bar{\zeta}\right) \\ &= \mathrm{Re}\left((\sin\theta + 4i\cos\theta)|\zeta'|^2\right) = \sin\theta|\zeta'|^2. \end{aligned}$$

□

[1] That is a configuration such that acceleration and position are parallel: $\nabla(\zeta^*) = \lambda\zeta^*$ for some scalar λ.

Later we will make use of the identity

$$\frac{d}{ds}|\zeta'|^2 = 2(\langle\zeta', \nabla\Omega\rangle - \sin\theta|\zeta'|^2) \tag{22}$$

Definition 1 The C^1 system $x' = g(x)$, defined in some open domain $D \subset \mathbb{R}^n$, is gradient-like with respect to a function $h(x)$ (also called a Lyapunov function) if h is strictly monotonic along solutions, which are not critical points.

A function satisfying these conditions is sometimes called a Lyapunov function. A stronger result than Proposition 1 is the following theorem.

Theorem 1 *The system* (13) *is gradient-like with respect to C in the regions*

$$\begin{aligned} \mathcal{S}^+ &= \{(\theta, \zeta, \zeta')|\theta > 0, \zeta \in \mathbb{C}\backslash\{-\mu_1, \mu_2\}, \zeta' \in \mathbb{C}\}, \\ \mathcal{S}^- &= \{(\theta, \zeta, \zeta')|\theta < 0, \zeta \in \mathbb{C}\backslash\{-\mu_1, \mu_2\}, \zeta' \in \mathbb{C}\}. \end{aligned}$$

Proof We consider the case $\theta > 0$. Using the (22) and by straightforward computation starting from the identity (21), $C' = 2\sin\theta|\zeta'|^2$

$$\begin{aligned} C''/2 &= \theta' \cos\theta|\zeta'|^2 + \sin\theta\frac{d}{ds}|\zeta'|^2 \\ &= \cos^2\theta|\zeta'|^2 + 2\sin\theta\left(\langle\zeta', \nabla\Omega\rangle - \sin\theta|\zeta'|^2\right) \\ &= (1 - 3\sin^2\theta)|\zeta'|^2 + 2\sin\theta\langle\zeta', \nabla\Omega\rangle \end{aligned} \tag{23}$$

Using again identity (22), and the last $C''/2 = (1 - 3\sin^2\theta)|\zeta'|^2 + 2\sin\theta\langle\zeta', \nabla\Omega\rangle$

$$\begin{aligned} C'''/2 = &-\sin\theta(1 + 3\cos 2\theta)|\zeta'|^2 + (2\cos 2\theta - \sin^2\theta)\langle\zeta', \nabla\Omega\rangle \\ &+ \sin\theta\langle\zeta', H\Omega\zeta'\rangle + 4\sin\theta\cos\theta\zeta'\Lambda\nabla\Omega + \sin\theta|\nabla\Omega|^2 \end{aligned} \tag{24}$$

where $H\Omega$ is the linear map defined by Hessian matrix at w and for $u = a+ib, v = c+id$ the identity $u\Lambda v \equiv ad - bc = \mathrm{Re}(\overline{iuv}) = \langle iu, v\rangle$ holds.

Now if (ζ, ζ') is not a critical point but $\zeta' = 0$ then by (21) and (23), we get $C' = 0$, $C'' = 0$ but from (24) $C''' = 4\sin\theta|\nabla\Omega|^2$. Since ζ is not a critical point then $\nabla\Omega(\zeta) \neq 0$ and $C''' > 0$, therefore C is increasing in the region $\mathcal{S}^+$. In a similar way we can prove that C is decreasing in the region $\mathcal{S}^-$. □

Remark 2 The function C in the Proposition 1 is analogous to Jacobi integral in the restricted circular three-body problem, *but not the same though*. In fact in the restricted circular three-body problem the Jacobi integral would be $2\bar{\Omega} - |\zeta'|^2 = 4\Omega - |\zeta'|^2$, instead of (20).

Remark 3 The proof of Theorem 1 applies in particular to the invariant system $\theta = \pi/2$, where C is increasing except at critical points. Similarly C is decreasing along non-critical solutions of the invariant system $\theta = -\pi/2$.

4 Dynamics on the boundary manifolds

It is sufficient to study the system at infinity for $\theta = \pi/2$:

$$\begin{aligned} \zeta' &= \beta, \\ \beta' &= -\beta + \nabla\Omega(\zeta). \end{aligned} \tag{25}$$

In fact, changing (ζ,β,s) by $(\zeta,-\beta,-s)$ transforms the asymptotic system for $\theta = \pi/2$ to the corresponding system for $\theta = -\pi/2$:

$$\begin{aligned} \zeta' &= \beta, \\ \beta' &= \beta + \nabla\Omega(\zeta). \end{aligned} \tag{26}$$

According to (18) and the Remark 1, the number of critical points are as follows.

Proposition 2 *There are exactly five critical points of system* (25) *corresponding to Lagrangian and Eulerian configurations of the zero mass particle with respect to the primaries.*

The linearization of (25) at a critical point $\zeta^*, \beta = 0$ has the form

$$\begin{pmatrix} \delta & \zeta \\ \delta & \beta \end{pmatrix}' = \begin{pmatrix} 0 & I \\ B & -I \end{pmatrix} \begin{pmatrix} \delta & \zeta \\ \delta & \beta \end{pmatrix} \tag{27}$$

where B is the Hessian of Ω at ζ^* and I is the 2×2 identity matrix.

Lemma 1 *Let e be an eigenvector of B with eigenvalue* κ, *then* $\lambda_\pm = \frac{1}{2}(-1 \pm \sqrt{1+4\kappa})$ *are eigenvalues of (27) with eigenvectors* $(e, \lambda_\pm e)^{\mathrm{T}}$.

Proof Let $(e,f)^{\mathrm{T}}$ be an eigenvector with eigenvalue λ for (27) then

$$\begin{pmatrix} 0 & I \\ B & -I \end{pmatrix} \begin{pmatrix} e \\ f \end{pmatrix} = \lambda \begin{pmatrix} e \\ f \end{pmatrix}$$

can be written as

$$\begin{aligned} f &= \lambda e, \\ Be - f &= \lambda f \end{aligned}$$

or substituting for f

$$Be = \lambda(\lambda + 1)e$$

therefore, if κ is an eigenvalue of B with eigenvector e then $\lambda^2 + \lambda = \kappa$, thus

$$\lambda_\pm = \frac{1}{2}\left(-1 \pm \sqrt{1+4\kappa}\right) \tag{28}$$

are the eigenvalues of (27) with eigenvectors $(e,f)^{\mathrm{T}} = (e, \lambda \pm e)^{\mathrm{T}}$.

The stability of the critical points of the restricted parabolic three-body problem can be investigated taking advantage the results obtained for the circular three-body as given in Szebehely (1967, pp. 141–230). In what follows we will use an overbar for quantities referred to Szebehely (1967).

The stability of the Lagrangian is given as follows.

Proposition 3 *The critical points of the invariant system* (25) *associated to equilateral triangle configurations* L_4, L_5 *are of the saddle type*

$$\dim W^u_{L_k} = \dim W^s_{L_k} = 2, \quad k = 4, 5.$$

Proof For the Lagrangian points, from Szebehely (1967, p. 144), the Hessian matrix is $B = 2\bar{B}$ where

$$\bar{B} = \begin{pmatrix} \frac{3}{4} & \pm 3\frac{3^{1/2}}{2}(\mu - \frac{1}{2}) \\ \pm 3\frac{3^{1/2}}{2}(\mu - \frac{1}{2}) & \frac{9}{4} \end{pmatrix}$$

$\mu = \mu_2$, and the positive sign corresponds to L_4, the negative sign to L_5, therefore the eigenvalues of B are twice those of $\bar{B}$. From Szebehely (1967, p. 144), it is read

$$\kappa_{1,2} = 3\left(1 \pm \sqrt{3\left(\mu_2 - \frac{1}{2}\right)^2 + \frac{1}{4}}\right)$$

(subindex 1 belongs to the positive sign, subindex 2 to the negative sign), since for $0 < \mu_2 < 1$, the radical varies between $\frac{1}{4}$ and 1, both roots $\kappa_{1,2}$ are positive. According to Lemma 1, the eigenvalues for the linearization of (27) are:

$$\text{for } \kappa_1 : \lambda_{\pm}^{(1)} = \frac{1}{2}\left(-1 \pm \sqrt{1 + 12\left(1 + \sqrt{3\left(\mu_2 - \frac{1}{2}\right)^2 + \frac{1}{4}}\right)}\right), \tag{29}$$

$$\text{for } \kappa_2 : \lambda_{\pm}^{(2)} = \frac{1}{2}\left(-1 \pm \sqrt{1 + 12\left(1 - \sqrt{3\left(\mu_2 - \frac{1}{2}\right)^2 + \frac{1}{4}}\right)}\right), \tag{30}$$

but for $\mu_2 \in (0, 1/2]$,

$$18 \leq 12\left(1 + \sqrt{3\left(\mu_2 - \frac{1}{2}\right)^2 + \frac{1}{4}}\right) < 24,$$

$$0 < 12\left(1 - \sqrt{3\left(\mu_2 - \frac{1}{2}\right)^2 + \frac{1}{4}}\right) \leq 6$$

then each pair of roots of (29) and (30) are real and have opposite signs, so the dimensions of the unstable and stable manifolds are two. □

In contrast to the restricted planar circular three-body problem, where the linear stability of the Lagrangian points depends on the Routh's value, for the restricted parabolic three-body problem, they are unstable for any value of the masses, μ_2.

Proposition 4 *The critical points of the invariant system* (25) *associated to collinear configurations* L_1 *to* L_3 *are of the saddle type. The dimensions of the unstable and stable manifolds are*

$$\dim W_{L_i}^u = 1, \quad \dim W_{L_i}^s = 3 \qquad i = 1, 2, 3.$$

Proof For the collinear configurations L_1 to L_3 the explicit expression of the coordinate x depends on a quintic equation but the Hessian matrix is diagonal $B = \mathrm{diag}(\kappa_1, \kappa_2)$ with $\kappa_1 > 0$ and $\kappa_2 < 0$ for all values of the masses (see Szebehely 1967). From the expression (28), it follows that for $\kappa_1 > 0$, the roots

$$\lambda_{\pm}^{(1)} = \frac{1}{2}\left(-1 \pm \sqrt{1 + 4\kappa_1}\right)$$

are real and have opposite signs. For $\kappa_2 < 0$, if $1 + 4\kappa_2 > 0$ then the two eigenvalues

$$\lambda_{\pm}^{(2)} = \frac{1}{2}\left(-1 \pm \sqrt{1 + 4\kappa_2}\right)$$

are real and negative, and if $1 + 4\kappa_2 < 0$ then the roots are complex with negative real part. In summary, there is one positive eigenvalue and three eigenvalues with negative real part, so $\dim W^u_{L_i} = 1$ and $\dim W^s_{L_i} = 3$, for $i = 1, 2, 3$. □

For completeness, we state the corresponding dimensions of the invariant manifolds for the invariant system $\theta = -\pi/2$.

Proposition 5 *The critical points of the invariant system $\theta = -\pi/2$ associated to equilateral triangle configurations L_4, L_5 are of the saddle type.*

$$\dim W^u_{L_k} = \dim W^s_{L_k} = 2 \quad k = 4, 5.$$

Proposition 6 *The critical points of the invariant system $\theta = -\pi/2$ associated to collinear configurations L_1 to L_3 are of the saddle type. The dimensions of the unstable and stable manifolds are*

$$\dim W^u_{L_i} = 3, \quad \dim W^s_{L_i} = 1, \qquad i = 1, 2, 3.$$

We end up this section by considering the linearization of the global system (16) at the critical points. For $\theta = \pm\pi/2$ it has the form

$$\begin{pmatrix} \delta\theta \\ \delta\zeta \\ \delta\beta \end{pmatrix}' = \begin{pmatrix} \mp & 0 & 0 \\ 0 & 0 & I \\ 0 & B & -I \end{pmatrix} \begin{pmatrix} \delta\theta \\ \delta\zeta \\ \delta\beta \end{pmatrix} \tag{31}$$

where the minus sign corresponds to $\theta = \pi/2$ and the plus sign corresponds to $-\pi/2$. Thus, the stable (resp. unstable) manifolds of the invariant system $\theta = \pi/2$ (resp. $\theta = -\pi/2$) alluded in Propositions 3, 4 (resp. 5, 6) add one dimension. This extra dimension comes out as the result of the homothetic orbits, which connect the critical points at $\theta = -\pi/2$ with the corresponding at $\theta = \pi/2$.

Theorem 2 *The dimensions of the stable/unstable manifolds of the critical points of the global system* (16) *are summarized in the following table.*

$\theta = \pi/2$	L_1 to L_3	L_4, L_5	$\theta = -\pi/2$	L_1 to L_3	L_4, L_5
W^s	*4*	*3*	W^s	*1*	*2*
W^u	*1*	*2*	W^u	*4*	*3*

5 Structure of parabolic escape orbits

Saari has shown Saari (1971) that when the particle separations are $\sim t^{2/3}$, without any further assumption on the value of the energy, the solutions for the n body problem asymptotically approach a central configuration as $t \to \infty$. However he could not ascertain whether the particles spiralled in or approached a fixed central configuration. Hulkower (1978) considered the three-body problem with zero energy. Since

Chazy's (1922) classification, this is the only case for the three-body problem where all the particles can separate like $t^{2/3}$: Hulkower proved that parabolic escapes tend asymptotically to a central configurations without "infinite" spin, that is, neither of the masses librates around the center of mass.

Let us characterize for the restricted parabolic problem, with reference to the pulsating coordinate system, the kind of motions where all the particles escape like $t^{2/3}$ in the original frame. Recall that z_{0i} denotes the relative position of the mass particle with respect to the i-th primary, $i = 1, 2$, and z_0 its position relative to the center of mass; the relative position of the primaries is $z(t)$ and is parabolic.

Proposition 7 *Let $(\zeta(s), \beta(s))$ be a solution such that $\zeta(s)$ tends to a non-collision configuration ζ^* as $s \to \infty$. Then $|z_{01}(t)|, |z_{02}(t)|, |z_0(t)| \sim t^{2/3}$ as $t \to \infty$. Moreover as $s \to \infty, \zeta'(s)$ remains bounded if and only if $\dot{z}0(t) \to 0$.*

Proof First, notice that $s \to \infty$ if and only if $t \to \infty$. This follows from the changes of variables

$$n(t - T) = \sigma + \frac{\sigma^3}{3}, \qquad \sigma = \sinh(s).$$

Let $z_1(t) = \mu_2 z(t), z_2(t) = -\mu_1 z(t)$ be the position of the primaries relative to the center of mass at origin, then

$$\begin{aligned}
|z_{10}| &= |z_0 - z_1| = |\zeta z - \mu_2 z| = |z||\zeta - \mu_2|, \\
|z_{20}| &= |z_0 - z_2| = |\zeta z + \mu_1 z| = |z||\zeta + \mu_1|, \\
|z_0| &= |z\zeta| = |z||\zeta|
\end{aligned}$$

therefore, if $\zeta(s) \to \zeta^* \neq \mu_2, -\mu_1$ as $s \to \infty$ then $|z_{10}(t)|, |z_{20}(t)|$ tend to infinity asymptotically to $|z(t)| \sim t^{2/3}$. To proof the second statement, apply the chain rule

$$\begin{aligned}
\dot{z}_0(t) &= \zeta\dot{z} + \frac{d\zeta}{dt} z = \zeta\dot{z} + \frac{d\zeta}{ds}\frac{ds}{dt} z \\
&= \zeta\dot{z} + \zeta' n q^{3/2} r^{-3/2} = \zeta\dot{z} + \zeta' n (\sigma^2 + 1)^{-3/2} \\
&= \zeta\dot{z} + \zeta' n \cosh^{-3}(s) = \zeta\dot{z} + \zeta' n \operatorname{sech}^3(s)
\end{aligned} \tag{32}$$

Now, as $s \to \infty, \dot{z} \to 0, \operatorname{sech}^3(s) \to 0$ and $\zeta(s) \to \zeta^*$, it follows that if $\zeta'(s)$ remains bounded then $\dot{z}_0(t) \to 0$. Suppose now that ζ' remains bounded, say $|\zeta'| \leq M$. Taking out ζ' from (32)

$$n\zeta' = (\dot{z}_0 - \zeta\dot{z})\cosh^3(s) = (\dot{z}_0 - \zeta\dot{z})(\sigma^2 + 1)^{3/2}$$

then

$$|\dot{z}_0| - |\zeta\dot{z}| \leq |\dot{z}_0 - \zeta\dot{z}| \leq \frac{nM}{(\sigma^2 + 1)^{3/2}}$$

thus

$$|\dot{z}_0| \leq |\zeta||\dot{z}| + \frac{nM}{(\sigma^2 + 1)^{3/2}} \to 0$$

since ζ is bounded and $\dot{z} \to 0$. □

Following Chazy's nomenclature, let us define parabolic escape orbits.

Definition 2 Let P^+ (resp. P^-) be the set of initial conditions leading to a motion of the infinitesimal mass $z_0(t)$, such that $z_0(t), \sim t^{2/3}$ and $\dot{z}_0(t) \to 0$ as $t \to \infty$ (resp. $t \to \infty$). We call P^+ (resp. P^-) the set of forward (resp. backward) escape parabolic orbits.

Of course the relationships of the relative positions and velocities

$$z_{01} = z_0 - \mu_2 z, \quad z_{02} = z_0 + \mu_1 z$$
$$\dot{z}_{01} = \dot{z}_0 - \mu_2 \dot{z}, \quad \dot{z}_{02} = \dot{z}_0 + \mu_1 \dot{z}$$

and for parabolic two-body motion

$$z(t) \sim t^{2/3}, \quad \dot{z}(t) \to 0 \qquad \text{as } t \to \pm\infty,$$

show that in fact, for parabolic escape orbits the three mutual distances are $\sim t^{2/3}$ and the relative velocities tend to zero. This is Chazy's definition of parabolic motions for the general three-body problem.

Theorem 3 *For the restricted parabolic three-body problem, the set of parabolic orbits P^+ is a union of immersed manifolds of dimensions 4 and 3; the four-dimensional manifold consists of orbits tending asymptotically to an equilateral triangle configuration and the three-dimensional manifold consists of orbits tending asymptotically to a collinear central configuration.*

Proof Proposition 7 shows that the stable manifolds of the critical points L_i, $i = 1, \ldots, 5$ are in fact parabolic orbits: $|\zeta(t)| \to \infty$ and $\dot{\zeta}(t) \to 0$ as $t \to +\infty$. The proof then follows from Theorem 2. □

6 A geometric criterion for elliptic–parabolic motion

In this section we recall Merman's criterion for elliptic-parabolic motion and introduce a new more geometrical criterion based on the topology of the Hill's regions.

In a series of papers, Merman (1953a, b, 1954b) gave sufficient conditions for preservation of hyperbolic–elliptic and parabolic-elliptic motion. In the restricted parabolic three-body problem, one of the criteria of elliptic–parabolic motion given by Merman (1953a) was used by N. G. Kocina to show a numerical example of elliptic–parabolic motion. We state one of such criteria due to Merman, then introduce a geometric criterion of such motions based on the Hill's regions.

According to Merman's (1954b) notation for the restricted parabolic three-body problem, $\mathbf{r}_{ij}$ denotes the relative position vector from m_i to m_j , where m_1, m_2 are the primaries and m_0 is the massless particle; φ denotes the angle between the the radius vector $\mathbf{r}_{21}$ and the x-axis.

Proposition 8 (*Merman*). *If at initial instant $t = 0$ we can choose constants $R > 0$ and $\alpha > 1$, such that*

$$r_{21}(0) \geq \alpha R, \quad r_{21}(0) \leq a_1 r_{20}(0),$$
$$\frac{1}{2a} = \frac{m_1}{r_{01}(0)} - \frac{|\dot{r}_{01}(0)|^2}{2} \geq \frac{m_1}{R} + B\frac{\varphi(0) - \sin\varphi(0)}{2nq^3}$$

where

$$a_1 \geq \frac{1}{1-\frac{1}{\alpha}}, \quad 2B = m_2\sqrt{2m_1 R}(2a_1^3 + a_1^2 + a_1)$$

then the motion is parabolic–elliptic for $t \to +\infty$ and $|\mathbf{r}_{01}(t)|$ is bounded superiorly.

Similar criterion exists for $t \to -\infty$, see Merman (1954b).

Kocina (1954) used the parametrization of the motion of the primaries given in Sect. 2. He integrated the equations of motion in the interval $\sigma \in [-12.2, 7.7]$ using Kowell's method. In the interval [2, 7.7] with a step size of 0.0125 and using the equations of motions with origin at the mass m_1, he was able to satisfy Merman's criterion for $\sigma = 7.7$. In the interval $[-12.275, 2]$, he is unable to satisfy Merman's criterion at $\sigma = -12.2$, he then estimates the variations of the osculatory elements from -12.2 to -14.200 and for this value of σ Merman's criterion could be satisfied. He considered the case of equal masses $m_1 = m_2 = 1, q = 10$, and $2nq^3 = 63.24556$ obtaining the results of the following table.

σ	R	α	a_1	B	$(2a)^{-1}$	$R^{-1} + B\frac{\varphi - \sin\varphi}{2nq^3}$
-14.200	400	5.066	1.246	94.289	0.0034518	0.0031902
7.7	200	3.014	1.497	104.48	0.010104	0.0097288

Merman's criterion is satisfied whenever the value in the sixth column is greater than the corresponding value of the seventh column.

In Fig. 1, we show the orbit found by Kocina (1954).

6.1 The geometric criterion

Using the results of Sect. 3.2, we now state a geometric criterion of capture using the Jacobian function C.

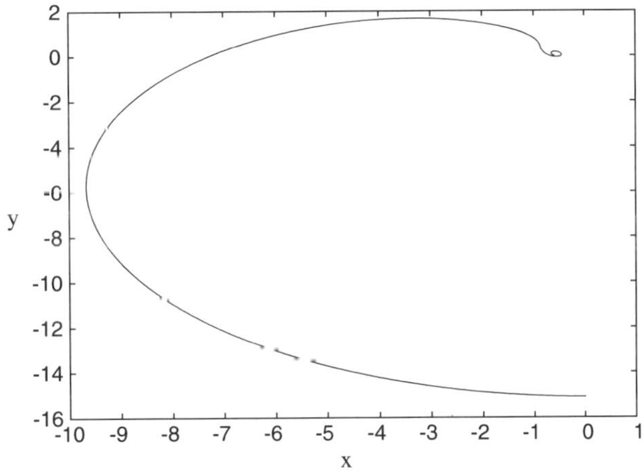

Fig. 1 Kocina's orbit

We first recall some basic facts derived from the gradient-like property. In order to avoid confusion we use the symbol C^*, with subscript * whenever we want to fix a value rather than the Jacobian function $C(\zeta, \zeta')$. The type of conditions in Theorem 4 will be shown to be sufficient.

Since $|\zeta'|^2 = 2\Omega(\zeta) - C \geq 0$, and C increases along solution when $s \to \infty$, when comparing the initial value of C, say $C_0' = C(\zeta(0), \beta(0))$ and its value at a later time $s > 0, C_s = C(\zeta(s), \beta(s))$ then $C_s > C_0'$, which in turns implies

$$2\Omega(\zeta(s)) \leq 2\Omega(\zeta(s)) - |\beta(s)|^2 \equiv C_s > C_0$$

or

$$\Omega(\zeta(s)) > \frac{C_0}{2}$$

for all $s > 0$. Therefore, each solution is "trapped" in its initial Hill's region $\Omega > C_0/2$. In fact the same argument shows that if $s_2 > s_1 > 0$ then

$$\Omega(\zeta(s_2)) > \frac{C_{s_1}}{2}.$$

Definition 3 *The Hill's region corresponding to the value C_* is*

$$M_{C_*} = \{\zeta \in \mathbb{C}\backslash\{-\mu_1, \mu_2\} | \Omega(\zeta) > C_*/2\}.$$

Its boundary ∂M_{C_*} will be called the zero–velocity curve.

The boundary of Hill's regions is given by the called zero–velocity curves since $\beta = \zeta' = 0$. The following proposition shows that a solution that enters region M_{C_0} forwards in time immediately leaves it.

Proposition 9 *Let $(\zeta(s), \beta(s))$ be a non-stationary solution of the global system* (16) *and let $C_0 = C(\zeta(0), \zeta'(0))$ If $\Omega(\zeta(0)) = C_0/2$ then*

$$(\Omega \circ \zeta)'(0) = 0 \quad \textit{but } (\Omega \circ \zeta)''(0) > 0.$$

In particular, if $\zeta(0) \in \partial M_{C_0}$, then Ω $(\zeta(s))$ has a local minimum at $s = 0$ and there exists $\delta > 0$, such that $\zeta(s) \in \mathrm{Int}(M_{C_0})$ for $0 < |s| < \delta$.

Proof The proof is similar to the gradient-like property. By the chain rule

$$\begin{aligned}
(\Omega \circ \zeta)' &= \langle \nabla\Omega, \zeta' \rangle, \\
(\Omega \circ \zeta)'' &= \langle H\Omega \cdot \zeta', \zeta' \rangle + \langle \nabla\Omega, \zeta'' \rangle \\
&= \langle H\Omega \cdot \zeta', \zeta' \rangle + \langle \nabla\Omega, -(\sin\theta - 4i\cos\theta)\zeta' + \nabla\Omega \rangle \\
&= \langle H\Omega \cdot \zeta', \zeta' \rangle - \sin\theta \langle \nabla\Omega, \zeta' \rangle + 4\cos\theta \nabla\Omega \Lambda \zeta' + |\nabla\Omega|^2
\end{aligned}$$

where $H\Omega$ is the adjoint map associated to the Hessian matrix. Evaluating at $s = 0$ it is obtained $(\Omega \circ \zeta)'(0) = 0$, but $(\Omega \circ \zeta)''(0) = |\nabla\Omega(\zeta(0))|^2$. □

We denote by C_i the value of the function C at the critical point L_i, $i = 1, 2, \ldots, 5$. A short computation shows that for equal masses the values are the following:

$$\frac{11}{2} = C_4 = C_5 < C_1 = C_3 = 6.913592448172306 \cdots < C_2 = 8. \tag{33}$$

For general masses, the critical values are shown in Fig. 2. In fact

$$2(3-\mu+\mu^2) = C_4 = C_5 < C_3 < C_1 < C_2$$

holds for $\mu \in (0, 1/2)$. In Figs. 3, 4 and 5 the Hill's regions for values nearby the critical values are shown.

Theorem 4 (C-criterion of capture for equal masses). *For a non-stationary solution* $(\zeta(s), \zeta'(s))$*:*

1. *If for some time, $s_0 > 0$ it is verified that $C(\zeta(s_0), \zeta'(s_0) \geq C_1$ then for $s \geq s_0$, either $(\zeta\ (s), \beta(s))$ is a capture orbit or the solution tends asymptotically to L_2.*
2. *If for some time s_0 it is verified that $C(\zeta(s_0), \zeta'(s_0) \geq C_2$ and $\zeta(s_0) \leq 0$ (resp. $\zeta\ (s_0) \geq 0$) then for $s \geq s_0$, $(\zeta(s), \beta(s))$ is a capture orbit by the left primary at $-\mu_2$ (resp. right primary at μ_1).*

Proof Let us proof the first statement: by Theorem 1, the set defined by the inequality $C \geq C_1$ is positively invariant and projects into configuration space to the nut-shaped region M_{C_1}, which contains the primaries and the configuration L_1. Since the flow is gradient-like then the solutions tends to a critical point or $C(\zeta(s), \zeta'(s))$ is unbounded

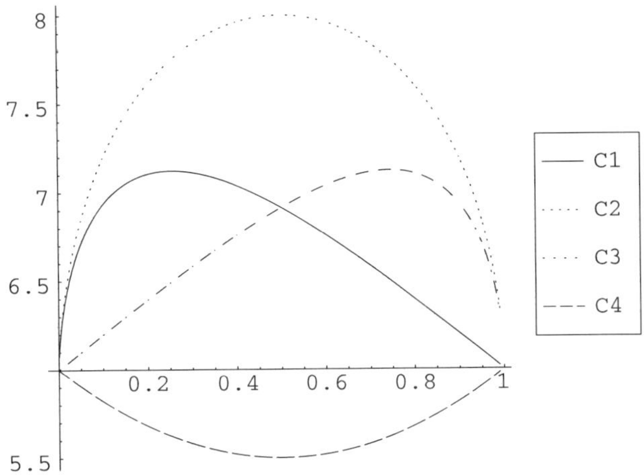

Fig. 2 The critical values C_i as function of the mass parameter μ in the range (0,1)

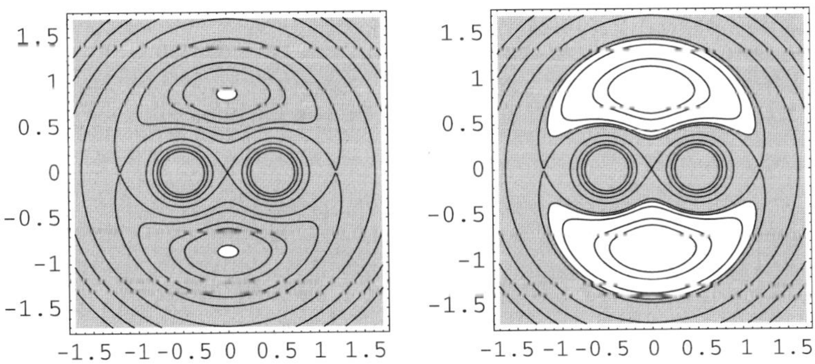

Fig. 3 Hill's region $\Omega \geq C_4/2 + \epsilon$ and $\Omega \geq C_4/2 + 2\epsilon$, for $\epsilon > 0$, small

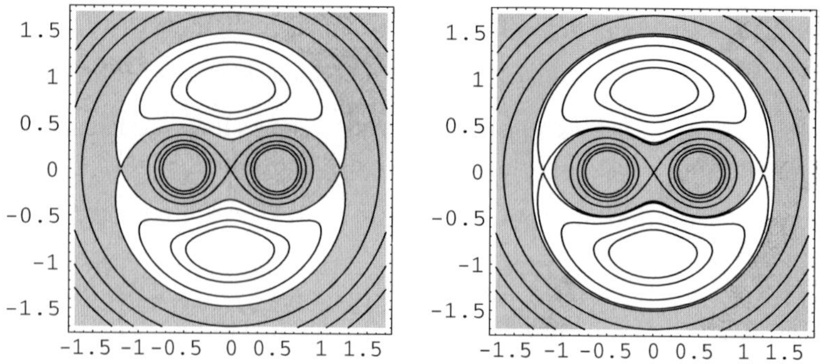

Fig. 4 Hill's region $\Omega \geq C_1/2$ and $\Omega \geq C_1/2+\epsilon$, for $\epsilon > 0$, small

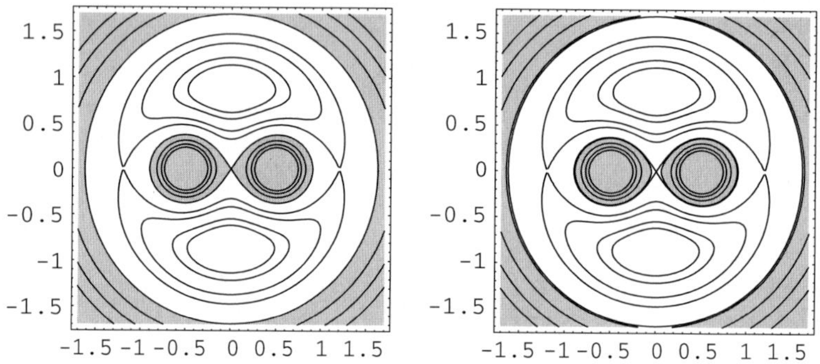

Fig. 5 Hill's region $\Omega \geq C_2/2$ and $\Omega \geq C_2/2+\epsilon$, for $\epsilon > 0$, small

from below in the last case $C(\zeta(s),\zeta'(s)) \to +\infty$; from its definition, $C = 2\Omega - |\zeta'|^2$, which this can only be true if $\Omega(\zeta(s)) \to \infty$, that is, if $\zeta(s)$ tends to collision with any of the boundaries. The second statement is proved similarly, except that in the case $C > C_2$ there are no critical points. □

Tracking back the reparametrizations and changes of variables one can translate this criterion to the original coordinates. The following table shows that the C-criterion $C > C_2 = 8$ is satisfied for Kocina's example at the extremes of the interval of integration.

σ	s	x	y	x'	y'	$\Omega(x,y)$	C
−14.2	−3.34763	−0.510113	−0.0291475	−7.06951	−0.685905	33.663389	16.8783
−12.2	−3.19626	−0.579429	−0.0198056	−2.34265	−2.78979	13.478215	13.6855

7 Heteroclinic connections

We have shown in Sect. 3.1 that critical points at $\theta = -\pi/2$ connect with those at $\theta = \pi/2$ along the homothetic orbits. Therefore, in order to fully understand the global dynamics of the problem, all possible heteroclinic connections among the critical point on the invariant system $\theta = \pi/2$ have to be determined (the invariant system $\theta = -\pi/2$ can be obtained by symmetry). This is a hard problem though, due to the high dimension of the phase space, which is four. Let us then consider the invariant system $\theta = \pi/2$; according to Fig. 3, the unstable manifold $W^u(L_4)$ (which is two-dimensional according to Theorem 5) "spreads out" in the shaded region of Fig. 3 with no further restriction, so we expect an open set of initial conditions relative to $W^u(L_4)$, which escape to infinity towards the upper shaded region. Also some solutions starting in $W^u(L_4)$ project to curves passing over the critical collinear configurations. A natural question is that if we can adjust the velocity in such a way that some such solution tends to a collinear configuration with asymptotic velocity tending to zero, that is, does $W^u(L_4)$ connect with any of $W^u(L_i)$, $i = 1,2,3$? We will give a partial answer for the equal mass case in Sect. 7.2. In the next section, we will study the final phase of the simplest case – the unstable manifolds associated to the collinear configurations. For this purpose, it is convenient to introduce collisions with the primaries in the same vein as if it were critical points. In what follows $(\zeta(s,\xi), \zeta'(s,\xi))$, will denote the solution of the invariant system (25), which passes through $\xi \in \mathbb{C}^2$ at $s = 0$.

7.1 Connections of the unstable manifolds associated to collinear configurations

In Fig. 4, the Hill's regions for $\Omega \geq C_1$ and $\Omega \geq C_1 + \epsilon$ for small positive ϵ are shown. The unstable manifold $W^u(L_1)$ (or its projection into configuration space) is trapped in the nut-shaped region contain-ing the primaries. Then, either $W^u(L_1)$ goes to collision with one of the primaries or connects with $W^u(L_2)$. According to Proposition 4, $W^u(L_1)$ is one-dimensional and as will be shown in Proposition 10, it lies completely on the plane $\mathrm{Im}(\zeta) = \mathrm{Im}(\zeta') = 0$ (i.e., it projects along the real axis $\mathrm{Im}(\zeta) = 0$), so necessarily one of the branches ends up in collision with the left binary.

In Fig. 5, the Hill's regions for $\Omega \geq C_2$ and $\Omega \geq C_2 + \epsilon$ for small positive ϵ are shown. The unstable manifold $W^u(L_2)$ (or its projection on the configuration space) is trapped into two disks containing the primaries. Then each of the two component of the one-dimensional manifold $W^u(L_2)$ (see Proposition 4) goes to collision with a different primary. Let us state this result formally.

Definition 4 The stable (unstable) set associated to binary collision with the primary μ_1 located at μ_2 is

$$W^{s,u}(B_{\mu_1}) = \{\xi = (\zeta_0, \zeta_0') \mid \exists s_n \to \pm\infty,\ \zeta(s_n,\xi) \to \mu_2\}.$$

The stable (unstable) set associated to binary collision with the primary μ_2 located at $-\mu_1$ is

$$W^{s,u}(B_{\mu_2}) = \{\xi = (\zeta_0, \zeta_0') \mid \exists s_n \to \pm\infty, \zeta(s_n,\xi) \to -\mu_1\}.$$

The stable (unstable) set to infinity

$$W^{s,u}(\infty) = \{\xi \in \mathbb{C}^2 | \zeta(s,\xi) \to \infty \ as \ s \to \pm\infty\}.$$

Proposition 10 *For any value of the masses the following connections occur. More precisely,*

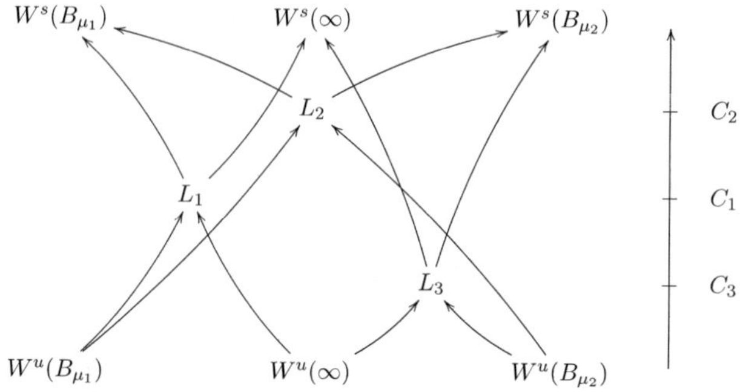

1. *One of the branches of the one-dimensional manifold* $W^u(L_1)$ (*see Proposition* 4) *ends up in double collision with the primary at* $-\mu_1$, *the other escapes to infinity, that is,* $\zeta(s) \to -\infty$ *along the real axis* $Im(\zeta) = 0$.
2. *One of the branches of the one-dimensional manifold* $W^u(L_3)$ (*see Proposition* 4) *ends up in double collision with the primary at* μ_2, *the other escapes to infinity, that is,* $\zeta(s) \to \infty$ *along the real axis* $Im(\zeta) = 0$.
3. *One of the branches of the one-dimensional manifold* $W^u(L_2)$ (*see Proposition* 4) *ends up in double collision with the primary at* $-\mu_1$, *the other ends up in double collision with the primary at* μ_2.

Proof It remains to show that the real axis is invariant in reduced systems (19) for all values of the masses, but it is a consequence of the symmetry

$$(\zeta,\beta,s) \to (\bar{\zeta},\bar{\beta},s) \tag{34}$$

□

Corollary 1 *For any value of the masses the connection diagram for the invariant system* (26) *is obtained from that of Proposition 10 by reversing the arrows and exchanging the superindex of stable for unstable and vice versa.*

For completeness we describe the phase portrait for the invariant system (25) restricted to the real axis $\mathrm{Im}(\zeta) = \mathrm{Im}(\zeta') = 0$. In that case, system (25) reduces to

$$x'' = -x' + \nabla\Omega(x,0) \tag{35}$$

where $x = \mathrm{Re}(\zeta)$ and

$$\Omega(x,0) = x^2 + \frac{2\mu_2}{|x+\mu_1|} + \frac{2\mu_1}{|x-\mu_2|}.$$

Proposition 11 *The phase portrait of system* (35) *is topologically conjugate to that shown in Fig.* 6.

Proof The potential function $\Omega(x,0)$ is positive, convex and tends to infinity when $x \to \pm\infty$ or $x \to -\mu_1, \mu_2$. In fact

$$\Omega_{xx}(x,0) = 2 + \frac{4\mu_2}{|x+\mu_1|} + \frac{4\mu_1}{|x-\mu_2|} > 0$$

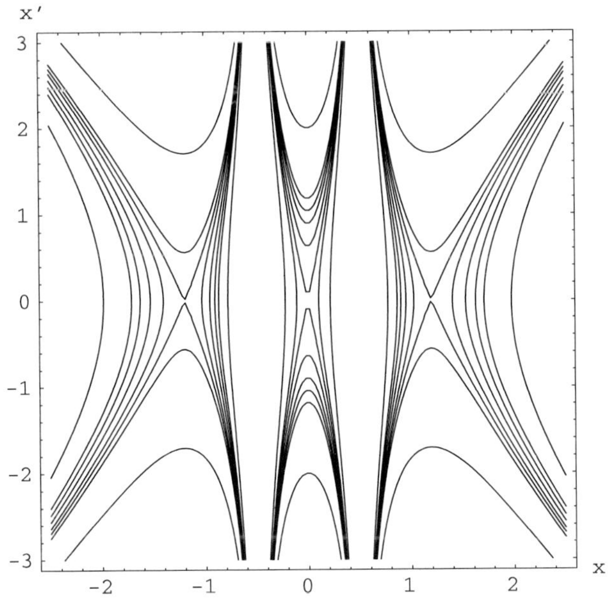

Fig. 6 Phase portrait of the Hamiltonian system (36)

The Hamiltonian system associated to (35) is

$$x'' = \nabla\Omega(x, 0) \tag{36}$$

Its phase portrait follows immediately from the form of the potential $\Omega(x, 0)$, and is shown in Fig. 6. System (35) has the same critical points but differs from (36) in the friction term $-\zeta'$, then the solutions of the actual system (35) crosses the level sets of Fig. 6, keeping the saddle character of the critical points. Therefore, the phase portrait is only slightly distorted from that in (6), or is topologically the same. In fact one can use the gradient flow of the Jacobian function $C(x, x', 0, 0) = x'^2/2 - \Omega(x, 0)$ to define an explicit topological conjugacy. □

7.2 The case of equal masses

For equal masses, an additional symmetry of (25) holds:

$$(\zeta, \beta, s) \to (-\zeta, -\beta, s)$$

composed with symmetry (34) lead to the new symmetry

$$(\zeta, \beta, s) \to (-\zeta, -\beta, s)$$

having $\mathrm{Re}(\zeta) = \mathrm{Re}(\beta) = 0$, as fixed points, therefore is invariant. System (25) restricted to this invariant plane has the form

$$y'' = -y' + \nabla\Omega(0, y) \tag{37}$$

where

$$\Omega(0, y) = y^2 + \frac{2}{\sqrt{\frac{1}{4} + y^2}}.$$

Proposition 12 *The phase portrait of system (37) is topologically conjugate to that shown in Fig. 7.*

Proof The idea is to compare solutions of (37) with the associated Hamiltonian system

$$y'' = \nabla\Omega(0, y) \tag{38}$$

whose phase portrait is shown in Fig. 8. There is a double heteroclinic connection among L_4 and L_5 contained in the level set $C = C_4 = 11/2$, which is destroyed by the friction term in (37) forcing a connection of L_4 and L_5 with L_2. Let us formalize the argument: the behavior of the orbits in Fig. 8 is consequence of the fact that the partial derivative of the potential function

$$\Omega_y = 2y\left(1 - \frac{1}{\sqrt{\frac{1}{4} + y^2}}\right)$$

has exactly three critical points: $y = \pm\sqrt{3}/2$ and $y = 0$. Since $C = 2\Omega - y'^2$, then at a critical point, Hess $C = \mathrm{diag}(\Omega_{yy}, -2)$; therefore minimum for Ω is obtained from the inequality $\Omega_{yy} > 0$, which implies that the critical point is a saddle for C. In the same way, we get that maximum for Ω is obtained from the inequality $\Omega_{yy} < 0$, which implies that the critical point is a maximum for C.

Since Ω has local minima at the Lagrangian configurations $y = \pm\sqrt{3}/2$ then the critical points $L_{4,5}$ are saddles for C with value C_4; Ω has local maximum at the collinear configuration $y = 0$ then the critical point L_2 is a local maximum with value C_2.

Now for system (37) the flow is gradient-like with respect to C, then the component D of the region $\{(y, y') \,|C_4 \leq C(0, 0, y, y') \leq C_2\}$, which projects into the bounded

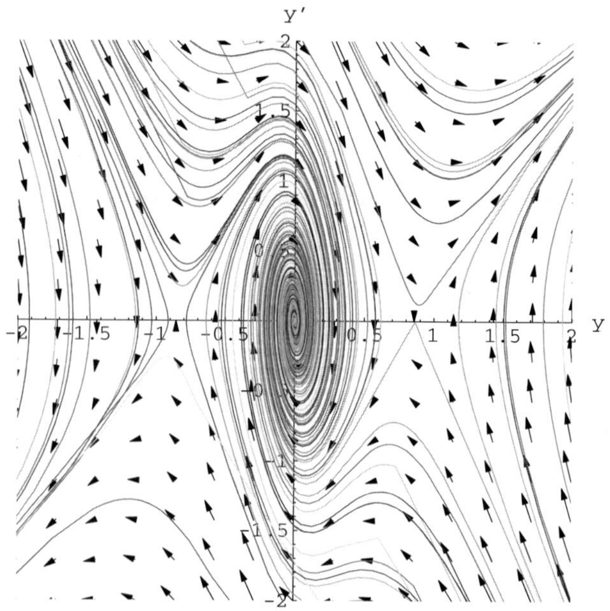

Fig. 7 Phase portrait of the Hamiltonian system (37)

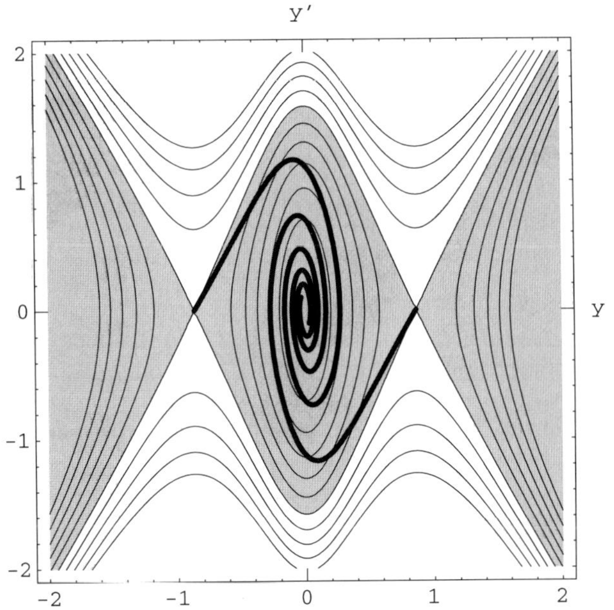

Fig. 8 Phase portrait of the Hamiltonian system (38) and unstable branches (*dashed*) of L_4, L_4 for system (37) connecting to L_2. The *shaded region* is M_{C_4}

component of the Hills region M_{C_4} containing the origin, is compact and positively invariant, then it follows that the unstable manifolds have to point onto the interior of D and then its ω-limit is L_2. □

Proposition 13 *For equal masses the following connections occur.*

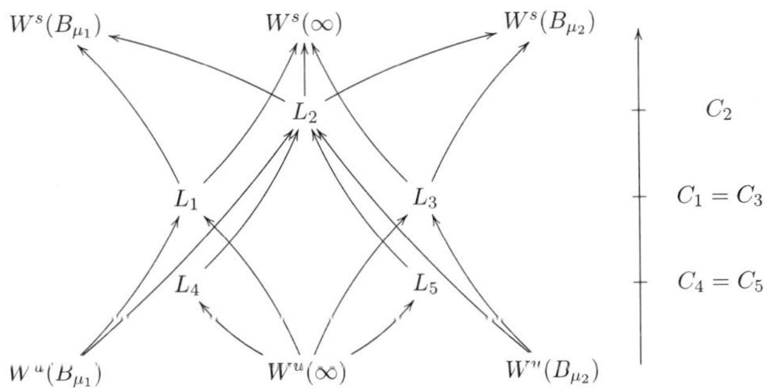

Corollary 2 *For equal the masses the connection diagram for the invariant system* (26) *is obtained from that of Proposition* 13 *by re-versing the arrows and exchanging the superindex of stable for unstable and vice versa.*

8 Dynamical consequences

From the connection diagrams given in Propositions 10, 13 and its Corollaries 1, 2 a number of dynamical consequences can be obtained, which we expect to be of some interest to astronomers. We enunciate some of them in the original coordinate system z_i and time scale t.

Recall the terminology introduced in Sect. 1.

Proposition 14 (Ejection and recapture). *For arbitrary masses of the primaries, there exist solutions of the restricted parabolic three-body problem* (6) *of the following type: the infinitesimal particle is emitted by one of the primaries, passes arbitrarily close to one of the collinear configurations L_i, $i = 1,2,3$, and is captured by the same primary.*

Proof Chose a value k of L_k, $k = 1,2,3$ and a binary B_{μ_j}, $j = 1,2$. From Corollary 1 and Proposition 10, there exists a connection from $W^u(B_{\mu_j})$ to L_k. A connection $L_k \dashrightarrow L_k$ corresponding to the homothetic orbit exists. Again from Proposition 10, a connection from L_k to B_{μ_j} exists leading to the following connecting diagram

$$\begin{array}{ccc} W^u(B_{\mu_j}) & & W^s(B_{\mu_j}) \\ \downarrow & & \uparrow \\ L_k & \dashrightarrow & L_k \end{array}$$

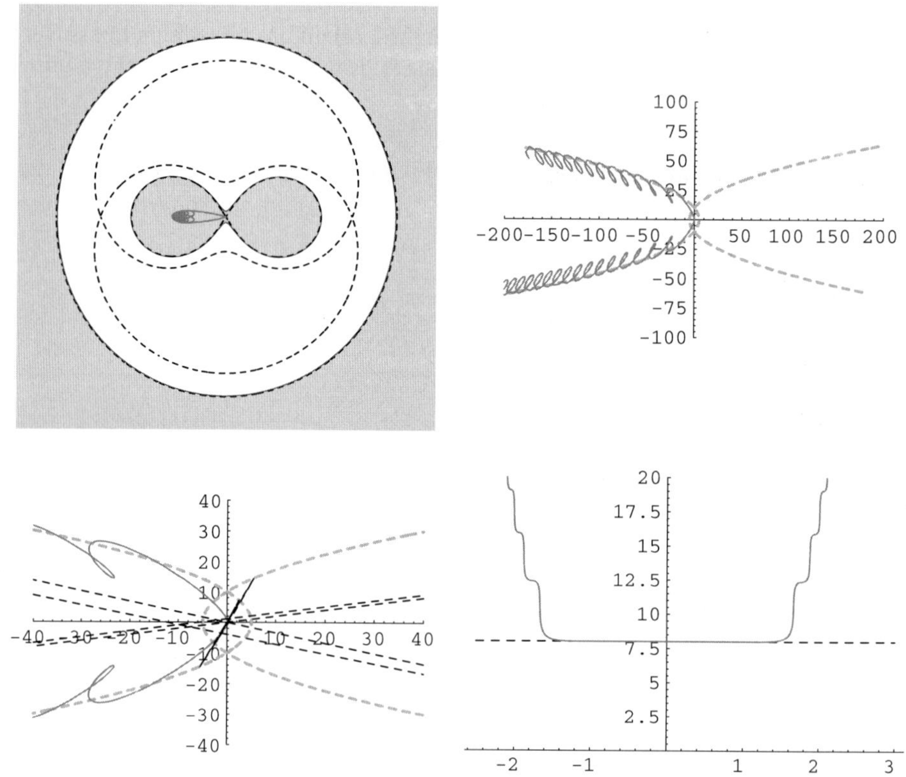

Fig. 9 An example of an orbit of emission–recapture in the rotating–pulsating frame (*left*) and in the inertial frame (*right*)

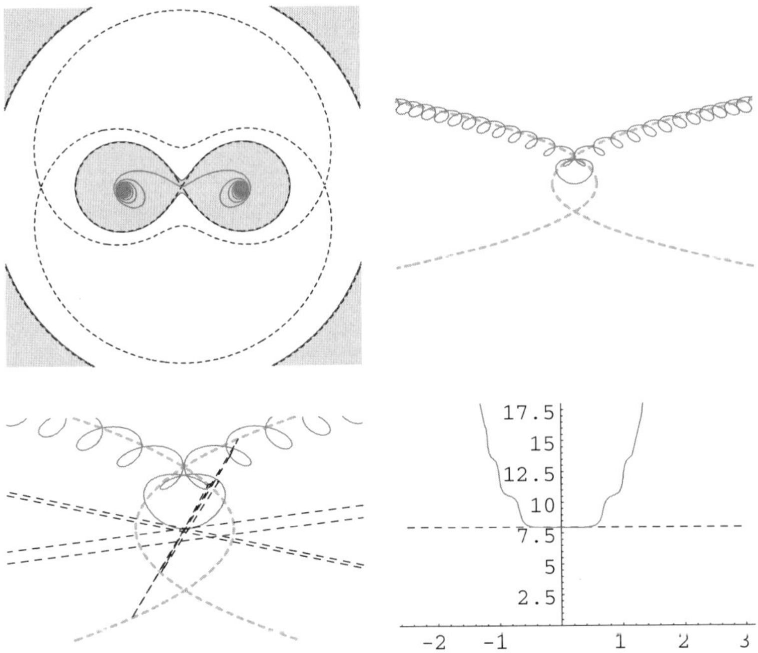

Fig. 10 An example of an orbit of exchange in the rotating pulsating frame (*left*) and in the inertial frame (*right*)

Now the *existence* of a solution, which realizes the path of the previous diagram is proved by tracking backwards and forwards a small ball around L_k, using hyperbolicity and continuity with respect to initial conditions in bounded domains of time.

Proposition 15 (Exchange). *For arbitrary masses of the primaries, there exist solutions of the restricted parabolic three-body problem* (6) *of the following type: the infinitesimal particle is emitted by one of the primaries, passes arbitrarily close to one of the collinear configurations* L_i, $i = 1, 2, 3$, *and is captured by the same primary.*

Proof Choose $j \neq k$. The proof is similar to the previous case, except that we can complete the final part of the connecting diagram with a different binary, leading to

$$\begin{array}{ccc} W^u(B_{\mu_j}) & & W^s(B_{\mu_k}) \\ \downarrow & & \uparrow \\ L_k & \dashrightarrow & L_k \end{array}$$

Proposition 16 (Emission and escape). *For arbitrary masses of the primaries, there exist solutions of the restricted parabolic three-body problem* (6) *of the following type: the infinitesimal particle is emitted by one of the primaries, passes arbitrary close to one of the collinear configuration* L_1 *or* L_3 *and escape to infinity.*

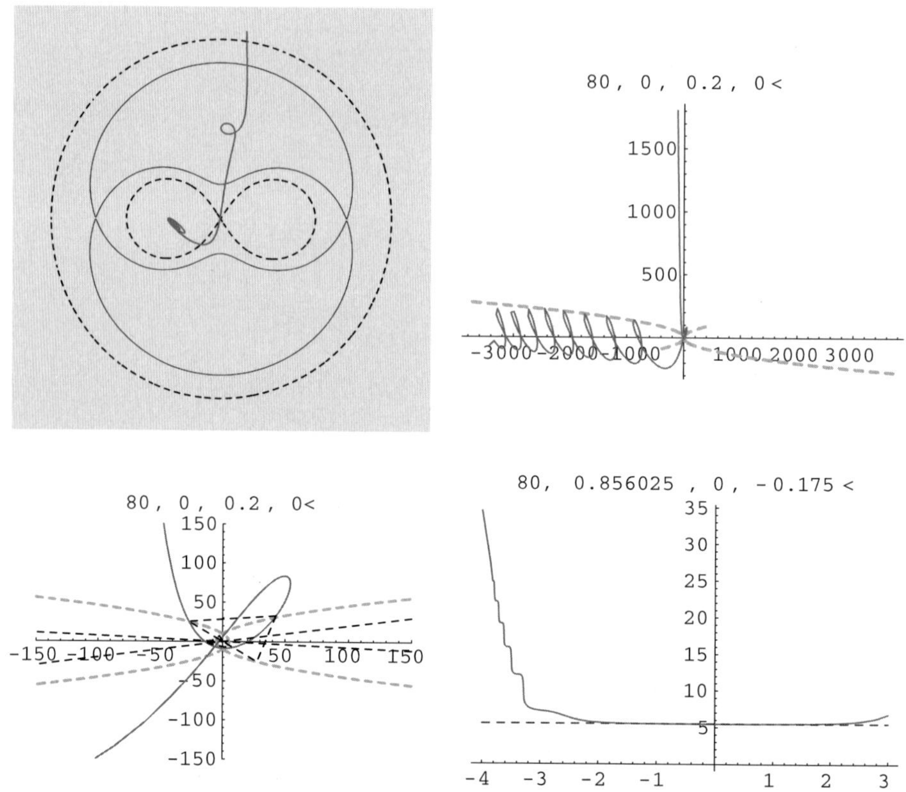

Fig. 11 An example of an emission and escape orbit with nearby passages to L_2 and L_4 before escape

Proof Choose $j = 1, 2$. The following connecting diagrams are obtained by the previous arguments.

$$\begin{array}{ccc} W^u(B_{\mu_j}) & & W^s(\infty) \\ \downarrow & & \uparrow \\ L_{1,3} & \dashrightarrow & L_{1,3} \end{array}$$

The following results apply for equal masses.

Proposition 17 (Visiting). *For equal masses of the primaries, there exist solutions of the restricted parabolic three-body problem (6) of the following type: the infinitesimal particle comes from infinity, passes a neighborhood of L_2 and then of L_4, and back to L_2 before escaping to infinity.*

Proof Choose $j = 1, 2$. The following connecting diagrams are obtained by the previous arguments. □

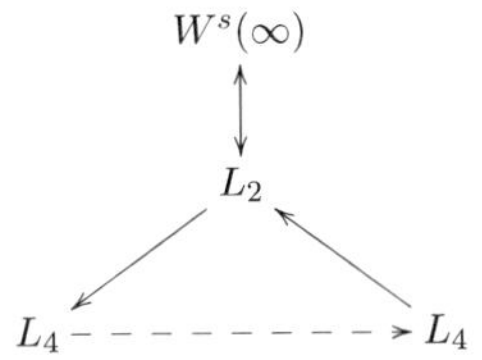

Proposition 18 (Visiting and capture). *For equal masses of the primaries, there exist solutions of the restricted parabolic three-body problem* (6) *of the following type: the infinitesimal particle comes from infinity, passes a neighborhood of L_2 and then of L_4, and back to L_2 before being captured by any of the binaries.*

Proof Choose $j = 1, 2$. The following connecting diagrams are obtained by the previous arguments.

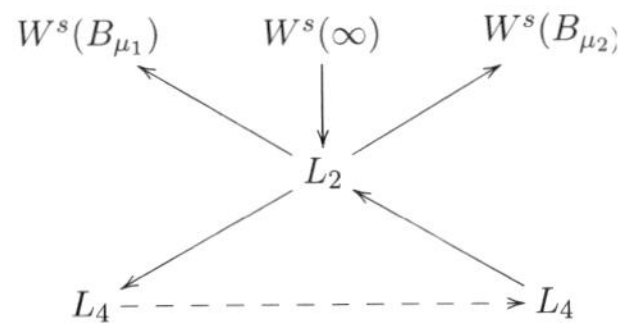

□

9 Numerical examples

In this section, we present some special orbits, which illustrate some of the theory presented here. In all following figures the orbits are shown in the rotating-pulsating frame (top–left) and in the inertial frame (top–right). A zoom of the same orbit in the inertial frame is at (bottom–left). At bottom–right the graph of the Jacobian function is a function of time s showing the gradient-like character of the flow. All initial conditions were taken at $s = 0$ with $\theta = 0$.

Figure 9 shows an orbit of emission and recapture. Figure 10 shows an exchange orbit. Finally in Fig. 11, an orbit emitted from one of the primaries with close passages near L_4 and L_2 before escaping to infinity.

Acknowledgements The second author is partially supported by DGES grant number BFM2002-04236-C02-02 and by a DURSI grant number 2001SGR 00173. The first and third authors were partially supported by SEP-CONACYT grant number SEP-2004-C01-47768.

References

Alvarez Ramírez, M., Delgado, J.: Blow up of the isosceles 3-body problem with an infinitesimal mass. Discrete Contin. Dyn. Syst. **9**(5), 1149–1173 (2003)
Broucke, R.: On the elliptic restricted three-body problem. J. Astronaut. Sci. **19**(6), 417–432 (1972)
Chazy, J.: Sur l'allure du mouvement dans le probleme des trois corps quand le temps croit indefniment. Ann. Ecole Norm. Sup. **39**(3), 29–130 (1922)
Devaney, Robert L.: Triple collision in the planar isosceles three-body problem. Invent. Math. **60**(3), 249–267 (1980)
ElBialy, M.S.: Triple collisions in the isosceles three body problem with small mass ratio. Z. Angew. Math. Phys. **40**(5), 645–664 (1989)
Hulkower, N.D.: The zero energy three body problem. Indiana Univ. Math. J. **27**(3), 409–447 (1978)
Kocina, N.G.: An example of motion in the restricted parabolic problem of three bodies. (Russian). Byull. Inst. Teoret. Astr. **5**, 617–622 (1954)

Lacomba, E.A., Bryant, J.: Contact structures for total collision and zero energy infinity manifolds in celestial mechanics. Proceedings of the IUTAM- ISIMM symposium on modern developments in analytical mechanics, vol. II (Torino, 1982). Atti Accad. Sci. Torino Cl. Sci. Fis. Mat. Nature. **117**(2), 563–568 (1983)

Llibre, J., Waldvogel, J.: Qualitative behaviour of the flow of the n-body problem in the zero energy level. New trends for Hamiltonian systems and celestial mechanics (Cocoyoc, 1994), pp. 275–288, Adv. Ser. Nonlinear Dynam., 8, World Sci. Publishing, River Edge, NJ (1996)

Martínez, R., Simó, C.: Qualitative study of the planar isosceles three-body problem. Celest. Mech. **41**(1–4), 179–251 (1987/88)

Martin, Monroe, H.: The restricted problem of three bodies. Trans. Amer. Math. Soc. **52**, 522–538 (1942)

McGehee, Richard: Triple collision in the collinear three-body problem. Invent. Math. **27**, 191–227 (1974)

Merman, G.A.: New criteria of hyperbolic–elliptic motion in the problem of three bodies (Russian). Akad. Nauk SSSR. Astr. Zurnal **30**, 332–339 (1953b)

Merman, G.A.: On Chazy's investigations in the problem of three bodies. Byull. Inst. Teor. Astron. Akad. Nauk SSSR. **5**, 594–605 (1954a)

Merman, G.A.: The restricted parabolic problem of three bodies (Russian). Byull. Inst. Teoret. Astr. **5**, 606–616 (1954b)

Moeckel, R.: Orbits near triple collision in the three-body problem. Indiana Univ. Math. J. **32**(2), 221–240 (1983)

Saari, D.G.: Expanding gravitational systems. Trans. Amer. Math. Soc. **156**, 219–240 (1971)

Saari, Donald G., Hulkower, Neal D.: On the manifolds of total collapse orbits and of completely parabolic orbits for the n-body problem. J. Differential Equations **41**(1), 27–43 (1981)

Simó, C., Susín, A.: Connections between critical points in the collision manifold of the planar three-body problem. The geometry of Hamiltonian systems, pp. 497–518. Berkeley, CA (1989), Math. Sci. Res. Inst. Publ., 22, Springer, Berlin, Heidelberg, New York (1991)

Szebehely, V.: Theory of orbits: the restricted problem of three bodies. Academic Press, New York (1967)

Celestial Mechanics and Dynamical Astronomy (2006) 95:201–212
DOI 10.1007/s10569-005-5663-7

ORIGINAL ARTICLE

Quaternions and the perturbed Kepler problem

Jörg Waldvogel

Received: 3 October 2005 / Revised: 24 November 2005 /
Accepted: 5 December 2005 / Published online: 17 August 2006

Abstract Quaternions, introduced by Hamilton (Philos. Mag. **25**, 489–495, 1844) as a generalization of complex numbers, lead to a remarkably simple representation of the perturbed three-dimensional Kepler problem as a perturbed harmonic oscillator. The paper gives an overview of this technique, including an outlook to applications in perturbation theories.

Keywords Kustaanheimo–Stiefel regularization · Quaternions · Perturbed Kepler problem · Birkhoff transformation

1 Introduction

A large branch of Celestial Mechanics is concerned with handling perturbations of the Kepler problem, described by a strongly non-linear differential equation for the vector $x \in \mathbb{R}^n, n = 2, 3$ as a function of time t,

$$\ddot{x} + \mu \frac{x}{r^3} = \varepsilon f(x,t), \quad r = \|x\|, \tag{1}$$

where x is the position vector of the moving particle with respect to the central body (with gravitational parameter $\mu > 0$), dots denote derivatives with respect to t, and $\varepsilon f(x,t)$ is a given small perturbation.

As will be demonstrated in the example below, the linearity of a problem leads to formally simple perturbation theories. Fortunately, there exist sets of variables in which the Kepler problem becomes linear; these are preferred variables for treating perturbed Kepler problems.

In this paper, we will revisit two closely related sets of variables that were introduced in order to regularize the collision singularity in the Kepler problem: the variables

J. Waldvogel (✉)
Seminar for Applied Mathematics,
Swiss Federal Institute of Technology ETH,
8092 Zurich, Switzerland
e-mail: waldvogel@sam.math.ethz.ch

introduced by Levi-Civita (1920), in which the planar Kepler problem appears as a harmonic oscillator in two dimensions, and the KS variables (Kustaanheimo 1964; Kustaanheimo and Stiefel 1965), in which the spatial Kepler problem appears as a harmonic oscillator in four dimensions. Both sets of variables have therefore the agreeable property of transforming the differential equations of the Kepler problem into a system of linear differential equations; they are therefore good variables for formulating theories of the perturbed Kepler problem. We will present a unified treatment of these two classical topics, using complex variables in the planar case and quaternions in the three-dimensional case. For completeness we mention that in a set of variables based on radial inversion in momentum space (see, e.g., Siegel and Moser 1971; or the summary in Celletti 2002) the Kepler problem also becomes linear.

The use of quaternions for the purpose of regularization of the Kepler problem in three dimensions has been contemplated before. In the comprehensive text by Stiefel and Scheifele (1971) the use of quaternions was taken into consideration (p. 286), but clearly rejected: *"Any attempt to substitute the theory of the KS matrix by the more popular theory of the quaternion matrices leads to failure or at least to a very unwieldy formalism"*. Almost simultaneously, Vivarelli (1994) and Vrbik (1994, 1995) demonstrated the usefulness of quaternions in this field. Here we will describe a new, elegant way of handling the three-dimensional case in complete analogy to the well-known planar case by introducing an unconventional conjugation of quaternions (see the definition in Eq. (24) below), first mentioned by Waldvogel (2006).

Perturbation theories of ordinary differential equations are comparatively simple for linear problems. Consider, e.g., the perturbed system

$$\dot{x}(t) + A(t)\,x(t) - b(t) = \varepsilon f(x,t), \quad x : t \in \mathbb{R} \mapsto x(t) \in \mathbb{R}^n \tag{2}$$

of linear differential equations, where $A(t)$ is a given time-dependent matrix. Equation (2) may formally be solved to arbitrary order by the series

$$x(t) = x_0(t) + \varepsilon\, x_1(t) + \varepsilon^2\, x_2(t) + \cdots,$$

where $x_k(t)$ satisfies the linear differential equation

$$\dot{x_k}(t) + A(t)\,x_k(t) = f_{k-1}(t), \quad k = 0, 1, 2, \ldots \tag{3}$$

Here $f_{-1}(t) := b(t)$, and $f_0(t), f_1(t), \ldots$ are defined as the coefficients of the formal Taylor series of $f(x,t)$ with respect to ε:

$$\sum_{k=0}^{\infty} \varepsilon^k f_k(t) = f(x_0(t) + \varepsilon\, x_1(t) + \varepsilon^2\, x_2(t) + \cdots, t)\,.$$

Note that the linear differential equations (3) are all of the type of the unperturbed problem $k = 0$; they only differ in their right-hand sides.

In Sect. 2, Levi-Civita's regularization procedure in complex notation will be summarized. Section 3 contains a brief introduction to quaternion algebra and states the KS transformation in quaternion notation. In Sect. 4, the main part of this paper, we develop a simple, concise way of transforming the spatial perturbed Kepler problem into a perturbed harmonic oscillator in four dimensions. As a byproduct, an elegant representation of the spatial Birkhoff transformation in quaternion notation will be given in Sect. 5.

2 The Levi-Civita transformation

Here we summarize the *three* steps necessary for regularizing (and thus "linearizing") the perturbed planar Kepler problem by Levi-Civita's transformation. Throughout this section, we use complex notation, i.e. instead of the vectors $x = (x_1, x_2)^T \in \mathbb{R}^2$, $f = (f_1, f_2)^T$ we use the corresponding complex numbers $\mathbf{x} = x_1 + i x_2 \in \mathbb{C}$, $\mathbf{f} = f_1 + i f_2 \in \mathbb{C}$.

2.1 First step: slow-motion movie

Instead of the physical time t a new independent variable τ, called the fictitious time, is introduced by the differential relation

$$\mathrm{d}t = r \cdot \mathrm{d}\tau, \qquad \frac{\mathrm{d}}{\mathrm{d}\tau}(\,) = (\,)'. \tag{4}$$

Therefore, the ratio $\mathrm{d}t/\mathrm{d}\tau$ of two infinitesimal increments is made proportional to the distance r; the movie is run in slow-motion whenever r becomes small. With the differentiation rules

$$\frac{\mathrm{d}}{\mathrm{d}t} = \frac{1}{r}\,\frac{\mathrm{d}}{\mathrm{d}\tau}, \qquad \frac{\mathrm{d}^2}{\mathrm{d}t^2} = \frac{1}{r^2}\,\frac{\mathrm{d}^2}{\mathrm{d}\tau^2} - \frac{r'}{r^3}\,\frac{\mathrm{d}}{\mathrm{d}\tau}.$$

Equation (1) transforms into

$$r\,\mathbf{x}'' - r'\,\mathbf{x}' + \mu\,\mathbf{x} = r^3\,\varepsilon\mathbf{f} \in \mathbb{C}. \tag{5}$$

2.2 Second step: conformal squaring

This part of Levi-Civita's regularization procedure consists of representing the complex physical coordinate $\mathbf{x}$ as the square $\mathbf{u}^2$ of a complex variable $\mathbf{u} = u_1 + i\,u_2 \in \mathbb{C}$,

$$\mathbf{x} = \mathbf{u}^2, \tag{6}$$

i.e. the mapping from the parametric plane to the physical plane is chosen as a conformal squaring. This is based on the observation that conformal squaring maps an origin-centered ellipse to a Keplerian ellipse with one focus at the origin (see Fig. 1). Equation (6) implies

$$r = |\,\mathbf{x}\,| = |\,\mathbf{u}\,|^2 = \mathbf{u}\,\bar{\mathbf{u}}, \tag{7}$$

and differentiation of Eqs. (6) and (7) yields

$$\mathbf{x}' = 2\,\mathbf{u}\,\mathbf{u}', \qquad \mathbf{x}'' = 2\left(\mathbf{u}\,\mathbf{u}'' + \mathbf{u}'^{\,2}\right) \in \mathbb{C}, \quad r' = \mathbf{u}'\,\bar{\mathbf{u}} + \mathbf{u}\,\bar{\mathbf{u}}'. \tag{8}$$

By substituting this into (5), cancelling two equal terms ($2\,r\,\mathbf{u}'^2$ and $2\,\mathbf{u}'\,\bar{\mathbf{u}}\,\mathbf{u}\,\mathbf{u}'$) and dividing by $\mathbf{u}$ we obtain

$$2\,r\,\mathbf{u}'' + (\,\mu - 2\,|\mathbf{u}'|^2\,)\,\mathbf{u} = r^2\,\bar{\mathbf{u}}\,\varepsilon\,\mathbf{f}. \tag{9}$$

Remark Obtaining initial values $\mathbf{u}(0) = \sqrt{\mathbf{x}(0)}$ requires the computation of a complex square root. This can conveniently be accomplished by means of the formula

$$\sqrt{\mathbf{x}} = \frac{\mathbf{x} + |\mathbf{x}|}{\sqrt{2\,(|\mathbf{x}| + \mathrm{Re}\,\mathbf{x})}}, \tag{10}$$

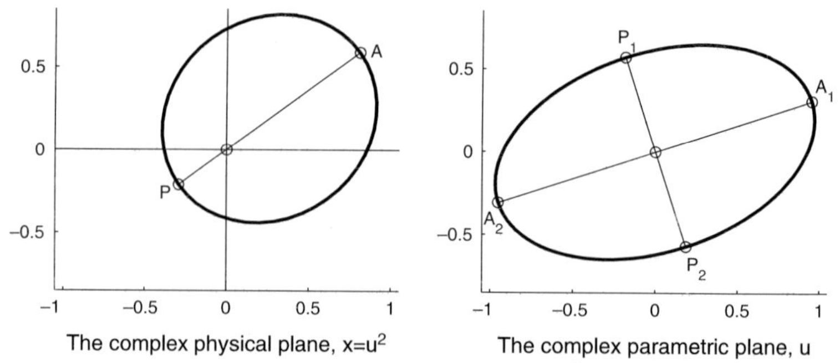

Fig. 1 The image of a (doubly covered) Keplerian ellipse with one focus at the origin of the physical plane (left) under the conformal square root is an ellipse centered at the origin of the parametric plane (right)

which reflects the observation that the complex vector $\sqrt{\mathbf{x}}$ has the direction of the bisector between $\mathbf{x}$ and the real vector $|\mathbf{x}|$; it holds in the range $-\pi < \arg(\mathbf{x}) < \pi$. The alternative formula

$$\sqrt{\mathbf{x}} = \frac{\mathbf{x} - |\mathbf{x}|}{i\sqrt{2\,(|\mathbf{x}| - \operatorname{Re}\mathbf{x})}}$$

holds in $0 < \arg(\mathbf{x}) < 2\pi$ and agrees with (10) in the upper half-plane; it therefore provides the analytic continuation of (10) into the sector $\pi \leqslant \arg(\mathbf{x}) < 2\pi$. Furthermore, it avoids a loss of accuracy near the negative real axis $\mathbf{x} < 0$.

2.3 Third step: fixing the energy

This step is simple for the unperturbed problem, $\varepsilon = 0$. Integrating the inner product of Eq. (1) and the vector $\dot{x} \in \mathbb{R}^2$ yields the well-known energy equation

$$\frac{1}{2}\,\|\,\dot{x}\,\|^2 - \frac{\mu}{r} = -h = \text{const}, \tag{11}$$

where the energy constant h is chosen such that $h > 0$ corresponds to an elliptic orbit. From (4) and (8) there follows (using again complex notation):

$$\dot{\mathbf{x}} = \frac{1}{r}\cdot 2\,\mathbf{u}\,\mathbf{u}', \qquad \frac{1}{2}\,|\,\dot{\mathbf{x}}\,|^2 = 2\,\frac{|\,\mathbf{u}'\,|^2}{r}, \tag{12}$$

and (11) implies

$$\mu - 2\,|\,\mathbf{u}'\,|^2 = r\,h\,. \tag{13}$$

Substituting this into (9) and dividing by r yields

$$2\,\mathbf{u}'' + h\,\mathbf{u} = 0, \tag{14}$$

a system of linear differential equations describing a harmonic oscillator in two dimensions with frequency $\omega = \sqrt{h/2}$.

In the perturbed case h of Eq. (11) is no longer a constant, but is a slowly varying function and satisfies the differential equation

$$\dot{h} = -\,\langle \dot{x},\, \varepsilon f\rangle \quad \text{or} \quad h' = -\,\langle x',\, \varepsilon f\rangle\,, \tag{15}$$

where $\langle x, y\rangle$ denotes the inner product of the vectors $x, y \in \mathbb{R}^2$. The energy equation (13) still holds, and instead of (14) we obtain

$$2\,\mathbf{u}'' + h\,\mathbf{u} = |\,\mathbf{u}\,|^2\,\bar{\mathbf{u}}\,\varepsilon\,\mathbf{f}\,. \tag{16}$$

Remark Equation (16), together with the second Eq. (15), describes a perturbed harmonic oscillator with slowly varying frequency; it may be transformed to constant frequency by introducing a new independent variable s proportional to the *osculating eccentric anomaly,* e.g., according to the differential relation

$$\mathrm{d}s = \sqrt{h}\,\mathrm{d}\tau\,.$$

This results in the system of differential equations

$$\frac{\mathrm{d}h}{\mathrm{d}s} = -\frac{r}{\sqrt{h}}\,\langle\dot{x}, \varepsilon f\rangle\,, \qquad 2\,\frac{\mathrm{d}^2\mathbf{u}}{\mathrm{d}s^2} + \mathbf{u} = \frac{r}{h}\,\bar{\mathbf{u}}\,\varepsilon\mathbf{f} - \frac{1}{h}\,\frac{\mathrm{d}h}{\mathrm{d}s}\,\frac{\mathrm{d}\mathbf{u}}{\mathrm{d}s}$$

(for more details regarding applications to perturbation theories, see Waldvogel 2006).

3 Quaternion algebra and the KS transformation

In this section we indicate how Levi-Civita's regularization procedure may be generalized to three-dimensional motion. The essential step is to replace the conformal squaring of Sect. 2.2 by the Kustaanheimo–Stiefel (KS) transformation. A preliminary version of this transformation using spinor notation was proposed by Kustaanheimo (1964); the full theory was developed in a subsequent joint paper (Kustaanheimo and Stiefel 1965); the entire topic is extensively discussed in the comprehensive text by Stiefel and Scheifele (1971). The relevant mapping from the 3-sphere onto the 2-sphere was discovered already by Hopf (1931) and is referred to in topology as the Hopf mapping.

Both the Levi-Civita and the KS regularization share the property of "linearizing" the equations of motion of the two-body problem. Quaternion algebra, introduced by Hamilton (1844), turns out to be very well suited as a tool for regularizing the three-dimensional Kepler motion, as was observed by Vivarelli (1994) and Vrbik (1994, 1995). Here we will present a new elegant way of extending the Levi-Civita regularization to three dimensions by means of quaternions.

3.1 Basics

Quaternion algebra is a generalization of the algebra of complex numbers obtained by using three independent "imaginary" units i, j, k. As for the single imaginary unit i in the algebra of complex numbers, the rules

$$i^2 = j^2 = k^2 = -1$$

are postulated, together with the non-commutative multiplication rules

$$i\,j = -j\,i = k\,, \quad j\,k = -k\,j = i\,, \quad k\,i = -i\,k = j\,.$$

Given the real numbers $u_l \in \mathbb{R}$, $l = 0, 1, 2, 3$, the object

$$\mathbf{u} = u_0 + i\,u_1 + j\,u_2 + k\,u_3 \tag{17}$$

is called a *quaternion* $\mathbf{u} \in \mathbb{U}$, where $\mathbb{U}$ denotes the set of all quaternions (in the remaining sections bold-face characters denote quaternions). The sum $iu_1+ju_2+ku_3$ is called the *quaternion part* of $\mathbf{u}$, whereas u_0 is naturally referred to as its real part. The above multiplication rules and vector space addition define the *quaternion algebra*. Multiplication is generally non-commutative; however, any quaternion commutes with a real:

$$c\,\mathbf{u} = \mathbf{u}\,c\,, \qquad c \in \mathbb{R}\,, \quad \mathbf{u} \in \mathbb{U}\,, \tag{18}$$

and for any three quaternions $\mathbf{u}$, $\mathbf{v}$, $\mathbf{w} \in \mathbb{U}$ the associative law holds:

$$(\mathbf{u}\,\mathbf{v})\,\mathbf{w} = \mathbf{u}\,(\mathbf{v}\,\mathbf{w})\,. \tag{19}$$

The quaternion $\mathbf{u}$ may naturally be associated with the corresponding vector $u = (u_0, u_1, u_2, u_3) \in \mathbb{R}^4$. For later reference we introduce notation for three-vectors in two important particular cases: $\vec{u} = (u_1, u_2, u_3) \in \mathbb{R}^3$ for the vector associated with the *pure quaternion* $\mathbf{u} = i\,u_1 + j\,u_2 + k\,u_3$, and $\underline{u} = (u_0, u_1, u_2)$ for the vector associated with the quaternion with a vanishing k-component, $\mathbf{u} = u_0 + i\,u_1 + j\,u_2$.

For convenience we also introduce the vector $\vec{\imath} = (i, j, k)$; the quaternion $\mathbf{u}$ may then be written formally as $\mathbf{u} = u_0 + \langle \vec{\imath}\,, \vec{u} \rangle$. For the two quaternion products of $\mathbf{u}$ and $\mathbf{v} = v_0 + \langle \vec{\imath}\,, \vec{v} \rangle$, we then obtain the concise expressions

$$\begin{aligned} \mathbf{u}\,\mathbf{v} &= u_0 v_0 - \langle \vec{u}, \vec{v} \rangle + \langle \vec{\imath}\,,\, u_0\,\vec{v} + v_0\,\vec{u} + \vec{u} \times \vec{v} \rangle, \\ \mathbf{v}\,\mathbf{u} &= u_0 v_0 - \langle \vec{u}, \vec{v} \rangle + \langle \vec{\imath}\,,\, u_0\,\vec{v} + v_0\,\vec{u} - \vec{u} \times \vec{v} \rangle\,, \end{aligned} \tag{20}$$

where $\times$ denotes the vector product. Note that the non-commutativity shows only in the sign of the term with the vector product.

The *conjugate* $\bar{\mathbf{u}}$ of the quaternion $\mathbf{u}$ is defined as

$$\bar{\mathbf{u}} = u_0 - i\,u_1 - j\,u_2 - k\,u_3\,; \tag{21}$$

then the *modulus* $|\mathbf{u}|$ of $\mathbf{u}$ is obtained from

$$|\mathbf{u}|^2 = \mathbf{u}\,\bar{\mathbf{u}} = \bar{\mathbf{u}}\,\mathbf{u} = \sum_{l=0}^{3} u_l^2\,. \tag{22}$$

As transposition of a product of matrices, conjugation of a quaternion product reverses the order of its factors:

$$\overline{\mathbf{u}\,\mathbf{v}} = \bar{\mathbf{v}}\,\bar{\mathbf{u}}\,. \tag{23}$$

3.2 The KS transformation with quaternions

Here, we will revisit KS regularization and present a new, elegant derivation of it, using quaternion algebra and an unconventional "conjugate" $\mathbf{u}^\star$ referred to as the *star conjugate* of the quaternion $\mathbf{u} = u_0 + i\,u_1 + j\,u_2 + k\,u_3$:

$$\mathbf{u}^\star := u_0 + i\,u_1 + j\,u_2 - k\,u_3\,. \tag{24}$$

The star conjugate of $\mathbf{u}$ may be expressed in terms of the conventional conjugate $\bar{\mathbf{u}}$ as

$$\mathbf{u}^\star = k\,\bar{\mathbf{u}}\,k^{-1} = -k\,\bar{\mathbf{u}}\,k\,;$$

however, it turns out that the definition (24) leads to a particularly elegant treatment of KS regularization. The following elementary properties are easily verified:

$$\begin{aligned} (\mathbf{u}^\star)^\star &= \mathbf{u}, \\ |\,\mathbf{u}^\star|^2 &= |\mathbf{u}|^2\,, \\ (\mathbf{u}\,\mathbf{v})^\star &= \mathbf{v}^\star\,\mathbf{u}^\star\,. \end{aligned} \tag{25}$$

Consider now the mapping

$$\mathbf{u} \in \mathbb{U} \longmapsto \mathbf{x} = \mathbf{u}\,\mathbf{u}^\star. \tag{26}$$

Star conjugation immediately yields $\mathbf{x}^\star = (\mathbf{u}^\star)^\star\,\mathbf{u}^\star = \mathbf{x}$; hence $\mathbf{x}$ is a quaternion of the form $\mathbf{x} = x_0 + i\,x_1 + j\,x_2$ which may be associated with the vector $\underline{x} = (x_0, x_1, x_2) \in \mathbb{R}^3$. From $\mathbf{u} = u_0 + i\,u_1 + j\,u_2 + k\,u_3$ we obtain

$$\begin{aligned} x_0 &= u_0^2 - u_1^2 - u_2^2 + u_3^2, \\ x_1 &= 2(u_0\,u_1 - u_2\,u_3), \\ x_2 &= 2(u_0\,u_2 + u_1\,u_3)\,, \end{aligned} \tag{27}$$

which is exactly the KS transformation in its classical form or – up to a permutation of the indices – the Hopf map. Therefore, we have

Theorem 1 *The KS transformation which maps* $u = (u_0, u_1, u_2, u_3) \in \mathbb{R}^4$ *to* $\underline{x} = (x_0, x_1, x_2) \in \mathbb{R}^3$ *is given by the quaternion relation*

$$\mathbf{x} = \mathbf{u}\,\mathbf{u}^\star\,,$$

where $\mathbf{u} = u_0 + i\,u_1 + j\,u_2 + k\,u_3$, $\mathbf{x} = x_0 + i\,x_1 + j\,x_2$.

Corollary 1 *The norms of the vectors* $\underline{x}$ *and* u *satisfy*

$$r := \|\underline{x}\| = \|u\|^2 = \mathbf{u}\,\bar{\mathbf{u}}\,. \tag{28}$$

Proof By appropriately combining the two conjugations and using the rules (18), (19), (22), (23) and (25) we obtain

$$\|\underline{x}\|^2 = \mathbf{x}\,\bar{\mathbf{x}} = \mathbf{u}\,(\mathbf{u}^\star\,\bar{\mathbf{u}}^\star)\,\bar{\mathbf{u}} = |\mathbf{u}^\star|^2\,|\mathbf{u}\,|^2 = |\mathbf{u}\,|^4 = \|u\|^4$$

from where the statement follows. □

3.3 The inverse map

Since the mapping (27) does not preserve the dimension its inverse in the usual sense does not exist. However, the present quaternion formalism yields an elegant way of finding the corresponding *fibration* of the original space $\mathbb{R}^4$. Being given a quaternion $\mathbf{x} = x_0 + i\,x_1 + j\,x_2$ with vanishing k-component, $\mathbf{x} = \mathbf{x}^\star$, we want to find all quaternions $\mathbf{u}$ such that $\mathbf{u}\,\mathbf{u}^\star = \mathbf{x}$. We propose the following solution in two steps:

First step: Find a particular solution $\mathbf{u} = \mathbf{v} = \mathbf{v}^\star = v_0 + i\,v_1 + j\,v_2$ which has also a vanishing k-component. Since $\mathbf{v}\,\mathbf{v}^\star = \mathbf{v}^2$ we may use Eq. (10), which was developed for the complex square root, also for the square root of a quaternion:

$$\mathbf{v} = \frac{\mathbf{x} + |\mathbf{x}|}{\sqrt{2\,(|\mathbf{x}| + x_0)}}.$$

Clearly, $\mathbf{v}$ has a vanishing k-component.

Second step: The entire family of solutions (the fiber corresponding to $\mathbf{x}$, geometrically a circle in $\mathbb{R}^4$ parametrized by the angle φ), is given by

$$\mathbf{u} = \mathbf{v} \cdot e^{k\varphi} = \mathbf{v}\,(\cos\varphi + k\sin\varphi)\,.$$

Proof $\mathbf{u}\,\mathbf{u}^\star = \mathbf{v}\,e^{k\varphi}\,e^{-k\varphi}\,\mathbf{v}^\star = \mathbf{v}\,\mathbf{v}^\star = \mathbf{x}\,.$ □

4 KS regularization with quaternions

In order to regularize the perturbed three-dimensional Kepler motion by means of the KS transformation it is necessary to look at the properties of the map (26) under differentiation.

The transformation (26) or (27) is a mapping from $\mathbb{R}^4$ to $\mathbb{R}^3$; it therefore, leaves one degree of freedom in the parametric space undetermined. In KS theory (Kustaanheimo and Stiefel, 1965; Stiefel and Scheifele, 1971), this freedom is taken advantage of by trying to inherit as much as possible of the conformality properties of the Levi-Civita map, but other approaches exist (e.g., Vrbik 1995). By imposing the "bilinear relation"

$$2\,(u_3\,\mathrm{d}u_0 - u_2\,\mathrm{d}u_1 + u_1\,\mathrm{d}u_2 - u_0\,\mathrm{d}u_3) = 0 \tag{29}$$

between the vector $u = (u_0, u_1, u_2, u_3)$ and its differential $\mathrm{d}u$ on orbits the tangential map of (27) becomes a linear map with an orthogonal (but non-normalized) matrix.

This property has a simple consequence on the differentiation of the quaternion representation (26) of the KS transformation. Considering the non-commutativity of the quaternion product, the differential of Eq. (26) becomes

$$\mathrm{d}\,\mathbf{x} = \mathrm{d}\,\mathbf{u} \cdot \mathbf{u}^\star + \mathbf{u} \cdot \mathrm{d}\,\mathbf{u}^\star\,, \tag{30}$$

whereas (29) takes the form of a commutator relation,

$$\mathbf{u} \cdot \mathrm{d}\,\mathbf{u}^\star - \mathrm{d}\,\mathbf{u} \cdot \mathbf{u}^\star = 0\,. \tag{31}$$

Combining (30) with the relation (31) yields the elegant result

$$\mathrm{d}\,\mathbf{x} = 2\,\mathbf{u} \cdot \mathrm{d}\,\mathbf{u}^\star\,, \tag{32}$$

i.e. the bilinear relation (29) of KS theory is equivalent with the requirement that the tangential map of $\mathbf{u} \mapsto \mathbf{u}\,\mathbf{u}^\star$ behaves as in a commutative algebra.

By using the tools collected in Section 3 together with Eq. (32) the regularization procedure outlined in Section 2 will now be carried out for the three-dimensional perturbed Kepler problem. Care must be taken to preserve the order of the factors in quaternion products. Exchanging two factors is permitted if one of the factors is real or if the factors are mutually conjugate. An important tool for simplifying expressions is regrouping factors of multiple products according to the associative law (19). In order to stress the simplicity of this approach, we present all the details of the formal computations.

4.1 First step in space: slow-motion movie

Let $\mathbf{x} = x_0 + i\,x_1 + j\,x_2 \in \mathbb{U}$ be the quaternion associated with the position vector $\underline{x} = (x_0, x_1, x_2) \in \mathbb{R}^3$; then the perturbed Kepler problem (1) is given by

$$\ddot{\mathbf{x}} + \mu\,\frac{\mathbf{x}}{r^3} = \varepsilon\,\mathbf{f}(\mathbf{x},t) \in \mathbb{U}\,, \qquad r = |\mathbf{x}|\,, \tag{33}$$

where $\mathbf{f}(\mathbf{x},t) = f_0(\mathbf{x},t) + i\,f_1(\mathbf{x},t) + j\,f_2(\mathbf{x},t) = \mathbf{f}^\star(\mathbf{x},t)$ is the quaternion associated with the perturbation $f(\underline{x},t) \in \mathbb{R}^3$.

The first transformation step calls for introducing the fictitious time τ according to Eq. (4), $dt = r \cdot d\tau$; the result is formally identical with Eq. (5),

$$r\,\mathbf{x}'' - r'\,\mathbf{x}' + \mu\,\mathbf{x} = r^3\,\varepsilon\,\mathbf{f} \in \mathbb{U}\,. \tag{34}$$

4.2 Second step: KS transformation with quaternions

Instead of the conformal squaring according to Eq. (6) we use the KS transformation (26),

$$\mathbf{x} = \mathbf{u}\,\mathbf{u}^\star\,, \qquad r := |\mathbf{x}| = \mathbf{u}\,\bar{\mathbf{u}}\,. \tag{35}$$

Differentiation by means of the commutator relation (31) yields

$$\mathbf{x}' = 2\,\mathbf{u}\,\mathbf{u}^{\star\prime}\,, \qquad \mathbf{x}'' = 2\,\mathbf{u}\,\mathbf{u}^{\star\prime\prime} + 2\,\mathbf{u}'\,\mathbf{u}^{\star\prime}\,, \quad r' = \mathbf{u}'\,\bar{\mathbf{u}} + \mathbf{u}\,\bar{\mathbf{u}}'\,. \tag{36}$$

Substitution of (35) and (36) into (34) results in the lengthy equation

$$(\mathbf{u}\,\bar{\mathbf{u}})\,(2\,\mathbf{u}\,\mathbf{u}^{\star\prime\prime} + 2\,\mathbf{u}'\,\mathbf{u}^{\star\prime}) - (\mathbf{u}'\,\bar{\mathbf{u}} + \mathbf{u}\,\bar{\mathbf{u}}')\,2\,\mathbf{u}\,\mathbf{u}^{\star\prime} + \mu\,\mathbf{u}\,\mathbf{u}^\star = r^3\,\varepsilon\,\mathbf{f}\,, \tag{37}$$

which is considerably simplified by observing that the second and third term—after applying the distributive law—compensate:

$$2\,(\mathbf{u}\,\bar{\mathbf{u}})\,\mathbf{u}'\,\mathbf{u}^{\star\prime} - 2\,\mathbf{u}'\,(\bar{\mathbf{u}}\,\mathbf{u})\,\mathbf{u}^{\star\prime} = 0\,.$$

Furthermore, by means of (18), (19) and (31) the fourth term of (37) may be simplified as follows:

$$-2\,(\mathbf{u}\,\bar{\mathbf{u}}')\,(\mathbf{u}\,\mathbf{u}^{\star\prime}) = -2\,\mathbf{u}\,(\bar{\mathbf{u}}'\,\mathbf{u}')\,\mathbf{u}^\star = -2\,|\,\mathbf{u}'|^2\,\mathbf{u}\,\mathbf{u}^\star\,.$$

By using this and left-dividing by $\mathbf{u}$ Equation (37) now becomes

$$2\,r\,\mathbf{u}^{\star\prime\prime} + (\,\mu - 2\,|\mathbf{u}'|^2\,)\,\mathbf{u}^\star = r^2\,\bar{\mathbf{u}}\,\varepsilon\,\mathbf{f} \tag{38}$$

in almost perfect formal agreement with Eq. (9) of the planar case.

4.3 Third step: fixing the energy in space

In formal agreement with the planar case the energy equation expressed by fictitious time is

$$\frac{1}{2\,r^2}\,|\mathbf{x}'|^2 - \frac{\mu}{r} = -h \quad \text{with} \quad h' = -\langle \underline{x}',\,\varepsilon\,\underline{f}\rangle\,. \tag{39}$$

From (36), (25), (28) we have

$$|\,\mathbf{x}'\,|^2 = \mathbf{x}'\,\bar{\mathbf{x}}' = 4\,\mathbf{u}\,(\,\mathbf{u}^{\star\prime}\,\bar{\mathbf{u}}^{\star\prime}\,)\,\bar{\mathbf{u}} = 4\,r\,|\,\mathbf{u}'\,|^2,$$

and the first equation of (39) becomes

$$\mu - 2\,|\,\mathbf{u}'\,|^2 \;=\; r\,h \tag{40}$$

in formal agreement with Eq. (13) found for the planar case. Substituting this into the star-conjugate of (38) and dividing by r yields the elegant final result

$$2\,\mathbf{u}'' + h\,\mathbf{u} \;=\; |\,\mathbf{u}\,|^2\,\varepsilon\,\mathbf{f}\,\bar{\mathbf{u}}^\star \tag{41}$$

a differential equation in perfect agreement with (16) for the planar case; however, it takes more than an educated guess to get the correct right-hand side.

5 The Birkhoff transformation

The topic of this section is not directly related to the preceding text; we add it here because the quaternion tools discussed before allow for an elegant representation of the spatial Birkhoff transformation.

This regularizing transformation was proposed by Birkhoff (1915), in order to regularize all singularities of the planar restricted three-body problem with a single transformation. Half a century later Stiefel and Waldvogel (1965) published a generalization of Birkhoff's transformation to three dimensions, using the KS transformation. Later these ideas were used by Waldvogel (1967a, b).

Here, we will first revisit the classical Birkhoff transformation (the same conformal map is known in aerodynamics as the Joukowsky transformation) and represent it as the composition of three elementary conformal mappings; this will then readily generalize to the spatial situation by means of quaternions.

Consider a rotating physical plane parametrized by the complex variable $\mathbf{y} \in \mathbb{C}$; for convenience we assume the fixed primaries of the restricted three-body problem to be situated at the points A, C given by the complex posititons $\mathbf{y} = -1$ and $\mathbf{y} = 1$, respectively (see Fig. 2). The complex variable of the parametric plane will be denoted by $\mathbf{v}$ and will be normalized in such a way that the primaries are mapped to $\mathbf{v} = -1$ or $\mathbf{v} = 1$, respectively.

The key observation is that Levi-Civita's conformal map (6), $\mathbf{u} \mapsto \mathbf{x} = \mathbf{u}^2$, not only regularizes collisions at $\mathbf{x} = 0$ but also analogous singularities at $\mathbf{x} = \infty$. This is seen

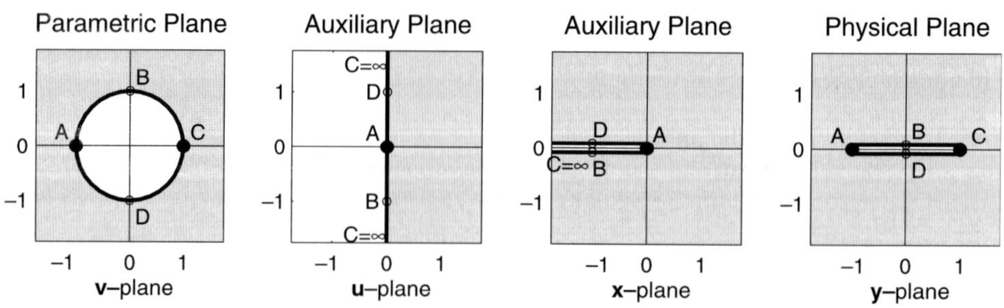

$$\mathbf{v} \mapsto \mathbf{u} = \frac{\mathbf{v}+1}{\mathbf{v}-1} \qquad \mathbf{u} \mapsto \mathbf{x} = \mathbf{u}^2 \qquad \mathbf{x} \mapsto \mathbf{y} = \frac{\mathbf{x}+1}{\mathbf{x}-1}$$

Fig. 2 The sequence of conformal maps generating the planar Birkhoff transformation by closing the complex planes to become Riemann spheres (by adding the point at infinity) and using inversions $\mathbf{x} = 1/\tilde{\mathbf{x}}$, $\mathbf{u} = 1/\tilde{\mathbf{u}}$.

Taking advantage of this fact, we first map the **v**-sphere to an auxiliary **u**-sphere by the Möbius transformation

$$\mathbf{v} \longmapsto \mathbf{u} = \frac{\mathbf{v}+1}{\mathbf{v}-1} = 1 + \frac{2}{\mathbf{v}-1}, \tag{42}$$

which takes the primaries A, C to the points $\mathbf{u} = 0$, $\mathbf{u} = \infty$, respectively. The Levi-Civita map (6) will leave these points invariant while regularizing collisions at A or C. Finally, the Möbius transformation

$$\mathbf{x} \longmapsto \mathbf{y} = \frac{\mathbf{x}+1}{\mathbf{x}-1} = 1 + \frac{2}{\mathbf{x}-1} \tag{43}$$

maps A, C to $\mathbf{y} = -1$ and $\mathbf{y} = 1$, respectively. The composition of the maps (42), (6), (43) yields

$$\mathbf{y} = \frac{\left(\frac{\mathbf{v}+1}{\mathbf{v}-1}\right)^2 + 1}{\left(\frac{\mathbf{v}+1}{\mathbf{v}-1}\right)^2 - 1} \quad \text{or} \quad \mathbf{y} = \frac{1}{2}\left(\mathbf{v} + \frac{1}{\mathbf{v}}\right), \tag{44}$$

the well-known map used by Joukowsky and Birkhoff.

In the spatial case, we choose $\mathbf{v}, \mathbf{u}, \mathbf{x}, \mathbf{y} \in \mathbb{U}$ to be quaternions, $\mathbf{x} = \mathbf{x}^\star$, $\mathbf{y} = \mathbf{y}^\star$ being quaternions with vanishing k-components associated with 3-vectors $\underline{x}$, $\underline{y}$. Then the mappings (42), (43), now being shifted inversions in four or three dimensions, are both conformal maps, in fact the only conformal maps existing in those dimensions, except for the translations, magnifications, and rotations. Composing these with the KS or Hopf map (35), $\mathbf{u} \mapsto \mathbf{x} = \mathbf{u}\,\mathbf{u}^\star$, yields

$$\mathbf{y} = 1 \;+\; (\mathbf{v}^\star - 1)\;(\mathbf{v} + \mathbf{v}^\star)^{-1}\;(\mathbf{v} - 1) \tag{45}$$

after a few lines of careful non-commutative algebra. This is easily split up into components by means of the inversion formula $1/\mathbf{v} = \bar{\mathbf{v}}/|\mathbf{v}|^2$; it agrees with the results of Stiefel and Waldvogel (1965) up to the sign of v_3. Both transformations regularize; the discrepancy is due to the different definition of the orientation in the inversions.

References

Birkhoff, G.D.. The restricted problem of three bodies. Rendiconti del Circolo Matematico di Palermo **39**, 1 (1915); Reprinted in Collected Mathematical Papers, vol. 1. Dover Publications, New York (1968)

Celletti, A.: The Levi-Civita, KS and radial-inversion regularizing transformations. In: Benest, D., Frœschlé, C. (eds.), Singularities in Gravitational Systems, Lecture Notes in Physics, pp. 25–48. Springer-Verlag, Berlin, Heidelberg, New York (2002)

Hamilton, W.R.: On quaternions, or a new system of imaginaries in algebra. Philos. Mag. **25**, 489–495 (1844)

Hopf, H.: Über die Abbildung der dreidimensionalen Sphare auf die Kugelfläche. Math. Ann. **104** (1931); Reprinted in Selecta Heinz Hopf, 38–63. Springer-Verlag, Berlin, Heidelberg, New York (1964)

Kustaanheimo, P.: Spinor regularization of the Kepler motion. Ann. Univ. Turku, Ser. A **73**, 1–7 (1964); Publ. Astr. Obs. Helsinki **102**

Kustaanheimo, P., Stiefel, E.L.: Perturbation theory of Kepler motion based on spinor regularization. J. Reine Angew. Math. **218**, 204–219 (1965)
Levi-Civita, T.: Sur la régularisation du problème des trois corps. Acta Math. **42**, 99–144 (1920)
Siegel, C.L., Moser, J.K.: Lectures on Celestial Mechanics, 290 pp. Springer-Verlag, Berlin, Heidelberg, New York (1971)
Stiefel, E.L., Scheifele, G.: Linear and Regular Celestial Mechanics, 301 pp. Springer-Verlag, Berlin, Heidelberg, New York (1971)
Stiefel, E.L., Waldvogel, J.: Généralisation de la régularisation de Birkhoff pour le mouvement du mobile dans l'espace à trois dimensions. C.R. Acad. Sc. Paris **260**, 805 (1965)
Vivarelli, M.D.: The KS transformation revisited. Meccanica **29**, 15–26 (1994)
Vrbik, J.: Celestial mechanics via quaternions. Can. J. Phys. **72**, 141–146 (1994)
Vrbik, J.: Perturbed Kepler problem in quaternionic form. J. Phys. A **28**, 193–198 (1995)
Waldvogel, J.: Die Verallgemeinerung der Birkhoff-Regularisierung für das räumliche Dreikörperproblem. Bull. Astronomique, Série 3, Tome II, Fasc. **2**, 295–341 (1967a)
Waldvogel, J.: The restricted elliptic three-body problem. In: Stiefel, E., Rössler, M., Waldvogel, J., Burdet, C.A. (eds.), Methods of Regularization for Computing Orbits in Celestial Mechanics. NASA Contractor Report NASA CR **769**, pp. 88–115 (1967b)
Waldvogel, J.: Order and chaos in satellite encounters. In: Steves, B.A., Maciejewski, A.J., Hendry, M. (eds.), Chaotic Worlds: From Order to Disorder in Gravitational N-Body Dynamical Systems, pp. 233-254. Springer, Dordrecht (2006)

Celestial Mechanics and Dynamical Astronomy (2006) 95:213–224
DOI 10.1007/s10569-006-9032-y

REVIEW ARTICLE

The 3:2 spin-orbit resonant motion of Mercury

Anne Lemaitre · Sandrine D'Hoedt · Nicolas Rambaux

Received: 29 November 2005 / Accepted: 18 May 2006 /
Published online: 15 August 2006

Abstract Our purpose is to build a model of rotation for a rigid Mercury, involving the planetary perturbations and the non-spherical shape of the planet. The approach is purely analytical, based on Hamiltonian formalism; we start with a first-order basic averaged resonant potential (including J_2 and C_{22}, and the first powers of the eccentricity and the inclination of Mercury). With this kernel model, we identify the present equilibrium of Mercury; we introduce local canonical variables, describing the motion around this 3:2 resonance. We perform a canonical untangling transformation, to generate three sets of action-angle variables, and identify the three basic frequencies associated to this motion. We show how to reintroduce the short-periodic terms, lost in the averaging process, thanks to the Lie generator; we also comment about the damping effects and the planetary perturbations. At any point of the development, we use the model SONYR to compare and check our calculations.

Keywords Mercury · Resonance spin-orbit · Hamiltonian formalism

1 Introduction

Two space missions are programmed to discover the planet Mercury: Messenger, launched by NASA in 2004, and BepiColombo, a cornerstone of the European Space Agency, scheduled for launch in 2013. Except for Mariner 10 mission, Mercury has not retained much attention from the scientific and spatial community; thanks to this unique experience, we know that Mercury has no atmosphere and a cratered, Moon-like surface, a small magnetic field and a relatively large iron-rich core.

A. Lemaitre (✉) · S. D'Hoedt · N. Rambaux
Département de mathematique, FUNDP,
Rempart de la Vierge, 8, 5000 Namur, Belgium
e-mail: anne.lemaitre@fundp.ac.be

S. D'Hoedt
e-mail: sandrine.dhoedt@fundp.ac.be

Mercury is a very special case, unique in our Solar System, blocked in a 3:2 spin-orbit resonance. Recently, with the new numerical techniques and the power of the nowadays computers, Correia and Laskar (2004) published very convincing results, showing that the probability of capture of Mercury in such a resonance 3:2 was higher than 0.5.

One of the purposes of the present space missions is to detect the presence of a liquid core, to measure its size and its composition and, ideally, to understand its evolution and its different stages.

To be successful, it should be based on a complete and precise model for the rotational motion, to be able to measure the internal structures and properties in a differential way.

Unfortunately, probably due to the lack of data and the complexity of the resonant rotation, complete analytical models of rotation, using classical perturbation techniques (applied for other planets or satellites), were not developed for Mercury. The interest of getting such a model for the space missions is obvious, even if besides, very sophisticated numerical integrations have to be performed. The analytical model of rotation can be used to analyze the effect of each contribution on the motion of the satellite, through the corresponding variational equations. Let us mention the coefficients of the gravitational field or the obliquity or the combined motion of the axes of figure and of the spin axis, as examples of contributions. The analytical methods allow to keep the unknown constants (like the higher zonal and tesseral coefficients) as parameters and to stress their influence on several variables.

Our purpose in this paper is to present the hypotheses and the main steps, in the building of such a model, the already confirmed calculations and the open questions, as well as the main contributions of other teams, working with the same philosophy.

In particular, the three *free* or *proper* frequencies have been detected and checked; the influence of a constant precession rate on these free librations is already quantified, several damping sources are identified, the ideal almost inertial reference frame is discussed and the motion of the obliquity is pointed out, in coupled and uncoupled coordinates systems.

We have, at each step, the opportunity of testing our results through a very performing numerical integration of the coupled rotational and orbital motions, of a planetary system of N-bodies : the model SONYR[1] developed by Rambaux and Bois (2004).

2 The model and the variables

To describe the spin-orbit motion of Mercury, we need four reference frames and two sets of canonical variables: the orbital ones and the rotational ones.

2.1 The reference frames

Let us choose and denote the reference frames of our problem, centered on the center of mass of Mercury:

[1] The approach of SONYR (Spin-Orbit N-bodY Relativistic) derives from the DSX formulation of the post-Newtonian theory of motion for a system of N arbitrary extended, weakly self-graviting, rotating and deformable bodies in mutual interactions. In particular, the SONYR model gives an accurate simultaneous integration of the spin-orbit motion of Mercury.

(a) $(\vec{X}_0, \vec{Y}_0, \vec{Z}_0)$ considered as an *inertial* frame (which can be the ecliptic plane, fixed at some epoch, or an appropriately chosen Laplacian plane, also fixed at some epoch).
(b) $(\vec{X}_1, \vec{Y}_1, \vec{Z}_1)$ the *orbital* frame with $\vec{Z}_1$ perpendicular to the orbit plane.
(c) $(\vec{X}_2, \vec{Y}_2, \vec{Z}_2)$ the *spin* frame, with $\vec{Z}_2$ pointing to the spin axis direction; the plane generated by $(\vec{X}_2, \vec{Y}_2)$ is the equatorial plane, with $\vec{X}_2$ chosen along the ascending node of the equatorial plane on the plane $(\vec{X}_0, \vec{Y}_0)$.
(d) $(\vec{X}_3, \vec{Y}_3, \vec{Z}_3)$ the *body* or the *figure* frame, with $\vec{Z}_3$ in the direction of the axis of greatest inertia and $\vec{X}_3$ in the direction of the axis of smallest inertia.

We introduce, for each couple of frames, the direction $\vec{N}_{ij}$, direction (length $= 1$) chosen along the ascending node of the plane generated by $(\vec{X}_j, \vec{Y}_j)$ on the plane $(\vec{X}_i, \vec{Y}_i)$; we can already notice that $\vec{X}_2 = \vec{N}_{02}$.

2.2 Orbital motion

For the orbital motion, we use Delaunay's elements,

$$(L_o, G_o, H_o, l_o, g_o, h_o)$$

with respect to the inertial reference frame, the subscript "o" meaning "orbital". Mercury is reduced to its center of mass, the capital letters designate the conjugated momenta associated to the angles l_o (mean anomaly), $g_o = \omega_o$ (argument of the pericenter) and $h_o = \Omega_o$ (longitude of the ascending node); they are classically defined as:

$$L_o = m\sqrt{\mu\, a}, \qquad G_o = L_o\sqrt{1-e^2}, \qquad H_o = G_o \cos i_o,$$

where e is the eccentricity, i_o the inclination and a is the semi-major axis of Mercury's orbit, μ is approximated by $G\,(m+M)$, where G is the universal constant of gravitation, m the mass of Mercury and M that of the Sun.

The present values of the eccentricity and inclination (e.g., with respect to the ecliptic frame) do not require the passage to non-singular orbital variables.

We use the terminology *orbital motion* to designate the orbital motion of the Sun about Mercury; however it is given, in most cases, by the orbital motion of Mercury about the Sun, with a change of sign in the mean anomaly. This is why very often we call it the orbital motion of Mercury.

The orbital motion is considered as known and given by an external theory; in the simplest case, we shall assume that it is Keplerian, which means that L_o, G_o, H_o, g_o, and h_o are constants and $\dot{l}_o = n_o = m^3\,\mu^2/L_o^3$; in the more complete cases, we assume that the six elements are known functions of time.

2.3 Rotational motion

For the rotational motion, following (Deprit 1967) we adopt the non singular Andoyer's variables $(\Lambda_1, \Lambda_2, \Lambda_3, \lambda_1, \lambda_2, \lambda_3)$ defined, for the angular variables, as:

$$\begin{aligned} \lambda_1 &= l + g + h, \\ \lambda_2 &= -l, \\ \lambda_3 &= -h, \end{aligned} \tag{1}$$

where h is the angle between $\vec{X}_0$ and $\vec{X}_2$, measured in the plane $(\vec{X}_0, \vec{Y}_0)$, g the angle between $\vec{X}_2$ and $\vec{N}_{23}$ measured in the plane $(\vec{X}_2, \vec{Y}_2)$, and l is the angle between $\vec{N}_{23}$ and $\vec{X}_3$ measured in the plane $(\vec{X}_3, \vec{Y}_3)$.

And for their conjugated momenta:

$$\begin{aligned} \Lambda_1 &= G, \\ \Lambda_2 &= G - L = G\,(1 - \cos J), \\ \Lambda_3 &= G - H = G\,(1 - \cos K), \end{aligned} \tag{2}$$

where G being the norm of the angular momentum, $L = G\cos J$ the projection of the angular momentum on $\vec{Z}_3$, and $H = G\cos K$ the projection of the angular momentum on $\vec{Z}_0$.

K is the angle (called the *inertial* obliquity) between $\vec{Z}_0$ and $\vec{Z}_2$ and J is the angle between $\vec{Z}_2$ and $\vec{Z}_3$.

The usually called *obliquity*, that we shall denote by θ, is the angle between the orbital plane and the spin plane, which means here between $\vec{Z}_2$ and $\vec{Z}_1$; the relation between θ and K is given by the formula:

$$\cos\theta = \cos i_o \cos K + \sin i_o \sin K \cos(h - h_o). \tag{3}$$

2.4 Non-resonant Hamiltonian formalism

Without any direct planetary perturbation, the Hamiltonian can be written:

$$H = H_{2\mathrm{B}} + T_{\mathrm{ROT}} + V_{\mathrm{G}}, \tag{4}$$

where

(a) $H_{2\mathrm{B}}$ is the energy of the two body problem simply given by:

$$H_{2\mathrm{B}} = -\frac{m^3\mu^2}{2L_o^2}. \tag{5}$$

(b) T_{ROT} is the rotational kinetic energy, which takes the following simplified form, thanks to the choice of Andoyer's variables (Deprit 1967):

$$T_{\mathrm{ROT}} = \frac{(\Lambda_1 - \Lambda_2)^2}{2I_3} + \frac{1}{2}(\Lambda_1^2 - (\Lambda_1 - \Lambda_2)^2)\left(\frac{\sin^2\lambda_2}{I_1} + \frac{\cos^2\lambda_2}{I_2}\right), \tag{6}$$

where I_1, I_2, I_3 are the inertia momenta with $I_1 < I_2 < I_3$.

(c) V_{G} is the gravitational potential due to the Sun attraction, in which the selected orbital motion of Mercury is inserted.

We use the diagonalized expression of the matrix of inertia, with I_1, I_2, and I_3 on the diagonal, which means that the third frame (the body frame) is chosen as reference here.

This potential V_{G} is developed (Kaula 1966) in terms of spherical harmonics $Y_{i,j}(\phi,\theta)$:

$$V_{\mathrm{G}} = -\mathrm{GM}\sum_{i,j} \frac{1}{r^{i+1}}\, Y_{i,j}(\phi,\theta), \tag{7}$$

where

$$Y_{i,j}(\phi,\theta) = P_{i,j}(\sin\phi)\,(C_i^j \cos j\theta + S_i^j \sin j\lambda), \tag{8}$$

where (r,ϕ,θ) represent the spherical coordinates (ϕ is the longitude and θ the latitude); the $P_{i,j}$ are Legendre's polynomials. However, for Mercury, the order of magnitude and even the signs of the coefficients C_i^j or S_i^j with $i > 2$ are unknown.

This is why V_{G} is limited to the order 2 spherical harmonics (the order 0 is involved in $H_{2\mathrm{B}}$ and the order 1 disappears by the choice of the reference frame). We denote this part of the potential by $V_{\mathrm{G},2}$, with an explicit expression:

$$V_{\mathrm{G},2} = -\frac{\mathrm{GM}}{a^3}\frac{a^3}{r^3}\left[\frac{C_2^0}{2}\left(2\bar{z}_3^2 - \bar{x}_3^2 - \bar{y}_3^2\right) + 3C_2^2\left(\bar{x}_3^2 - \bar{y}_3^2\right)\right], \tag{9}$$

where r is the distance between the Sun's and Mercury's centers of mass, $(\bar{x}_3, \bar{y}_3, \bar{z}_3)$ designate the normalized coordinates of the vector pointing to the Sun, in the frame $(\vec{X}_3, \vec{Y}_3, \vec{Z}_3)$.

C_2^0 and C_2^2 are the only two remaining coefficients; they can be rewritten in terms of the moments of inertia, and in terms of dimensionless coefficients J_2 and C_{22}:

$$C_2^0 = \frac{I_1 + I_2 - 2I_3}{2} = -mR_{\mathrm{e}}^2 J_2,$$

$$C_2^2 = \frac{I_2 - I_1}{4} = mR_{\mathrm{e}}^2 C_{22},$$

where R_{e} is Mercury's equatorial radius. The relation with our canonical set of rotational variables $(\Lambda_1, \Lambda_2, \Lambda_3, \lambda_1, \lambda_2, \lambda_3)$ and orbital variables $(L_{\mathrm{o}}, G_{\mathrm{o}}, H_{\mathrm{o}}, l_{\mathrm{o}}, g_{\mathrm{o}}, h_{\mathrm{o}})$ is performed through the following cascade of rotations:

$$\begin{pmatrix} \bar{x}_3 \\ \bar{y}_3 \\ \bar{z}_3 \end{pmatrix} = R_3(l)R_1(J)R_3(g)R_1(K)R_3(h)R_3(-h_{\mathrm{o}})R_1(-i_{\mathrm{o}})R_3(-g_{\mathrm{o}}) \begin{pmatrix} \cos v_{\mathrm{o}} \\ \sin v_{\mathrm{o}} \\ 0 \end{pmatrix},$$

where v_{o} is the true anomaly, R_i are rotation matrices, where their subscript is the index of the rotation axis, and, in terms of our new variables, $l = -\lambda_2$, $g = \lambda_1 + \lambda_2 + \lambda_3$, and $h = -\lambda_3$.

The last step is the expansion of v_{o} and $(\frac{a}{r})^3$, in terms of e and l_{o}, up to a given degree of e.

2.5 Resonant Hamiltonian

In a first approach, if we forget about the shape of Mercury ($V_{\mathrm{G}} = 0$), if we consider that the $\vec{Z}_3 = \vec{Z}_2$ (which means $\Lambda_2 = 0$ and $J = 0$) and that the motion of Mercury is Keplerian, we obtain a completely uncoupled problem, in terms of two degrees of freedom, (λ_1, Λ_1) and $(l_{\mathrm{o}}, L_{\mathrm{o}})$, with $H = H_{2\mathrm{B}} + \frac{\Lambda_1^2}{2I_3}$.

The following frequencies are associated to the orbital and rotational motions:

$$\dot{\lambda}_1 = \frac{\Lambda_1}{I_3}, \tag{10}$$

$$\dot{l}_{\mathrm{o}} = \frac{m^3\,\mu^2}{L_{\mathrm{o}}^3} = n_{\mathrm{o}}. \tag{11}$$

For Mercury, these two frequencies are in resonance 3:2, which means that, at the exact resonance, and for a given Keplerian orbit for Mercury, we can deduce a specific value (called $G^\star$) for the norm of the angular momentum:

$$\Lambda_1 = G^\star = \frac{3}{2} n_o I_3. \tag{12}$$

In a second approach, let us now introduce the complete T_{ROT}, and $V_{\mathrm{G},2}$ which introduce the coupling between both motions. A canonical transformation between $(\lambda_1, l_o, \Lambda_1, L_o)$ and $(\sigma, l_o, \Lambda_1, \Lambda_o)$ introduces the 3:2 resonance and transforms a problem with two fast angles to a problem with one slow variable (σ) and one fast variable l_o. The variables with subscripts 2 and 3 for the rotation, and the other Delaunay's elements do not change.

$$\sigma = \lambda_1 - \frac{3}{2} l_o \qquad \text{and} \qquad \Lambda_o = L_o + \frac{3}{2}\Lambda_1. \tag{13}$$

We can express the Hamiltonian in terms of these new canonical variables, and obtain, in a schematic way:

$$\begin{aligned} H = {} & H_{2\mathrm{B}}(\Lambda_o, \Lambda_1) + T_{\mathrm{ROT}}(\lambda_2, \Lambda_1, \Lambda_2) \\ & + V_{\mathrm{G},2}(\sigma, \lambda_2, \lambda_3, \Lambda_1, \Lambda_2, \Lambda_3, l_o, g_o, h_o, \Lambda_o, G_o, H_o). \end{aligned} \tag{14}$$

As it is explained in details in D'Hoedt and Lemaitre (2004a), we can now perform a canonical transformation to get rid of the short periodic terms (containing l_o) in $V_{\mathrm{G},2}$, always considered as an expansion in powers of e. After this averaging process, we can write:

$$\begin{aligned} \langle H \rangle = {} & H_{2\mathrm{B}}(\Lambda_o, \Lambda_1) + T_{\mathrm{ROT}}(\lambda_2, \Lambda_1, \Lambda_2) \\ & + \langle V_{\mathrm{G},2} \rangle(\sigma, \lambda_2, \lambda_3, \Lambda_1, \Lambda_2, \Lambda_3, -, g_o, h_o, \Lambda_o, G_o, H_o), \end{aligned} \tag{15}$$

where all the variables should be replaced by averaged variables.

Let us remind that: $\langle V_{\mathrm{G},2} \rangle = \frac{1}{2\pi} \int_0^{2\pi} V_{\mathrm{G},2} \, \mathrm{d}l_o$, for the first-order of the averaging process.

If we stop the expansion of $\langle V_{\mathrm{G},2} \rangle$ to degree 2 in eccentricity, we can again simplify the problem; in this truncated potential $\langle V_{\mathrm{G},2,2} \rangle$, the variables σ, λ_2, and λ_3 always appear with the same combinations of h_o and g_o. These combinations can be isolated and presented as a new canonical transformation:

$$\begin{array}{ll} \sigma_1 = \sigma - h_o - g_o, & \Lambda_1, \\ \sigma_2 = \lambda_2 & \Lambda_2, \\ \sigma_3 = \lambda_3 + h_o & \Lambda_3, \\ g_o & G'_o = G_o + \Lambda_1, \\ h_o & H'_o = H_o + \Lambda_1 - \Lambda_3 \end{array}$$

and the result is:

$$\mathcal{K} = H_{2\mathrm{B}} + T_{\mathrm{ROT}} + \langle V_{\mathrm{G},2,2} \rangle(\sigma_1, \sigma_2, \sigma_3, \Lambda_1, \Lambda_2, \Lambda_3, \Lambda_o, G'_o, H'_o). \tag{16}$$

This Hamiltonian $\mathcal{K}$ is called the *kernel* (with only three angles); it is related to the complete averaged Hamiltonian through:

$$\langle H \rangle = \mathcal{K} + R(\sigma_1, \sigma_2, \sigma_3, \Lambda_1, \Lambda_2, \Lambda_3, -, g_o, h_o, \Lambda_o, G_o, H_o), \tag{17}$$

where R is the (averaged) perturbation (with still five angles).

2.6 Equilibrium

Vanishing the partial derivatives of the Hamiltonian $\mathcal{K}$, with respect to $\sigma_1, \sigma_2, \sigma_3, \Lambda_1, \Lambda_2$, and Λ_3, several equilibria can be calculated (D'Hoedt et al. 2006); in particular the present status of Mercury, is characterized by: $(\sigma_1, \sigma_2, \sigma_3) = (0, 0, 0), J = 0, K = i_o$.

This value of K of the equilibrium can be slightly changed by introducing a constant precession rate for the node (h_o) and for the pericenter (g_o). The variation is of the order of 1.6 Arcmin, using the present orbit of Mercury. This value coincides with the results of Rambaux and Bois (2004).

Let us call the values at the equilibrium: $(\sigma_1, \sigma_2, \sigma_3) = (0, 0, 0)$, $(\Lambda_1, \Lambda_2, \Lambda_3) = (\Lambda_1^\star, 0, \Lambda_3^\star)$.

2.7 Generalized Cassini's laws

Let us show that our equilibrium checks the generalized Cassini's laws, as for our hypotheses, these laws (taken from Beletsky 2000) are based on a Keplerian orbit for Mercury, plus constant precessions of the node and of the pericenter.

(a) *The body rotates uniformly around a central principal axis of inertia and the angular velocity of this rotation is equal or close to $\frac{3}{2}n$.* At the equilibrium, $\vec{Z_2} = \vec{Z_3}$ because $J = 0$; so the body rotates about the third axis of inertia. Because $\sigma_3 = 0$, $h = h_o$ and using that $\sigma_1 = l + g + h - \frac{3}{2}l_o - g_o - h_o = 0$ we deduce that:

$$\frac{d(l+g)}{dt} = \frac{3}{2}\dot{l}_o - \dot{g}_o = \frac{3}{2}\dot{l}_o + \epsilon \text{ with } \epsilon = 0 \text{ in the pure Keplerian motion.}$$

(b) *The spin axis of the body and the normal to the orbital plane make a constant angle.* At the equilibrium, K is a constant, and we can easily deduce that the corresponding value of the obliquity θ is also constant.

(c) *The spin axis of the body ($\vec{Z_2}$), the normal to the orbital plane ($\vec{Z_1}$), and the axis of precession of the orbit ($\vec{Z_0}$) lie in one plane.* This is trivial in the pure Keplerian case because $K = i_o$ and $\vec{Z_1} = \vec{Z_2}$; it is also true with the precessions, and can be shown with a few algebra.

(d) *Each time the body passes at the pericenter of its orbit, one of the principal axes of inertia orthogonal and the radius vector of the pericenter lie at equal distances from the line of nodes.* At the equilibrium, $\sigma_1 = 0 = \sigma_3$; the passage through the pericenter means that $l_o = 0$. Putting all these conditions together we deduce that, at the pericenter: $l+g = g_o$, where $l+g$ is the angular distance between the line of nodes ($\vec{N_{01}}$) and the axis $\vec{X_3}$ when $J = 0$, and g_o is the angular distance between ($\vec{N_{01}}$) and the pericenter.

3 The short-periodic terms

A first category of short-periodic terms is present in the model, even for a strictly Keplerian model: the terms corresponding to the orbital period of 88 days of Mercury. To perform (at first-order) the averaging transformation and determine its generator, we have to manipulate the different parts of the Hamiltonian, and introduce an expansion of Λ_1 about $G^\star$: $\Lambda_1 = G^\star + X$: let us come back to the non averaged Hamiltonian H, and develop it around $G^\star$:

$$\mathcal{H}_{2B}(\Lambda_o, \Lambda_1) = \mathcal{H}_{2B}(\Lambda_o, G^\star) + \alpha_{2B} X + \mathcal{B}, \tag{18}$$

$$\begin{aligned} \mathcal{T}_{ROT}(\sigma_2, \Lambda_1, \Lambda_2) &= \frac{(G^\star + X)^2}{2I_3} + \cdots \\ &= \frac{G^{\star 2}}{2I_3} + \alpha_{ROT} X + \beta_{ROT}, \end{aligned} \tag{19}$$

where $\mathcal{B}$ and β_{ROT} contain all the terms of $\mathcal{H}_{2B}$ and $\mathcal{T}_{ROT}$ with powers of X higher than 1. We have: $\alpha_{2B} + \alpha_{ROT} = 0$ thanks to the definition of $G^\star$ and we can drop the constant term $G^{\star 2}/2I_3$.

Consequently, to start the averaging process, we have (forgetting about the constant terms):

$$\begin{aligned} \mathcal{H} &= \mathcal{H}_{2B} + \mathcal{T}_{ROT} + \mathcal{V}_{G,2} \\ &= \mathcal{H}_{2B}(\Lambda_o, G^\star) + \mathcal{B} + \beta_{ROT} + \mathcal{V}_{G,2} \\ &= \mathcal{H}_{2B}(\Lambda_o, G^\star) + \mathcal{P}, \end{aligned} \tag{20}$$

where $\mathcal{P}$ contains all the perturbing terms of $\mathcal{H}$. $\mathcal{H}_{2B}$ is selected as the zeroth-order (unperturbed) part of the Hamiltonian and is denoted by H_0^0 in Lie triangle formalism; the perturbation $\mathcal{P}$ is, in the same formalism, called H_1^0 (see, e.g., Meyer and Hall 1992).

The homological equation for the first-order averaging Lie process can be written as follows:

$$H_0^1 = H_1^0 + [H_0^0; W_1] = H_1^0 - \frac{\partial H_0^0}{\partial \Lambda_o} \frac{\partial W_1}{\partial l_o}, \tag{21}$$

where H_0^1 is chosen as the mean value of H_1^0,

$$H_0^1 = \langle H_1^0 \rangle = \langle \mathcal{B} \rangle + \langle \beta_{ROT} \rangle + \langle \mathcal{V}_{G,2} \rangle = \mathcal{B} + \beta_{ROT} + \langle \mathcal{V}_{G,2} \rangle.$$

We can deduce the generator of this transformation, W_1, by a simple integration with respect to l_o:

$$H_1^0 - H_0^1 = n_o \frac{\partial W_1}{\partial l_o}, \text{ where } n_o = \frac{m^3 \mu^2}{L_o^3} \text{ and here } L_o = \Lambda_o - \frac{3}{2} G^\star. \tag{22}$$

If we cut $\mathcal{V}_{G,2}$ to the second power in eccentricity ($\mathcal{V}_{G,2,2}$) we have already more than 700 terms in the generator W_1.

Thanks to this generator, we can write (at first-order) the relationship between averaged and non-averaged variables and momenta:

$$\sigma_i = \bar{\sigma}_i + \frac{\partial W_1}{\partial \Lambda_i} \qquad \Lambda_i = \bar{\Lambda}_i - \frac{\partial W_1}{\partial \sigma_i}, \quad i = 1, 2, 3. \tag{23}$$

In particular, using the averaged values corresponding to the equilibrium of the kernel, we can get, for a given time, their corresponding non-averaged values, to start a numerical integration, for example. This is the case when we compare our results with those of the model SONYR, to avoid a systematic shift between the two approaches. The amplitudes of those short-periodic terms (about Cassini's state) are of the order of a few arcseconds.

Theoretically, we can push the averaging transformation to higher orders and calculate the corresponding generators; another improvement would be to add an iterative process between the averaging transformation and its reciprocal so to find a fixed point of the processus, and again get better initial conditions for numerical tests. This is not yet implemented.

4 The coupled motion

Let us come back to the kernel model $\mathcal{K}$ and let us find the proper frequencies of the rotational model. Mercury is not blocked at the exact 3:2 resonance, but performs a small libration about this equilibrium. To measure the amplitude and the period of this libration, we perform a translation to the equilibrium, and introduce relative coordinates, denoted by ξ_i or η_i, to reproduce this oscillation.

We introduce the following local cartesian coordinates:

$$\begin{aligned}
\sqrt{2\Lambda_1}\cos\sigma_1 &= \sqrt{2\Lambda_1^\star} + \xi_1, & \sqrt{2\Lambda_1}\sin\sigma_1 &= 0 + \eta_1,\\
\sqrt{2\Lambda_2}\cos\sigma_2 &= 0 + \xi_2, & \sqrt{2\Lambda_2}\sin\sigma_2 &= 0 + \eta_2,\\
\sqrt{2\Lambda_3}\cos\sigma_3 &= \sqrt{2\Lambda_3^\star} + \xi_3, & \sqrt{2\Lambda_3}\sin\sigma_3 &= 0 + \eta_3.
\end{aligned}$$

Expanding the kernel in powers of ξ_i and η_i, we obtain, as first approximation, a quadratic form with the following terms:

$$\mathcal{K} = a\,\xi_1^2 + b\,\xi_2^2 + 2c\,\xi_1\xi_3 + d\,\xi_3^2 + e\,\eta_1^2 + f\,\eta_2^2 + 2g\,\eta_1\eta_3 + h\,\eta_3^2 + \cdots, \tag{24}$$

where we clearly see the coupling between variables number 1 and 3.

At this equilibrium, the numerical values for the coefficients are: $a = 39.126449$, $b = 0.002931$, $2c = -0.000511$, $d = 0.002961$, $e = 0.001003$, $f = 0.009875$, $h = 0.002946$, and $2g = 0.000171$.

This coupling could be solved by introducing another canonical transformation (Henrard and Lemaitre 2005) to untangle these variables:

$$\begin{aligned}
\xi_1 &= \tilde{\xi}_1\cos\phi - \tilde{\xi}_3\sin\phi, & \eta_1 &= \rho\,(\tilde{\eta}_1\cos\phi + \tilde{\eta}_3\sin\phi),\\
\xi_2 &= \tilde{\xi}_2, & \eta_2 &= \tilde{\eta}_2,\\
\xi_3 &= \rho\,(-\tilde{\xi}_1\sin\phi + \tilde{\xi}_3\cos\phi) & \eta_3 &= \tilde{\eta}_1\sin\phi + \tilde{\eta}_3\cos\phi
\end{aligned}$$

with $\rho^2 = \frac{ag+hc}{ce+gd} = (1038.52951)^2$ and $\sin 2\phi = \frac{2c\rho}{a+\rho^2 d} = -1.64163\times 10^{-4}$. It is followed, for $i = 1, 2, 3$, by a usual scaling transformation (the scaling parameter is called s_i^2) and a passage to action-angle transformation:

$$2J_i = s_i^2\,\tilde{\xi}_i^2 + \frac{1}{s_i^2}\,\tilde{\eta}_i^2 + \cdots \tag{25}$$

so to get

$$K = \omega_1 J_1 + \omega_2 J_2 + \omega_3 J_3. \tag{26}$$

The three proper frequencies ω_1, ω_2, and ω_3 correspond to periods of 15.86, 583.99, and 1065.05 years if we do not take into account the precession rates in the calculation of the equilibrium; a difference of a few years appears in the third period if we introduce the precession (D'Hoedt et al. 2006). Those three values are in agreement with those of SONYR, with or without the precession (Rambaux and Bois 2004).

The second degree of freedom (D'Hoedt and Lemaitre 2004b) (the *free wobble*) is linked to the motion of the third axis of figure $\vec{Z}_3$ about the spin axis $\vec{Z}_2$; it is connected (through the second term of the kinetic energy of rotation T_{ROT}) to the difference between I_1 and I_2, and consequently to the $C_{2,2}$ measure.

Let us concentrate on the coupling on the other two motions (called *free libration in longitude* for the first and *free precession* for the third).

The variables $\tilde{\xi}_3$ and $\tilde{\xi}_1$ follow dissociated classical harmonic oscillator differential equations.

The very small value of the angle ϕ means that $\tilde{\xi}_1 \simeq \xi_1$ and $\tilde{\xi}_3 \simeq \rho\xi_3$, which explains that the frequencies ω_1 and ω_3 are very often associated to (ξ_1, η_1) and (ξ_3, η_3), forgetting about the uncoupling transformation.

These solutions depend on initial conditions (positions and velocities) of the rotational motion. Different sets of initial conditions have been tested by Yseboodt and Margot (2005); the amplitudes of the motion of the obliquity in particular, linked and dominated by the motion of our variables number 3, could be reduced or amplified by an important factor, only by adjusting the initial conditions.

If planetary motions are considered, after averaging, they will introduce forced contributions on these free motions, with apparition of the very long periods (frequencies $\dot{\Omega}_p$, for the nodes and $\dot{\omega}_p$ for the pericenters) of the planetary motions:

$$\ddot{\xi}_i + \gamma_i \xi_i = \mathcal{F}(\dot{\Omega}_p, \dot{\omega}_p), \quad i = 1,2,3. \tag{27}$$

Bills and Comstock (2005) performed a study on the obliquity variations of Mercury over 10 millions years. They used a linear spin model to describe the orientation of Mercury's pole and a given secular variation model of the orbit. They found that the obliquity remains small with values of 2–4 Arcmin with a typical period of 825 millions of years.

4.1 Damping effects

It has been shown by Peale (2005a) that Mercury had suffered different non conservative forces like tidal forces or core-mantle frictions: on periods of time of about 10^5 years, the damping has played a very efficient role, and reduced the amplitudes of the free motions to very small values. Only the planetary frequencies should still be relevant on their motions, after millions of years of damping.

5 The inertial reference plane

In our first calculations, we have always considered the inertial reference frame as the ecliptic one, frozen at some epoch. The main reason is that we can use orbital elements for Mercury in a well known and easy way; for example, the inclination of the orbit of Mercury on the ecliptic is about 7°, a small but non-zero value, which is convenient for our approach. Furthermore, the same conviviality was expected for the introduction of the direct planetary perturbations, calculated in general with respect to an ecliptic frame at some epoch.

However, at several times, we have noted that the *Laplacian plane* has been quoted as the ideal inertial plane for Mercury's rotation. Very recently, two papers (Peale 2005b; Yseboodt and Margot 2005) presented different calculations and parameters for this specific plane.

We would like to insist on the need for a common definition of the Laplacian plane; it is different for each planet, is linked to the specific perturbations chosen for the model and is moving (secularly) with time.

The idea is the following: let us imagine a non-Keplerian orbital motion for Mercury, which means here a precession of the ascending node Ω_o and a variation of the inclination i_o. For some periods of time, these two motions (node and inclination) could be considered as linear, with constant given rates μ_Ω and μ_i.

We define a reference frame $(\vec{X}, \vec{Y}, \vec{Z})$ with respect to which the orbital frame $(\vec{X}_1, \vec{Y}_1, \vec{Z}_1)$ will have a *constant* inclination. The position of this frame (called Laplacian) with respect to the initial reference frame $(\vec{X}_0, \vec{Y}_0, \vec{Z}_0)$ is given by:

$$\begin{aligned} \vec{X} &= \cos\nu\, \vec{X}_0 + \sin\nu\, \vec{Y}_0, \\ \vec{Y} &= -\sin\nu \cos I\, \vec{X}_0 + \cos\nu \cos I\, \vec{Y}_0 + \sin I\, \vec{Z}_0, \\ \vec{Z} &= \sin\nu \sin I\, \vec{X}_0 - \cos\nu \sin I\, \vec{Y}_0 + \cos I\, \vec{Z}_0, \end{aligned} \tag{28}$$

where $\sin I = \frac{\mu_i}{N}$ and $\nu = \Omega_o + \frac{\pi}{2}$ with $N^2 = \mu_\Omega^2 + \mu_i^2$.

With those values, it is easy to show that the orbital plane precesses on the Laplacian one with a velocity equal to N, and has a "constant" inclination $\hat{i}_o$ given by: $\sin \hat{i}_o = \sqrt{\sin^2 I + \cos^2 I \sin^2 i_o}$.

Of course, if we are looking for a good inertial frame for a space mission, which means 100 years, a mean Laplacian frame could be considered as fixed. On the opposite, if we are interested in the long-period evolution of the spin-orbit resonance, this Laplacian plane has a motion, which has to be taken into account. This motion comes from the non-Keplerian properties of the orbit of Mercury, which clearly means from the planetary perturbations. Two different timescales are still present: the short-periodic terms, due to the orbital motions of the planets, and the long-periodic terms, coming from the secular motions. Following Yseboodt and Margot (2005) or Peale (2005b), we can call them *instantaneous* or mean Laplacian frames and their determination is different: on 10^4 or 10^5 years, using numerical ephemerides for the calculation of the rates, and on 10^6 or more years, using long-periodic analytical variations of the orbital elements.

For those long periods of time, e, i_o, g_o, and h_o are functions of time and not constants anymore. Peale (2005b) introduced this slow motion in the definition of the equilibrium (which obviously depends on these quantities) by an adiabatic invariant: the action corresponding to the momentum Λ_3. He showed that a body initially at (or near) the equilibrium (Cassini's state) is going to move slowly with time, staying near the equilibrium during the evolution.

6 The planetary perturbations

Indirect planetary perturbations have been introduced by Yseboodt and Margot (2005) to test the dependence of the amplitudes of the free motions on the planetary perturbations, using the secular development of Murray and Dermott (1999) for a 3:2 spin-orbit resonance.

It is obvious that the indirect planetary perturbations are more relevant than the direct ones, as it has already been showed by the introduction of a constant rate of precession and its influence on the proper modes of rotation, and by the slow variations of the Laplacian frame.

Again, we have to consider the introduction of the indirect effects in two different contexts: for a space mission, we need a very precise non-Keplerian orbit for Mercury, based on short-periodic terms and probably expressed in polynomials of time, valid for maximum a few hundred years; on the opposite for a model of rotation of Mercury, we need an averaged orbit, with very long-periodic terms, given by a complete analytical work or a synthetic theory, with a validity of millions of years.

It is not always easy to determine which terms are slow or not, with respect to the three proper periods of rotation: the orbital period of Mercury of 88 days is obviously short, the secular motions of the nodes and pericenters are surely slow, but is the orbital period of Jupiter (about 12 years) still short with respect to the first proper period of 15.86 years? Consequently, the introduction of these *short* and *long* periods in the analytical theory could be performed in different ways and is not an easy task.

References

Beletsky, V.V.: Essays on the Motion of Celestial Bodies. Birkhauser Verlag (2000)

Bills, B.G., Comstock, R.I.: Forced obliquity variations of Mercury. J. Geophys. Res. **110**, E04006 (2005)

Correia, A., Laskar, J.: Mercury's capture into the 3:2 spin-orbit resonance as a result of its chaotic dynamics. Nature **429**, 848–850 (2004)

Deprit, A.: Free rotation of a rigid body studied in the phase plane. Am. J. Phys. **35**(5), 424–428 (1967)

D'Hoedt, S., Lemaitre, A.: The spin-orbit resonant rotation of Mercury: a two degree of freedom Hamiltonian model. Celest. Mech. Dynam. Astron. **89**, 267–283 (2004a)

D'Hoedt, S., Lemaitre, A.: The spin-orbit resonance of Mercury: a Hamiltonian approach. In: Kurtz, DW (ed.) Proceedings of the International Astronomical Union 196, pp. 263–270. (2004b)

D'Hoedt, S., Lemaitre, A., Rambaux, N.: 2006, Note on Mercury's rotation: The four equilibria of the Hamiltonian model, Celest. Mech. Dynam. Astron. **96** (2006) in press.

Henrard, J., Lemaitre, A.: The untangling transformation. Astron. J. **130**, 2415–2417 (2005)

Kaula, W.M.: Theory of Satellite Geodesy: Applications of Satellites to Geodesy. Blaisdell Publishing, NY (1996)

Meyer, K.R., Hall, G.R.: Introduction to Hamiltonian Dynamical Systems and the N-Body Problem. Applied Mathematical Sciences, vol. 90. Springer-Verlag, Berlin (1992)

Murray, C.D., Dermott, S.F.: Solar System Dynamics. Cambridge University Press, Cambridge (1999)

Peale, S.J.: The free precession and libration of Mercury. Icarus **178**, 4–18 (2005a)

Peale, S.J.: The proximity of Mercury's spin to Cassini's state 1 from adiabatic invariance. Icarus **181**, 338–347 (2005b)

Rambaux, N., Bois, E.: Theory of the Mercury's spin-orbit motion and analysis of its main librations. Astron. Astrophy. **413**, 381–393 (2004)

Yseboodt, M., Margot, J.L.: Evolution of Mercury's obliquity. Icarus **181**, 327–337 (2005)

Celestial Mechanics and Dynamical Astronomy (2006) 95:225–244
DOI 10.1007/s10569-006-9007-z

ORIGINAL ARTICLE

Symmetric and asymmetric librations in extrasolar planetary systems: a global view

John D. Hadjidemetriou

Received: 15 November 2005 / Revised: 6 January 2006 /
Accepted: 20 February 2006 / Published online: 15 August 2006

Abstract We present a global view of the resonant structure of the phase space of a planetary system with two planets, moving in the same plane, as obtained from the set of the families of periodic orbits. An important tool to understand the topology of the phase space is to determine the position and the stability character of the families of periodic orbits. The region of the phase space close to a stable periodic orbit corresponds to stable, quasi periodic librations. In these regions it is possible for an extrasolar planetary system to exist, or to be trapped following a migration process due to dissipative forces. The mean motion resonances are associated with periodic orbits in a rotating frame, which means that the relative configuration is repeated in space. We start the study with the family of symmetric periodic orbits with nearly circular orbits of the two planets. Along this family the ratio of the periods of the two planets varies, and passes through rational values, which correspond to resonances. At these resonant points we have bifurcations of families of resonant elliptic periodic orbits. There are three topologically different resonances: (1) the resonances $(n+1){:}n$, (2:1, 3:2, …), (2) the resonances $(2n+1){:}(2n-1)$, (3:1, 5:3, …) and (3) all other resonances. The topology at each one of the above three types of resonances is studied, for different values of the sum and of the ratio of the planetary masses. Both symmetric and asymmetric resonant elliptic periodic orbits exist. In general, the symmetric elliptic families bifurcate from the circular family, and the asymmetric elliptic families bifurcate from the symmetric elliptic families. The results are compared with the position of some observed extrasolar planetary systems. In some cases (e.g., Gliese 876) the observed system lies, with a very good accuracy, on the stable part of a family of resonant periodic orbits.

Keywords Periodic orbits · Resonances · Extrasolar planetary systems

J. D. Hadjidemetriou (✉)
Department of Physics, University of Thessaloniki,
54124 Thessaloniki, Greece
e-mail: hadjidem@auth.gr

1. Introduction

In recent years, it was discovered that our Solar System is not the only planetary system in the universe. Up to the present there are 146 observed extrasolar planetary systems, with 18 of them having two or more planets. In many planetary systems with two planets close to each other, the two planets are in mean motion resonance. Examples are: HD 82943 (Israelian et al. 2001; Mayor et al. 2004), Gliese 876 (Marcy et al. 2001; Rivera and Lissauer 2001), at the 2:1 resonance and 55Cnc at the 3:1 resonance (Marcy et al. 2002). Some of these systems have large eccentricities and are evidently stable.

There are different approaches to the study of the dynamical evolution of a planetary system and on the mechanisms that stabilize the system, or generate chaotic motion and instability: Beaugé et al. (2003, 2006), Ferraz-Mello et al. (2003), Gozdjiewski et al. (2002), Ji et al. (2003a, b), Malhotra (2002), and Lee and Peale (2002, 2003). In these papers different methods have been applied as the averaging method, direct numerical integrations of orbits, or various numerical methods which provide indicators for the exponential growth of nearby orbits. In this way the regions where stable motion exists have been detected, in the orbital elements space.

The purpose of this paper is to present a global view of the structure of the phase space of a planetary system with two planets, moving in the plane, as obtained from the set of the families of periodic orbits. In this way, we can detect the regions where stable librations could exist. These will be the regions where a real planetary system could exist in nature. As we shall see, stable regions corresponding to elliptic orbits of the two planets with relatively large eccentricities are associated with mean motion resonances. We remark that stable motion could also exist far from resonances, if the eccentricities are small. This latter motion is close to a stable periodic orbit of the *circular family* of periodic orbits. We also remark that it is possible to have stable motion far from a periodic orbit, but we believe that in this latter case the two planets are not close to each other, so that their gravitational interaction is not very significant.

We found that the gravitational interaction between the planets is not negligible, even for very small planetary masses (compared to the mass of the star). In some cases it even dominates the attraction from the star. In this latter case the geometry of the planetary orbits is such that at some epoch the two planets come so close to each other that their gravitational interaction dominates the attraction from the star. After that, the two planets are trapped in a close binary revolving around the sun. The gap in the lower right quarter in Fig. 11, at the collision area, is generated by this mechanism. For this reason the model we shall use in this study is the *general three body problem*, for planar motion, with the star as the more massive body and the two planets with small, but not negligible masses.

A very good tool to study the structure of the phase space is to find the position and the stability character of the periodic orbits or, equivalently, the fixed points of the Poincaré map on a surface of section. To make this clear, think of a two-dimensional Poincaré map on a surface of section. The position and the stability character of the fixed points (periodic orbits) play a crucial role on the structure of the phase space. A different position of the fixed points and/or a different stability character will change completely the topology of the phase space. This means that the fixed points/periodic orbits provide the framework of the phase space, although they are a set of measure zero.

Stable quasi periodic librations are expected to exist close to a stable periodic orbit. So, although the periodic orbits are a set of measure zero, they can be used as a tool to provide the regions of the phase space (or, equivalently, the space of orbital elements), where stable, quasi periodic librations exist. These stable librations are not periodic motion, but are close to an exact periodic motion. A real planetary system could exist at these regions. The periodic motion in whose vicinity we have stable librations is of two kinds: either non resonant, with small planetary eccentricities, or resonant, where the two planets are in a mean motion resonance. In this latter case the eccentricities are relatively large.

We remark that a periodic orbit is represented as a fixed point in the four dimensional phase space of the Poincaré map. A perturbed periodic orbit is defined by a four-dimensional deviation vector, and this makes the study of the properties of motion close to a periodic orbit quite complicated. For example, a possible perturbation could be to shift the planet P_2 on its orbit, or to rotate the orbit of P_2, at $t = 0$, leaving all other elements the same. Or, we can change the value of the planetary eccentricities from those of the exact periodic motion, and/or the angle of apsides, leaving the semimajor axes the same, so that the resonance is not changed. Although a complete study along these lines is not available, we believe that several quasi periodic librations, as for example the cases studied by Nelson and Papaloizou (2002) are motions in the vicinity of exact periodic orbits.

It can be proved (Hadjidemetriou 1975) that families of periodic orbits in the planar general three body problem exist, in a *rotating* frame which will be defined in Sect. 2. This means that in a periodic orbit the *relative* configuration is repeated in phase space. In particular, in the planetary three body problem (one big body, the star and two small bodies, the planets) the periodic orbits are of two types:

- Non-resonant periodic orbits with nearly circular orbits of the two planets.
- Resonant periodic orbits with nearly elliptic orbits of the two planets.

The circular orbits are all symmetric but the elliptic orbits may be symmetric or asymmetric. Close to a stable periodic orbit there exists a region of stable librations, and it is at these regions that a planetary system could be trapped.

There exist families of elliptic periodic orbits for every mean motion resonance. As we will show, there are three topologically different resonant cases:

- The resonances of the form $(n + 1){:}n$, (2:1, 3:2, ...).
- The resonances $(2n + 1){:}(2n - 1)$, (3:1, 5:3, ...).
- All other resonances, (5:2, 7:3, 8:3, ...).

A global view of the resonant families of elliptic periodic orbits will be computed for each one of the above resonance types. There exist both *symmetric* and *asymmetric* families. As we shall see, the ratio of the planetary masses plays an important role on the stability and the existence of asymmetric families of periodic orbits. The sum of the masses of the planets also plays an important role on the stability and the existence of families of resonant periodic orbits. The stability of a symmetric periodic orbit depends, all other things being the same (semimajor axes, eccentricities), on the phase of the two planets, that is on whether the line of apsides are aligned or antialigned and on the position of the two planets at perihelion of aphelion at some epoch.

The properties of motion close to a periodic orbit are studied by considering a Poincaré map on a surface of section. Close to a stable periodic orbit we have stable librations and the motion in phase space takes place on a torus. On the contrary, close to an unstable periodic orbit we have irregular, chaotic, motion and in many cases the system disrupts into a binary system (the star and one planet) and an escaping planet.

The position of some real extrasolar planetary systems is compared with the above mentioned regions of stable librations.

2. The rotating frame

The model that we will use is the general planar three body problem, for the star, which is the more massive body, and the two planets, with small, but not negligible masses. From this point on, the star will be called the *sun* and the planets will be denoted as P_1 and P_2. The indices 1 and 2 are for the inner and the outer planet, respectively, at least initially, because in a chaotic motion the roles of inner and outer planet may change. For intersecting planetary orbits, the index 1 will refer to the body with the smaller semimajor axis.

In the inertial frame where the center of mass of the system is at rest, we have four degrees of freedom. We define now a rotating frame of reference, whose origin is at the center of mass of the *sun* and the planet P_1, the x-axis is the line *sun*-P_1, the positive direction being from the *sun* to P_1, and the y-axis is perpendicular to the x-axis. As coordinates we use the position x_1 of P_1 on the x-axis, the coordinates (x_2, y_2) of P_2 on the xOy plane and the angle θ of the x-axis with a fixed direction in the inertial frame (Fig. 1). The xOy frame is a *non-uniformly rotating* frame. The Lagrangian of the system in the above mentioned coordinates x_1, x_2, y_2, θ is

$$\mathcal{L} = \frac{1}{2}(m_1 + m_0)\left\{q(\dot{x}_1^2 + x_1^2\dot{\theta}^2) + \frac{m_2}{m}\left[\dot{x}_2^2 + y_2^2 + \dot{\theta}^2(x_2^2 + y_2^2)\right.\right.$$
$$\left.\left. + 2\dot{\theta}(x_2\dot{y}_2 - \dot{x}_2 y_2)\right]\right\} - V, \tag{1}$$

where

$$V = -\frac{Gm_0m_1}{r_{01}} - \frac{Gm_0m_2}{r_{02}} - \frac{Gm_1m_2}{r_{12}} \tag{2}$$

and

$$m = m_0 + m_1 + m_2, \quad q = m_1/m_0. \tag{3}$$

G is the gravitational constant and r_{01}, r_{02} and r_{12} are the distances between the *sun* and P_1, the *sun* and P_2 and P_1P_2, respectively. We note that the angle θ is ignorable, so besides the energy (or Jacobi) integral there also exists the angular momentum integral, $p_\theta = \partial\mathcal{L}/\partial\dot{\theta} =$ constant,

$$p_\theta = (m_0 + m_1)\left\{\dot{\theta}\left[qx_1^2 + \frac{m_2}{m}(x_2^2 + y_2^2)\right] + \frac{m_2}{m}(x_2\dot{y}_2 - \dot{x}_2 y_2)\right\}. \tag{4}$$

By making use of the angular momentum integral, we can reduce the system to *three* degrees of freedom, in the variables x_1, x_2, and y_2. In this way, the study is restricted to the rotating frame xOy, which means that the relative motion of the three bodies in the rotating frame can be studied, independently of the motion of the system xOy in the inertial frame. The Lagrangian of the reduced system in the rotating frame is the *Routhian function* $\mathcal{R}$,

$$\mathcal{R} = \frac{1}{2}(m_1 + m_0)\left\{q\dot{x}_1^2 + \frac{m_2}{m}(\dot{x}_2^2 + y_2^2) - \frac{\left[\frac{p_\theta}{m_1+m_2} - \frac{m_2}{m}(x_2\dot{y}_2 - \dot{x}_2 y_2)\right]^2}{q\dot{x}_1^2 + \frac{m_2}{m}(x_2^2 + y_2^2)}\right\},$$
$$-V. \tag{5}$$

Note that the angular momentum p_θ appears as a fixed parameter.

From this point on, we shall restrict our study to the rotating frame only. In order to avoid duplication of the results, we must fix the units of mass, length and time, by considering three

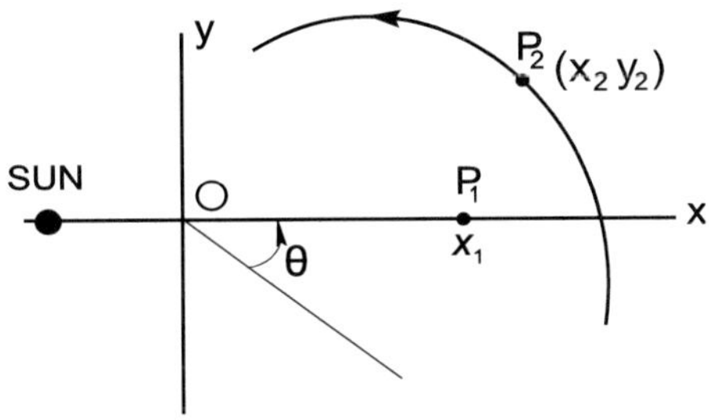

Fig. 1 The rotating frame xOy. The planet P_1 moves on the x-axis and the planet P_2 moves in the xOy plane. The angle θ defines the orientation of the rotating frame

normalizing conditions:

$$m_0 + m_1 + m_2 = 1, \quad G = 1, \quad p_\theta : \text{fixed}.$$

This means that the unit of mass is the total mass of the system, the gravitational constant is taken equal to unity and the constant angular momentum is given a fixed value (note that the angular momentum appears as a fixed parameter in the differential equations in the rotating frame).

It can be proved (Hadjidemetriou 1975) that *families of periodic orbits* exist in the rotating frame xOy. In particular, we are interested in periodic orbits of the planetary type, that we shall study in the next section.

3. Periodic orbits of the planetary type

We consider a three-body system with the *sun* S as the main body and two small, but gravitationally interacting, planets P_1 and P_2. As we mentioned above, families of periodic orbits exist in the rotating frame xOy. A periodic orbit can be *symmetric* or *asymmetric* with respect to the rotating x-axis. We remind that in all cases P_1 moves on the x-axis and P_2 in the xOy plane. In a symmetric periodic orbit, for a certain epoch, for example $t = 0$, it is $\dot{x}_1 = 0$ and $\dot{x}_2 = 0$. This implies that at $t = 0$ the planet P_2 crosses perpendicularly the x-axis and at that time the planet P_1 is temporarily at rest on the x-axis. So, the non zero initial conditions of a periodic orbit are: $x_{10}, x_{20}, \dot{y}_{20}$, which means that a family of symmetric periodic orbits can be represented by a one-dimensional curve in the three-dimensional space of initial conditions $x_{10}, x_{20}, \dot{y}_{20}$. The orbits for which we cannot find an epoch for which the above conditions are satisfied, are asymmetric.

There are two different types of periodic orbits of the *planetary type*:

- Periodic orbits with the two planets describing *nearly circular* orbits. In general, the system is non resonant. This is seen in Fig. 2. We remark that the planetary orbits are exactly circular in the unperturbed case, and *nearly circular* if the gravitational interaction between the planets is introduced. All these orbits are symmetric with respect to the x-axis of the rotating frame.
- Periodic orbits with the two planets describing nearly *elliptic orbits*. In the unperturbed case the two planetary orbits are exactly elliptic. We remark that, contrary to the circular case, all elliptic periodic orbits are resonant, corresponding to a *mean motion resonance*. We can distinguish between symmetric and asymmetric resonant elliptic periodic orbits.

  - Symmetric elliptic periodic orbits: the two planets describe elliptic orbits whose lines of apsides are aligned or antialigned, and in addition, when P_1 is at perihelion or aphelion, P_2 is at that time also at perihelion or aphelion (Fig. 3a). This means that there

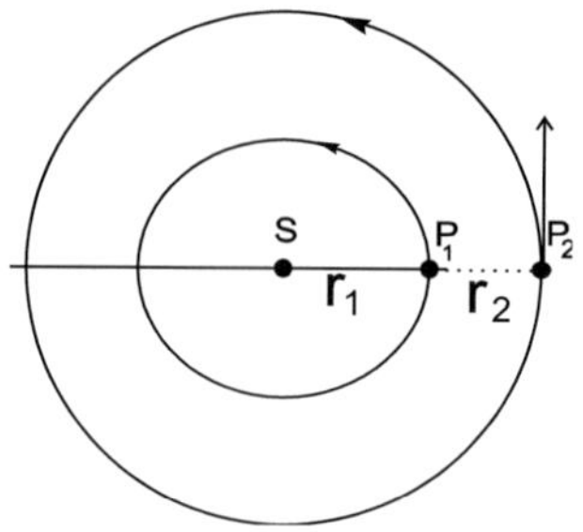

Fig. 2 The circular motion of the two planets P_1 and P_2. The motion is periodic in the rotating frame xOy, for any value of the planetary semimajor axes

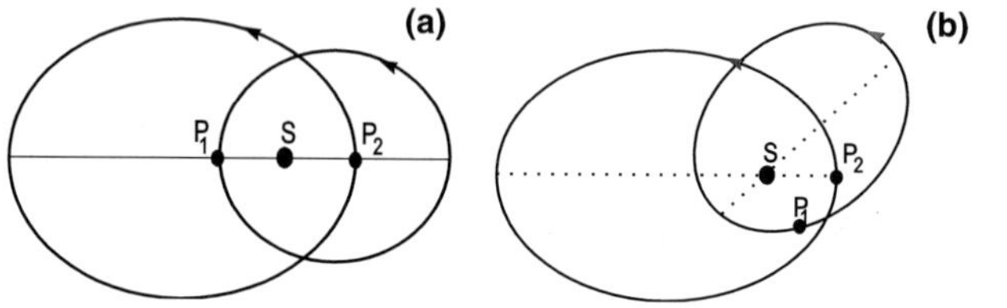

Fig. 3 (a) The elliptic motion of the two planets P_1 and P_2 in the symmetric configuration. (b) The elliptic motion of the two planets in the asymmetric configuration

are four different initial configurations, which are equivalent in pairs, i.e. position at $t = 0$ and $t = T/2$, T being the period, depending on the particular resonance.

- Asymmetric elliptic periodic orbits: The lines of apsides of the two planets are not in the same line (Fig. 3b).

4. Families of periodic orbits

It can be proved that families of periodic orbits of the planetary type exist, in a rotating frame, in the general three-body problem (Hadjidemetriou 1976). In order to obtain a global view of the resonant and non resonant periodic orbits, we will start with the unperturbed problem (zero planetary masses) and circular planetary orbits, and study how things change when we give masses to the planets.

4.1. The unperturbed family of circular orbits

Consider the two planets P_1 and P_2 describing circular orbits in the plane (Fig. 2). It is evident that the motion is periodic in the rotating frame xOy defined in Sect. 1, for all values of the semimajor axes (radii) of the two planets. The planet P_1 moves on the x-axis (in this particular case it is fixed) and the planet P_2 moves in the xOy plane. This motion is *symmetric* with respect to the x-axis, because for some time, say $t = 0$, it is $\dot{x}_1 = 0$ (this is always the case) and at that time the planet P_2 crosses perpendicularly the x-axis, $\dot{x}_2 = 0$. These periodic orbits belong to a monoparametric family of symmetric periodic orbits. To make this clear, let us fix the value of the semimajor axis a_1 of P_1 and vary the semimajor axis a_2 of P_2. In this way, we obtain a continuous family of periodic orbits, along which the ratio a_1/a_2 of the semimajor axes, or, equivalently, the ratio n_1/n_2 of the frequencies varies. Evidently n_1/n_2 passes through rational values, which means that on this single family of periodic orbits there exist resonant orbits, corresponding to mean motion resonance between the planets.

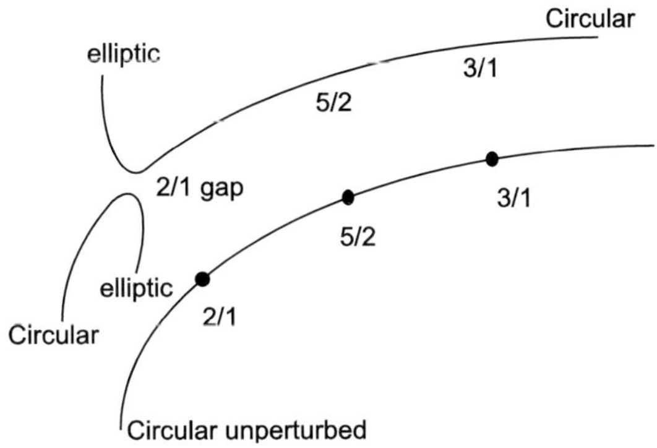

Fig. 4 The unperturbed family of circular orbits (lower curve) with three representative resonances, 2:1, 3:1, and 5:2 (schematically). The upper curves represent the evolution of the single circular family, when the gravitational interaction between the planets is introduced. A gap appears at the 2:1 and two families of 2:1 resonant elliptic orbits bifurcate from the gap

4.2. Continuation of the circular family to non zero planetary masses

We give now small, but not negligible masses to the two planets. It can be proved (Hadjidemetriou 1976) that all the unperturbed circular periodic orbits are continued, with the same period, as periodic orbits with nearly circular planetary orbits, in the rotating frame xOy. These orbits are symmetric with respect to the x-axis. The continuation is not possible only at the resonances of the form $(n+1){:}n$, i.e, 2:1, 3:2, 4:3,... At these resonances a *gap* appears, as we show schematically in Fig. 4, using the 2:1 resonance as a representative example. At such a gap, we have a bifurcation of two families of 2:1 resonant *elliptic* periodic orbits, as we shall see in the following.

4.3. Evolution of the stability on the family of circular orbits

It is obvious that all the orbits of the circular unperturbed family are orbitally stable, because the two planetary orbits are uncoupled Keplerian orbits. We have, in the rotating frame, three degrees of freedom and consequently, we have three pairs of unit eigenvalues, which are all on the unit circle in the complex plane (for the theory see Hadjidemetriou 2006). One pair of eigenvalues is equal to one, because of the existence of the energy integral. The other two pairs of eigenvalues are complex conjugate on the unit circle and are obtained, by Keplerian theory, to be equal to

$$\lambda_{1,2} = e^{\pm i\phi}, \qquad \lambda_{3,4} = e^{\pm i\phi}, \qquad \phi = \frac{2\pi}{1 - n_2/n_1}, \tag{6}$$

where n_1, n_2 are the frequencies of the two planets in their orbits (in the inertial frame). We see that $\lambda_1 = \lambda_3$ and $\lambda_2 = \lambda_4$, i.e. there is a double eigenvalue equal to $e^{i\phi}$ and a double complex conjugate eigenvalue equal to $e^{-i\phi}$ (Fig. 5a).

As we mentioned in the previous section, the ratio n_1/n_2 varies along the family of periodic orbits of the unperturbed problem and consequently passes through rational values, which correspond to mean motion resonance between the two planets. This means that the double eigenvalue $\lambda_1 = \lambda_3$ (and its complex conjugate) move on the unit circle, as we move on the unperturbed family of circular periodic orbits. From Eq. 6, we see that the ratio $\phi/2\pi$ takes rational values at all resonant orbits on the family, and, in general, the double eigenvalue $\lambda_1 = \lambda_3$ is not at $+1$ or -1.

Let us now switch on the planetary masses, introducing the gravitational interaction between the two planets. The double eigenvalue $\lambda_1 = \lambda_3$ (and its complex conjugate) split into two separate eigenvalues and move in the complex plane, but must obey the Hamiltonian

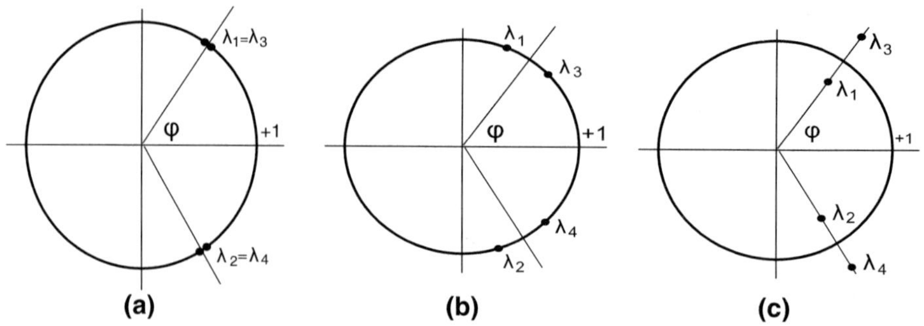

Fig. 5 (a) The two double pairs of eigenvalues $\lambda_1 = \lambda_3$ and $\lambda_2 = \lambda_4$ of the unperturbed problem on the unit circle in the complex plane. (b) The two double eigenvalues split and remain on the unit circle when the gravitational interaction of the two planets is taken into account. The stability is preserved. (c) The evolution of the four eigenvalues outside the unit circle, generating complex instability. This cannot happen in a planetary system

properties: they must be arranged in reciprocal pairs and also in complex conjugate pairs. This means that we have two possibilities: either to move *on* the unit circle, as shown in Fig. 5b, *preserving the stability*, or move outside the unit circle, as shown in Fig. 5c (complex conjugate pairs: (λ_1, λ_2), (λ_3, λ_4), reciprocal pairs (λ_1, λ_4), (λ_2, λ_3)), *generating complex instability*. It can be proved however, by making use of the theory of Krein (Hadjidemetriou 1982), that the evolution to complex instability, when $\phi \neq 0$ or π (the eigenvalues are not at ± 1), is impossible in a planetary system with two planets moving in the plane in the same direction. In fact the proof is stronger. It is proved that *there does not exist* a Hamiltonian perturbation to generate complex instability in such a system.

The only case where the perturbation can generate instability is when two eigenvalues meet at $+1$ or at -1, on the unit circle in the complex plane, as the corresponding orbit moves on the family of circular orbits. The first case, meeting of the eigenvalues at $+1$, corresponds to a resonance of the form $(n + 1) : n$, i.e., 2:1, 3:2, ... and the second case, meeting of the eigenvalues at -1, corresponds to a resonance of the form $(2n + 1) : (2n - 1)$, i.e., 3:1, 5:2, ... But in the first case the continuation is not possible (see the gap in Fig. 4), so we will consider the case for the resonances 3:1, 5:3, ...

Let us consider the region on the perturbed circular family close to the 3:1 resonance. The two double pairs of complex conjugate eigenvalues $\lambda_1 = \lambda_3$ and $\lambda_2 = \lambda_4$ have split on the unit circle to two separate complex conjugate pairs (λ_1, λ_2) and (λ_3, λ_4) (Fig. 5b) and each pair moves toward the point -1. When they meet at -1, the eigenvalues may move *outside* the unit circle, on the real axis, and thus generate instability, as predicted by the theory of Krein (Hadjidemetriou 1982). This is seen in Fig. 6. It may happen that only one pair of eigenvalues, (λ_1, λ_2), goes outside the unit circle, and the other, (λ_3, λ_4), staying on the unit circle, or both pairs of eigenvalues go outside the unit circle. This depends on the magnitude of the masses, as we shall see in the following. After we cross the 3:1 resonance, the eigenvalues return again on the unit circle, restoring the stability.

From the above, we see that the only region on the family of *circular* periodic orbits which is unstable is the region close to resonances of the form 3:1, 5:3,...All other resonant and non-resonant cases are stable.

4.4. Families of symmetric and asymmetric elliptic periodic orbits

The periodic motion of a planetary system with *elliptic* orbits of the two planets is necessarily resonant, with the two planets locked in a mean motion resonance. All these orbits belong

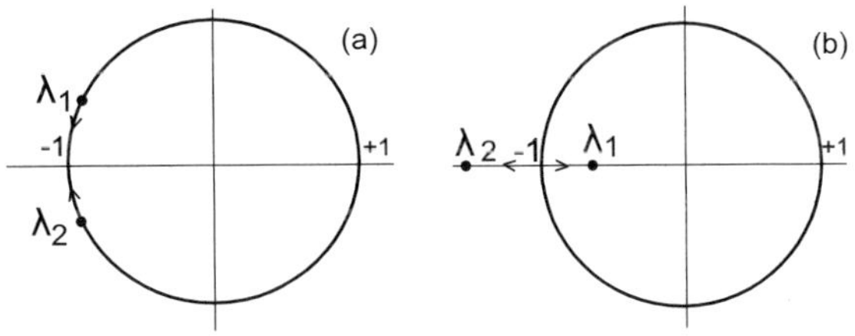

Fig. 6 (a) The position of the complex conjugate eigenvalues λ_1, λ_2 on the unit circle close to the 3:1 resonance. (b) The evolution of the eigenvalues λ_1, λ_2 outside the unit circle when the orbit crosses the 3:1 resonance

to monoparametric families, along which the eccentricities of the two planetary orbits vary, and, in the asymmetric case, the difference $\Delta\varpi$ between the angle of apsides also varies. In order to obtain a global view of the families of elliptic periodic orbits, we will start with the families that bifurcate from the family of circular periodic orbits. Next, we shall study the elliptic families that bifurcate from the above mentioned elliptic families. We remark however that apart from these two categories of elliptic families, there also exist families of elliptic orbits that appear as independent from them. Some of these latter families may be the continuation of families of the *elliptic* restricted three body problem, by giving mass to the massless body.

As we will show in the following, the orbits of the circular family are all symmetric with respect to the x-axis of the rotating frame. The families of elliptic orbits that bifurcate from the circular family are also symmetric. However, the orbits of the elliptic families that bifurcate from the elliptic families are all asymmetric. The independent elliptic families are either symmetric or asymmetric.

Consider a family of periodic orbits, either circular or elliptic. A bifurcation of a new family of periodic orbits is only possible from that orbit on the family which has an eigenvalue equal to $+1$ (Hadjidemetriou 2006). We remark that this is a double unit eigenvalue, because the eigenvalues are always in complex conjugate pairs. We call this orbit a *critical* orbit. This critical orbit may be the point on the family where we have a change from stability to instability (or vice versa), but it can also be, on the circular family, any resonant $p{:}q$ periodic orbit, if it is described several times, as we show in the following.

We will consider separately the bifurcations from the circular family and from the elliptic families.

4.4.1. *Bifurcation from the circular family*

We shall show now that from each *resonant* region on the family of circular periodic orbits we have a bifurcation of families of *elliptic* periodic orbits. There are two elliptic families bifurcating from a $p{:}q$ resonant periodic orbit if $p{:}q \neq 3{:}1, 5{:}3, \ldots$ and two, or four families if $p{:}q = 3{:}1, 5{:}3, \ldots$, as we shall prove below. All these families are symmetric with respect to the rotating x-axis.

As we mentioned in Sect. 4.3, there are three pairs of eigenvalues: one unit pair, due to the existence of the energy integral and two more complex conjugate pairs, on the unit circle. In the unperturbed case (zero planetary masses) these two complex conjugate pairs coincide,

$$\lambda_1 = \lambda_3 = e^{+i\phi}, \qquad \lambda_2 = \lambda_4 = e^{-i\phi}, \qquad \phi = \frac{2\pi}{1 - n_2/n_1}, \tag{7}$$

where n_1/n_2 is the ratio of the mean motions of the two planets. This ratio is not, in general, rational. It is at the rational (resonant) points that a bifurcation of families of *elliptic* periodic orbits is possible. In addition, a bifurcation of a family of elliptic orbits is possible at all

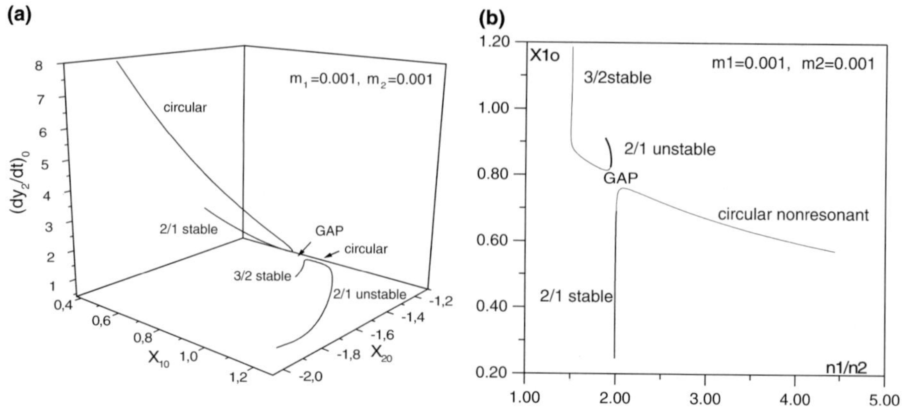

Fig. 7 (a) The circular and the elliptic families of periodic orbits at the 2:1 resonance, for a fictitious system with $m_0 = 0.998$, $m_1 = 0.001$, and $m_2 = 0.001$. The gap is clearly seen. (b) The same system in the space $(n_1{:}n_2)$-x_{10} The 3:2 resonant family is also shown

points on the circular family where we have a change from stability to instability, or vice versa. This latter case happens at the resonances of the form 3:1, 5:3,...In this case, we may have a bifurcation of two or of four families of resonant elliptic periodic orbits.

There are three topologically different resonances:

- The resonances of the form $n_1/n_2 = (n+1)/n$, $n = 1, 2, \ldots$
  These are the resonances 2:1, 3:2, ... At these resonances, the periodic orbits are not continued as circular periodic orbits, but instead, a gap appears on the continued circular family and at both ends of the gap we have the beginning of a family of resonant elliptic periodic orbits, corresponding to the $(n+1) : n$ resonance (Fig. 3). A typical example is shown in Fig. 7, for a planetary system with masses $m_0 = 0.998$, $m_1 = 0.001$, and $m_2 = 0.001$. In Fig. 7a, we show the circular family of symmetric periodic orbits, in the space of initial conditions $x_{10}x_{20}\dot{y}_{20}$, the gap at the 2:1 resonance and the bifurcation from the gap of two families of elliptic symmetric periodic orbits. In Fig. 7b, we show the same diagram in the space $(n_1 : n_2)$-x_{10}. The gap at the 2:1 resonance and the elliptic resonant periodic orbits are clearly seen. In Fig. 7b the 3:2 resonant elliptic family is also shown.
- The resonances of the form $n_1{:}n_2 = (2n+1){:}(2n-1)$, $n = 1, 2, \ldots$
  These are the resonances 3:1, 5:3,... As we mentioned in Sect. 4.3, instability is generated in a small region close to the above resonances. As shown schematically in Fig. 6, only one pair of eigenvalues may go outside the unit circle, generating *simple instability*, or both pairs may go outside the unit circle, generating *double instability*. Whether we have the first or the second case, it depends on the magnitude of the perturbation, i.e. on the total mass of the planets. This is shown in Fig. 8, where we plot the *stability index* along the family of circular orbits, close to the 3:1 resonance, for two different planetary masses. The stability index corresponds to a complex conjugate pair of eigenvalues and it is defined as

$$b_1 = \lambda_1 + \lambda_2, \qquad b_2 = \lambda_3 + \lambda_4.$$

  This means that we have stability if

$$|b_i| \leq 2 \quad (i = 1, 2).$$

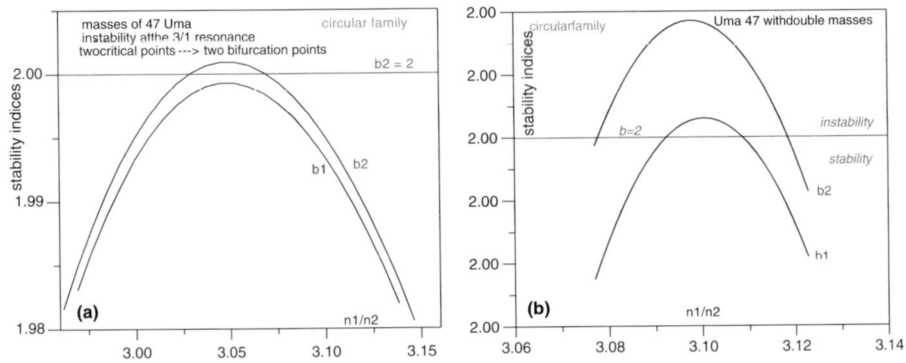

Fig. 8 (a) The stability index along the circular family, close to the 3:1 resonance, in the space $n_1 : n_2$ *stability indices* for the masses of the system 47 UMa, $m_0 = 0.998$, $m_1 = 0.001$, and $m_2 = 0.001$. Two critical points exist. (b) The same as in (a), but with double planetary masses. Four critical points exist now

In Fig. 8a, we use the (normalized) masses of the system 47 UMa, $m_0 = 0.996942$, $m_1 = 0.002354$, and $m_2 = 0.000704$, (Fisher et al. 2002) as an example and in Fig. 8b the double masses. In the first case only one stability index becomes unstable for a small region close to the 3:1 resonance, but as the perturbation (the masses) increases, both stability indices become unstable. At the points where the stability changes, the stability index is equal to $+2$, which means that the eigenvalues are equal to $e^{\pm i\pi} = -1$. If now this orbit is described twice, the eigenvalue becomes equal to $(e^{\pm i\pi})^2 = +1$, and consequently the orbit is critical and a bifurcation of a family of elliptic periodic orbits is possible. The period of the elliptic orbits is equal to twice the period of the circular orbit from which it bifurcates.

In the former case we have two critical points, and consequently two elliptic 3:1 resonant periodic orbits. In the latter case there are four critical points and consequently four families of resonant 3:1 elliptic periodic orbits. All these families are symmetric with respect to the rotating x-axis and differ in the initial phase only, i.e. on whether the lines of apsides are aligned or antialigned and on the position of the two planets at perihelion or at aphelion at $t = 0$. Along these families the planetary eccentricities vary, starting with zero values.

- All other resonances $n_1{:}n_2 \neq (n+1){:}n$, $(2n+1){:}(2n-1)$, $n = 1, 2, \ldots$
  These are resonances of the form 5:2, 7:3, 5:1, 8:3, ... As we mentioned in Sect. 4.3, the stability is preserved when we go from the perturbed to the unperturbed problem, which means that the eigenvalues $\lambda_1 = \lambda_3$ and $\lambda_2 = \lambda_4$ split, but stay *on* the unit circle. Let us assume now that we approach a p/q resonance ($n_1/n_2 = p/q$, $p/q \neq (n+1)/n, (2n+1)/(2n-1)$) on the perturbed family. The eigenvalues λ_1, λ_3 move on the unit circle and cross the point

$$e^{i2\pi \frac{q}{q-p}},$$

one after the other. So, there are *two orbits* on the perturbed family of periodic orbits, close to each other, and close to the $p{:}q$ resonance, which have a pair of eigenvalues λ_1, λ_3 (or λ_2, λ_4) equal *exactly* to

$$e^{\pm i2\pi \frac{q}{q-p}}. \tag{8}$$

This is not a critical orbit (it does not have a unit pair of eigenvalues). However, it can become critical if this orbit is described $|q-p|$ times, because its eigenvalues are equal to

$$\lambda_i^{|q-p|} = e^{\pm i2\pi q} = +1.$$

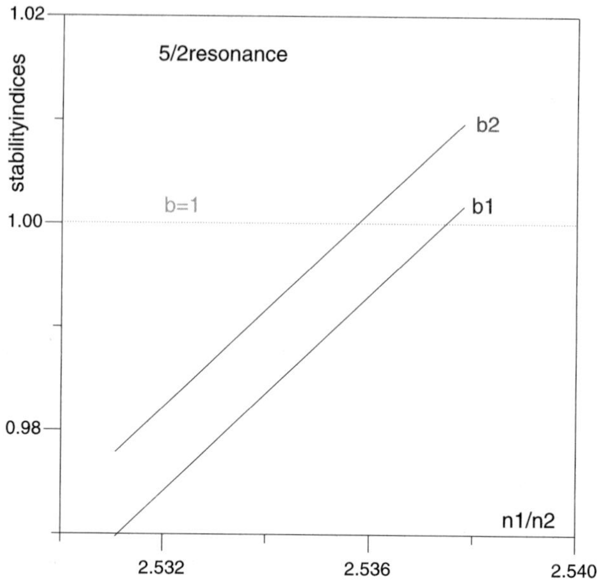

Fig. 9 The two stability indices close to the 5:2 resonance along the perturbed family of circular orbits. They cross the value $b = 1$, corresponding to the pair of eigenvalues $e^{\pm i2\pi/3}$

At this point we have a bifurcation of a p:q family of resonant elliptic periodic orbits, along which the planetary eccentricities vary, starting from zero values. The period along this resonant family is $|p-q|$ times the period of the corresponding circular resonant periodic orbit. Since there are two such points, there are *two* p:q resonant families of elliptic periodic orbits that bifurcate from the circular family at the region close to the p:q resonance. As an example, we consider the region close to the 5:2 resonance on the family of circular periodic orbits. There exist two distinct periodic orbits, close to each other, with eigenvalues $\lambda_{1,2} = e^{\pm 2i\pi/3}$ and $\lambda_{3,4} = e^{\pm 2i\pi/3}$, respectively (stability indices $b_1 = +1$, $b_2 = +1$, respectively). From each one there is a bifurcation of a 5:2 resonant elliptic symmetric family of periodic orbits, with period three times the period of the circular orbit from which they bifurcate. This is shown in Fig. 9, for the masses of 47 UMa, used also in the 3:1 resonance above. These orbits differ in the phase only and the planetary eccentricities vary, starting with zero values.

4.4.2. Bifurcation from the elliptic families

As we mentioned before, there is a bifurcation of a family of periodic orbits from each orbit of a family which is critical, i.e. it has a pair of eigenvalues equal to $+1$. This is the case where we have a change of stability to instability, or vice versa, as we have already seen in the 3:1 resonance case on the family of circular orbits. Let us consider now a family of *elliptic* periodic orbits that bifurcates from the circular family. It may happen that the stability along the family changes. At this point there is a bifurcation of a new family of *elliptic* periodic orbits. The numerical computations have shown that all such elliptic families are *asymmetric*. We remark that in the unperturbed case the planetary orbits of an elliptic family are uncoupled elliptic Keplerian orbits in resonance, which implies that there are two pairs of eigenvalues

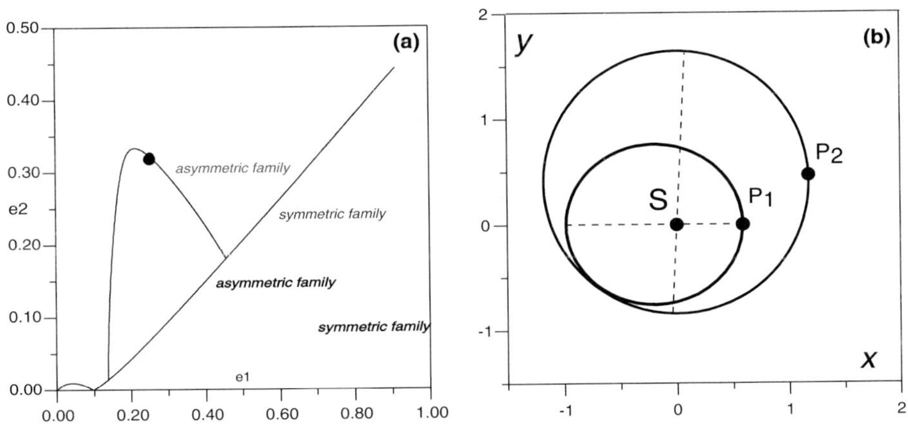

Fig. 10 (a) The family of symmetric elliptic periodic orbits, in the eccentricity space, for the masses of the Io-Europa satellites of Jupiter, and the bifurcation of an asymmetric family starting from one end of the unstable region and ending to the other end of the unstable region. (b) The asymmetric orbit indicated on the asymmetric family of (a)

equal to +1 (in addition to the third unit pair which always exists, due to the existence of the energy integral). When the perturbation is applied one, or both, of these unit pairs may go outside the unit circle in the complex plane, on the real axis, and thus generate instability. This depends mainly on the ratio m_1/m_2 of the planetary masses, as we shall show by some examples that follow.

An example of bifurcation of an asymmetric family of elliptic periodic orbits, for the masses of the Io-Europa satellites of Jupiter at the 2:1 resonance, is given in Fig. 10a. The normalized masses are $m_0 = 0.999928$, $m_1 = 0.000047$, and $m_2 = 0.000025$. Note that $m_1 > m_2$. On the symmetric family there is an unstable region and from both ends of it, at the critical points, there is a bifurcation of an elliptic 2:1 resonant asymmetric family of periodic orbits. It turns out that these two families coincide, as shown in Fig. 10a. In Fig. 10b, we present a representative asymmetric orbit, corresponding to the one indicated in Fig. 10a. We remark that the unstable region on the symmetric elliptic family appears because it is $m_1 > m_2$. If it were $m_1 < m_2$, the whole symmetric family is stable and consequently there are no critical points from which to bifurcate a new family of elliptic periodic orbits. The critical mass ratio, beyond which an unstable region appears is $m_1/m_2 = 0.97$ (Ferraz-Mello et al. 2003).

Finally we remark, that apart from all the above mentioned, symmetric or asymmetric families of resonant elliptic periodic orbits, there exist also such families that do not bifurcate from the above mentioned families. Some of them are the continuation from the *elliptic* restricted problem to the general problem, if we give mass to the massless body.

5. Factors that affect the stability of a planetary system

Families of periodic orbits for the masses of some observed extrasolar planetary systems were studied in several papers: Hadjidemetriou (2002), Hadjidemetriou and Psychoyos (2003), Psychoyos and Hadjidemetriou (2005), and Voyatzis and Hadjidemetriou (2005, 2006), studying symmetric and asymmetric periodic orbits for the systems Gliese 876, HD 82943,

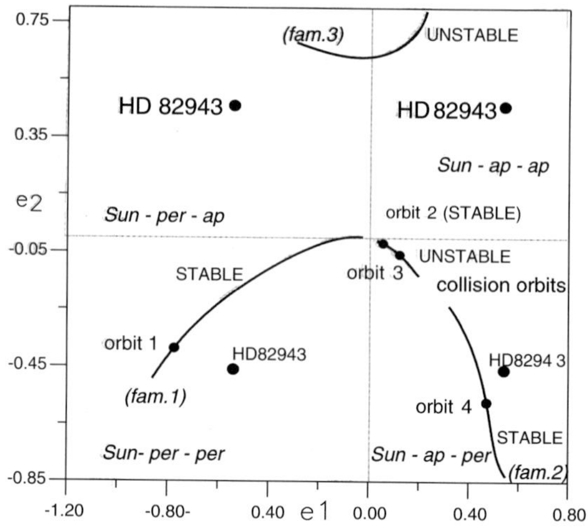

Fig. 11 The three families of symmetric periodic orbits at the 2:1 resonance in the eccentricity space, for the masses of HD 82943 given by Israelian et al. (2001). A positive eccentricity denotes position at aphelion and a negative eccentricity position at perihelion. The positions of HD 82943, for the four different initial configurations are also shown

at the 2:1 resonance, 55CnC at the 3:1 resonance, 47 UMa at the 5:2 resonance and also for the Io-Europa satellites of Jupiter at the 2:1 resonance. Stationary solutions close to mean motion resonance, which in fact correspond to periodic orbits, have been studied by Beauge et al. (2003, 2006). It is not the object of the present paper to review these results, but to use some of them as examples to make clear the factors that affect the stability of a planetary system and to show the stability regions close to a linearly stable periodic orbit, by making use of the Poincaré map on a surface of section. In particular, we will start with the families of periodic orbits corresponding to the masses of the system HD 82943. Although the elements of this system have been revised several times, the fit given by Israelian et al. (2001), that we shall use here, is enough for our purpose. The elements of this fit are: $m_0 = 1.05 M_{\mathrm{SUN}}$, $m_1 \sin i = 0.88\,\mathrm{MJ}$, $m_2 \sin i = 1.63\,\mathrm{MJ}$, $a_1 = 0.73\,\mathrm{AU}$, $a_2 = 1.16\,\mathrm{AU}$, $e_1 = 0.54 \pm 0.05$, and $e_2 = 0.41 \pm 0.08$. In Fig. 11, we present three families of symmetric periodic orbits, for the normalized masses corresponding to the above system, in the space $e_1 e_2$, using as a convention $e_i > 0$ to denote position at aphelion and $e_i < 0$ position at perihelion. There is a stable family, *family* 1 where the line of apsides are aligned and the two planets are both at perihelion, at $t = 0$, and another family, *family* 2, which starts as stable and becomes unstable at a region where close encounters between the two planets take place. We remark that the whole *family* 1 is stable because $m_1 < m_2$. After this collision region, at relatively large eccentricities, the family is again stable. Along this family, the lines of apsides are antialigned and at $t = 0\,P_1$ is at aphelion and P_2 at perihelion. These two families bifurcate from the circular family at the region where the 2:1 resonant gap appears. A third family, *family* 3, independent from these two also exists, and is unstable. In this diagram, we also show the position of the system HD 82943, with its elements as given above, for all four initial different configurations and we shall comment on their stability in the next sections.

The behavior of the system when a periodic orbit is perturbed, is studied by the Poincaré map on a surface of section. As we shall see in the following, stable librations exist close to a stable periodic orbit and chaotic motion appears for larger deviations from the exact periodic motion of a stable orbit or for perturbations close to an unstable periodic orbit. The Poincaré map is defined on the surface of section (Fig. 12)

$$H = h = \text{constant}, \quad y_2 = 0, \quad \dot{y}_2 > 0$$

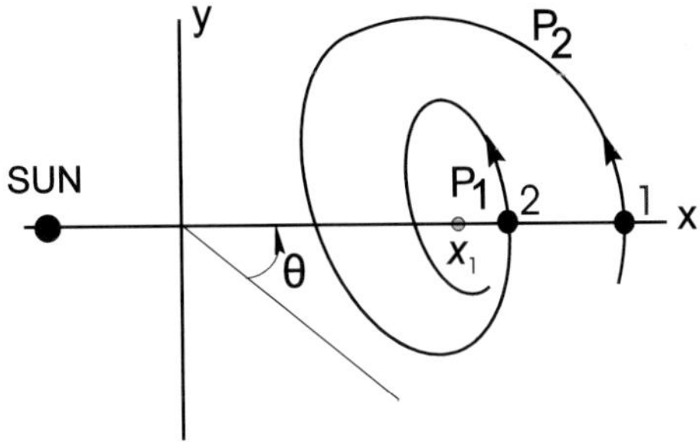

Fig. 12 The Poincaré map on a surface of section, defined at the times when P_2 crosses the x-axis in the positive sense

and is in the four-dimensional space $x_1, \dot{x}_1, x_2$, and $\dot{x}_2$. In some examples that we shall show below, we give the projections in some coordinate plane of the phase space of the Poincaré map.

5.1. Deviation from the symmetry of a stable periodic orbit: region of stability

In Fig. 13, we consider a typical orbit on the *family* 1, *orbit* 1, and study its stability, by shifting the position of P_2 on its orbit at $t = 0$, thus destroying the symmetry. This orbit corresponds to the configuration where the line of apsides are aligned and both planets are at perihelion, at $t = 0$. We note that for a range of deviations from the exact periodic motion, the perturbed motion takes place on a well defined torus. Although the phase space of the Poincaré map is four dimensional and consequently the deviations from the periodic motion that we consider do not cover all possible directions, it is clear that close to a stable periodic orbit there exists a region of stable librations, with small amplitude, close to the 2:1 resonance.

The same is repeated for a typical orbit on the *family* 2, *orbit* 4, lying on the stable part beyond the collision area (Fig. 14). The planetary orbits are antialigned and P_1 lies at aphelion while P_2 at perihelion, at $t = 0$. We note that although the eccentricities are large and the planetary orbits intersect, there exists a region of stable librations.

5.2. Comparison between a linearly stable and an unstable orbit

We shall study here the importance of the linear stability or instability of a periodic orbit, with respect to the properties of the phase space close to such an orbit. As examples we shall use two periodic orbits on the *family* 2, with small eccentricities, one stable and the other unstable. As in the previous case, we perturb the system by destroying the symmetry by shifting the planet P_2 on its orbit. Close to the stable orbit (Fig. 15) we have stable librations, and the motion takes place on a well defined torus. On the contrary, close to the unstable periodic orbit (Fig. 16), the motion is finally chaotic. It is worth noting that for a relatively long time interval the motion looks as a stable libration, on a torus, but later it becomes chaotic.

5.3. Dependence of the stability on the phase

We consider here the motion corresponding to the four different configurations of the extrasolar system HD 82943 (old fit mentioned above), for the four different phases shown in Fig. 11. The behavior is shown in Fig. 17. Note that the configurations: (alignment, P_1, P_2 at perihelion) and (antialignment, P_1 at aphelion, P_2 at perihelion) are stable, while the other two configurations, (alignment, P_1, P_2 at aphelion) and (antialignment, P_1 at perihelion, P_2

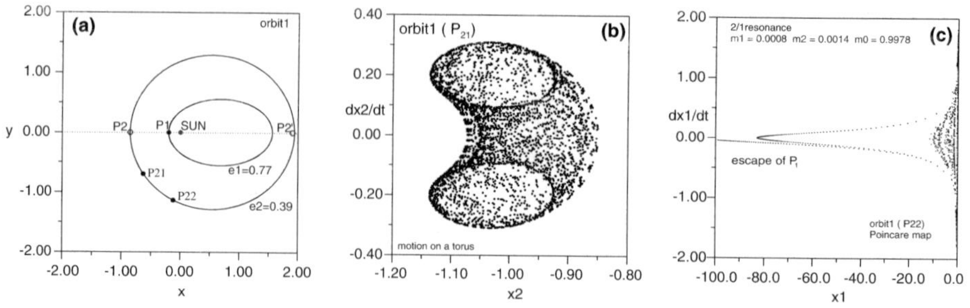

Fig. 13 (a) The stable *orbit* 1 and the two shifted positions, P_{21} and P_{22} of P_2 on its orbit. (b) The Poincaré map: P_2 at P_{21}, resulting to stable librations. (c) The Poincaré map: P_2 at P_{22}, resulting to chaotic motion and disruption of the system

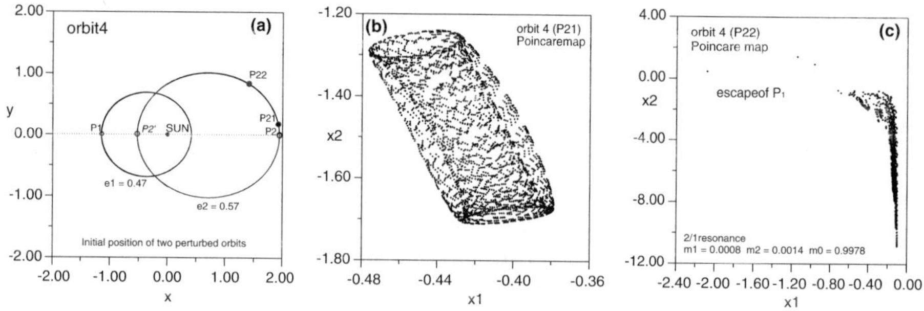

Fig. 14 (a) The stable *orbit* 4 and the two shifted positions, P_{21} and P_{22} of P_2 on its orbit. (b) The Poincaré map: P_2 at P_{21}, resulting to stable librations. (c) The Poincaré map: P_2 at P_{22}, resulting to chaotic motion and disruption of the system

at aphelion) are unstable. Note that the stable configurations are close to a stable family of periodic orbits.

6. Discussion

From all the previous examples it is clear that the position and the stability character of the periodic orbits plays an essential role in defining the structure of the phase space. A real extrasolar planetary system can exist in nature, if it lies in that region of the phase space which is close to a stable branch of a family of periodic orbits. We remark that all families of periodic orbits with non zero eccentricities are resonant, corresponding to a mean motion resonance between the planets. This means that resonances play an important role on the stability of a planetary system with large eccentricities, when the two planets are close to each other, so that their gravitational interaction is important. A phase protection mechanism appears in resonant cases, for some phases (alignment or antialignment, position of the planets at perihelion or aphelion), in such a way that we can have stable planetary systems even with large eccentricities and intersecting planetary orbits. Thus, all other things being the same (masses, eccentricities, and semimajor axes), the phase plays an important role on the stability of a resonant orbit. In addition, comparing Figs. 14 and 16, that correspond to two orbits with the same phase, one stable and one unstable, respectively, we have an indication that the increase of the planetary eccentricities may stabilize the system, because it makes the planets to avoid close encounters.

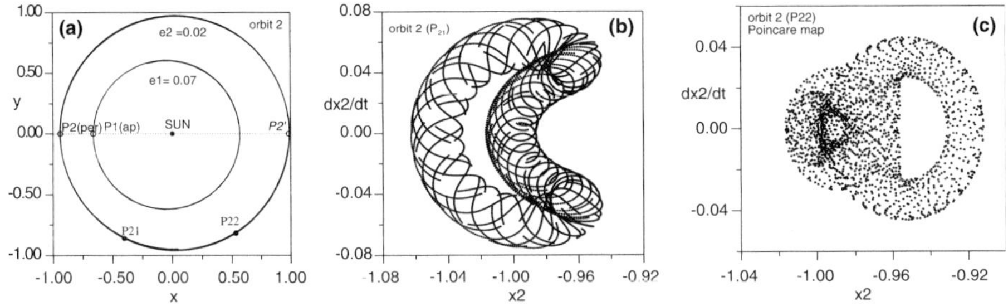

Fig. 15 (a) The stable *orbit* 2 and the two shifted positions, P_{21} and P_{22} of P_2 on its orbit. (b) The Poincaré map: P_2 at P_{21}, resulting to stable librations. (c) The Poincaré map: P_2 at P_{22}, resulting also to stable librations

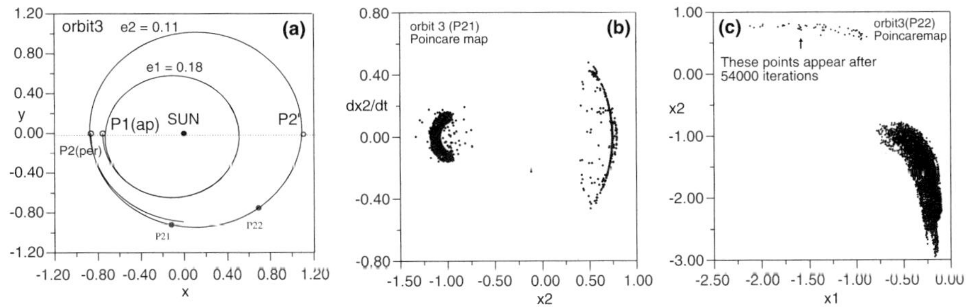

Fig. 16 (a) The unstable *orbit* 3 and the two shifted positions, P_{21} and P_{22} of P_2 on its orbit. (b) P_2 at P_{21}, resulting to chaotic motion. For a long time the motion is trapped on a 'torus' (the *dark region*, with the shape of the torus of figure **a** but later it deviates from this torus and the motion is chaotic. (c) P_2 at P_{22}, resulting to chaotic motion and disruption of the system

The resonances are not the only cases where stable motion exists. All non resonant periodic orbits on the *circular* family of periodic orbits are stable (with the exception of a small region close to the 3/1, 5/3,... resonances) and consequently stable librations are expected in their vicinity. In this case however, the planetary eccentricities are small.

Another factor that affects the topology of the phase space, and consequently the stability regions for a planetary system, is the ratio m_1/m_2 of the planetary masses and also the sum $m_1 + m_2$ of the planetary masses, because it affects the appearance or not of families of periodic orbits and also the stability regions on a family of periodic orbits. As an example, consider the *family* 1 at the 2:1 resonance (Fig. 11). For $m_1 < m_2$, the whole family is stable, but when the ratio of the planetary masses changes, $m_1 > m_2$, an unstable region appears (Fig. 10) and we have a bifurcation of a new family of asymmetric periodic orbits (the critical mass ratio beyond which an unstable region appears on the family is $m_1/m_2 = 0.97$). This changes the topology and the stability properties of the phase space.

The total mass of the planetary masses, which is a measure of the perturbation, plays also an important role in some resonances, and in particular, in the resonances 3:1, 5:3, ... As we showed in Sect. 4.4, we have a bifurcation of two families of elliptic orbits if the sum $m_1 + m_2$ is small, but we have four elliptic families if this sum is larger. This means that the topology of the phase space, and consequently the dynamical properties of the system depend critically on the sum $m_1 + m_2$ of the masses close to the 3:1 resonance.

As an example to show how the knowledge of the resonant structure of the phase space may help us to locate the positions of stable planetary systems, we present in Fig. 18a the families of periodic orbits for the masses of the system Gliese 876. The stable *family* 1 exists,

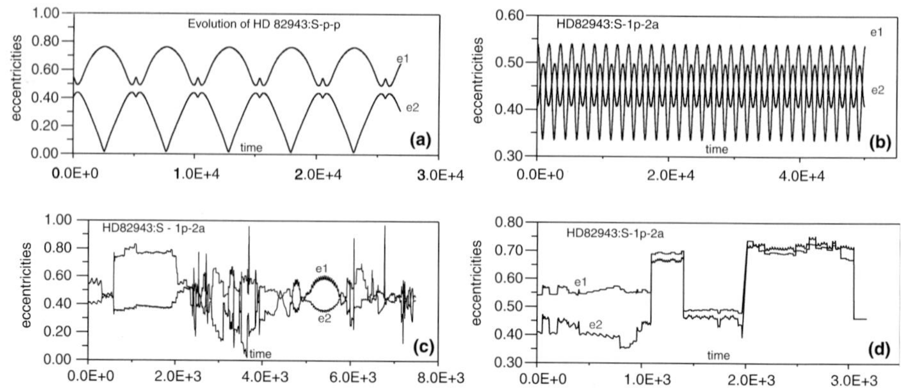

Fig. 17 The evolution of the eccentricities of the system HD 82943 for the four different positions shown in Fig. 11. (a) alignment, P_1, P_2 at perihelion: ordered motion. (b) antialignment, P_1 at aphelion, P_2 at perihelion: ordered motion. (c) alignment, P_1, P_2 at apocenter: chaotic motion. (d) antialignment, P_1 at perihelion, P_2 at aphelion: chaotic motion

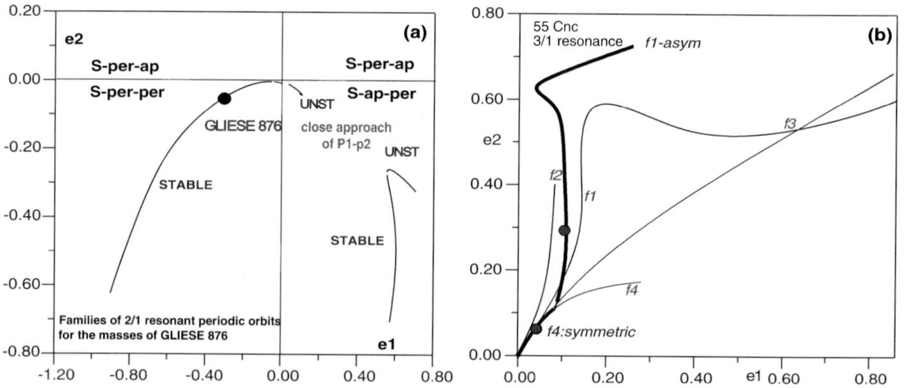

Fig. 18 (a) The 2:1 resonant families for the masses of Gliese 876, and the position of the observed system on the *family* 1. (b) The four different families, $f1$, $f2$, $f3$, $f4$ of 3:1 resonant periodic orbits. Only one of them, $f4$, denoted by the thicker line, is stable up to a point. A new, stable asymmetric family, $f1$-*asym*, bifurcates from this critical point. The planetary system, under migration, moves on the stable part of $f4$ and then changes route and follows the family $f1$-*asym*

as in the case of HD 82943 (Fig. 11), and we see that the elements of the observed system (from the web page of Schneider 2006), $a_1 = 0.13\,\mathrm{AU}$, $a_2 = 0.21\,\mathrm{AU}$, $e_1 = 0.27$, and $e_2 = 0.025$ lie almost exactly on the stable *family* 1, as it is shown in Fig. 18a.

Finally, we remark that the knowledge of the resonant structure of the phase space is useful in interpreting the results on the evolution of a planetary system under migration, when dissipative forces act. From the work on migration by Ferraz-Mello et al. (2003), Peale and Lee (2002), Lee and Peale (2003), and Lee (2004), we see that when a planetary system starts from an arbitrary position, it is trapped on a *stable* family of periodic orbits, and as the migration continues, it moves *on* the stable family. It is worth noting that if, as the system evolves on a stable family, there is a change from stability to instability, from which another stable family bifurcates, the system changes route and follows the new, stable, family. This is shown in Fig. 18b, for a system close to the 3:1 resonance, studied by Ferraz-Mello et al. (2003). There are four different families of resonant symmetric periodic orbits, $f1$, $f2$, $f3$, and $f4$ (Voyatzis and Hadjidemetriou 2006), and only one family, $f4$, is stable at its lower part. The system is trapped on this stable branch and moves up to the point where the family becomes unstable. From that point on, we have a bifurcation of a stable asymmetric 3:1 resonant family of periodic orbits, family $f1$-*asym*, and the system continues to migrate on this latter stable family.

From all the above, it is clear that the knowledge of the resonant structure of the phase space, as determined by the network of families of periodic orbits, is very useful in understanding the dynamics of a planetary system and interpreting the results obtained by different methods.

Acknowledgements I thank Dr. C. Beauge whose remarks helped me improve the presentation of the paper. This work was supported by the research program *Pythagoras II*, Nr.80831 of the Greek Ministry of Education and the E.U.

References

Beauge, C., Ferraz-Mello, S., Michtchenko, T.: Extrasolar planets in mean-motion resonance: apses alignment and asymmetric stationary solutions. ApJ **593**, 1124 (2003)

Beauge, C., Michtchenko, T., Ferraz-Mello, S.: Stationary solutions in resonant extrasolar systems. MNRAS, **365**, 1160–1170 (2006)

Beauge, C., Callegari, N., Ferraz-Mello, S., Michtchenko, T.: Resonances and stability of extra-solar planetary systems. In: Knezevic Z., Milani A. (eds.) Dynamics of Populations of Planetary Systems. Cambridge University Press, Cambridge, p. 3

Ferraz-Mello, S., Beaugé, C., Michtchenko T.: Evolution of migrating planet pairs in resonance, Cel. Mech. Dyn. Astr. **87**, 99–112 (2003)

Fischer, D., Marcy, G., Buttler, P., Laughlin, G., Vogt, S.: A second planet orbiting 47 UMa. ApJ **564**, 1028 (2002)

Gozdziewski, K., Bois, E., Maciejewski, A.: Global dynamics of the Gliese 876 planetary system. MNRAS **332**, 839 (2002)

Hadjidemetriou, J.D.: The continuation of periodic orbits from the restricted to the general three-body problem. Cel. Mech. Dyn. Astr. **12**, 155–174 (1975)

Hadjidemetriou, J.D.: Families of periodic planetary type orbits in the three-body problem and their stability. Astrophys. Sp. Sci. **40**, 201–224 (1976)

Hadjidemetriou, J.D.: On the relation between resonance and instability in planetary systems. Cel. Mech. Dyn. Astr. **27**, 305–322 (1982)

Hadjidemetriou, J.D.: Resonant periodic motion and the stability of extrasolar planetary systems. Cel. Mech. Dyn. Astr. **83**, 141–154 (2002)

Hadjidemetriou, J.D.: Periodic orbits in gravitational systems. In: Proceedings of the Cortina meeting 2004, 2006

Hadjidemetriou, J.D., Psychoyos, D.: Dynamics of extrasolar planetary systems: 2/1 resonant motion. In: G. Contopoulos and N. Voglis (eds.), Lecture Notes in Physics: Galaxies and Chaos, Vol. 626, pp. 412–432, Springer-Verlag, Berlin (2003)

Israelian, G., Santos, N., Mayor, M., Rebolo, R.: Evidence for planet engulfment by the star HD82943. Nature **411**, 163 (2001)

Ji, J., Liu, L., Kinoshita, H., Zhou, J., Nakai, H., Li, G.: The librating companions in HD 37124, HD 12661, HD 82943, 47 Uma and GJ 876: alignment or antialignment? ApJ **591**, L57–L60 (2003a)

Ji, J., Kinoshita, H., Liu, L., Li, G., Nakai, H.: The apsidal antialignment of the HD 82943 system. Cel. Mech. Dyn. Astron. **87**, 113–120 (2003b)

Lee, M.H., Peale, S.: Dynamic and origin of the 2:1 orbital resonances of the GJ 876 planets. ApJ **567**, 596–609 (2002)

Lee, M.H., Peale, S.J.: Extrasolar planets and mean motion resonances. In Deming, D., Seager, S. (eds) Scientific Frontiers in Research of Extrasolar planets. ASP, **197** (2003)

Lee, M.H.: Diversity and origin of 2:1 orbital resonance in extrasolar planetary systems. ApJ **611**, 517 (2004)

Malhotra, R.: A dynamical mechanism for establishing apsidal resonance. ApJ **575**, L33–36 (2002)

Marcy, G., Butler, P., Fischer, D., Vogt, S., Lissauer, J., Rivera, E.: A Pair of resonant planets orbiting GJ 876, ApJ **556**, 296 (2001)

Marcy, G.W., Butler, R.P., Fischer, D.A., Laughlin, G., Vogt, S.S., Henry, G.W., Pourbaix, D.: A planet at 5AU around 55Cnc. ApJ **581**, 1375–1388 (2002)

Mayor, M., Udry, S., Naef, D., Pepe, F., Queloz, D., Santos, N.C., Burnet, M.: The CORALIE survey for southern extra-solar planets. XII. Orbital solutions for 16 extra-solar planets discovered with CORALIE. A&A **415**, 291 (2004)

Nelson, R., Papaloizou, J.C.B.: Possible commensurabilities among pairs of extrasolar planets. MNRAS **333**, 25–30 (2002)

Peale, S., Lee, M.: Extrasolar planets and the 2:1 orbital resonances, In DDA 33rd Meeting, BAAS **34**, 933 (2002)

Psychoyos, D., Hadjidemetriou, J.D.: Dynamics of 2/1 resonant extrasolar systems. Application to HD82943 and Gliese876. Cel. Mech. Dyn. Astr. **92**, 135–156 (2005)

Rivera, E.J., Lissauer, J.J.: Dynamical models of the resonant pair of planets orbiting the star GJ 876, ApJ **558**, 392–402 (2001)

Schneider, J.: http://www.obspm.fr/encycl/catalog.html, (2006)

Voyatzis, G., Hadjidemetriou, H.D.: Symmetric and asymmetric librations in planetary and satellite systems at the 2/1 resonance. Cel. Mech. Dyn. Astr. **93**, 263–294 (2005)

Voyatzis, G., Hadjidemetriou, H.D.: Symmetric and asymmetric 3:1 resonant periodic orbits: an application to the 55Cnc extra-solar system. Cel. Mech. Dyn. Astr. this issue (2006)

Celestial Mechanics and Dynamical Astronomy (2006) 95:245–257
DOI 10.1007/s10569-006-9011-3

ORIGINAL ARTICLE

The influence of mutual perturbations on the eccentricity excitation by jet acceleration in extrasolar planetary systems

F. Namouni · J. L. Zhou

Received: 24 November 2005/Accepted: 13 March 2006 /
Published online: 15 August 2006

Abstract We study the influence of mutual planetary perturbations on the process of eccentricity excitation by jet acceleration suggested by Namouni (Astron. J. **130**, 280–294). Modeling the jet's action by a constant-direction acceleration, we solve the linear secular equations of the combined planetary perturbations and the jet acceleration of the host star for a two-planet system. The effects of the acceleration's strength, relative mass ratio and the relative distance of the two planets are investigated. The model is applied to the extrasolar planetary systems of HD 108874, 47 Uma, and HD 12661.

Keywords Extrasolar planets · Stellar jets · Eccentricity excitation

1 Introduction

Ten years after the discovery of the first Jupiter-like planet around 51 Pegasi, the number of extrasolar planets is steadily on the rise. As of this writing, we know of 173 planets and 18 multiplanet systems (see the extrasolar planetary encyclopaedia at http://www.obspm.fr/planets). Half of these planets have orbital eccentricities in excess of 0.28. Such large eccentricities are surprising as it is thought that planets form in a protoplanetary cloud on nearly circular orbits much like the planets of the solar system. In effect, these observations seem to have moved the paradigm of planetary formation away from the solar system. Possible explanations for the large eccentricities include the perturbations that follow planet–planet scattering (Marzari and Weidenschilling 2002; Ford et al. 2003), the tidal

F. Namouni (✉)
CNRS, Observatoire de la Côte d'Azur, BP 4229, 06304 Nice, France
email: namouni@obs-nice.fr

J. L. Zhou
Department of Astronomy, Nanjing University, Nanjing 210093, China
e-mail: zhoulj@nju.edu.cn

interaction between the gas disk and the planets (Goldreich and Tremaine 1981), Kozai's secular eccentricity cycles (Holman et al. 1997; Terquem and Papaloizou 2002; Takeda and Rasio 2005), the eccentricity's pumping during planetary pair migration in mean motion resonance (Chiang et al. 2002; Lee and Peale 2002), the perturbations by stellar encounters (Zakamska and Tremaine 2004) and the outcome of a stellar-like relaxation that occurred if planets formed by gravitational instability (Papaloizou and Terquem 2001). The merits and drawbacks of most of these mechanisms are well documented (Tremaine and Zakamska 2004). A main recurring criticism is that when these mechanisms work they do so for specific systems. This lack of generality suggests that in reality various combinations of these mechanisms are at work in all systems. While satisfactory, such a conclusion does not explain (1) why the solar system is such an outlier, (2) what gives rise to apsidal alignment in non-resonant systems, and (3) what the origin of the similarity of the eccentricity distributions of extrasolar planets and spectroscopic binary stars is (Stepinski and Black 2001).

Looking for possible mechanisms that can explain the states of both solar and extrasolar planets, Namouni (2005) (hereafter N05) reasoned that the similarity in the eccentricity distribution of extrasolar planets and spectroscopic binary stars may be a hint at a common excitation mechanism. If this were true then such a mechanism must depend weakly on the local dynamics of the companion. The description of the corresponding physical process would therefore, depend weakly on the companion's position and velocity. The simplest possible process satisfying this condition is a constant acceleration. To complete the description of the physical process, the direction of its acceleration too has to be independent of the planetary plane. A possible choice is the star's rotation axis, which naturally hints at stellar jets and star-disk mass loss as natural mechanisms for such an acceleration.

Typical stellar jet accelerations of magnitudes $A \sim 10^{-11}$ to $10^{-13}\,\mathrm{km\,s^{-2}}$ are capable of exciting planetary eccentricities. The acceleration's parameters are best known for multiple systems because mutual gravitational perturbations cause the planetary orbits to precess. If the secular frequencies associated with these precessions are larger than the jet acceleration's excitation frequency, the efficiency of eccentricity excitation is hampered. To verify the validity of the eccentricity excitation by jet acceleration, it is necessary to quantify its relative importance with respect to planetary perturbations in specific multiple systems. Such an approach has been applied in N05 to the Jupiter–Saturn and υ Andromedae binary system using direct numerical integration of the equations of motion. It was shown that the excitation by jet acceleration was able to explain the observed states of these systems.

In this paper, we develop a fast semi-analytic approach to study the effect of mutual gravitational perturbations in a two-planet system whose host star is subject to an acceleration by a material jet. We assume that the jet's intrinsic precession frequency is small compared to its excitation frequency $n_{\mathrm{A}} = 3|A|/na$ where n and a are the mean motion and semi-major axis of the planet, respectively. This puts the jet resonance studied in N05 much closer to the star than the location of the planetary region allowing us to ignore the complex issues of eccentricity damping interior to the resonance and resonant eccentricity driving. We also assume that the two-planets are not in mean motion resonance.

The paper is organized as follows: in Sect. 2, we lay down the analytical model based on secular perturbation theory applied to a system consisting of two planets and a star accelerated by a jet. In Sect. 3, we compare the secular semi-analytical model to the direct integration of the equations of motion. The effects of the acceleration's strength, the planetary mass ratio, and relative distance are discussed in Sects. 4–6, respectively. The secular model is applied to the planets of HD 108874, 47 Uma, and HD 12661 in Sect. 7, where we derive the amplitudes of the secular eigenmodes and determine the minimum acceleration strength required for efficient excitation by a slowly-precessing acceleration. Section 8 contains a summary.

2 Linear perturbations

We consider the system made up of two planets orbiting a host star that is accelerated by an asymmetric jet. It has been shown in N05 that the jet's acceleration perturbs a planetary orbit through the potential:

$$\langle R\rangle_{\text{jet}} = -\tfrac{3}{2}a\,\mathbf{A}\cdot\mathbf{e} = -\tfrac{3}{2}A\,ae\,\sin(\varpi-\Omega)\sin I, \tag{1}$$

where $\mathbf{e} = \mathbf{v}\times\mathbf{h}/G(m+M) - \mathbf{x}/|\mathbf{x}|$ is the eccentricity vector of magnitude e, which is a constant of the Keplerian motion and can be expressed in terms of the position $\mathbf{x}$, the velocity $\mathbf{v}$, the angular momentum $\mathbf{h}$, the gravitational constant G, and the masses of the star M and that of its planetary companion m. The potential also depends on the orbital semi-major axis of the planet, a. The last expression is obtained by choosing the z-axis along $\mathbf{A}$; in this case, ϖ, Ω, and I are the longitude of pericenter, the longitude of ascending node, and the inclination of the orbit with respect to $\mathbf{A}$.

The two planets are assumed not to be in mean motion resonance. This allows us to use the classical secular perturbation potential of first-order in the masses and second-order in eccentricity and inclination to describe the long-term motion of the two planets. Its expression is given as (see, e.g. Murray and Dermott 1999):

$$\begin{aligned}\langle R\rangle_{\text{planets}} = &\frac{1}{8}\left[2\alpha\partial+\alpha^2\partial^2\right]b_{1/2}^{(0)}(e_1^2+e_2^2) - \frac{1}{8}b_{3/2}^{(1)}(I_1^2+I_2^2)\\ &+\frac{1}{4}\left[2-2\alpha\partial-\alpha^2\partial^2\right]b_{1/2}^{(1)}e_1e_2\cos(\varpi_1-\varpi_2)\\ &+\frac{1}{4}\alpha b_{3/2}^{(1)}I_1I_2\cos(\Omega_1-\Omega_2),\end{aligned} \tag{2}$$

where α is the ratio of the inner planet's semi-major axis to that of the outer planet a_1/a_2, ∂ is the partial derivative with respect to α, and $b_{1/2}^{(0)}$, $b_{1/2}^{(1)}$ and $b_{3/2}^{(1)}$ are the usual Laplace coefficients.

The evolution equations for the eccentricity and inclination of the two planets are obtained from the Laplace–Lagrange equations for the the eccentricity and inclination vectors ($h_i = e_i\sin\varpi_i$, $k_i = e_i\cos\varpi_i$, $p_i = I_i\sin\Omega_i$, $q_i = I_i\cos\Omega_i$, $i = 1$ or 2) as:

$$\begin{aligned}\dot{h}_i &= \frac{1}{n_ia_i^2}\frac{\partial}{\partial k_i}\left[\frac{Gm_j}{a_2}\langle R\rangle_{\text{planets}}+\langle R\rangle_{\text{jet}}^{(i)}\right],\\ \dot{k}_i &= -\frac{1}{n_ia_i^2}\frac{\partial}{\partial h_i}\left[\frac{Gm_j}{a_2}\langle R\rangle_{\text{planets}}+\langle R\rangle_{\text{jet}}^{(i)}\right],\\ \dot{p}_i &= \frac{1}{n_ia_i^2}\frac{\partial}{\partial q_i}\left[\frac{Gm_j}{a_2}\langle R\rangle_{\text{planets}}+\langle R\rangle_{\text{jet}}^{(i)}\right],\\ \dot{q}_i &= -\frac{1}{n_ia_i^2}\frac{\partial}{\partial p_i}\left[\frac{Gm_j}{a_2}\langle R\rangle_{\text{planets}}+\langle R\rangle_{\text{jet}}^{(i)}\right],\end{aligned} \tag{3}$$

where $\langle R\rangle_{\text{jet}}^{(i)}$ is the jet acceleration's potential at the location of planet (i). For consistency with the validity domain of the second-order planetary perturbation potential, we substitute I for $\sin I$ in the expression (1).

Introducing the vector $\mathbf{E} = (h_1, h_2, k_1, k_2, p_1, p_2, q_1, q_2)$, the equations of motion (3) take the simple form:

$$\dot{\mathbf{E}} = L \cdot \mathbf{E}, \tag{4}$$

where the matrix L is given as:

$$L = \begin{bmatrix} 0 & 0 & G_{11} & G_{12} & \nu_{\rm A}^{(1)} & 0 & 0 & 0 \\ 0 & 0 & G_{21} & G_{22} & 0 & \nu_{\rm A}^{(2)} & 0 & 0 \\ -G_{11} & -G_{12} & 0 & 0 & 0 & 0 & \nu_{\rm A}^{(1)} & 0 \\ -G_{21} & -G_{22} & 0 & 0 & 0 & 0 & 0 & \nu_{\rm A}^{(2)} \\ -\nu_{\rm A}^{(1)} & 0 & 0 & 0 & 0 & 0 & -G_{11} & G_{11} \\ 0 & -\nu_{\rm A}^{(2)} & 0 & 0 & 0 & 0 & G_{22} & -G_{22} \\ 0 & 0 & -\nu_{\rm A}^{(1)} & 0 & G_{11} & -G_{11} & 0 & 0 \\ 0 & 0 & 0 & -\nu_{\rm A}^{(2)} & -G_{22} & G_{22} & 0 & 0 \end{bmatrix} \tag{5}$$

with:

$$G_{11} = \frac{n_1}{4} m_2 \alpha^2 b_{3/2}^{(1)}, \qquad G_{22} = \frac{n_1}{4} m_1 \alpha^{5/2} b_{3/2}^{(1)}, \tag{6}$$

$$G_{12} = -\frac{n_1}{4} m_2 \alpha^2 b_{3/2}^{(2)}, \qquad G_{21} = -\frac{n_1}{4} m_1 \alpha^{5/2} b_{3/2}^{(2)}, \tag{7}$$

$$\nu_{\rm A}^{(i)} = \frac{3A}{2n_i a_i}. \tag{8}$$

To find the eccentricity and inclination expressions, we have to diagonalize the system (5) and use the initial orbital state of the two planets in terms of the vector $\mathbf{E}$ to identify its projections on the eigenvectors. As we are interested primarily in planetary systems on initially circular coplanar orbits, the acceleration's inclination I_0 with the planetary orbital plane implies:

$$\mathbf{E}(t=0) = I_0\ (0,\ 0,\ 0,\ 0,\ 0,\ 0,\ 1,\ 1). \tag{9}$$

Before using the solution of the secular system, we remark that: first, the addition of the acceleration term in the form of the new terms $\nu_{\rm A}^{(i)}$, lifts the degeneracy of the secular two-body problem. Indeed if $\nu_{\rm A}^{(i)} = 0$ (i.e. there is no external acceleration), the matrix (5) can be split up into two separate blocs for the eccentricity and inclination leading, respectively, to two eigenmodes for the eccentricity and one eigenmode for the inclination. The latter is simply related to the arbitrary choice of a reference plane for the inclination. In contrast, the jet acceleration, the eccentricity and inclination variations are mixed. The second remark concerns the scaling of the eccentricity and inclination solution. Since the differential system representing the planets' eccentricities and inclinations is linear and since the initial state is proportional to I_0, the solutions for e and I will be proportional to I_0. This tells us that the solution is valid for any inclination provided that it is covered by the second-order approximation. As we shall see in the next section, the integration of the full equations of motion for an acceleration with $I_0 = 30°$ shows that the agreement with the secular solution is reasonable. That is why we will not pay much attention to the actual amplitude of I_0 and quote mostly e/I_0 and I/I_0. This is consistent with the main goal of the paper as it was shown in N05 that a conservative constant-direction acceleration yields periodic oscillations of maximum amplitude sin I_0 (I_0 in the second-order approximation). Mutual perturbations cause the orbits to precess and tend to reduce the maximum eccentricity, it is therefore, natural to ascertain the effect of mutual planetary perturbations by scaling the eccentricity and inclination to I_0.

3 Comparison with the integration of the full equations of motion

To illustrate the competition between mutual planetary perturbations and the eccentricity excitation by jet acceleration, N05 used a direct integration of the full equations of motion of the Jupiter–Saturn system. The set up was such that Jupiter and Saturn had circular coplanar orbits with their current semi-major axes. The inclination of the perturbing acceleration was 30° and two different acceleration strengths, $A = 2 \times 10^{-12}$ and $2 \times 10^{-14}\,\mathrm{km\,s^{-2}}$ corresponding to the locations of the Keplerian boundary a_{kplr} at 10^3 and 10^4 AU, respectively. The Keplerian boundary corresponds to the location where the jet's acceleration balances the gravitational acceleration by the host star. For a stationary jet, objects with orbital semi-major axes beyond a_{kplr} are lost to the interstellar medium. In terms of a_{kplr}, the acceleration is given as:

$$|A_0| \simeq 2 \times 10^{-12} \left(\frac{M+m}{M_\odot}\right) \left(\frac{10^3\,\mathrm{AU}}{a_{\mathrm{kplr}}}\right)^2 \mathrm{km\,s^{-2}} \tag{10}$$

and the corresponding excitation period, $T_\mathrm{A} = 2\pi/n_\mathrm{A}$, is given as:

$$T_\mathrm{A} \simeq 10^6 \left(\frac{M+m}{M_\odot}\right)^{1/2} \left(\frac{a_{\mathrm{kplr}}}{10^3\,\mathrm{AU}}\right)^2 \left(\frac{1\,\mathrm{AU}}{a}\right)^{1/2} \text{years.} \tag{11}$$

In Fig. 1, we compare the results of the numerical integrations using the full equations of motion with the solution of the secular systems (4)–(9) for an acceleration of Keplerian boundary

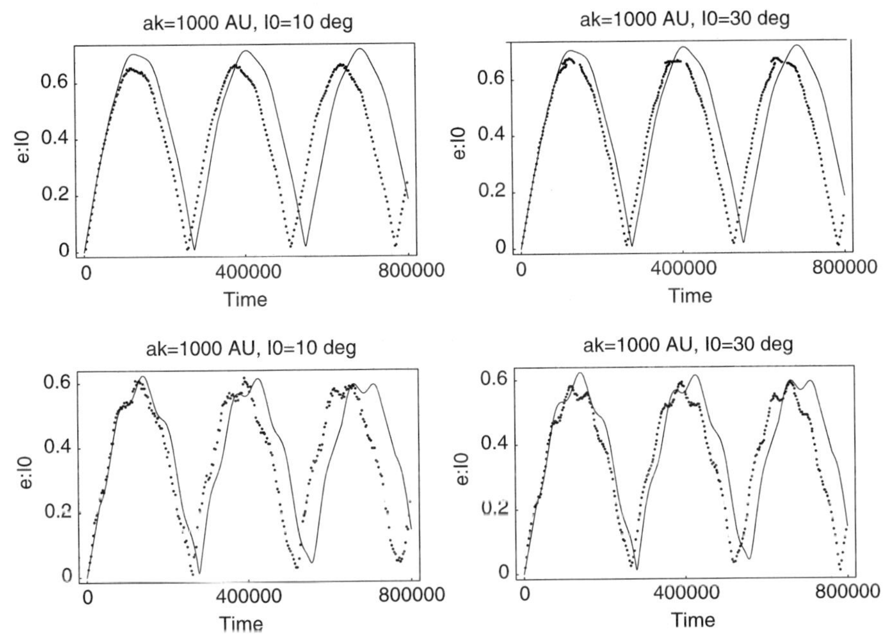

Fig. 1 The ratio of the eccentricity to the inclination as a function of time for Jupiter (*upper panels*) and Saturn (*lower panels*). The *solid lines* are obtained from the solution of the linear systems (4)–(9). The *dotted lines* are obtained from the direct integration of the full equations of motion, where Jupiter and Saturn have zero initial eccentricities. The two columns correspond to the two jet angles 10 and 30°. In all cases, the acceleration is $A = 2 \times 10^{-12}\,\mathrm{km\,s^{-2}}$ corresponding to $a_{\mathrm{kplr}} = 1000\,\mathrm{AU}$. Note that the *dotted lines*, we divided the eccentricity by sin I_0 instead of I_0 for consistency with the full (non-linear) solution derived in N05

$a_{\mathrm{kplr}} = 1000\,\mathrm{AU}$. As explained in the previous section, the ratio of the eccentricity e to the jet inclination I_0 in the case of the secular system and to sin I_0 for the numerical integrations is the relevant measure of excitation. We note three features: (1) the linear solution is a good approximation for relatively high eccentricities, ~0.3, and for relatively large inclinations, ~30°. (2) The excitation period is somewhat overestimated for both small and large inclinations while (3) the amplitudes are somewhat underestimated for both inclinations. The agreement is only slightly better for smaller inclinations.

A stronger acceleration with a Keplerian boundary at $a_{\mathrm{kplr}} = 500\,\mathrm{AU}$ reveals similar features as seen in Fig. 2. Only in this case, the amplitudes of Jupiter are better reproduced than those with a weaker acceleration. However, this good fit is lost for larger times as it can be seen on Saturn's amplitudes.

The better agreement for Jupiter and the loss of accuracy with time suggest that the Jupiter–Saturn interaction combined to the jet acceleration is not purely secular in time, when observed over longer excitation cycles. This in turn suggests that mean motion resonances, which are neglected in the secular approach affect the eccentricity evolution over a time longer than the excitation time scale. In order to confirm this hypothesis, we study the previous Jupiter–Saturn configuration but use smaller masses for the two planets. Figure 3 shows the eccentricity excitation in such cases. The better agreement of the two approaches is already visible if we reduce Saturn's mass. An almost perfect agreement can be obtained if both planets—Jupiter and Saturn—have smaller masses and the jet angle is small too.

We conclude that our method yields good results over the first excitation cycles when non-secular long-period perturbations do not have a pronounced effect on the eccentricity excitation. Such an agreement is sufficient for our purposes because as it has been shown in

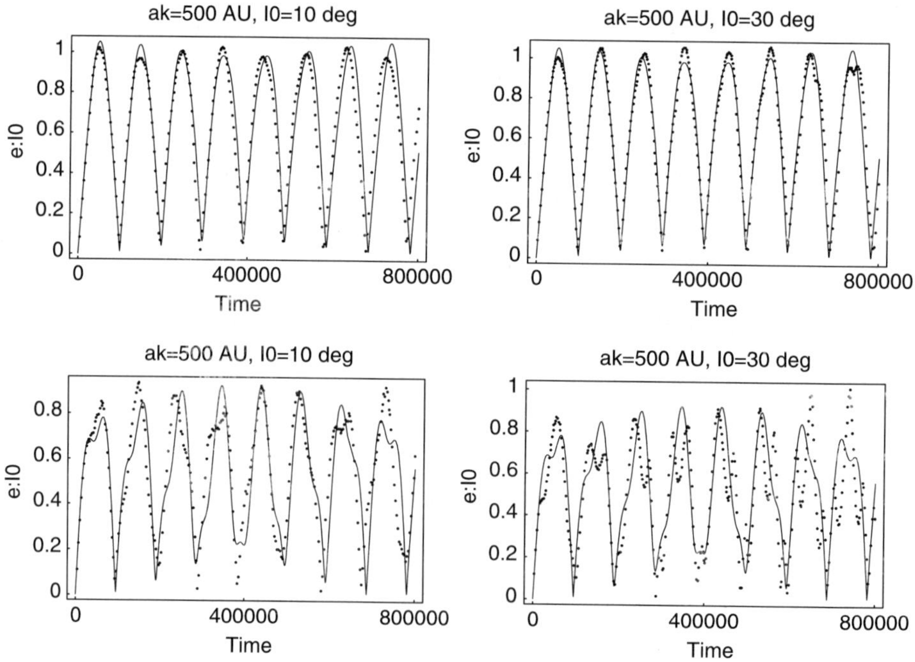

Fig. 2 Same as Fig. 1, but for a stronger acceleration $A = 8\times10^{-12}\,\mathrm{km\,s^{-2}}$ corresponding to $a_{\mathrm{kplr}} = 500\,\mathrm{AU}$. The upper panels correspond to Jupiter and the lower panels to Saturn

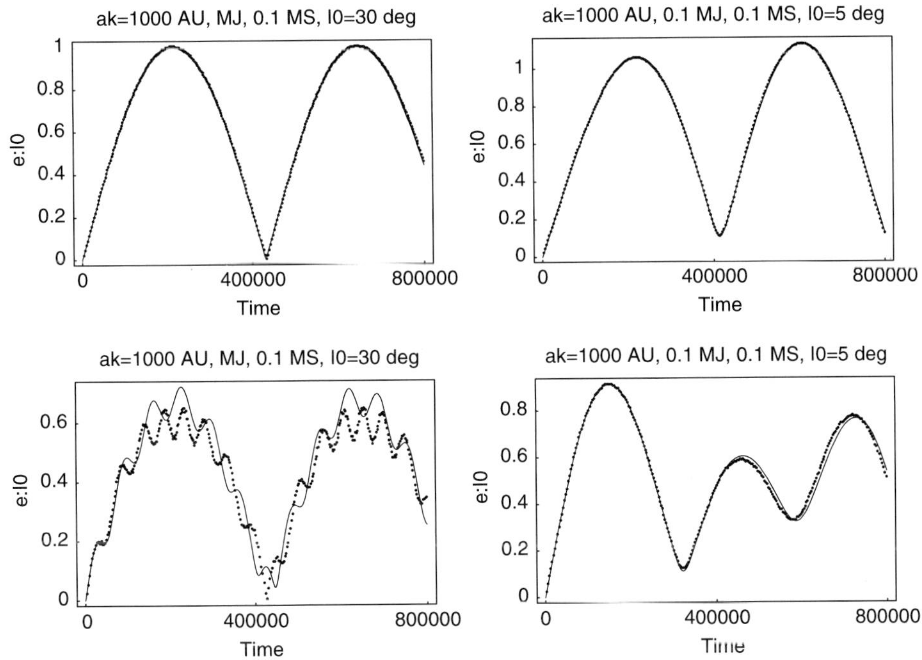

Fig. 3 The ratio of the eccentricity to the inclination as a function of time for two configurations. In the first configuration (*left column*), Saturn (*lower plot*) has a tenth of its actual mass; Jupiter's mass is set to its current value (*upper plot*). The jet angle is 30° and the Keplerian boundary is at $a_{\rm kplr} = 1000\,\rm AU$. In the second configuration (*right columns*), both Jupiter (*upper plot*) and Saturn (*lower plot*) have a tenth of their actual masses. The jet angle is chosen as 5° to enhance the agreement with the direct numerical integrations; the Keplerian boundary is at $a_{\rm kplr} = 1000\,\rm AU$

N05 (1) the modeling of a realistic time dependent jet amounts to a simple convolution of the secular oscillations that we have described and the time window of the jet episode, and (2) typical action times for stellar jets are usually smaller than half the excitation time. The linear secular method is therefore, satisfactory to assess the influence of mutual planetary perturbations that are mainly secular over a single oscillation cycle.

4 Effect of the acceleration's strength

To characterize the relative importance of mutual gravitational perturbations and jet eccentricity excitation, we examine the dependence of the fundamental secular eigenmodes and the corresponding amplitudes as a function of the accelerations strength. For the acceleration strength, we take the more physical parameter $a_{\rm kplr}$ that shows for instance where a putative disk is truncated by the jet's action. We also note that henceforth we shall not refer to the jet inclination because we normalize the eccentricity of both planets to I_0 (see Sect. 2). In Fig. 4, we illustrate the variation of the secular eigenperiods with $a_{\rm kplr}$. There are four eigenperiods, three of which tend to the free eigenmodes of the Jupiter–Saturn System: $T_1 = 375,000$ years, $T_2 = 60,000$ years, and $T_3 = 51,000$ years (N05, Murray and Dermott 1999). The last frequency T_4 is that of the jet excitation; it varies by two orders of magnitude as $a_{\rm kplr}$ varies from $\sim 10^2$ to 10^3 AU as roughly expected from expression (11). Writing the eccentricity vectors of the two planets as:

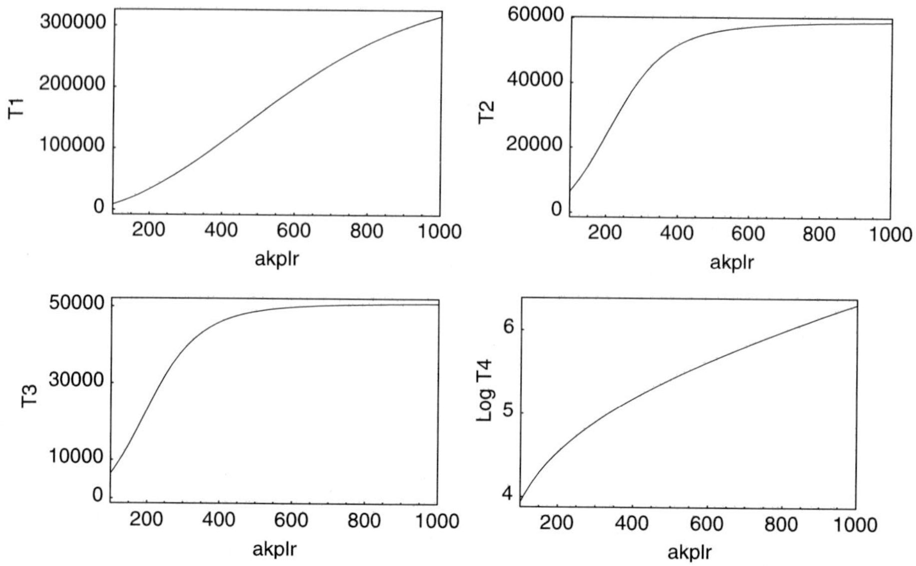

Fig. 4 Fundamental secular eigenperiods of the Jupiter–Saturn system as a function of the acceleration's strength represented by a_{kplr} (AU). The eigenperiods tend to the well known free periods as the acceleration gets weaker

$$\mathbf{e}_i = I_0 \sum_{k=1}^{4} \mathbf{e}_{ik}, \tag{12}$$

where $i = 1$ for the inner planet (Jupiter) and 2 for the outer planet (Saturn) and the sum runs over the four eigenmodes, we show the amplitudes of the eccentricity vectors e_{ik} for Jupiter in Fig. 5 and Saturn in Fig. 6. We note that the dominant eccentricity modes for Jupiter are 1 and 4 while all modes contribute to the eccentricity of Saturn.

5 Effect of the planets' mass ratio

We ascertain the effect of the planets' mass ratio by varying the mass of the Jupiter–Saturn system. We consider two cases: first, Saturn is left unchanged while Jupiter's mass is varied between 0.1 and 1 M_{J}; the semi-majors axes are unchanged. The acceleration strength corresponds to a Keplerian boundary at $a_{\mathrm{kplr}} = 500$ AU. The eigenmode amplitudes are shown in Fig. (7) for both the inner planet (e_1) and Saturn (e_2) as a function of their mass ratio. We observe the expected trend that the smaller the inner planet, the larger the excited eigenmodes. Fixing Jupiter's mass and varying Saturn's leads to the eigenmode amplitudes shown in Fig. 8. We note the degeneracy of the eigenmodes as the outer planet's mass decreases (mass ratio of inner to outer $\sim$35). We remark however, that these eigenmodes should be used to obtain an estimate of the maximum eccentricity and do not exactly reproduce the eccentricity evolution beyond the first excitation cycle. This has already been encountered in Sect. 3, where we explained that the effect of mean motion resonances may cause departures from the purely secular evolution of the linear system (see also Fig. 2).

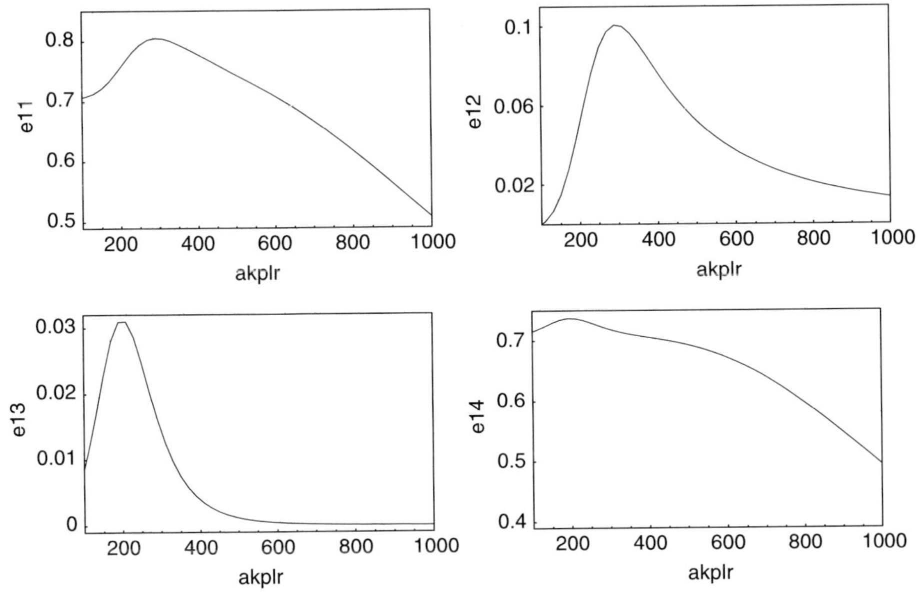

Fig. 5 Amplitudes of the secular eigenmodes of Jupiter as a function of the acceleration's strength represented by a_{kplr} (AU). Note that the amplitudes have been divided by I_0 (see Sect. 2)

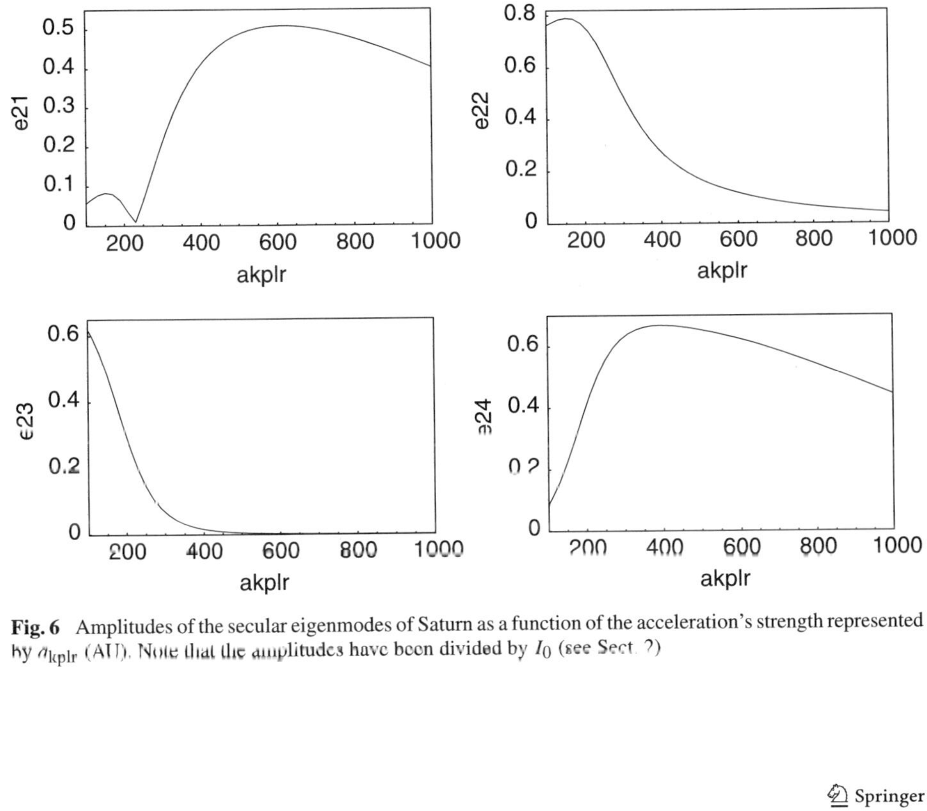

Fig. 6 Amplitudes of the secular eigenmodes of Saturn as a function of the acceleration's strength represented by a_{kplr} (AU). Note that the amplitudes have been divided by I_0 (see Sect. 2)

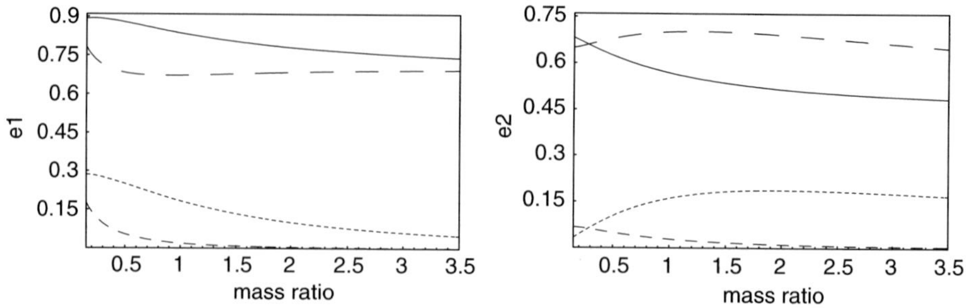

Fig. 7 Amplitudes of the secular eigenmodes of Saturn (*right panel*) interacting with an inner planet (*left panel*) at Jupiter's semi-major axis whose mass varies from 0.1 to 1 M_J. The horizontal axis shows the mass ratio of the inner planet to the outer planet. The amplitudes correspond to the eigenmodes 1 (*solid*), 2 (*short-dashed*), 3 (*medium-dashed*) and 4 (*long-dashed*)

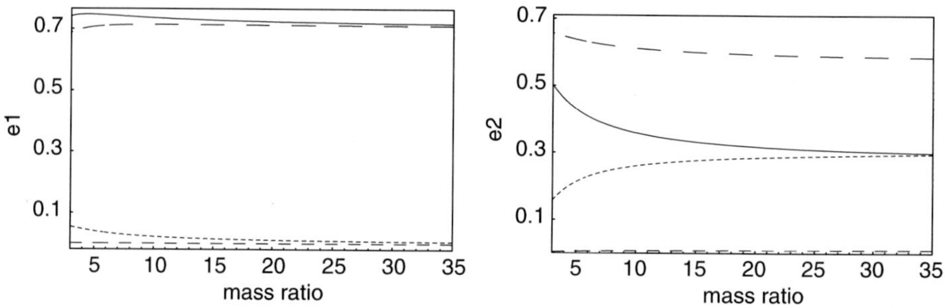

Fig. 8 Amplitudes of the secular eigenmodes of Jupiter (*left panel*) interacting with an outer planet (*right panel*) at Saturn's semi-major axis whose mass varies from 0.1 to 1 M_S. The horizontal axis shows the mass ratio of the inner planet to the outer planet. The amplitudes correspond to the eigenmodes 1 (*solid*), 2 (*short-dashed*), 3 (*medium-dashed*), and 4 (*long-dashed*)

6 Effect of the planets' relative distance

The effect of mutual interactions is diminished by an increasing distance between the two planets. Equivalently, one can choose a weaker acceleration that has a smaller Keplerian boundary radius, a_{kplr}. This is recovered in the linear model as it is shown in Fig. 9 for an acceleration with a Keplerian boundary at 500 AU. In this example, Saturn's semi-major axis was kept at its current value while Jupiter's semi-major axis was varied in the range set by the 2:1 and the 4:1 resonances. We note a curious mode amplitude exchange for Saturn at $\alpha = 0.475$. An important conclusion that we can draw from the previous two sections is that an acceleration with a Keplerian boundary at 500 AU while displaying the correct trends as the mass ratio or the semi-major axis vary, is relatively strong so as to excite the two planets' eccentricities. Excitation does actually occur only if the direction of acceleration represented by the factor I_0 (or more precisely sin I_0), that appears in the eccentricity expression (12), is not too small.

7 Application to HD 108874, 47 Uma, and HD 12661

To illustrate the use of the secular method in the case of extrasolar planets, we determine the eigenmode amplitudes as a function of the acceleration's strength represented by the radius

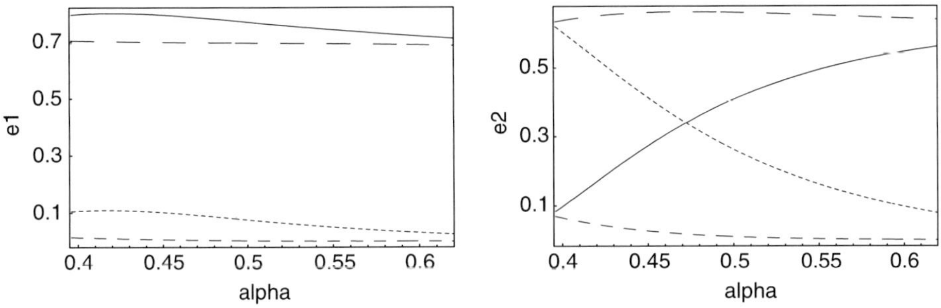

Fig. 9 Amplitudes of the secular eigenmodes of Jupiter (*left panel*) and Saturn (*right panel*) as a function of the semi-major axis ratio. Saturn's semi-major axis is set at its current value 9.2525 and Jupiter's varies. The amplitudes correspond to the eigenmodes 1 (*solid*), 2 (*short-dashed*), 3 (*medium-dashed*), and 4 (*long-dashed*)

of the Keplerian boundary for the two-planet systems of HD108874 (Vogt et al. 2005), 47 Uma (Fischer et al. 2002a) and HD 12661 (Fischer et al. 2002b). The corresponding orbital data are given in Table 1 (Appendix) and the amplitudes in Fig. 10.

Assuming a reference inclination of $I_0 = 20°$ ($\sin I_0 = 0.34$), our excitation criterion for these systems is that the maximum value of all eigenmode amplitudes be larger than the observed eccentricity values. Such a criterion is not the most conservative nor is it the most relaxed one. The latter comes from the fact that the eccentricity of each planet is the sum (with the corresponding phases) of the various modes and is always likely to be larger than the largest mode amplitude.

For the chosen criterion, the HD 108874 planets are different, while the inner planet requires only a small eccentricity and therefore satisfies the criterion quite well, HD 108874c with an eccentricity of 0.25 barely meets the criterion for a strong acceleration of Keplerian boundary $a_{\mathrm{kplr}} = 150\,\mathrm{AU}$. The planets of 47 Uma require stronger accelerations because of their larger masses (and hence stronger gravitational interaction), while those of HD 12661 with their low-eccentricity values satisfy the criterion for $a_{\mathrm{kplr}} \leq 350\,\mathrm{AU}$. In view of this analysis, it is important to bear in mind that we have assumed a constant-direction acceleration throughout this paper, which in reality represents slowly-precessing jets only. The fact that the planets of 47 Uma, for instance, do not meet the excitation criterion does not imply that the jet acceleration model is not able to explain the planets' orbital configuration. It has been shown in N05 that the resonance between the jet's precession frequency and the excitation frequency n_{A} is able to excite the eccentricities to unity even for nearly perpendicular jets. In this instance, it is interesting to note that HD 108874b has a small eccentricity indicating the possibility of eccentricity damping that is characteristic of the orbital region inside the resonance radius. This suggests that the jet resonance is located outside 1 AU and that the more eccentric outer planet lies in the resonance region.

8 Summary

We have presented a simple and fast method to determine the effect of a jet induced acceleration on a two-planet system by applying linear secular perturbation theory. We combined the Laplace–Lagrange theory of first-order in the masses and second-order in the eccentricities secular potential with the secular potential of a jet acceleration derived in N05. This has enabled us (1) to verify the general excitation trends as the acceleration's strength, the planetary mass ratio and the relative semi-major axes vary and (2) to estimate of the accelera-

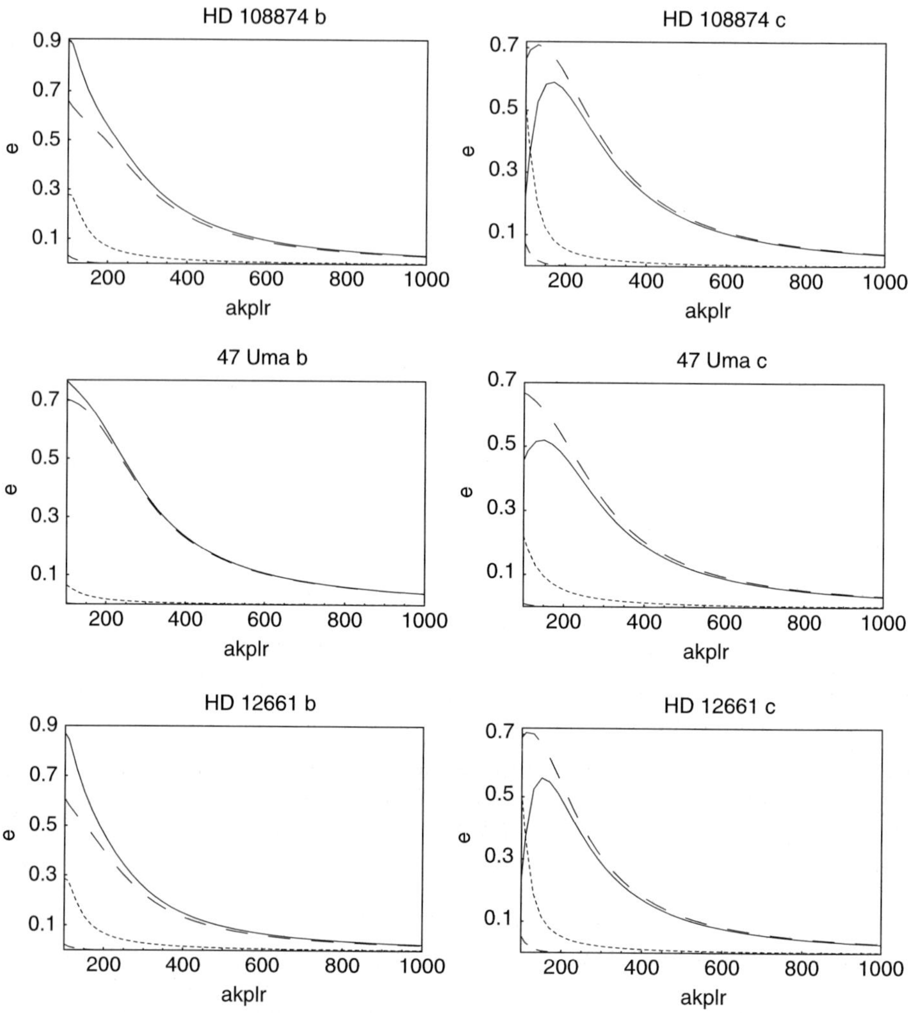

Fig. 10 Amplitudes of the secular eigenmodes of the companions to HD 108874, 47 Uma, and HD 12661 as a function of the radius of the Keplerian boundary. The amplitudes correspond to the eigenmodes 1 (*solid*), 2 (*short-dashed*), 3 (*medium-dashed*), and 4 (*long-dashed*)

tion strengths required in the extrasolar systems of HD 108874, 47 Uma, and HD 12661. We found that the first two systems were likely excited by the resonance of the jet's precession frequency with its natural excitation frequency that is known to yield maximum eccentricity values near unity regardless of the jet inclination angle. The eccentricities of the planets around HD 12661 can be explained by the acceleration of a slowly precessing jet modeled with the linear theory. Extensions of this work include generalizing the linear secular analysis to any number of planets, and also the coupling of the secular dynamics to specific mean motion resonances. The treatment of the jet resonance however, can not be dealt with in the linear regime and requires the Kozai potential instead of that in Eq. (2).

Acknowledgements The authors thank the referee for helpful comments. FN thanks Alessandra Celletti and the Celmec IV organizing committee for their invitation and the Programme National de Planétologie for partial financial support. JLZ was supported by the Natural Science Foundation of China (Grant 10233020) and the program for New Century Excellent Talents in Universities of China (Grant 04-0468). Part of this work was done while JLZ was on a CNRS–NSFC exchange programme in the Observatoire de la Côte d'Azur; he thanks Christiane Froeschlé, Elena Lega, Claude Froeschlé and the Cassiopée Department for their hospitality.

Appendix

Table 1 Orbital elements and masses of the companions to HD 108874, 47 Uma, and HD 12661 (Fischer et al. 2002a, b; Vogt et al. 2005)

Companion	m/M_J sin i	a	e	$M_*/M_\odot$
HD 108874b	1.36	1.051	0.07	1
HD 108874c	1.018	2.68	0.25	1
47 Uma b	2.54	2.09	0.35	1.03
47 Uma c	0.76	3.73	0.2	1.03
HD 12661b	2.3	0.83	0.06	1.07
HD 12661c	1.57	2.56	0.1	1.07

References

Chiang, E.I., Fischer, D., Thommes, E.: Excitation of orbital eccentricities of extrasolar planets by repeated resonance crossings. Astrophys. J. **564**, L105–L109 (2002)

Fischer, D.A., Marcy, G.W., Butler, R.P., Laughlin, G., Vogt, S.S.: A second planet orbiting 47 ursae majoris. Astrophys. J. **564**, 1028–1034 (2002a)

Fischer, D.A., Marcy, G.W., Butler, R.P., Vogt, S.S., Henry, G.W., Pourbaix, D., Walp, B., Misch, A.A., Wirght, J.T.: A planetary companion to HD 40979 and additional planets orbiting HD 12661 and HD 385291. Astrophys. J. **586**, 1394–1408 (2002b)

Ford, E.B, Rasio, F.A., Yu, K.: Dynamical instabilities in extrasolar planetary systems. In: Deming, D. Seager S. (eds.) Scientific Frontiers in Research on Extrasolar Planets ASP Conf. Series **294**, 181–188 (2003)

Goldreich, P., Tremaine, S.: The origin of the eccentricities of the rings of uranus. Astrophys. J. **243**, 1062–1075 (1981)

Holman, M., Touma, J., Tremaine, S.: Chaotic variations in the eccentricity of the planet orbiting 16 Cyg B. Nature **386**, 254–256 (1997)

Lee, M.H., Peale S.J.: Dynamics and origin of the 2:1 orbital resonances of the GJ 876 planets. Astrophys. J. **567**, 596–609 (2002)

Marzari, F., Weidenschilling, S.J.: Eccentric extrasolar planets: the jumping Jupiter model. Icarus **156**, 570–579 (2002)

Murray, C.D., Dermott, S.F.: Solar system dynamics. Cambridge University Press, Cambridge (1999)

Namouni, F.: On the origin of the eccentricities of extrasolar planets. Astron. J. **130**, 280–294 (2005)

Papaloizou, J.C.B., Terquem, C.: Dynamical relaxation and massive extrasolar planets. MNRAS **325**, 221–230 (2001)

Stepinski, T.F., Black, D.C.: On orbital elements of extrasolar planetary candidates and spectroscopic binaries. Astron. Astrophys. **371**, 250–259 (2001)

Takeda, G., Rasio, F.: High orbital eccentricities of extrasolar planets induced by the Kozai mechanism. Astrophys. J, **512**, 1001–1010 (2005)

Terquem, C., Papaloizou, J.C.B.: Dynamical relaxation and the orbits of low-mass extrasolar planets. MNRAS **332**, L39–L43 (2002)

Tremaine, S., Zakamska, N.: Extrasolar planet orbits and eccentricities. In: Holt, S.S. Deming, D. (eds.) The Search for Other Worlds, AIP Conf. Proc. **713**, 243–252 (2004)

Vogt, S.S., Butler, R.P., Marcy, G.W., Fischer, D.A., Henry, G.W., Laughlin, G., Wright, J.T., Johnson, J.A.: Five new multicomponent planetary systems. Astrophys. J. **632**, 638–658 (2005)

Zakamska, N., Tremaine, S.: Excitation and propagation of eccentricity disturbances in planetary systems. Astron. J. **128**, 869–877 (2004)

Celestial Mechanics and Dynamical Astronomy (2006) 95:259–271
DOI 10.1007/s10569-006-9005-1

ORIGINAL ARTICLE

Symmetric and asymmetric 3:1 resonant periodic orbits with an application to the 55Cnc extra-solar system

George Voyatzis · **John D. Hadjidemetriou**

Received: 3 November 2005 / Revised: 16 January 2006 /
Accepted: 5 February 2006 / Published online: 15 August 2006

Abstract We study the dynamics of 3:1 resonant motion for planetary systems with two planets, based on the model of the general planar three body problem. The exact mean motion resonance corresponds to periodic motion (in a rotating frame) and the basic families of symmetric and asymmetric periodic orbits are computed. Four symmetric families bifurcate from the family of circular orbits of the two planets. Asymmetric families bifurcate from the symmetric families, at the critical points, where the stability character changes. There exist also asymmetric families that are independent of the above mentioned families. Bounded librations exist close to the stable periodic orbits. Therefore, such periodic orbits (symmetric or asymmetric) determine the possible stable configurations of a 3:1 resonant planetary system, even if the orbits of the two planets intersect. For the masses of the system 55Cnc most of the periodic orbits are unstable and they are associated with chaotic motion. There exist however stable symmetric and asymmetric orbits, corresponding to regular trajectories along which the critical angles librate. The 55Cnc extra-solar system is located in a stable domain of the phase space, centered at an asymmetric periodic orbit.

Keywords Extra-solar planetary systems · Resonances · Periodic orbits

1 Introduction

The general three body problem can be considered as a good model for studying the dynamics of two-planet extra-solar planetary systems. To date, about 12% of the observed extra-solar systems have two or more planets (e.g., see Schneider 2005). In many cases (but few confirmed) the two planets are in mean motion resonance, with relatively large eccentricity values.

G. Voyatzis (✉) · J. D. Hadjidemetriou
Department of Physics, University of Thessaloniki
54124 Thessaloniki, Greece
e-mail: voyatzis@auth.gr

J. D. Hadjidemetriou
e-mail: hadjidem@auth.gr

An efficient way for studying the resonant dynamics of a planetary system and detecting stable and unstable regions in phase space is to determine the periodic configurations of the system, i.e. the periodic orbits in an appropriate rotating frame of reference, and the corresponding stability (e.g. Hadjidemetriou 2002; Psychoyos and Hadjidemetriou 2005a). In this way, we obtain a chart of the phase space indicating where a resonant planetary system could exist. These are the stable configurations to which a planetary system could be trapped under a migration process (Lee and Peale 2002; Ferraz-Mello et al., 2003, Beauge et al. 2006). Such stable configurations are associated with stable apsidal corotations and stable asymmetric librations as has been shown first by Beaugé et al. (2003).

There are three topologically different resonances: (1) The first order resonances $(n+1):n$, i.e. 2:1, 3:2, …,(2) The resonances $(2n+1):(2n-1)$, i.e. 3:1, 5:3, …and (3) all other resonances, i.e. 5:2, 7:3, 8:3,. … In the first-order resonances the unperturbed circular periodic orbits (zero planetary masses) are not continued to m_1, $m_2 > 0$, and we have a gap on the family of circular periodic orbits, from which two resonant families of elliptic periodic orbits are generated. At the 5:2 resonance (and all similar resonances, i.e. 7:3, 8:3, etc.) the continuation of the family of circular orbits, from zero to nonzero masses, is possible. The stability is preserved and we have two points on the perturbed circular family from which there bifurcate families of resonant 5:2 elliptic periodic orbits. Voyatzis and Hadjidemetriou (2005) studied the periodic orbits and examined the dynamics of the 2:1 resonant motion. Psychoyos and Hadjidemetriou (2005b) studied the 5:2 mean motion resonance. The 3:1 resonant planetary dynamics is different from the above two types of resonant motion. The continuation is possible, but now an unstable region appears on the perturbed circular family of periodic orbits close to the 3:1 resonance. At this region we have a bifurcation of four 3:1 resonant families of elliptic periodic orbits, as we shall show in the following.

In the present paper, we study the 3:1 resonant dynamics by considering the approach of determining bifurcations and periodic orbits as mentioned above. We mainly use in our computations the values for the planetary masses that correspond to the companions B and C of the extra-solar system 55Cnc (Marcy et al., 2002; McArthur et al., 2004), which is at the 3:1 resonance. Numerical simulations that indicate the 3:1 resonant dynamics of the system can be found in Ji et al. (2003), Zhou et al. (2004) and Marzari et al. (2005). Stable symmetric and asymmetric configurations have been calculated by Beaugé et al. (2003) and possible resonance capture is shown in Ferraz-Mello et al. (2003). In the next section we discuss briefly the model and the possible periodic configurations. In Section 3, we present the families of periodic orbits and discuss the dynamics of the phase space at the 3:1 resonance, for the masses that correspond to the 55Cnc planetary system. We also study how the 3:1 resonant stable periodic configurations are affected when the planetary masses are changed. Finally, in Section 4, we study the stability of the 55Cnc system.

2 The model and the periodicity conditions

The dynamical model that we use is the general planar three-body problem of planetary type. The system consists of a star of mass m_0 and two planets, denoted as $\mathcal{P}_1$ and $\mathcal{P}_2$, of masses $m_1 \ll m_0$ and $m_2 \ll m_0$, respectively. The short-term evolution of the system is described by nearly Keplerian motion of the two planets, with semimajor axes a_i, eccentricities e_i, mean anomalies M_i, longitudes of pericenter ϖ_i and periods T_i (the index $i = 1, 2$ refers to the planet $\mathcal{P}_i$). The $\mathcal{P}_1$ is the "inner" planet and $\mathcal{P}_2$ is the "outer" planet in the sense that $T_1 < T_2$, although the planetary orbits may intersect.

The center of mass of the planetary system is considered as fixed in an inertial frame, and the study is made in a *nonuniformly rotating* frame of reference xOy, whose x-axis is the line defined by the star and the inner planet $\mathcal{P}_1$. The origin O is the center of mass of these two bodies and the y-axis is perpendicular to the x-axis. In this rotating frame $\mathcal{P}_1$ moves on the x-axis and $\mathcal{P}_2$ in the xOy plane. The system can be studied as a system of three degrees of freedom (Hadjidemetriou, 1975) and possesses the fundamental symmetry Σ:

$$x \to x, \quad y \to -y, \quad t \to -t.$$

We fix the units of mass, length and time by considering normalized values of the masses such that $m_0 + m_1 + m_2 = 1$, the gravitational constant, G, is equal to unity and also by keeping a fixed value of the angular momentum of the system.

Our study contributes to the dynamics of the 3:1 mean motion resonance. In this case it is $T_2/T_1 \approx 3$ and the corresponding resonant or critical angles are defined as

$$\theta_1 = \lambda_1 - 3\lambda_2 + 2\varpi_1, \qquad \theta_2 = \lambda_1 - 3\lambda_2 + 2\varpi_2,$$

where λ_i is the mean longitude of the planet $\mathcal{P}_i$. Since the motion is studied in a rotating frame, it is useful to consider the half of the difference of the critical angles, i.e. the angle $\Delta\varpi = \frac{1}{2}(\theta_1 - \theta_2) = \varpi_1 - \varpi_2$, which provides directly the angle between the apsidal lines of the two planets.

The periodic orbits, which are studied, are periodic with respect to the rotating frame defined above. Such a periodic orbit corresponds to a motion in the inertial plane where the *relative* configuration of the planets is periodically repeated. The system *is not*, in general, periodic in the inertial frame.

The periodic orbits that are invariant under the symmetry Σ are called symmetric. In a symmetric periodic orbit the planet $\mathcal{P}_2$ starts from the x-axis perpendicularly and at that time $\dot{x}_1 = 0$, and after some time $t = T/2$, T being the period, the planet $\mathcal{P}_2$ crosses again the x-axis perpendicularly and at that time it is $\dot{x}_1 = 0$. The symmetry implies that $\Delta\varpi = 0°$ or $180°$, i.e. the lines of apsides are either aligned or antialigned, respectively. Additionally, there exists a moment that when one planet is at perihelion (or aphelion) the other planet is also at perihelion or aphelion, i.e. is $\theta_i = 0°$ or $180°$, $i = 1, 2$.

An asymmetric periodic orbit is not invariant under Σ. It is mapped by Σ to another periodic orbit, called *mirror image*, which has the same elements a_i and e_i as the original one but opposite values in M_i and ϖ_i. The periodicity conditions can be defined in two different, but equivalent, options. In the first option the planet $\mathcal{P}_2$ starts from the x-axis (nonperpendicularly) and the planet $\mathcal{P}_1$ is not at rest on the x-axis. After a time $t = T$, when $\mathcal{P}_2$ crosses again the x-axis, the planets $\mathcal{P}_1$,$\mathcal{P}_2$ have the same initial position and velocity as at $t = 0$. In the second option we start, at $t = 0$, at the moment when $\mathcal{P}_1$ has zero velocity on the x-axis, which means that $\mathcal{P}_1$ is either at perihelion or at aphelion ($\mathcal{P}_2$ is not on the x axis), and after a time $t = T$, when $\mathcal{P}_1$ has again zero velocity on the x-axis, the planets $\mathcal{P}_1$,$\mathcal{P}_2$ have the same initial position and velocity as at $t = 0$. In both of the above cases the critical arguments θ_1 and θ_2, cannot take the value $0°$ or $180°$ at the same moment, i.e. the lines of apsides of the planets do not coincide and/or the planets do not pass from the pericenter or apocenter position simultaneously.

3 Families of 3:1 resonant periodic orbits

In our study, we shall consider the 55Cnc planetary system for the star and the companion planets B and C. We use the following normalized values for the masses (Schneider 2005):

$$m_0 = 0.99903, \qquad m_1 = 0.00078, \qquad m_2 = 0.00019.$$

The periods of the two companions were estimated to be 14.7 and 43.9 days, respectively, i.e. they are in a 3:1 mean motion resonance. The above values are slightly different from those given by McArthur et al. (2004) in the sense that they do not cause structural changes in the dynamics of the system.

Our study starts with the computation of the family of circular periodic orbits of the system and the determination of the critical periodic orbits with respect to the linear stability. These critical periodic orbits are the bifurcation points for the families of resonant elliptic periodic orbits.

3.1 The family of circular orbits

The unperturbed motion with zero masses and zero eccentricities of both planets is periodic in the rotating frame for *any* value of the radii. Thus by fixing the radius of the inner planet we can form a family of periodic orbits with parameter the radius of the outer planet. The ratio n_2/n_1 of the mean motion of the two planets varies along the family. If we switch on the masses, these periodic orbits are continued as periodic orbits in the rotating frame, which correspond to almost circular orbits in the inertial frame and form the *family C*. The continuation is not possible at the resonances 2:1, 3:2, ..., where a gap appears. In the particular study, we focus in the part of the *family C* where $n_2/n_1 \approx 3$. This is presented in Figure 1.

It can be proved (Hadjidemetriou 1982), that a region of instability appears on the family of circular orbits, close to the 3:1 resonance. We remark that, we have three degrees of freedom (in the rotating frame) and consequently, we have three pairs of eigenvalues. The orbit is stable only in the case where all the eigenvalues are on the unit circle, in the complex plane. One pair of eigenvalues is always on the unit circle, at +1, due to the existence of the energy (Jacobi) integral. The other two pairs are free to move on the unit circle, preserving the stability, or to move outside the unit circle and generate instability. The section of the family C cloce to the 3:1 resonance starts as stable, as we see in Figure 1. As we proceed along the family, from $n_1/n_2 = 3.008$ to $n_1/n_2 = 3.024$, one pair of eigenvalues goes outside

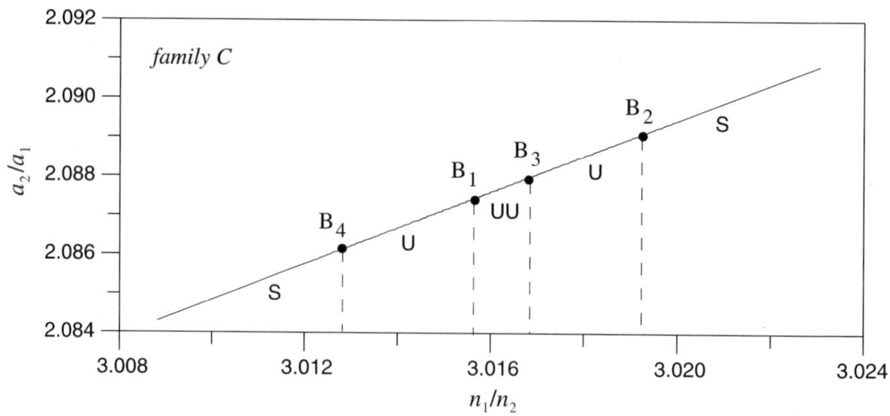

Fig. 1 The *family C* of circular periodic orbits near the 3:1 mean motion resonance. It is presented in the projection plane where the horizontal and vertical axis correspond to the ratio of mean motions and the ratio of semimajor axes of the planets, respectively. The type of linear stability is denoted by the symbols "S" for stable orbits, "U" for simply unstable orbits and "UU" for doubly unstable orbits. The four critical orbits (or bifurcation points) are denoted by $B_1, \ldots, B_4$

the unit circle at the point B_4, thus generating instability, while the other pair is still on the unit circle. As we proceed further along the family, the second pair of eigenvalues goes also outside the unit circle, starting from point B_1, and ending at the point B_3, where again in goes back on the unit circle. The first pair is still outside, up to the point B_2, where it goes back on the unit circle, thus restoring the stability. Conclusively, the region B_1B_3 is doubly unstable (two pairs of eigenvalues outside the unit circle) and the regions B_4B_1 and B_2B_3 are simply unstable (only one pair of eigenvalues outside the unit circle). The points B_1, B_2, B_3, and B_4 are the critical points, where the stability changes character, and it is from each of these points that we have bifurcation of a family of elliptic periodic orbits. All these orbits are symmetric.

3.2 The families of symmetric periodic orbits

There are four families of resonant elliptic periodic orbits, that bifurcate from the critical points B_1, B_2, B_3, and B_4, called *families* S_i, $i = 1, \ldots, 4$, where i denotes the corresponding bifurcation point B_i of the *family C*. The configuration of each of these families at $t = 0$ and at $t = T/2$ is given in Table 1. The characteristic curves of these families are presented in the space of the "signed eccentricities" $e_1^* = e_1 \sin \Delta\varpi$ and $e_2^* = e_2 \sin \theta_1$ (note that $\Delta\varpi$ and θ_1 are either 0° or 180°). In this presentation, shown in Figure 2, each one of the four families is placed in a particular quarter of the plane $e_1^* - e_2^*$.

The *family* S_1 is unstable. It starts with simply unstable orbits. The orbits become doubly unstable in the segment between the points B_{12} and B_{13}, which denote critical orbits. Starting with initial conditions close to an orbit of the *family* S_1 we have chaotic evolution and the critical angles $\Delta\varpi$ and θ_1 rotate. Chaos becomes more obvious as the planetary eccentricities increase.

The *family* S_2 starts from the circular family and consists initially of simply unstable orbits. The particular configuration permits the planets to come very close to each other and we have a collision orbit. After the collision the family is continued with stable periodic orbits, where the corresponding planetary orbits intersect.

The *family* S_3 starts with unstable orbits. The type of stability changes at the high-eccentricity value $e_1 = 0.74$ (and $e_2 = 0.59$) and the periodic orbits of the family become stable after this point. Around these stable orbits regular librating motion, with respect to the critical angles, is obtained. The point B_{31} where the stability changes is a bifurcation point for a family of stable asymmetric orbits (see paragraph 3.3).

The *family* S_4 is the only family that starts with stable orbits. The stable segment extends up to the point B_{41}, at $e_1 = 0.085$ and $e_2 = 0.110$. For higher eccentricities the orbits become unstable. The phase space region around the low-eccentricity stable periodic orbits consists of invariant tori, the orbits are regular and the critical angles $\Delta\varpi$ and θ_1 librate around the value of 180°.

Table 1 Configurations at $t = 0$ and $t = T/2$ obeyed by the 3:1 resonant symmetric periodic orbits of the families S_1–S_4

S_1:	$\mathcal{P}_2$(per)	–	Star	–	$\mathcal{P}_1$(apo)	→	$\mathcal{P}_1$(per)	–	Star	–	$\mathcal{P}_2$(apo)
S_2:	Star	–	$\mathcal{P}_1$(per)	–	$\mathcal{P}_2$(apo)	→	$\mathcal{P}_2$(per)	–	$\mathcal{P}_1$(apo)	–	Star
S_3:	Star	–	$\mathcal{P}_1$(apo)	–	$\mathcal{P}_2$(apo)	→	$\mathcal{P}_2$(per)	–	$\mathcal{P}_1$(per)	–	Star
S_4:	$\mathcal{P}_1$(per)	–	Star	–	$\mathcal{P}_2$(per)	→	$\mathcal{P}_2$(apo)	–	Star	–	$\mathcal{P}_1$(apo)

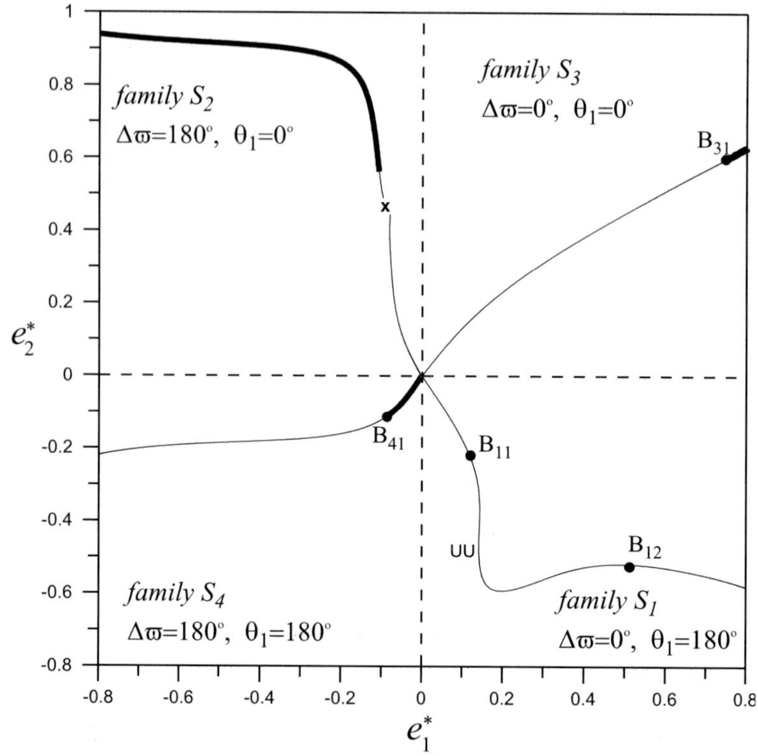

Fig. 2 The four families of symmetric periodic orbits presented in the plane of the "signed" eccentricities $e_1^*-e_2^*$. All families start from circular orbits at (0,0). The value of the critical angles $\Delta\varpi$ and θ_1 along each family is indicated. The bold line segments indicate stable orbits while the thin ones correspond to unstable orbits. The orbits between the critical orbits B_{11} and B_{12} are doubly unstable. The symbol × denotes a collision orbit

3.3 The families of asymmetric periodic orbits

The critical orbits B_{11}, B_{12}, B_{31}, and B_{41} on the symmetric families S_i, shown in Figure 2, are bifurcation points for new families. We found that all families that bifurcate from the above critical orbits consist of *asymmetric* periodic orbits. The corresponding characteristic curves are presented in the plane of eccentricities e_1–e_2 shown in Figure 3.

The asymmetric *family* A_1 starts from the critical point B_{11} of the symmetric family S_1 and terminates at the critical point B_{12} of the same family (Figure 3a). All orbits of the *family* A_1 are doubly unstable and the orbits close to the asymmetric periodic orbits are chaotic. The critical angle $\Delta\varpi$ corresponds to the value 0° at the critical points while it varies along the family increasing up to 35°. The critical angle θ_1 is 180° at the edges of the family and along the family reaches the maximum value of 222° (Figure 4a).

The second family of asymmetric orbits, the *family* A_3, bifurcates from the critical point B_{31} of the *family* S_3 (Figure 3b). This family starts as stable and is directed towards to lower values of the eccentricity e_1. At $e_1 \approx 0.13$ the family turns towards to higher eccentricity values e_1 and e_2. The stable character of the asymmetric orbits holds up to the critical orbit B_{32}. At this point the orbits become unstable and a new family of asymmetric orbits, the *family* A'_{31} bifurcates from this critical point. All orbits of A'_{31} are unstable. The critical angles

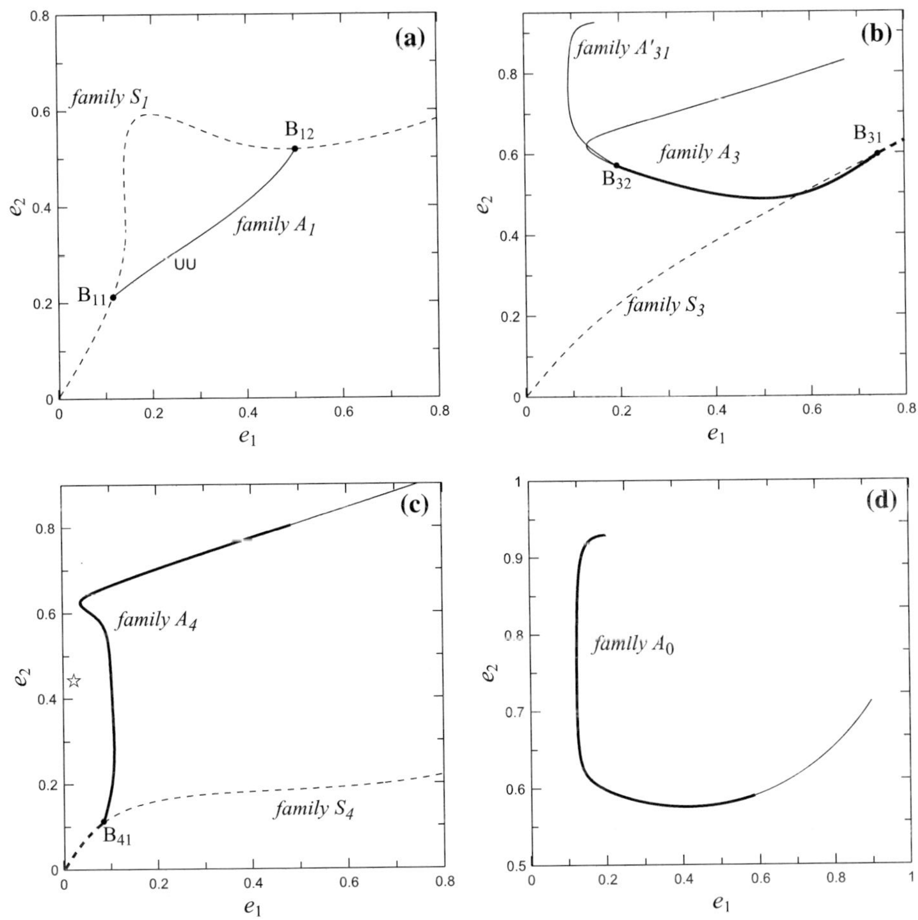

Fig. 3 The families of asymmetric periodic orbits (solid curves) presented in the plane of eccentricities $e_1 - e_2$. The associated symmetric families are indicated by dashed curves. The bold phase segments indicate stable orbits while the thin ones correspond to unstable orbits. The star in (c) indicates the position of the 55Cnc planetary system

$\Delta\varpi$ and θ_1 that correspond to the bifurcation point B_{31} are equal to $0°$, while at about the bifurcation point B_{32} they take their maximum value, $93°$ and $121°$, respectively (Figure 4b). We found that a relatively large region of phase space extended around the stable asymmetric periodic orbits shows regular motion. Along these orbits the critical angles librate around values different than $0°$ or $180°$.

The asymmetric *family* A_4 starts from the critical orbit B_{41} of the family S_4. Initially and up to $e_1 \approx 0.5$, it is $e_2 \approx$ const. (see Figure 3c). The apsidal difference $\Delta\varpi$ is almost constant ($\sim 250°$) along this part of the *family* A_4. This is shown in Figure 4c, which presents the variation of the critical angles along the family. The major part of the family consist of stable orbits and the phase space consist of regular librating resonant motion. The family becomes unstable for high eccentricity values ($e_1 \approx 0.5$ and $e_1 \approx 0.8$) but we have not succeed in calculating the family that bifurcates from this critical point.

Finally, we present the *family* A_0 of asymmetric periodic orbits in Figure 3d. This family does not bifurcate from a symmetric family. It consists of one segment of unstable orbits and

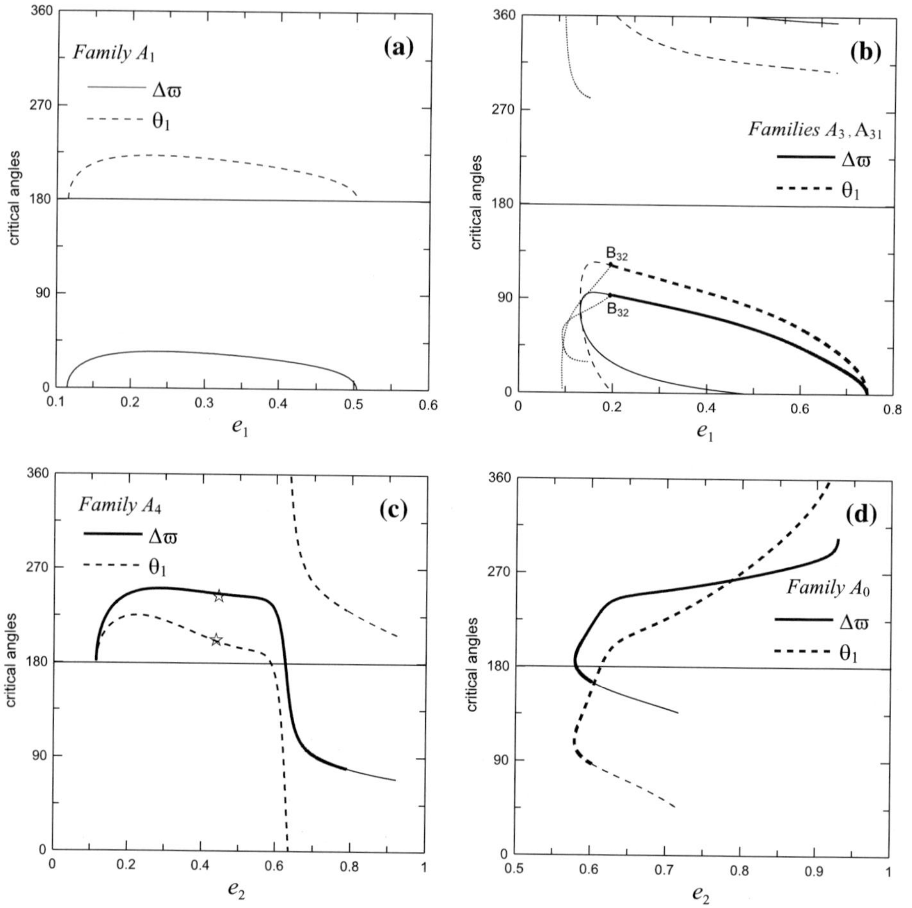

Fig. 4 The variation of the critical angles $\Delta\varpi$ (solid curves) and θ_1 (dashed curves) along the *families* of asymmetric periodic orbits presented in Figure 3. It is convenient to use as family parameter the value e_1 for the cases (a) and (b) and the value e_2 for the cases (c) and (d). Thin and bold lines indicate unstable and stable periodic orbits, respectively. The star in panel (c) indicates the position of the 55Cnc planetary system

one segment of stable orbits. It is remarkable that there exist stable orbits for $e_2 > 0.9$. Such a kind of family has been also found in the 2:1 resonant motion (Voyatzis and Hadjidemetriou 2005).

3.4 The families S_4 and A_4 for various mass values

The S_4 and A_4 are the only families of periodic orbits associated with stable resonant librations for low and moderate values of the planetary eccentricities. Also, as it is discussed in the next section, these families are associated with the dynamics of 55Cnc system. Our computations showed that these families exist for a large range of planetary masses. In the following, we present two numerical experiments in order to clarify the dependence of the families on the planetary masses.

In the first numerical experiment we multiply the planetary mass values, given in the beginning of Section 2, by a factor a and we recalculate the families. As it is shown in Figure 5a, when the planetary masses decrease, the families are not affected significantly with respect

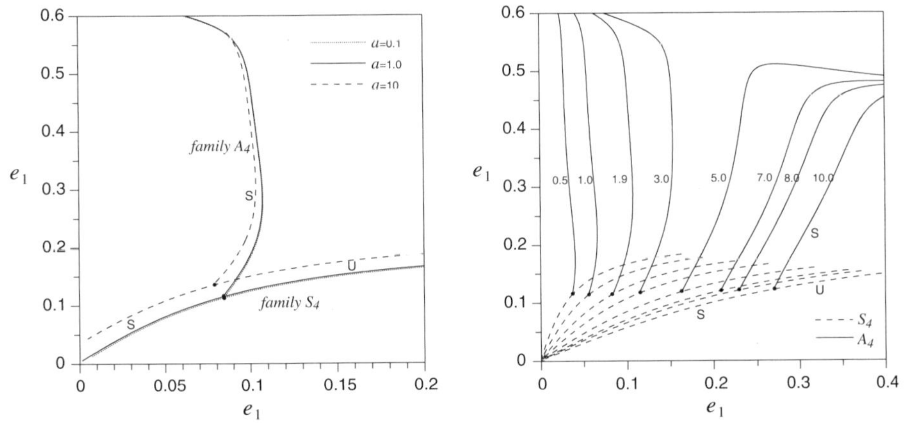

Fig. 5 Characteristic curves of the symmetric *family* S_4 and the asymmetric one A_4 for various values of planetary masses. (a) constant mass ratio, $m_1 = a \times 8 \times 10^{-4}, m_2 = a \times 2 \times 10^{-4}$. (b) $m_1 = 8 \times 10^{-4} = \text{const.}$ and $m_2 = \mu \times 10^{-4}$, where the value of μ is indicated beside each curve

to their location in the plane of eccentricities and their stability. When the planetary masses increase, the bifurcation point for the asymmetric family shifts only slightly to a lower value for the eccentricity e_1 and to a higher value for the eccentricity e_2. The most important consequence is that the unstable part of the family A_4 (see Figure 3c) increases and extends towards to lower eccentricity values as the masses increase. Thus, by increasing the masses by a factor of 10, the asymmetric orbits of the *family* A_4 become unstable for $e_2 > 0.5$.

In the second numerical experiment we fix the mass of the inner planet to the value $m_1 = 8 \times 10^{-4}$ and we set $m_2 = \mu \times 10^{-4}$, where we let μ to vary. The computed families for various values of μ are shown in Figure 5b. When the mass of the outer planet decreases, the *family* A_4 starts from lower e_1 values. The opposite case holds if m_2 increases. Note that the asymmetric *family* A_4 still exists for $m_2 > m_1$. These results are in a good qualitative agreement with those obtained from the averaged model and the migration scenario proposed by Beaugé et al. (2003) and Ferraz-Mello et al. (2003), respectively. It should be noted that the bifurcation point for the asymmetric family is located to $e_2 \approx 0.11$ for any value of μ or, equivalently, any planetary mass ratio. This characteristic feature for the 3:1 resonance has been indicated first by Ferraz-Mello et al. (2003).

4 Dynamical aspects and stability of the 55Cnc system

The observed position of the 55Cnc planetary system (Schneider 2005) is indicated by a star in Figures 3c and 4a and corresponds to $e_1 = 0.02$, $e_2 = 0.44$, $\Delta\varpi = 245°$, and $\theta_1 = 200°$. These elements are close to those given by McArthur et al. (2004). It is evident that this position is associated with the asymmetric periodic orbits of the *family* A_4 and, in particular, it seems to be located near the asymmetric orbit at $e_1 = 0.1$, $e_2 = 0.44$. The family A_4 is linearly stable at the particular domain, suggesting that a six-dimensional region of the phase space nearby the orbits of the family A_4 corresponds mainly to quasiperiodic motion or, equivalently, to regular trajectories for longterm evolution.

The width of the stable region can be determined numerically by examining a large number of trajectories. Particularly, for the initial conditions of the 55Cnc system, mentioned

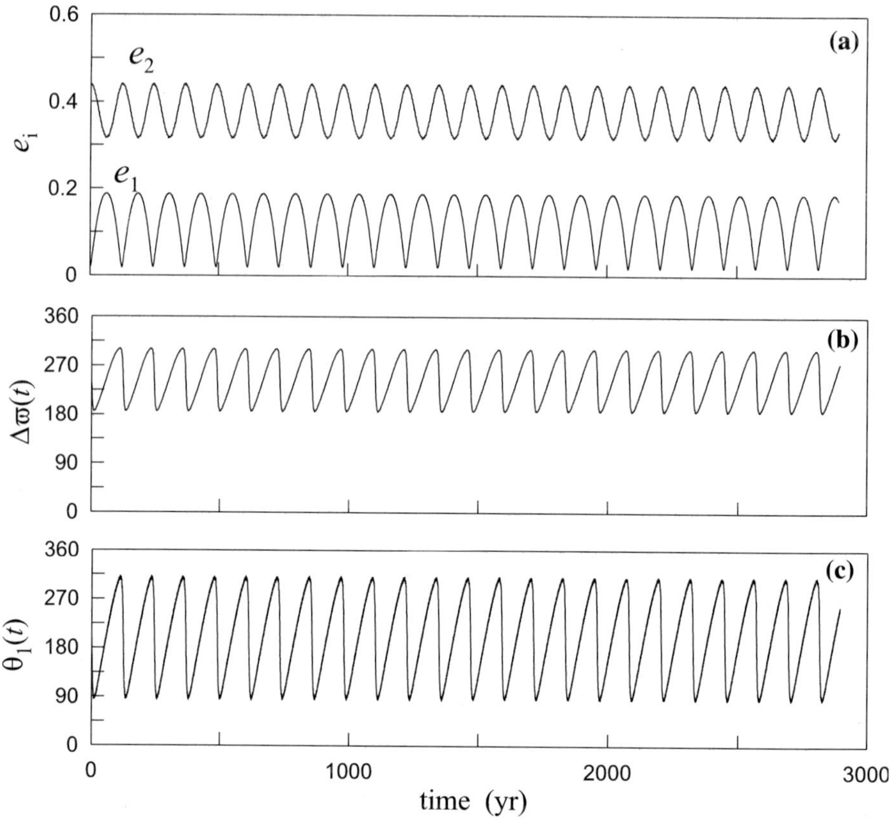

Fig. 6 The variation of the eccentricity values e_1 and e_2 (panel (a)) and the critical angles $\Delta\varpi$ and θ_1 (panels (b),(c)) along the trajectory associated with the 55Cnc planetary system. The initial conditions are $e_1(0) = 0.02$, $e_2(0) = 0.44$, $\Delta\varpi(0) = 245°$ and $\theta_1(0) = 200°$

above, and in the framework of the three body problem, we obtain regular evolution. The variation of the planetary eccentricities and the critical angles are presented in Figure 6. The semimajor axes of the planets are almost constant during the trajectory evolution but, as it is shown in Figure 6a, the eccentricities oscillate with relatively large amplitude. The large amplitude oscillations of the eccentricities have been also obtained and discussed by Zhou et al. (2004) and Beaugé et al. (2006). The critical angles $\Delta\varpi$ and θ_1 librate around the values 250 and 200°, respectively (Figure 6b, c). The period of libration is about 125 years for the particular case. Such stable librations have been also observed in the numerical simulations by Ji et al. (2003) and correspond to the stable motion of type '**a**' found by Zhou et al. (2004).

By examining a large number of trajectories, using Poincaré sections, we found that regular motion exists in a wide region of initial conditions around the stable asymmetric orbits of family A_4. In Figure 7, we present Poincaré maps on the surface of section $y_2 = 0$, $\dot{y}_2 > 0$, projected on the plane $x_2 - \dot{x}_2$. The panel (a) corresponds to the initial conditions of the 55Cnc system, mentioned above. The motion is regular on a torus presented in the section by two "islands". The corresponding periodic orbit close to this quasiperiodic orbit is of multiplicity two, i.e. intersects the plane of section twice per period and these fixed points are the centers of the "islands" in the four-dimensional space of the Poincaré map. In the

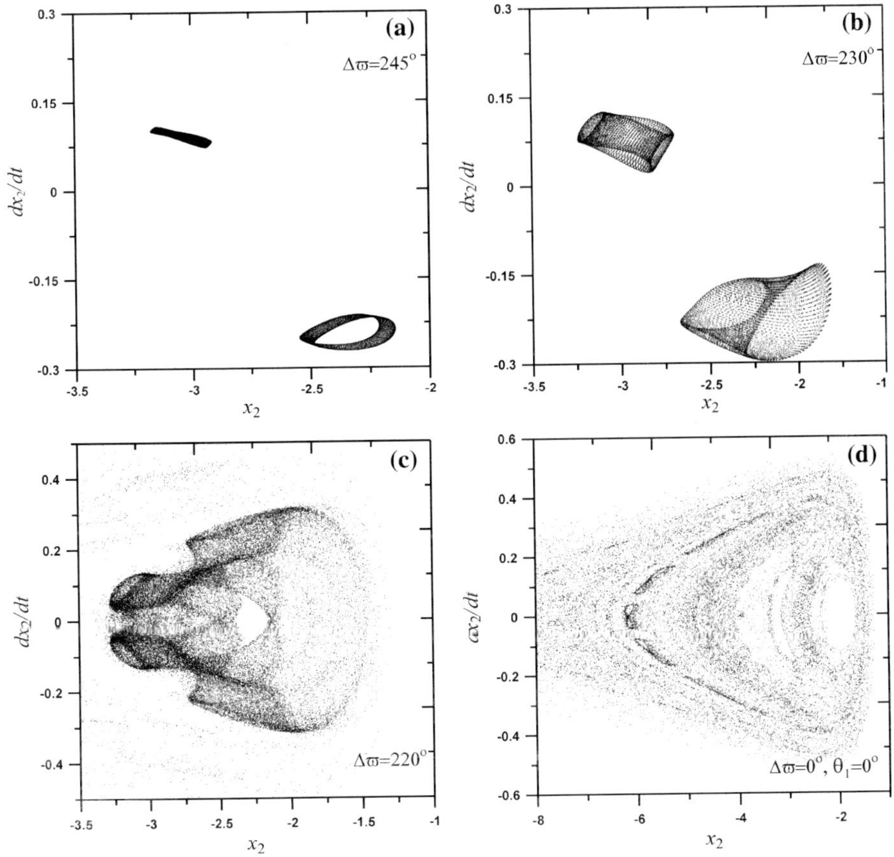

Fig. 7 Projection planes $x_2 - \dot{x}_2$ of the Poincare section $y_2 = 0$, $\dot{y}_2 > 0$ of the trajectory with initial conditions $e_1(0) = 0.02$, $e_2(0) = 0.44$ and (a) $\Delta\varpi(0) = 245°$, $\theta_1(0) = 200°$, (b) $\Delta\varpi(0) = 230°$, (c) $\Delta\varpi(0) = 220°$, $\theta_1(0) = 200°$ and (d) $\Delta\varpi(0) = 0°$, $\theta_1(0) = 0°$

panels (b) and (c), we present the trajectories corresponding to the same initial conditions as in panel (a), but we change the initial apsidal angle $\Delta\varpi$. For $\Delta\varpi = 230°$ (i.e. 15° deviation from the value of the corresponding periodic orbit) we observe still regular islands and librations of the critical angles. For larger deviations ($\Delta\varpi < 225°$) the trajectories show stickiness. Namely, the planetary orbits seem to evolve regularly in the resonant region for a long time interval. After this interval, the trajectory enters a wide chaotic region and the semimajor axis and the eccentricity of the outer planet increase rapidly. Such a trajectory is shown in Figure 7c and corresponds to $\Delta\varpi = 220°$. The points of the Poincaré map spread irregularly in a wide domain after about 15 K years. The evolution of the semimajor axes and eccentricities along this trajectory is shown in Figure 8a and illustrates the stickiness of the trajectory and the escape of the outer planet $\mathcal{P}_2$. The stickiness time decreases as $\Delta\omega$ decreases and for $\Delta\omega < 200°$ the chaotic motion becomes apparent after few iterations of the Poincaré map.

From the above results, we can see that the asymmetry, which is imposed by the periodic orbits of the family A_4, stabilizes the planetary system. Sufficient deviations from such an asymmetric configuration result to strongly chaotic motion. In the particular case the sym-

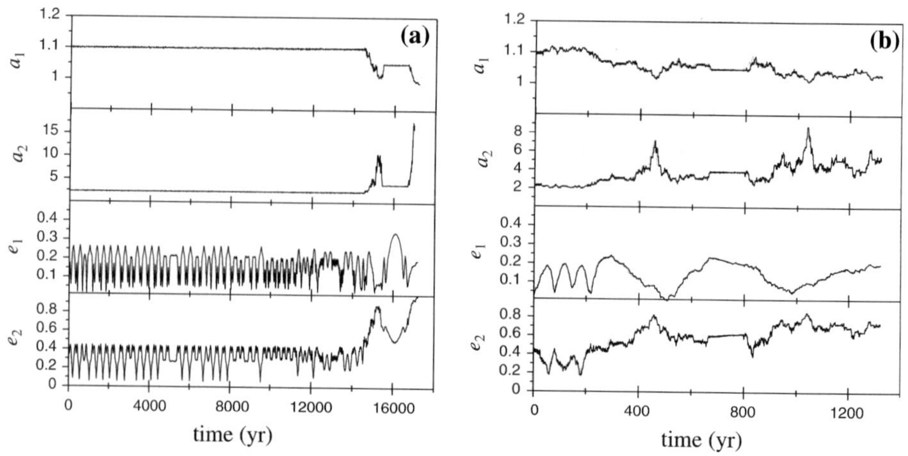

Fig. 8 (a) The evolution of the semimajor axes and eccentricities of the planetary orbits that corresponds to the sticky trajectory shown in Figure 7(c). (b) The same as in (a) for the chaotic trajectory of Figure 7(d), which corresponds to the symmetric initial configuration $\Delta\varpi(0) = \theta_1(0) = 0°$.

metric configurations are also unstable. All the trajectories with initial elements $e_1 = 0.02$, $e_2 = 0.44$, $\Delta\varpi(0) = 0$ or $180°$ and $\theta_1(0) = 0$ or $180°$, correspond to chaotic motion. An example is presented in Figure 7(d) and 8(b) for $\Delta\varpi(0) = \theta_1(0) = 0°$.

Acknowledgements We thank Dr Beaugé for useful comments. This work was supported by the research programme *Pythagoras II*, Nr.80831 of the Greek Ministry of Education and the E.U.

References

McArthur, B.E., Endl, M., Cochran, W.D., Benedict, F.G., Fischer, D.A., Marcy W.G., Butler, R.P., Naef, D., Mayor, M., Queloz, D., Udry, S., Harrison, E.: Detection of a Neptune-mass planet in the ρ^1 Cancri system using the Hobby-Eberly Telescope, ApJ **614**, L81–L84 (2004)

Beaugé, C., Ferraz-Mello, S., Michtchenko, T.: Extrasolar planets in mean-motion resonance: apses alignment and asymmetric stationary solutions, ApJ **593**, 1124–1133 (2003)

Beaugé, S., Michtchenko, T., Ferraz-Mello, C.: Planetary migration and extrasolar planets in the 2/1 mean motion resonance, MNRAS **365**, 1160–1170 (2006)

Ferraz-Mello, S., Beaugé, C., Michtchenko T.: Evolution of migrating planet pairs in resonance, Cel. Mech. Dyn. Astr. **87**, 99–112 (2003)

Hadjidemetriou, J.D.: The continuation of periodic orbits from the restricted to the general three-body problem, Cel. Mech. **12**, 155–174 (1975)

Hadjidemetriou, J.D.: On the relation between resonance and instability in planetary systems, Cel. Mech. **27**, 305–322 (1982)

Hadjidemetriou, J.D.: Resonant periodic motion and the stability of extrasolar planetary systems, Cel. Mech. Dyn. Astr. **83**, 141–154 (2002)

Ji, J., Kinoshita, H., Liu, L., Li, G.: Could the 55Cnc planetary system be in the 3:1 MMR?, ApJ **585**, L139–L142 (2003)

Marcy, G.W., Butler, R.P., Fischer, D.A., Laughlin, G., Vogt, S.S., Henry, G.W., Pourbaix, D.: A planet at 5AU around 55Cnc, ApJ **581**, 1375–1388 (2002)

Marzari, F., Scholl H., Tricarico, P.: Frequency map analysis of the 3/1 resonance between planets b and c in the 55Cnc system, A&A **442**, 359–364 (2005)

Lee, M.H., Peale, S.: Dynamic and origin of the 2:1 orbital resonances of the GJ 876 planets, ApJ **567**, 596–609 (2002)

Psychoyos, D., Hadjidemetriou, J.D.: Dynamics of 2/1 resonant extrasolar systems. Application to HD82943 and Gliese876, Cel. Mech. Dyn. Astr. **92**, 135–156 (2005a)

Psychoyos, D., Hadjidemetriou, J.D.: Dynamics of extrasolar systems at the 5/2 resonance: application to 47 UMa, in *Proceedings of IAU Colloquium*, No. 197, pp. 55–62 (2005b)
Schneider, J. http://www.obspm.fr/encycl/catalog.html, September 2005
Voyatzis, G., Hadjidemetriou, H.D.: Symmetric and asymmetric librations in planetary and satellite systems at the 2/1 resonance, Cel. Mech. Dyn. Astr. **93**, 265–296 (2005)
Zhou, L.Y., Lehto, H.J., Sun, Y.S., Zheng, J.Q.: Apsidal corotation in mean motion resonance: the 55Cnc system as an example, MNRAS **350**, 1495–1502 (2004)

Celestial Mechanics and Dynamical Astronomy (2006) 95:273–285
DOI 10.1007/s10569-006-9031-z

ORIGINAL ARTICLE

Estimations of orbital parameters of exoplanets from transit photometry by using dynamical constraints

Zsolt Sándor

Received: 30 November 2005 / Revised: 27 March 2006 /
Accepted: 16 May 2006 / Published online: 15 August 2006

Abstract The probability of the detection of Earth like exoplanets may increase in the near future after the launch of the space missions using the transit photometry as observation method. By using this technique only the semi-major axis of the detected planet can be determined, and there will be no information on the upper limit of its orbital eccentricity. However, the orbital eccentricity is a very important parameter, not only from a dynamical point of view, since it gives also information on the climate and the habitability of the Earth-like planets. In this paper a possible procedure is suggested for confining the eccentricity of an exoplanet discovered by transit photometry if an already known giant planet orbits also in the system.

Keywords Exoplanets · Planetary transit · Three-body problem · Stability · Chaos detection

1 Introduction

In the last decade, following the discovery of the first extrasolar planet around 51 Pegasi (Mayor and Queloz 1995), more than 168 exoplanets have been observed. The detection of exoplanets is of great importance, since they form planetary systems around their hosting stars, and by studying the main properties of these exoplanetary systems, the characteristics, formation and evolution of the Solar System can be treated and understood from a wider perspective. However, the above picture is rather ideal than complete yet, since the exoplanets observed until now are mainly Jupiter-like gas giants. This is a consequence of the fact that by using radial-velocity measurements, which is the most effective ground-based observing technique, it is not possible to detect Earth-like planets yet. (We note, however, that the detection of a

Z. Sándor (✉)
Department of Astronomy, Eötvös University,
Pázmány Péter stny. 1/A, H-1117 Budapest, Hungary
e-mail: Zs.Sandor@astro.elte.hu

$\sim 8M_\oplus$ planet by Rivera et al. (2005) indicates that radial velocity searches might be able to detect Earth-like planets in the future.)

The most compelling question related to exoplanetary research is the detection of Earth-like planets. Beside their importance in testing and improving formation theories of planetary systems, a major question is their habitability. If an Earth-like planet orbits in the habitable zone of a star, there could be chances of appearing (water based) life on its surface. On the habitable zone we mean that region around a star, where liquid water can exist on the surface of a planet. For details see in Kasting et al. (1993).

In order to find Earth-like planets, there are space missions in construction and planning phase. Such a mission is COROT (sponsored by CNES, ESA and other countries) to be launched in 2006, and the Kepler Mission (NASA) with a launch in 2008. These missions will use the transit photometry as detection technique, which is based on measuring the periodic dimming of a star's light intensity caused by an unseen transiting planet. Observations performed by these instruments will provide the semi-major axis a of the transiting planet calculated from Kepler's third law by measuring the period T of the transits, and assuming that the mass m_* of the hosting star is known. Since m_* is known only with limited accuracy, an uncertainty appears in a. (We note that m_* can be determined by spectroscopic observations and by stellar model calculations.) The error appearing in a and in the eccentricity e of the transiting planet due to the uncertainties in stellar mass will be estimated in Sect. 3.2.

In this paper we present a procedure which can be used to estimate the orbital eccentricity and inclination of a transiting planet, if (i) we can measure the duration of the transit, and (ii) there is another (giant) planet in the system. We shall see that an equation can be derived, which connects the mass and the radius R of the star, the semi-major axis a, the eccentricity e, the argument of the periastron ω, the inclination i of the transiting planet, and the duration τ of the transit. In this equation there are three unknowns, e, ω, and i. By fixing i, the corresponding (ω, e) pairs can be visualized as curves on the ω–e parameter plane. Thus the problem is underdetermined and there is no way to give an upper limit for e. We note that during one revolution of the transiting planet, in principle there are two minima in the *light curve* of the star. The first minimum occurs when the planet moves in front of the star, the second appears when the star is between the observer and the planet. The maximum of the light curve can be observed when both the star and the planet are visible. By using the second minimum observation, the eccentricity of the transiting planet can be estimated. However, the maximum and the second minimum of the light curve can not be seen in the case of an Earth-like planet, since its contribution to the whole light flux is not detectable. (The light curve is the variation of the light intensity of the star as a function of the orbital phase of the transiting planet.)

On the other hand, as suggested by planetary formation scenarios, we expect that beside Earth-like planets Jupiter-like giant planets can also be found in the majority of planetary systems. Having discovered an Earth-like planet around a star, by using complementary techniques (as observations by Space Interferometry Mission and ground-based Doppler spectroscopy) other more massive planets can be identified in the system, and their orbital parameters can be determined too. (Although the above scenario sounds very optimistic, it is one of the scientific goals of Kepler Mission, see http://kepler.nasa.gov/basis/goals.html.)

The presence of one or more additional giant planets (beside the transiting one) results in that, that both ordered and chaotic regions appear in the phase space of the system. If the phase trajectory of the Earth-like planet is in an ordered region of the phase space, the motion of the planet is stable for arbitrary long times. If the initial conditions of its orbit are in a chaotic region of the phase space, the motion of the planet can become unstable after some time. In this paper we try to exclude those orbital parameters of the transiting planet, which result in chaotic motion. We shall demonstrate that in some cases it is possible to determine an upper limit for the eccentricity and a lower limit for the inclination of the transiting planet. We stress that the eccentricity is a very important orbital parameter not only from a dynamical point of view, but also in studying the habitability and climatic variations of the Earth-like planet.

The paper is organized as follows: first we derive an equation between the duration of the transit and some important parameters of the star and the transiting body, then we solve this equation numerically. After examining the solutions of this equation, we map the stability structure of the system assuming the presence of a giant planet. Then we determine lower limits for the inclination and an upper bound for the eccentricity of the transiting planet depending on the eccentricity and the semi-major axis of the known giant planet.

2 An equation connecting the parameters of a star, a transiting planet and the transit

In this section we shall derive an equation between the orbital parameters of the transiting planet, the star's mass, and the duration of the transit from the geometry of the transit.

Let us suppose that the star's disc is a circle with radius R, and a planet is moving in a front of this disc with an average velocity v_{tr}. If the duration of the transit is denoted by τ and the length of the path of the transiting planet is d (see Fig. 1), the following approximation holds:

$$v_{\text{tr}} = \frac{d}{\tau}. \tag{1}$$

We note that according to Kepler's second law, the velocity of the planet is changing during the transit (except in the case of circular orbits), however this change is

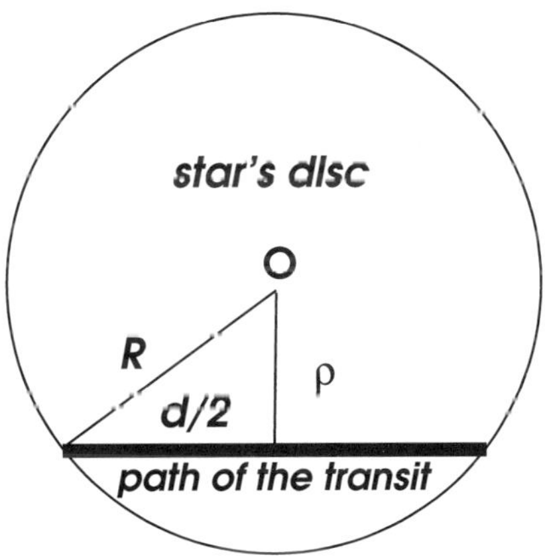

Fig. 1 The transit of a planet in front of the stellar disc. The straight sections denoted by R, ρ, and $d/2$ form a pythagorean triangle

negligible for moderate eccentricities. From the triangle in Fig. 1

$$R^2 = \rho^2 + \left(\frac{d}{2}\right)^2. \tag{2}$$

From Eq. 2 the length of the transit's path d can be expressed as

$$d = 2\sqrt{R^2 - r^2\cos^2 i}, \tag{3}$$

where, according to Fig. 2,

$$\rho = r\cos i, \tag{4}$$

i being the inclination (e.g. the angle between the orbital plane and the tangent plane to the celestial sphere), and r the distance between the center of the star and the planet at the moment of the middle of the transit.

By using the well known formula for r:

$$r = \frac{a(1-e^2)}{1+e\cos v}, \tag{5}$$

(where a is the semi-major axis, e is the eccentricity, and v is the true anomaly of the transiting planet), and Eqs. 1 and 3, the average transiting velocity of the planet (v_{tr}) can be written as

$$v_{\mathrm{tr}} = \frac{2}{\tau}\sqrt{R^2 - \left[\frac{a(1-e^2)}{1+e\cos v}\right]^2\cos^2 i}\,, \tag{6}$$

where v is the true anomaly of the transiting planet at the middle of the transit.

On the other hand, v_{tr} can also be approximated on the basis of the two-body problem. In the coordinate system (ξ, η), in which the axes of the orbital ellipse are on the axes ξ and η, the components of the orbital velocity vector are (Murray and Dermott 1999):

$$\begin{aligned}\dot{\xi} &= -\sqrt{\tfrac{\mu}{p}}\sin v,\\ \dot{\eta} &= \sqrt{\tfrac{\mu}{p}}(e+\cos v),\end{aligned} \tag{7}$$

where $p = a(1-e^2)$ is the parameter of the ellipse and $\mu = k^2(m_* + m_p)$, m_* and m_p being the stellar and planetary masses respectively, and k is the Gaussian constant of

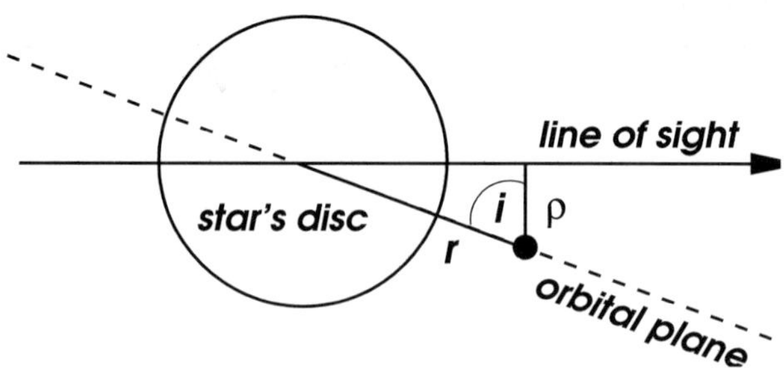

Fig. 2 Side-view of the transit, where r is the distance of the planet from the star's center and i is the inclination of its orbital plane

gravity. Let (x, y) denote the coordinate system in which the x-axis is the projection of the line of sight (e.g. the line connecting the center of the star to the observer) to the orbital plane of the transiting planet. From Fig. 3 it can be seen that the system (ξ, η) is just a rotation of the system (x, y) by an angle ω, which is the argument of the periastron of the transiting planet. Thus in the coordinate system (x, y) formulae (7) transform as

$$\begin{aligned} \dot{x} &= \dot{\xi} \cos\omega - \dot{\eta} \sin\omega, \\ \dot{y} &= \dot{\xi} \sin\omega + \dot{\eta} \cos\omega. \end{aligned} \tag{8}$$

Studying Fig. 3 one can find that the average velocity v_{tr} of the transiting planet is almost equal to $\dot{y}$, which is the velociy of the planet at the middle of the transit. Then by using Eqs. 7 and 8 we find

$$v_{\mathrm{tr}} \approx \dot{y} = -\sqrt{\frac{\mu}{p}} \sin v \sin\omega + \sqrt{\frac{\mu}{p}} (e + \cos v) \cos\omega. \tag{9}$$

Studying again Fig. 3, it is also true that at the middle of the transit

$$v + \omega = 360^\circ, \tag{10}$$

thus the average orbital velocity of the transiting planet is

$$v_{\mathrm{tr}} = \sqrt{\frac{\mu}{p}} (1 + e\cos\omega). \tag{11}$$

Combining Eqs. 6, 10, and 11 we obtain a relation between the orbital parameters of the transiting planet and the duration of the transit:

$$\sqrt{\frac{\mu}{a(1-e^2)}} (1 + e\cos\omega) - \frac{2}{\tau} \sqrt{R^2 - \left[\frac{a(1-e^2)}{1+e\cos\omega}\right]^2 \cos^2 i} = 0. \tag{12}$$

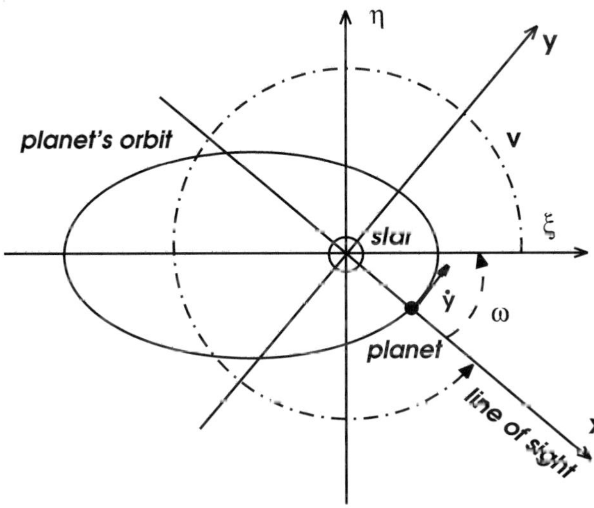

Fig. 3 The transit as viewed from above. At the middle of the transit v_{tr} is nearly equal to $\dot{y}$. The coordinate system (ξ, η) is the rotation of the coordinate system (x, y) by ω

This equation has the form

$$f(a, e, i, \omega, \mu, R, \tau) = 0, \tag{13}$$

where the unknown quantities are the eccentricity e, the inclination i, and the argument of the periastron ω. The other quantities, such as the semi-major axis a, the mass parameter (μ), the radius of the star (R), and the duration of the transit (τ) are supposed to be known.

3 Analysis of Eq. 12

In this section we present and analyse the solutions of Eq. 12 by using a model system. We shall see how the orbital eccentricity e of the transiting planet depends on its argument of periastron ω, or in other words, on the orbital position of the planet during the transit. We also study the sensitivity of the solutions to the uncertainties of the mass of the hosting star.

3.1 Solution of Eq. 12

According to the last paragraph of the previous section, the unknown quantities in Eq. 12 are the inclination i, the argument of periastron ω, and the eccentricity e of the transiting planet. Thus by fixing i, Eq. 12 can be solved numerically, and the (ω, e) pairs of the solution can be represented as a curve on the ω–e parameter plane.

In order to study the solutions of Eq. 12, we give specific values for the parameters in Eq. 12. Let us assume that the mass of the transiting planet is 1 Earth-mass, and it revolves around a 1 Solar-mass star with radius $R = 6.96 \times 10^8$ m, in an elliptic orbit characterized by $a = 1$ AU, $e = 0.1$, $i = 89.95°$. We consider two cases: (i) $\omega = 30°$ and (ii) $\omega = 130°$. It can be calculated easily that in these cases the durations of the transits are $\tau = 0.488029$ day and $\tau = 0.563743$ day, respectively.

By observing transits caused by the above planet, we can measure their duration τ and period T, from which the semi-major axis a can be calculated. In the first case $\tau = 0.488029$ day, and for different values of i the corresponding ω–e curves are plotted in Fig. 4 for $e < 0.5$. (We note that it is possible to study larger values of e as well, but for our demonstrational purpose we restrict ourselves for $e < 0.5$.) We also mark the real (ω, e) solution by a filled circle on the curve corresponding to $i = 89.95°$, but as we can see, there is no way to restrict efficiently the infinite set of solutions. The only restriction is that the solutions can not be chosen from the region above the ω–e curve corresponding to $i = 90°$.

In the second case corresponding to $\tau = 0.563743$ day, the ω–e curves are plotted in Fig. 5. It can be seen that if the transit happens nearly at the apastron, the set of the solutions of Eq. 12 is more limited than in the first case. Only those (ω–e) pairs satisfy Eq. 12, which are in the region above the $i = 90°$ curve. The real solution is also marked by a filled circle on the curve $i = 89.95°$.

Equation 12 has an infinite set of solutions formed by pairs of (ω, e) values. If only the duration of the transit is known, it is not possible to choose which (ω, e) pair represents the real parameters of the transiting planet.

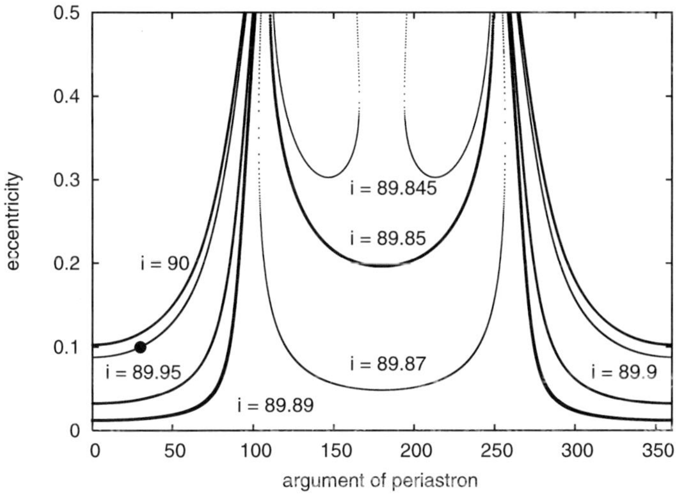

Fig. 4 Solutions of Eq. 12 for different inclinations when $\tau = 0.488029$ day. The original solution, which results in the above τ, is marked with a filled circle at $\omega = 30°$, $e = 0.1$, and $i = 89.95°$

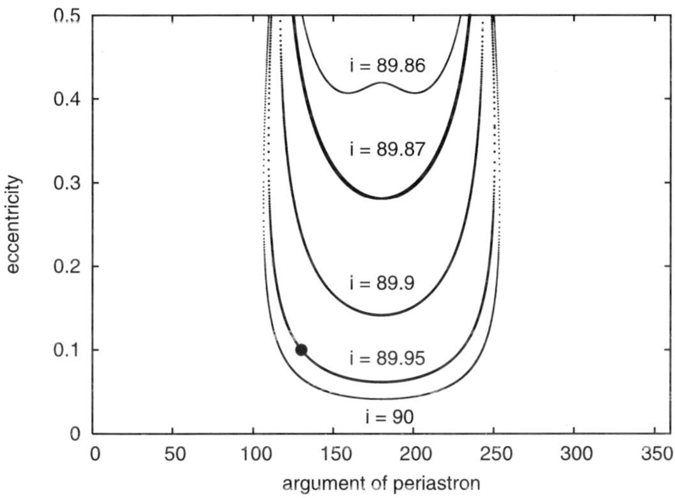

Fig. 5 Solutions of Eq. 12 for different inclinations when $\tau = 0.563743$ day. The original solution, which results in the above τ, is marked with a filled circle at $\omega = 130°$, $e = 0.1$, and $i = 89.95°$

3.2 Sensitivity of the solutions of Eq. 12 to the stellar mass error

As mentioned in the Introduction, the mass of the hosting star is known only with limited accuracy. This uncertainty in the stellar mass affects the semi-major axis of the transiting planet and causes an error in the eccentricity estimates. The semi-major axis of the transiting planet can be calculated from Kepler's third law

$$\frac{a^3}{T^2} = \frac{k^2}{4\pi^2}(m_* + m_p), \tag{14}$$

where T is the period of the transits, m_* is the mass of the hosting star, and m_p is the mass of the transiting planet, respectively (k is the Gaussian gravitational constant). In the case of Earth-like planets $m_p << m_*$, so m_p can be neglected. Denoting the errors in a and m_* by δa and δm_*, respectively, and expanding Eq. 14 around a and m_* up to first order we obtain:

$$3a^2\delta a = \frac{k^2T^2}{4\pi^2}\delta m_*. \tag{15}$$

Dividing Eq. 15 by a^3 and using again Eq. 14 (by neglecting m_p) we find for the relative errors

$$3\frac{\delta a}{a} = \frac{\delta m_*}{m_*}. \tag{16}$$

By using Eqs. 16 and 12 we calculated numerically the relative errors in the eccentricity of the transiting planet as a function of ω for the specific case when $i = 89.95°$ and $\tau = 0.488029$ day. Then we determined the maximum values $(\delta e/e)_{\max}$ of the relative error in eccentricity for different values of $\delta m_*/m_*$. We found that for $\delta m_*/m_* = 3\%$ $(\delta e/e)_{\max} = 12\%$, for $\delta m_*/m_* = 6\%$ $(\delta e/e)_{\max} = 25\%$, and for $\delta m_*/m_* = 10\%$ $(\delta e/e)_{\max} = 38\%$.

4 A possible confinement of the eccentricity of the transiting planet

4.1 The case of one additional giant planet

In this section we shall investigate the case when, beside the newly discovered planet, an already known giant planet also orbits around the hosting star. The presence of such a planet makes the problem non-integrable and both ordered and chaotic regions can be found in the phase space of the system. We suppose that the most probable orbital solutions of the transiting planet are those, which emanate from the ordered regions of the phase space. Such values of the orbital parameters of the transiting planet which would result in chaotic behaviour are unlikely, since in long terms the orbit of the planet could be unstable, thus these solutions should be avoided. We expect that the presence of a second (giant) planet represents a dynamical constraint, which reduces the infinit set of solutions of Eq. 12, and gives an upper limit for the maximum eccentricity of the transiting planet. We shall also demonstrate that by studying the solution-curves of Eq. 12 together with the stability structure of the ω–e plane, a lower bound for the inclination can also be determined. In what follows we shall investigate the stability in the (ω–e) plane within the framework of the planar general three-body problem.

In order to map the stability properties of the (ω–e) plane we used the Relative Lyapunov Indicator (RLI) (Sándor et al. 2000, 2004). The initial ω and e values are chosen from the intervals $e \in [0, 0.5]$ and $\omega \in [0°, 360°]$ with $\Delta e = 0.025$ and $\Delta\omega = 2°$. The initial value of the semi-major axis of the transiting planet is always $a = 1$ AU, while its true anomaly is calculated according to Eq. 10 as $v = 360° - \omega$ (see also Fig. 3). The masses of the planets are 1 Jupiter-mass and 1 Earth-mass, respectively, and the mass of the hosting star is 1 Solar-mass.

For each pair of the initial (ω, e) values we assign the RLI of the corresponding orbit calculated for 500 periods of the transiting planet. If the RLI is small

($\sim 10^{-12} - 10^{-13}$), the corresponding orbit is ordered and stable for very long time. If the RLI $\sim 10^{-11} - 10^{-9}$ the orbit is weakly chaotic. In practical sense such an orbit could be (Nekhoroshev) stable for very long terms as well, however, it can not be stable for arbitrary long time. Thus the regions characterized by these RLI values can already be the birth places of unstable orbits. Orbits having larger RLI $\sim 10^{-8} - 10^{-5}$, are strongly chaotic, and they will be unstable after certain time. In our stability maps the ordered regions are denoted by light, the weakly chaotic regions by grey, and the strongly chaotic regions by dark shades.

In what follows we consider the cases where the parameters of the known giant planet having 1 Jupiter mass are the following: $a_1 = 2.0$ AU, $e_1 = 0.2$, and 0.3. We fix the angular elements of the giant planet to $\lambda = \omega = 0°$. In Figs. 6 and 7 we show the dynamical structure of the ω–e parameter plane for the two different values of e_1. In these figures we also plot the solution curves of Eq. 12 by using $\tau = 0.488029$ day. From Fig. 6 it can be seen that there are two upper bounds for the eccentricity of the transiting planet depending on whether the transit occurs near the periastron,

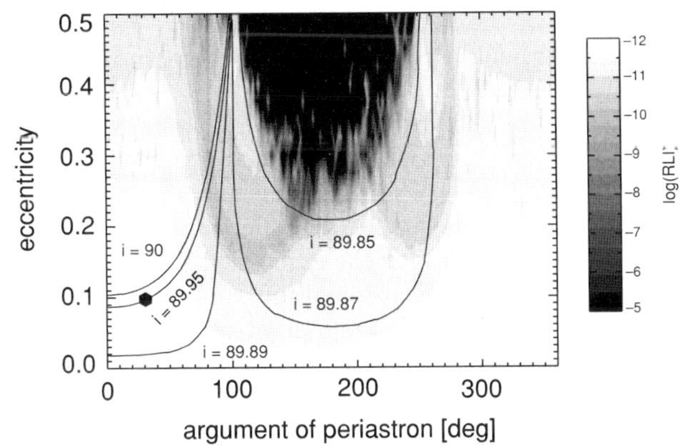

Fig. 6 The stability map of the ω–e parameter plane, when $a_1 = 2.0$ AU and $e_1 = 0.2$. The ω–e curves for different i are also plotted when $\tau = 0.488029$ day

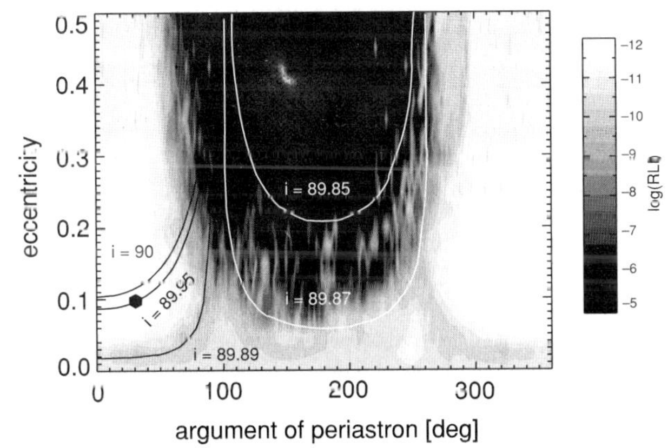

Fig. 7 The stability map of the ω–e parameter plane, when $a_1 = 2.0$ AU and $e_1 = 0.3$. The ω–e curves for different i are also plotted when $\tau = 0.488029$ day

or near the apastron. If the transit is near the periastron $\omega < 80°$, the upper limit of the eccentricity is $e < 0.27$, since the ω–e curves cross the chaotic region around this value. If the transit happens around the apastron $\omega \in [150°, 220°]$, the upper limit of the transiting planet's eccentricity is $e < 0.22$. In this case a lower limit can also be given for the inclination, $i > 89.85°$. The real solution is marked (as a filled circle) on the curve corresponding to $i = 89.95°$.

If the eccentricity of the giant planet is $e_1 = 0.3$, see Fig. 7, the maximum upper limit of the transiting planet's eccentricity is $e < 0.18$. However, in this case there exists a lower limit $e > 0.05$ as well. If the transit took place around the periastron the corresponding ω and e values would result in weakly chaotic orbits. A lower bound of the inclination in this case is $i > 89.89°$. Among the two values of the giant planet's eccentricity, this latter represents a more effective dynamical constraint for the orbital parameters of the transiting planet, which are $a = 1.0$ AU, $e = 0.1$, $\omega = 30°$, and $i = 89.95°$.

Next we shall study the cases when the transit takes place around the apastron $\omega = 130°$, resulting in a transit's duration $\tau = 0.563743$ day. We used the same orbital parameters of the giant planet as in the previous case, $a_1 = 2.0$ AU, $e_1 = 0.2$ and 0.3 respectively. Figure 8 shows the situation when $e_1 = 0.2$. The maximum upper limit of the eccentricity of the transiting planet is $e < 0.24$ corresponding to $\omega \approx 220°$. The lowest bound of its inclination is $i > 89.89°$.

Finally, when $e_1 = 0.3$, as shown in Fig. 9, there are no stable solutions for the orbital parameters of the transiting planet.

We have also investigated cases, when the semi-major axis of the giant planet was smaller or larger than 2 AU. If a_1 is smaller, a smaller e_1 is enough to result in an effective dynamical constraint. If a_1 is larger, the eccentricity of the giant planet should be larger as well for an efficient dynamical constraint.

In the above cases we assumed that the giant planet and the transiting planet revolved in the same plane. This, based on Solar System examples, seems to be a reasonable assumption, however, the orbital planes of the planets may differ slightly from each other. Thus we performed numerical simulations by using a small mutual inclination $I = 5°$ between the orbital planes as well. Comparing the correspond-

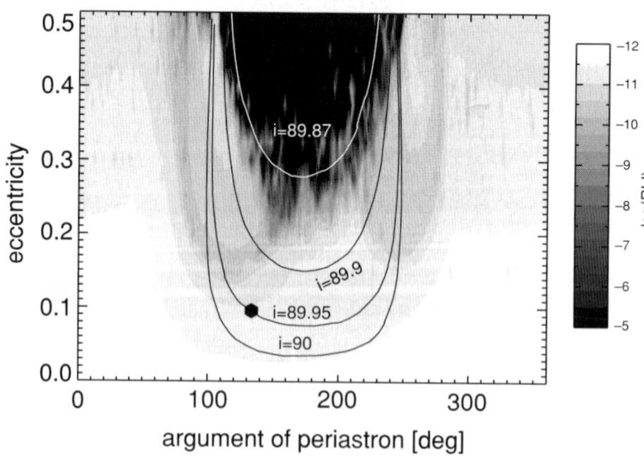

Fig. 8 The stability map of the ω–e parameter plane, when $a_1 = 2.0$ AU and $e_1 = 0.2$. The ω–e curves for different i are also plotted when $\tau = 0.563743$ day

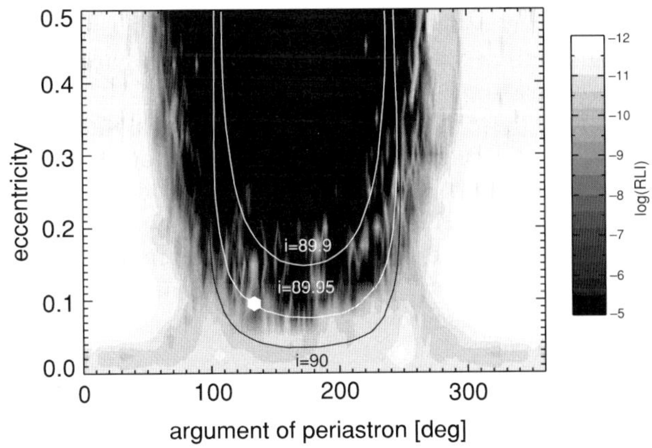

Fig. 9 The stability map of the ω–e parameter plane, when $a_1 = 2.0$ AU and $e_1 = 0.3$. The ω–e curves for different i are also plotted when $\tau = 0.563743$ day

ing ω–e parameter planes we have not found any significant differences between the planar and the spatial cases with small mutual inclination.

4.2 The case of two additional giant planets

Among the extrasolar giant planets with long observational baselines, there is a high rate of multiple planet systems. We expect that this rate will increase further with the accumulation of the observational data. Therefore we studied a case, when beside a transiting Earth-like planet there are two additional giant planets.

The inner giant planet's initial orbital parameters are the same as in Sect. 4.1. The outer giant planet, having 1 M_{Jup}, moves in an orbit characterized by $a_2 = 4$ AU, $e_2 = 0.05$ in the same plane as the transiting and the inner giant planet. Figure 10 shows the stability properties of the $\omega - e$ parameter plane of the transiting planet.

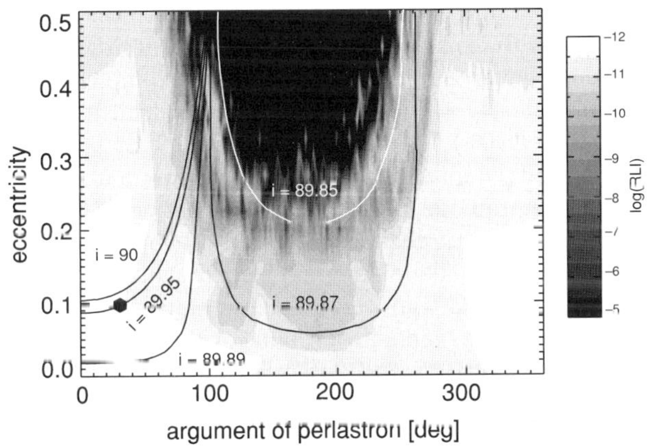

Fig. 10 The stability map of the ω–e parameter plane, when two giant planets are in the system. The semi-major axes and the eccentricities of the giants are $a_1 = 2.0$ AU, $e_1 = 0.2$, $a_2 = 4.0$ AU, and $e_2 = 0.05$. The ω–e curves for different i are also plotted when $\tau = 0.488029$ day

In order to study the effect of the outer giant planet on the ω–e parameter plane, we should compare Fig. 10 either to Fig. 8 or to Fig. 6. In these figures the initial conditions for the Earth-like and the inner giant planet are the same.

The main structure of these figures is very similar to each other, however in the case of the additional outer giant planet the strongly chaotic region on the ω–e parameter plane is more enhanced. This means that the dynamical constraint on the transiting planet could be more efficient in the case of two giant planets. Of course the strongness of the dynamical constraint depends on many parameters of the outer giant planet. To explore the complete parameter space is beyond the scope of this paper, it will be the subject of a forthcoming research.

5 Conclusions

The detection of Earth-like extrasolar planets by using ground based spectroscopic methods is beyond the present capabilities of observational astronomy. In the near future space instruments will be launched such as COROT and Kepler Mission which are devoted to observe such planets by using transit photometry.

In this paper we addressed the question whether it is possible to determine the orbital elements of Earth-like planets discovered by transit photometry if, apart from the period, the duration of the transit can be measured too. We supposed that the mass and the radius of the hosting star is known. We derived an equation, which connects the stellar and planetary masses, the duration of the transit, the semi-major axis, the eccentricity, the argument of periastron and the inclination of the transiting planet. By fixing the inclination, this equation contains two unknown variables, the argument of periastron ω and eccentricity e of the transiting planet. Thus the solutions for different inclinations can be represented as curves on the ω–e parameter plane.

In the last section of the paper we assumed that beside the transiting Earth-like planet additional giant planets orbit around the star as well. This assumption is quite reasonable if we accept the formation theories of planetary systems supporting the simultaneous presence of both rocky, Earth-like and gaseous, Jupiter-like planets. Since the detection of giant planets is possible by radial velocity measurements, we assumed their orbital parameters to be known. (We note that from radial velocity measurements only $m_p \sin i_p$ is known, where m_p and i_p are the mass and inclination of the giant planet. However, based on Solar System examples, we can assume that the giant planets orbit nearly in the same plane as the transiting Earth-like planet, so its inclination is $i_p \approx 90°$.) By using the framework of the general three-body problem, we investigated the influence of the known giant planet(s) on the transiting Earth-like planet's ω–e parameter plane. We found that on the ω–e parameter plane beside ordered domains chaotic regions appeared as well, where in long terms the motion of the transiting Earth-like planet may become unstable. Assuming that chaotic behaviour for an observed transiting planet is unlikely, we could determine an upper limit for the eccentricity, and a lower limit for the inclination.

In a future work we plan to extend our studies to investigate systematically the stability structure of the (ω–e) parameter plane for various values of the giant planet's semi-major axis, eccentricity, and inclination. Since the mass of the hosting star and several orbital parameters of the giant planets are known only with a limited accuracy from the radial velocity observations (see for instance Ford et al. 2005), we also plan to follow the propagation of these errors through the method presented in this paper.

In our future investigations we intend to consider cases of more massive transiting planets as well.

Acknowledgements This work has been supported by the Hungarian Scientific Research Fund (OTKA) under the grants D048424 and T043739. The author thanks the valuable help of Dr. C. Beaugé and Prof. B. Érdi in improving the manuscript. The useful comments of the unknown referees are also acknowledged.

References

Ford, E.B.: Quantifying the uncertainty in the orbits of extrasolar planets. Astron. J. **129**, 1706–1717 (2005)

Kasting, J.F., Whitmire, D.P., Reynolds, R.T.: Habitable zones around main sequence stars. Icarus **101**, 108–128 (1993)

Mayor, M., Queloz, D.: A Jupiter-mass companion to a solar-type star. Nature **378**, 355–359 (1995)

Murray, C.D., Dermott, S.: Solar System Dynamics. CUP, Cambridge (1999)

Rivera, E.J., Lissauer, J.J., Butler, R.P., Marcy, G.W., Vogt, S.S., Fischer, D.A., Brown, T.M., Laughlin, G., Henry, G.W.: A $\sim 7.5 M_{\oplus}$ planet orbiting the nearby star, GJ 876. Astrophys. J. **634**, 625–640 (2005)

Sándor, Zs., Érdi, B., Efthymiopoulos, C.: The phase space structure around L4 in the restricted three-body problem. Cel. Mech. Dynam. Astron. **78**, 113–123 (2000)

Sándor, Zs., Érdi, B., Széll, A., Funk, B.: The relative Lyapunov indicator: An efficient method of chaos detection. Cel. Mech. Dynam. Astron. **90**, 127–138 (2004)

Celestial Mechanics and Dynamical Astronomy (2006) 95:287–297
DOI 10.1007/s10569-005-5911-x

ORIGINAL ARTICLE

Critical inclination in the main problem of a massive satellite

S. Breiter · **A. Elipe**

Received: 20 October 2005 / Revised: 11 December 2005 /
Accepted: 14 December 2005 / Published online: 17 August 2006

Abstract The classical problem of the critical inclination in artificial satellite theory has been extended to the case when a satellite may have an arbitrary, significant mass and the rotation momentum vector is tilted with respect to the symmetry axis of the planet. If the planet's potential is restricted to the second zonal harmonic, according to the assumptions of the main problem of the satellite theory, two various phenomena can be observed: a critical inclination that asymptotically tends to the well known negligible mass limit, and a critical tilt that can be attributed to the effect of transforming the gravity field harmonics to a different reference frame. Stability of this particular solution of the two rigid bodies problem is studied analytically using a simple pendulum approximation.

Keywords Analytical methods · Critical inclination · Rigid body rotation

1 Introduction

The phenomenon of the critical inclination in artificial satellite theory has been known since 1950's. An in-depth overview and an extensive list of related papers can be found in Coffey et al. (1986, 1994). According to the first order theory, the secular perturbations in the argument of pericentre g vanish if the orbit's inclination I is the root of

$$1 - 5\cos^2 I = 0. \tag{1}$$

In the second order approximation, the phenomenon clearly becomes a resonance; with $J_2 > 0$ and all other harmonics neglected two kinds of frozen orbits appear: stable ones with $g \in \{90°, 270°\}$ and unstable with $g \in \{0°, 180°\}$.

S. Breiter (✉)
Astronomical Observatory of A. Mickiewicz University,
Słoneczna 36, PL 60 286 Poznań, Poland
e-mail: Breiter@amu.edu.pl

A. Elipe
Grupo de Mecánica Espacial, Universidad de Zaragoza,
50009 Zaragoza, Spain

The critical inclination was discovered and studied under the assumptions that the satellite's mass M_0 is negligible with respect to the planet's mass M_p. After a minor modification of the gravity parameter, the theory remains valid even for a significant mass of the satellite; this requires, however, that the torques, affecting the spin axes of both bodies, are neglected. It is not unreasonable to expect that the critical inclination in artificial satellite theory is a limiting case of some more general phenomenon occurring in the problem of two rigid bodies. Thus, in the present paper we consider the extended main problem of the artificial satellite where two restrictions have been suppressed: a spherical satellite may have an arbitrary mass (even much larger than the oblate "planet") and the "planet" may rotate around the axis that is tilted with respect to its symmetry axis. In order to avoid any suggestions concerning the mass ratio of the two bodies, we will use the term "central body" or "primary" instead of the "planet". Throughout the paper we use the formalism and, to large extent, the notation adopted from the fundamental paper of Kinoshita (1972).

Using the system of a nonspherical body and a homogenous sphere as a first approximation to the full problem of two rigid bodies is quite common in the binary asteroids modeling. On can find it in papers by Scheeres (2001, 2004a), Breiter et al. (2005), and—partially—in (Scheeres 2002, 2004b; Koon et al. 2004).

2 Reference frames and variables

Let us introduce two basic reference frames, both having their origins at the centre of mass of an axially symmetric central body O: the *body frame* $Oxyz$, with the basis unit vectors $\hat{\mathbf{x}}$, $\hat{\mathbf{y}}$, $\hat{\mathbf{z}}$, and the *fixed frame* $OXYZ$ with the $\hat{\mathbf{X}}$, $\hat{\mathbf{Y}}$, $\hat{\mathbf{Z}}$ unit vectors. Let $\hat{\mathbf{z}}$ be directed along the shortest axis of the central body, $\hat{\mathbf{x}}$ towards an arbitrary point on the equator, and $\hat{\mathbf{y}} = \hat{\mathbf{z}} \times \hat{\mathbf{x}}$. The total momentum of the system $\boldsymbol{\Gamma}$ (the sum of the central body angular momentum $\mathbf{G}_1$ and of the orbital angular momentum of the sphere $\mathbf{G}$) is constant, i.e.

$$\boldsymbol{\Gamma} = \mathbf{G}_1 + \mathbf{G} = \mathrm{I}_\mathrm{p}\, \boldsymbol{\omega} + \mathbf{R} \times \mathbf{P} = \mathbf{const}, \tag{2}$$

where $\boldsymbol{\omega}$ is the angular velocity vector of the rotating central body, $\mathbf{R}$ and $\mathbf{P}$ are the relative position of the sphere and its canonically conjugate momentum respectively, and I_p is the matrix of inertia

$$I_\mathrm{p} = \begin{pmatrix} A & 0 & 0 \\ 0 & A & 0 \\ 0 & 0 & C \end{pmatrix}. \tag{3}$$

Following Kinoshita (1972) we choose the $\hat{\mathbf{Z}}$ vector of the fixed frame directed along $\boldsymbol{\Gamma}$, the $\hat{\mathbf{X}}$ vector—orthogonal to $\hat{\mathbf{Z}}$—directed to some arbitrary fixed point, and $\hat{\mathbf{Y}} = \hat{\mathbf{Z}} \times \hat{\mathbf{X}}$.

Two sets of canonical variables are required in our problem: one for the rotation of the central body and one for the orbital motion of the sphere with respect to the central body. The rotation will be described by means of the Serret–Andoyer variables (Deprit and Elipe 1993) with the momenta

G_1—the magnitude of the rotation angular momentum,

L_1—the projection of $\mathbf{G}_1$ on the polar axis Oz

$$L_1 = \mathbf{G}_1 \cdot \hat{\mathbf{z}} = G_1 \cos J_1, \tag{4}$$

H_1—the projection of $\mathbf{G}_1$ on the fixed axis OZ

$$H_1 = \mathbf{G}_1 \cdot \hat{\mathbf{Z}} = G_1 \cos I_1, \tag{5}$$

and their conjugate angles g_1, ℓ_1, h_1. The geometrical meaning of the angles can be easily deduced from the rotations sequence required to transform any position vector, say $\mathbf{R}$, with the components expressed in the fixed frame, to the body frame; the resulting vector $\mathbf{r}$ will have the components

$$\mathbf{r} = R_3(\ell_1)\, R_1(J_1)\, R_3(g_1) R_1(I_1)\, R_3(h_1)\, \mathbf{R}, \tag{6}$$

where R_1 and R_3 are the usual matrices of rotation

$$R_1(\alpha) = \begin{pmatrix} 1 & 0 & 0 \\ 0 & \cos\alpha & \sin\alpha \\ 0 & -\sin\alpha & \cos\alpha \end{pmatrix}, \quad R_3(\alpha) = \begin{pmatrix} \cos\alpha & \sin\alpha & 0 \\ -\sin\alpha & \cos\alpha & 0 \\ 0 & 0 & 1 \end{pmatrix}. \tag{7}$$

An appropriate figure can be found in Kinoshita (1972, Figs. 1 and 5).

Orbital motion can be described in terms of the Delaunay variables, consisting of the mean anomaly ℓ, argument of pericentre g, longitude of the ascending node h (all measured in the fixed frame), and of their conjugate momenta

$$L = m\,\sqrt{\mu\, a} = m\, n\, a^2, \tag{8}$$

$$G = ||\mathbf{G}|| = L\,\eta = m\, n\, a^2 \sqrt{1 - e^2}, \tag{9}$$

$$H = \mathbf{G} \cdot \hat{\mathbf{Z}} = G\, \cos I, \tag{10}$$

where

$$m = \frac{M_\mathrm{p}\, M_0}{M_\mathrm{p} + M_0}, \quad \mu = k^2\,(M_\mathrm{p} + M_0), \tag{11}$$

M_0 being the mass of the orbiting sphere, k—the Gaussian gravity constant, a—orbital major semi-axis, e—orbital eccentricity, and n is the mean motion. Using the true anomaly f, we can express the position vector $\mathbf{R}$ of the sphere as

$$\mathbf{R} = R_3(-h)\, \mathrm{R}_1(-I)\, R_3(-f - g)\; (r, 0, 0)^\mathrm{T}\,. \tag{12}$$

Equations (6) and (12) allow to express the position of the sphere with respect to the body frame $\mathbf{r}$ as a function of the Delaunay and Serret–Andoyer variables.

According to the well known properties of the system (Kinoshita 1972), the choice of $\hat{\mathbf{Z}} = \hat{\mathbf{\Gamma}}$ implies important consequences:

1. Orbital plane and the plane normal to $\mathbf{G}_1$ intersect along the common line of the nodes on the *invariable plane* OXY, and

$$h_1 = h - \pi\,. \tag{13}$$

2. Momenta G_1, H_1, G, H are not independent, because of

$$G^2 - H^2 = G_1^2 - H_1^2, \quad H + H_1 = \alpha = \text{const.} \tag{14}$$

or, equivalently,

$$G \sin I - G_1 \sin I_1 = 0, \quad G \cos I + G_1 \cos I_1 = \alpha = \text{const.} \tag{15}$$

This means that given the value of α, the motion of L, G, L_1, G_1 and of their conjugate angles can be studied separately from H, h, H_1, and h_1; the number of degrees of freedom is effectively reduced from 6 to 4.

3 Hamiltonian and equations of motion

The Hamiltonian function of the system is

$$\mathcal{H} = \frac{G_1^2 - L_1^2}{2\,A} + \frac{L_1^2}{2\,C} - \frac{m^3\,\mu^2}{2\,L^2} + J_2\,\frac{m\,\mu\,a_{\rm p}^2}{r^3}\,P_2(\sigma), \tag{16}$$

where $a_{\rm p}$ is the equatorial radius of the primary, P_j is the Legendre polynomial of degree j, and $\sigma = \hat{\mathbf{r}} \cdot \hat{\mathbf{z}}$ is the sine of latitude of the sphere in the body frame.

The moments of inertia are related to the second zonal harmonic coefficient

$$C - A = J_2 M_{\rm p} a_{\rm p}^2. \tag{17}$$

Substituting this relation and expanding the Hamiltonian in power series of J_2, we may partition $\mathcal{H}$ into

$$\mathcal{H} = \mathcal{H}_0 + J_2\,\mathcal{H}_1 + \tfrac{1}{2}J_2^2\,\mathcal{H}_2 + O\left(J_2^3\right), \tag{18}$$

$$\mathcal{H}_0 = \frac{G_1^2}{2\,C} - \frac{m^3\,\mu^2}{2\,L^2}, \tag{19}$$

$$\mathcal{H}_1 = \frac{M\,a_{\rm p}^2\,(G_1^2 - L_1^2)}{2\,C^2} + \frac{m\,\mu a_{\rm p}^2}{r^3}\,P_2(\sigma), \tag{20}$$

$$\mathcal{H}_2 = \frac{M^2\,a_{\rm p}^4\,(G_1^2 - L_1^2)}{C^3}. \tag{21}$$

The unperturbed Hamiltonian $\mathcal{H}_0$ describes the system physically equivalent to the problem of two spheres: all momenta are integrals of motion and all angles are constant, save for

$$\dot{\ell} = \frac{\partial \mathcal{H}_0}{\partial L} = \frac{m^3\,\mu^2}{L^3} \equiv n, \tag{22}$$

$$\dot{g}_1 = \frac{\partial \mathcal{H}_0}{\partial G_1} = \frac{G_1}{C} \equiv n_1. \tag{23}$$

These two frequencies are present in the definition of the Lie derivative (Deprit 1969)

$$L_0 F = (F;\mathcal{H}_0) = n\,\frac{\partial F}{\partial l} + n_1\,\frac{\partial F}{\partial g_1}, \tag{24}$$

where F is any function and $(;)$ stands for the canonical Poisson bracket.

4 Rigid body potential in the inertial frame

Before we proceed to the perturbation treatment, let us look at the spheroid from another perspective. The usual treatment of the orbital motion amounts to transforming the position of the sphere from the inertial frame XYZ to the central body frame xyz. As the alternative, we propose to do the opposite, transforming the central body potential to the inertial frame XYZ. However, making use of the property (13), we transform the potential to the *nodal reference frame* X^*Y^*z with the axis OX^* directed towards the ascending node of the sphere's orbit.

Using theorems about the transformation of Legendre functions under rotation we obtain

$$P_2(\sigma) = C^*_{2,0} P_2(\sigma^*) + \sum_{j=1}^{2} P_{2,j}(\sigma^*) \left(C^*_{2,j} \cos j\lambda^* + S^*_{2,j} \sin j\lambda^* \right), \tag{25}$$

where $\sigma^* = \hat{\mathbf{R}} \cdot \hat{\mathbf{Z}}$ is the sine of latitude with respect to the invariant plane and λ^* is the longitude measured from the orbital ascending node. Introducing symbols $s_{I1} = \sin I_1$, $c_{I1} = \cos I_1$ and $c_1 = \cos J_1$, we can express the transformed potential coefficients as

$$C^*_{2,0} = P_2(c_{I1})P_2(c_1) - \frac{1}{3} P_{2,1}(c_{I1})P_{2,1}(c_1) \cos g_1 + \frac{1}{12} P_{2,2}(c_{I1})P_{2,2}(c_1) \cos 2g_1, \tag{26}$$

$$C^*_{2,1} = -\frac{c_{I1}}{3} P_{2,1}(c_1) \sin g_1 + \frac{s_{I1}}{6} P_{2,2}(c_1) \sin 2g_1, \tag{27}$$

$$C^*_{2,2} = -\frac{1}{24} P_{2,2}(c_{I1})P_2(c_1) - \frac{1}{18} P_{2,1}(c_{I1})P_{2,1}(c_1) \cos g_1 - \frac{1 + c^2_{I1}}{24} P_{2,2}(c_1) \cos 2g_1, \tag{28}$$

$$S^*_{2,1} = \frac{1}{3} P_{2,1}(c_{I1})P_2(c_1) - \frac{1 - 2c^2_{I1}}{3} P_{2,1}(c_1) \cos g_1 - \frac{1}{18} P_{2,1}(c_{I1})P_{2,2}(c_1) \cos 2g_1, \tag{29}$$

$$S^*_{2,2} = -\frac{s_{I1}}{6} P_{2,1}(c_1) \sin g_1 - \frac{c_{I1}}{12} P_{2,2}(c_1) \sin 2g_1. \tag{30}$$

Thus the second zonal harmonic of the potential in the X^*Y^*Z frame becomes

$$V^*_2 = \frac{m\,\mu\,a_p^2}{r^3} \sum_{j=0}^{2} P_{2,j}(\sigma^*) \left(J_2 C^*_{2,j} \cos j\lambda^* + J_2 S^*_{2,j} \sin j\lambda^* \right). \tag{31}$$

Assuming the unperturbed rotation with constant I_1, J_1 and n_1, we can see the rigid spheroid in the new reference frame as an object continuously changing its shape due to the time-dependence of $C^*_{2,j}$ and $S^*_{2,j}$ coefficients.

Looking forward to the application of averaging technique, let us ask about an average potential of the transformed body. Rejecting all terms that are periodic functions of g_1 we obtain

$$\bar{C}^*_{2,0} = P_2(c_{I1})P_2(c_1) = \tfrac{1}{4} \left(2 - 3s^2_{I1} \right) \left(2 - 3s^2_1 \right), \tag{32}$$

$$\bar{C}^*_{2,2} = -\frac{1}{24} P_{2,2}(c_{I1})P_2(c_1) = -\frac{s^2_{I1}}{8} \left(2 - 3s^2_1 \right), \tag{33}$$

$$\bar{S}^*_{2,1} = \frac{1}{3} P_{2,1}(c_{I1})P_2(c_1) = \frac{s_{I1}c_{I1}}{2} \left(2 - 3s^2_1 \right), \tag{34}$$

as the only nonvanishing terms. As it should be expected, if the rotation state is the shortest axis mode and the angular momentum $\boldsymbol{G}_1$ is normal to the invariant plane ($s_1 = \sin J_1 = 0$ and $s_{I1} = 0$), we recover the original potential with $C^*_{2,0} = \bar{C}^*_{2,0} = 1$

and all remaining coefficients equal to zero. But even more noteworthy is the fact that all the averaged coefficients have a common factor $P_2(c_1)$. Thus, if

$$\sin J_1 = \sqrt{\frac{3}{2}}, \tag{35}$$

the averaged potential vanishes. In further discussion we will call *the critical tilt* each of the two values $J_1 \approx 54^\circ\!.74$ and $J_1 \approx 125^\circ\!.26$, for which the potential of a rotating spheroid in the nodal frame differs from the point mass potential only by purely periodic terms in the quadrupole approximation.

5 First order normalization

Using the classical Lie transformation method to normalize the Hamiltonian function of our problem (Deprit, 1969), we will obtain the first order normalized Hamiltonian $\mathcal{K}_1$ as the sum of these terms of $\mathcal{H}_1$ that belong to the kernel of L_0 and so do not depend on the mean variables ℓ and g_1, i.e.

$$L_0 \mathcal{K}_1 = 0, \tag{36}$$

and $\mathcal{H}_1 - \mathcal{K}_1$ is purely periodic. This approach, however, requires two important assumptions:

1. We exclude all resonance that might occur due to the commensurability of n and n_1.
2. We assume the fast rotation with sufficiently large ratio G_1/G, because otherwise we obtain yet another kind of resonance leading to abnormally large amplitudes of g_1-dependent terms in the generator $\mathcal{W}_1$ defined through

$$L_0 \mathcal{W}_1 = \mathcal{H}_1 - \mathcal{K}_1. \tag{37}$$

If none of these restrictions is violated, we obtain the new Hamiltonian $\mathcal{K}_1$ as

$$\begin{aligned}\mathcal{K}_1 &= \frac{1}{2} M_\mathrm{p} a_\mathrm{p}^2 n_1^2 s_1^2 + \frac{m\,\mu\, a_\mathrm{p}^2}{a^3} \left[\bar{C}^*_{2,0} \left\langle \frac{a^3\, P_2(\sigma^*)}{r^3} \right\rangle \right. \\ &\left. + \bar{S}^*_{2,1} \left\langle \frac{a^3\, P_{2,1}(\sigma^*)}{r^3} \sin\lambda^* \right\rangle + \bar{C}^*_{2,2} \left\langle \frac{a^3\, P_{2,2}(\sigma^*)}{r^3} \cos 2\lambda^* \right\rangle \right].\end{aligned} \tag{38}$$

The average values in Eq. (38) are obtained by elementary quadratures, taking into account

$$\langle F(f) \rangle = \frac{1}{2\pi} \int_0^{2\pi} \frac{F\, r^2}{a^2 \eta}\, df, \tag{39}$$

and recalling that in the nodal frame

$$\begin{pmatrix} \sqrt{1-(\sigma^*)^2}\, \cos\lambda^* \\ \sqrt{1-(\sigma^*)^2}\, \sin\lambda^* \\ \sigma^* \end{pmatrix} = R_1(-I)\, R_3(-f-g) \begin{pmatrix} 1 \\ 0 \\ 0 \end{pmatrix}. \tag{40}$$

So we obtain

$$\left\langle \frac{a^3\, P_2(\sigma^*)}{r^3} \right\rangle = -\frac{2-3\,s^2}{4\,\eta^3}, \tag{41}$$

well known from the classical satellite theory, and

$$\left\langle \frac{a^3\,P_{2,1}(\sigma^*)}{r^3}\,\sin\lambda^* \right\rangle = \frac{3\,s\,c}{2\,\eta^3}, \tag{42}$$

$$\left\langle \frac{a^3\,P_{2,2}(\sigma^*)}{r^3}\,\cos 2\lambda^* \right\rangle = \frac{3\,s^2}{2\,\eta^3}, \tag{43}$$

where $s = \sin I$ and $c = \cos I$.

Collecting all results, we have the new Hamiltonian

$$\mathcal{K} = \mathcal{K}_0 + J_2\mathcal{K}_1 + O\left(J_2^2\right), \tag{44}$$

$$\mathcal{K}_0 = \frac{G_1^2}{2\,C} - \frac{m^3\mu^2}{2\,L^2}, \tag{45}$$

$$\mathcal{K}_1 = \frac{M_{\mathrm{p}}\,a_{\mathrm{p}}^2\,G_1^2\,s_1^2}{C^2} + \frac{a_{\mathrm{p}}^2 m\mu}{4a^3\eta^3}\left(-\bar{C}^*_{2,0}(2-3s^2) + 6s\left(\bar{S}^*_{2,1}\,c + \bar{C}^*_{2,2}\,s\right)\right). \tag{46}$$

Introducing the mutual inclination J as an angle between $\mathbf{G}$ and $\mathbf{G}_1$

$$J = I_1 + I, \tag{47}$$

we can simplify the final form of $\mathcal{K}_1$ that becomes

$$\mathcal{K}_1 = \frac{M\,a_{\mathrm{p}}^2\,G_1^2\,s_1^2}{C^2} - \frac{a_p^2 m\mu}{2a^3\eta^3}\,P_2(c_1)P_2(\cos J). \tag{48}$$

In the limiting case of the shortest axis rotation and negligible mass M_0, when $m \to M_0$, $J = I$, and $J_1 = 0$, the second term of (48) becomes the classical satellite theory Hamiltonian (Brouwer 1959). The Hamiltonian $\mathcal{K}_1$ depends only on Andoyer and Delaunay momenta that are prime integrals of the normalized system. It means that all Andoyer and Delaunay angles are linear functions of time.

6 Critical inclination and critical tilt

The frequency of the mean argument of pericentre in the inertial reference frame can be obtained from $\mathcal{K}_1$

$$\dot{g} = J_2\,\frac{\partial\mathcal{K}_1}{\partial G} = J_2\,\frac{3n\,a_{\mathrm{p}}^2}{8\,a^2\,(1-e^2)}\,P_2(c_1)\,\Psi(I,J), \tag{49}$$

where

$$\Psi(I,J) = 1 + 3\,\cos 2J + 2c\,s^{-1}\sin 2J. \tag{50}$$

There are two possibilities that lead to the situation when $\dot{g}=0$. The first one is $P_2(c_1) = 0$; this condition is satisfied at the critical tilt defined in Sect. 4. Its meaning is clearly understandable: secular perturbations in the argument of pericentre vanish because the potential differs from the Keplerian case only by short-periodic terms. Recalling that $\cos J_1 = L_1/G_1$ we observe, that $L_1 = \mathrm{const}$ not only as a mean variable, but it is also the prime integral of the original system due to the absence of ℓ_1 in the Hamiltonian $\mathcal{H}$, as a consequence of the axial symmetry of the central body.

The second condition for $\dot{g} = 0$—entirely independent on the first one—is obviously $\Psi = 0$. As we will see, it generalizes the classical critical inclination condition. In purely geometrical terms, the critical inclination I is a function of the mutual inclination J (Fig. 1a), or of the obliquity I_1 (Fig. 1b). The relation between the obliquity and mutual inclination is shown in Fig. 1c. The classical limit is recovered when the obliquity $I_1 = 0$, i.e. when $J = I$ (the dashed line in Fig. 1a). In this case $\Psi = 2\,(1-5\,c^2)$ and the critical inclination values are $I \approx 63\overset{\circ}{.}43$ and $I \approx 116\overset{\circ}{.}37$.

One should remember, however, that the geometry of I, I_1 and J is a function of momenta G and G_1. If we substitute Eq. (15) into the expression of Ψ, the critical inclination becomes a function of one physical parameter: the ratio G/G_1. This dependence is shown in Fig. 1d.

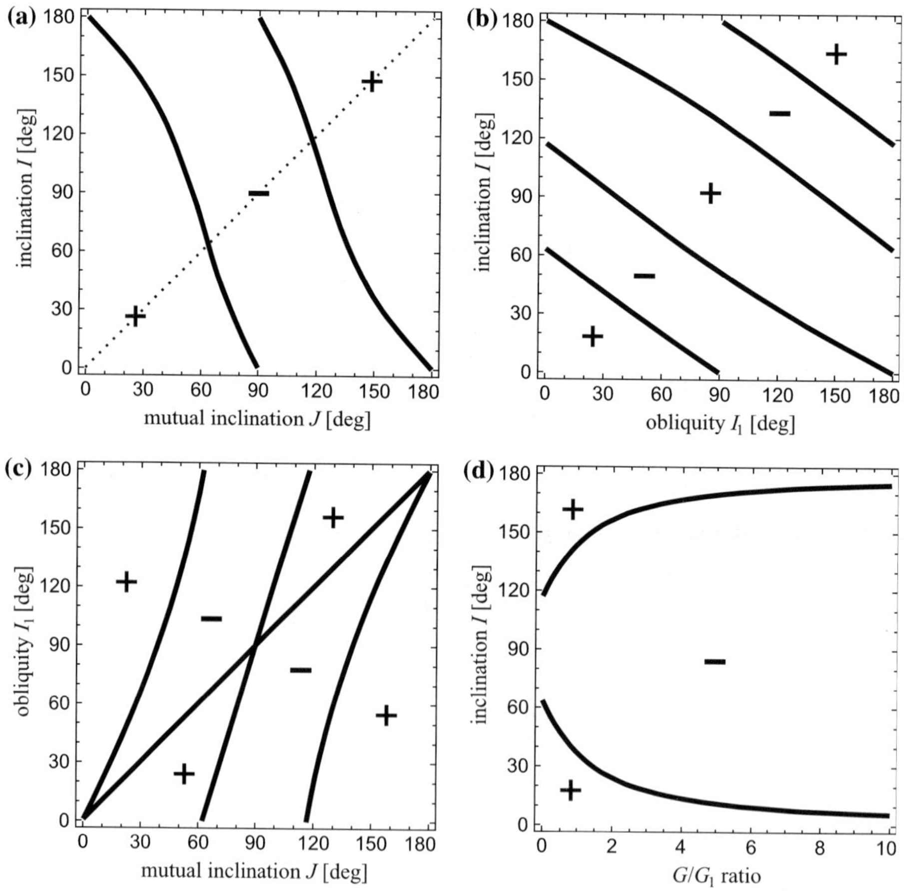

Fig. 1 Critical inclination in the first-order approximation. Plus and minus signs indicate the sign of secular perturbations in g when $P_2(c_1) > 0$

7 Stability of critical inclination

Within the first order approximation presented in Sect. 6, the critical inclination on a (g, G) phase plane is a line of critical points $G = \text{const}$. Similarly to the classical, negligible mass satellite problem, it is only at the second order of perturbation theory that we start observing isolated critical points with complementary stability indices, i.e. we pass from the critical inclination phenomenon to the notion of frozen orbits. But in our present generalization, reaching the second order becomes a real challenge: it is not feasible in a closed form, the perturbing potential is much more complicated, and the number of parameters is higher than in the small satellite case. For these reasons we present very limited results concerning the stability of the frozen orbits, obtained within a crude approximation. The second order normalization was performed according to the Lie–Deprit method (Deprit, 1969), with

$$\mathcal{K}_2 = \langle \mathcal{H}_2 + (\mathcal{H}_1 + \mathcal{K}_1; \mathcal{W}_1) \rangle , \tag{51}$$

but all short-periodic terms were a priori rejected before the evaluation of the Poisson bracket and ultimately we reduced the Hamiltonian

$$\mathcal{K} = \mathcal{K}_0 + J_2 \mathcal{K}_1 + \frac{J_2^2}{2} \mathcal{K}_2 \tag{52}$$

to a standard pendulum model

$$\mathcal{K}_{\mathrm{p}} = \frac{\mathcal{A}}{2} \hat{G}^2 + \mathcal{B} \cos 2g, \tag{53}$$

where

1. $\mathcal{A}$ is $J_2\, \partial^2 \mathcal{K}_1 / \partial G^2$ evaluated at the values of G/G_1 and I that satisfy $\Psi = 0$,
2. $\mathcal{B}$ is the coefficient of the $\cos 2g$ term in the $\frac{1}{2} J_2^2 \mathcal{K}_2$, evaluated at the values of G/G_1 and I that satisfy $\Psi = 0$, and truncated at the second power of the orbital eccentricity,
3. $\hat{G} = G - G_{\mathrm{cr}}$, where G_{cr} is the value of orbital momentum that satisfies $\Psi = 0$.

Even with so strong approximations, the pendulum Hamiltonian $\mathcal{K}_{\mathrm{p}}$ contains hundreds of terms and is too long to be quoted in this paper. However, one should be aware that some terms of $\mathcal{B}$ contain possibly resonant denominators $n \pm n_1$, $n \pm 2n_1$, and $2n \pm n_1$. Both n and n_1 are positive by definition, hence in the subsequent discussion the resonances $n \approx n_1$, $n \approx 2n_1$, and $2n \approx n_1$ have to excluded.

The approximate Hamiltonian $\mathcal{K}_{\mathrm{p}}$ leads to the equations of motion with two pairs of equilibria. E1 – $(g \in \{0, 180°\}, \hat{G} = 0)$, and E2 – $(g \in \{90°, 270°\}, \hat{G} = 0)$ with a complementary stability. The equilibrium E1 is stable (and E2—unstable) if $\mathcal{B}/\mathcal{A} < 0$. It can be easily verified that

$$\operatorname{sgn} \mathcal{A} = \operatorname{sgn} P_2(c_1), \tag{54}$$

hence the stability problem is reduced to the study of the sign of $\mathcal{B}$. On the other hand, according to Eq. (54) we will have to avoid the values of J_1 that are too close to the critical tilt $P_2(c_1) = 0$. The sign of $\mathcal{B}$ is determined by two parameters: the ratios G/G_1 and n/n_1.

The results are presented in Figs. 2 and 3. Thick dashed lines in these figures indicate the resonances between n and n_1 where the present analysis is not valid. Fig. 2 presents the case of an "average oblate" body when $P_2(c_1) > 0$. In Figure 2a one sees the case with no tilt ($J_1 = 0$). When $G/G_1 = 0$, we recover the classical satellite case of stable E2 orbits regardless of the n/n_1 ratio. But when the orbital momentum is not negligible, some these orbits may become unstable. However, when the tilt is significant (Fig. 2b) even the $G/G_1 = 0$ may lead to the occurrence of unstable E2 orbits; the situation is actually equivalent to the problem of critical inclination (or frozen orbits) around a significantly three-axial oblate body (i.e. one with J_2 comparable to the $J_{2,2}$ coefficient, but both considered small). Figure 3 presents two cases with $P_2(c_1) < 0$, ie. for an "average prolate" body. We have intentionally avoided the values of J_1 close to the critical tilt, where the pendulum approximation breaks down because of the vanishing $\mathcal{A}$.

As for the critical tilt phenomenon, it is not of the resonance type, because there are no isolated unstable critical points on the (ℓ_1, L_1) plane.

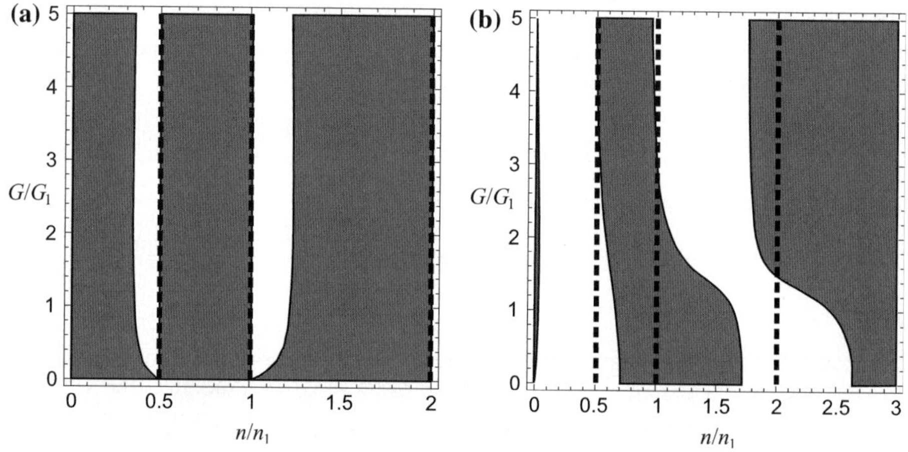

Fig. 2 Stability of frozen orbits when $P_2(c_1) > 0$, for two sample values of tilt: $J_1 = 0$ (**a**) and $J_1 = 45°$ (**b**). E2 equilibria are stable inside the grey zones

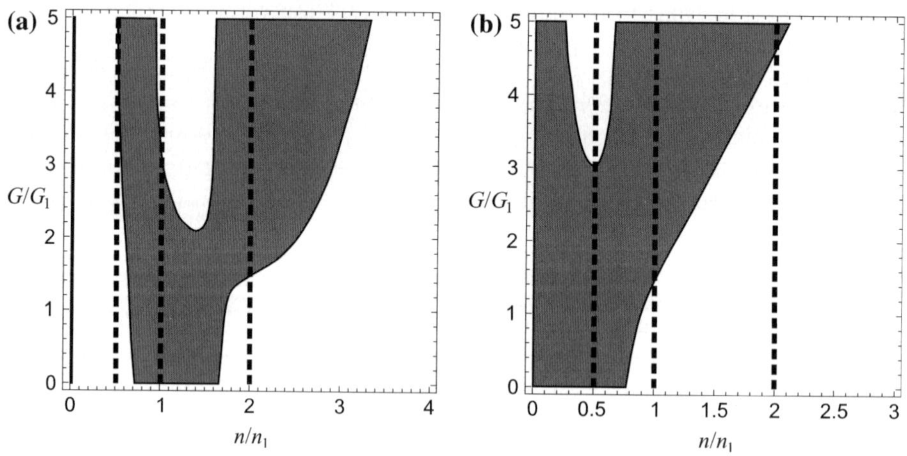

Fig. 3 Stability of frozen orbits when $P_2(c_1) < 0$, for two sample values of tilt: $J_1 = 60°$ (**a**) and $J_1 = 90°$ (**b**). E1 equilibria are stable inside the grey zones

8 Conclusions

The problem of frozen line of the apsides in the spheroid and sphere case becomes much more profound than its artificial satellite counterpart if we suppress at least one of the usual approximations: short axis rotation and negligible mass of the satellite. Without the short axis rotation assumption, the problem becomes actually equivalent to the motion around a three-axial body ("tesseral resonances" appear) and is accompanied by the phenomenon of critical tilt. On the other hand, increasing the satellite's mass we shift the critical inclination value towards 0 (prograde motion) or 180° (retrograde motion). Further studies concerning the stability of frozen orbits are still required, because the pendulum approximation and $O(e^2)$ truncation that were used in this paper should rather be treated as nothing more than a "quick look" model. Obviously, introducing more harmonics in the potential of the central body may change the stability of frozen orbits as it happens in the artificial satellite theory (Coffey et al. 1994). Nevertheless, the growing number of known binary asteroids with comparable masses of their components indicates that the problems discussed in this paper are more than "academic problems".

Acknowledgements The research financed by the Polish State Committee of Scientific Research (KBN) grant 1 P03D 020 27 and by the Spanish Ministry of Science and Technology Project ESP2005-07107.

References

Breiter, S., Melendo, B., Bartczak, P., Wytrzyszczak, I.: Synchronous motion in the Kinoshita problem. Application to satellites and binary asteroids. Astron. Astrophys. **437**, 753–764 (2005)

Brouwer, D.: Solution of the problem of artificial satellite theory without drag. Astron. J. **64**, 378–397 (1959)

Coffey, S.L., Deprit, A., Miller, B.R.: The critical inclination in artificial satellite theory. Celest. Mech. **39**, 365–406 (1986)

Coffey, S.L., Deprit, A., Deprit, E.: Frozen orbits for satellites close to an Earth-like planet. Celest. Mech. Dynam Astron. **59**, 37–72 (1994)

Deprit, A.: Canonical transformation depending on a small parameter. Celest. Mech. **1**, 12–30 (1969)

Deprit, A., Elipe, A.: Complete reduction of the Euler–Poinsot problem. J. Astron. Sci. **41**, 603–628 (1993)

Kinoshita, H.: First-order perturbations of the two finite body problem. Publ. Astron. Soc. Japan **24**, 423–457 (1972)

Koon, W.S., Marsden, J.E., Ross, S.D., Lo, M., Scheeres, D.J.: Geometric mechanics and the dynamics of asteroid pairs. Ann. N.Y. Acad. Sci. **1017**, 11–38 (2004)

Scheeres, D.J.: Changes in rotational angular momentum due to gravitational interactions between two finite bodies. Celest. Mech. Dynam. Astron. **81**, 39–44 (2001)

Scheeres, D.J.: Stability in the full two-body problem. Celest. Mech. Dynam. Astron. **83**, 155–169 (2002)

Scheeres, D.J.: Stability of relative equilibria in the full two-body problem. Ann. N.Y. Acad. Sci. **1017**, 81–94 (2004a)

Scheeres, D.J.: Bounds on rotations periods of disrupted binaries in the full 2-body problem. Celest. Mech. Dynam. Astron. **89**, 127–140 (2004b)

Celestial Mechanics and Dynamical Astronomy (2006) 95:299–326
DOI 10.1007/s10569-006-9027-8

ORIGINAL ARTICLE

Long-term effects of the Galactic tide on cometary dynamics

Marc Fouchard · Christiane Froeschlé · Giovanni Valsecchi · Hans Rickman

Received: 12 December 2005 / Revised: 11 March 2006 /
Accepted: 28 April 2006 / Published online: 15 August 2006

Abstract We introduce a model for integrating the effects of Galactic tides on Oort cloud comets, which involves two procedures, according to the values of the osculating semi-major axis a and eccentricity e. Ten simulations of the dynamics of 10^6 comets over 5 Gyr are performed using this model. We thus investigate the long-term effects of the Galactic tide with and without a radial component, the effects of the local density of the Galactic disk, and those of the Oort constants. Most of the results may be understood in terms of the integrability or non-integrability of the system. For an integrable system, which occurs for moderate semi-major axes with or without radial component, the dynamics is explained by periodic variation of the cometary perihelion, inducing the depletion of the outer region of the Oort cloud, a constant flux from the inner region after 500 Myr, and the quick formation of a reservoir of comets with argument of perihelion near 26.6°. When the system is non-integrable, the efficiency of the tide in reducing the cometary perihelion distance is enhanced both by replenishing the Oort cloud domain from which comets are sent toward the planetary system, and by reducing the minimal value that the perihelion distance may

M. Fouchard (✉)
INAF-IASF, via Fosso del Cavaliere 100, I-00133 Roma, Italy
IMCCE/SYRTE, Observatoire de Paris, 77 av. Denfert-Rochereau, 75014 Paris, France
e-mail: marc.fouchard@iasf-roma.inaf.it

Ch. Froeschlé
Observatoire de la Côte d'Azur, UMR 6202, Bv. de l'Observatoire, B.P. 4229, 06304 Nice cedex 4, France
e-mail: froesch@obs-nice.fr

G. Valsecchi
INAF-IASF, via Fosso del Cavaliere 100, I-00133 Roma, Italy
e-mail: giovanni.valsecchi@iasf-roma.inaf.it

H. Rickman
Uppsala Astronomical Observatory, Box 515, 75120 Uppsala, Sweden
e-mail: hans@astro.uu.se

reach. No effects of varying the Oort constants were observed, showing that the flat rotation curve is a satisfactory approximation in Oort cloud dynamics.

Keywords Galactic dynamics · Numerical model · Long-period comets · Oort cloud

1 Introduction

Since the famous work of Oort (1950), it has been generally accepted that long-period (LP) comets appear to originate in a Sun-centered spherical reservoir extending to $\sim 2 \times 10^5$ Astronomical Unit (AU). The mechanisms that influence the Oort cloud comets and produce the observed LP comets (i.e., place Oort cloud comets on orbits that lead into the inner planetary region) include the perturbations due to passing stars and the Galactic tide. Sporadically, the passage of the Solar System through a Giant molecular cloud may cause an upheaval of the entire cloud—however, due to the rarity of the latter type of event and the uncertainties around it (Bailey 1983), it is generally omitted from studies of Oort cloud dynamics.

Both stellar and Galactic perturbations act continuously, but they behave differently. The Galactic tide is a quasi-integrable perturbation that causes long-term variations of cometary orbital elements, including the perihelion distance, while stellar perturbations can be considered as stochastic events. These tend to randomize the orbital distribution of the Oort cloud comets, thus delivering some of them into the planetary region.

It has been shown (e.g., Heisler 1990; Matese and Lissauer 2002) that the main part of the steady flux of Oort cloud comets towards the observable region is due to the Galactic tide, while stellar perturbations may cause strong, temporary enhancements called comet showers (Hills 1981). Thus, for a general understanding of the origin of LP comets, it is important to model both the Galactic tide and stellar encounters with a relevant accuracy. Moreover, given that the dynamics of Oort cloud comets involves very long-time scales ($\sim 10^8$–10^9 years) and that the number of test comets must be very large in order to get reliable statistical data, the Galactic tide should clearly be modeled at the lowest possible computational cost.

Many studies have been devoted to the qualitative characteristics of the Galactic tide (Heisler and Tremaine 1986; Matese and Whitman 1989, 1992; Breiter et al. 1996; Brasser 2001; Breiter and Ratajczak 2005), while quantitative studies have been performed mostly through integrations over short-time spans (most recently, Matese and Lissauer 2004 and Dybczyński 2005). Long-term simulations of Oort cloud dynamics have either treated a small number of comets or made use of simplifications in the dynamical models (Wiegert and Tremaine 1999; Levison et al. 2001; Neslušan and Jakubík 2005).

In this paper, we investigate the effects of the Galactic tide, making use of an improved version of the model developed in Fouchard (2004) and Fouchard et al. (2005) (hereafter Papers I and II, respectively). In Sect. 2, the improved model is presented. Based on this model, calculations are performed and compared to numerical integrations, first considering only the normal component of the Galactic tide (Sect. 3.1). In Sect. 3.2, both normal and radial components are considered, and the long-term effects of the radial component[1] on Oort cloud comets are evidenced.

[1] For simplicity, we use the term *radial* to denote the entire in-plane component, although in general this has one radial part directed toward the Galactic center and one orthogonal to this direction.

Since the value of the local Galactic density $\rho_\odot$ is not well known (e.g., Holmberg and Flynn 2000), we study the effects of varying the strength of the normal component of the Galactic tide by considering different values of this density (Sect. 4.1). Moreover, since the Oort constants A and B are not known with great accuracy either (e.g., Olling and Merrifield 1998; Mignard 2000; Olling and Dehnen 2003), we compare calculations using different values of these constants, thereby effectively varying the radial component (Sect. 4.2). Finally, in Sect. 5, we summarize our conclusions and discuss a few aspects of the relevance of our model for the real Oort cloud.

2 The hybrid model

The model developed in Papers I and II can be called a *hybrid* since, according to the osculating values of the cometary semi-major axis a and eccentricity e, it involves two different procedures: (1) numerical integration of the Newtonian equations of motion in Cartesian coordinates, and (2) mappings obtained by averaging the Hamiltonian equations written, according to the osculating value of e, in Delaunay's variables or using the Matese elements defined in Sect. 2.3.

2.1 The Cartesian model

Following Paper I, two different heliocentric frames are considered. The first is a *rotating frame* with $\hat{x}'$-axis in the radial direction pointing toward the Galactic center, the $\hat{y}'$-axis pointing transversely along the local circular velocity, and $\hat{z}$ completing a right-handed system. The *fixed frame* $(\hat{x}, \hat{y}, \hat{z})$ is such that it coincides with the rotating frame $(\hat{x}', \hat{y}', \hat{z})$ at time $t = 0$ while keeping its axial directions fixed (see Fig. 1 of Paper II).

Let us define Ω_0 as the angular velocity about the Galactic centre, assuming the Sun to follow a circular orbit (since the motion of the Sun around the Galaxy is clockwise in both our frames, Ω_0 is negative, i.e., the vector is directed along $-\hat{z}$). If ϕ_r is an angle in the Galactic plane measured in the rotating frame from $\hat{x}'$, and ϕ the corresponding angle measured in the fixed frame from $\hat{x}$ at time t, we have the relation: $\phi = \phi_r + \Omega_0 t$ (see Fig. 1 of Paper II). All the final results will be presented in the fixed frame.

The force $\mathbf{F}$ per unit mass acting on a test particle orbiting the Sun under the influence of the Galactic tide is given by Heisler and Tremaine (1986):

$$\begin{aligned} \mathbf{F} = \; & - \frac{\mu M_\odot}{r^3}\mathbf{r} + (A-B)(3A+B)x'\hat{x}' - (A-B)^2 y'\hat{y}' \\ & - \left[4\pi\mu\rho_\odot - 2\left(B^2 - A^2\right)\right] z\hat{z}, \end{aligned} \tag{1}$$

where x', y', z are the coordinates of the comet in the rotating frame, $\mathbf{r}$ is the Sun–comet vector of length r, A and B are the Oort constants, and $\rho_\odot$ is the local density of the Galactic disk in the solar neighbourhood. The unit of mass is the solar mass ($M_\odot = 1$), the unit of time is the year, and the unit of length is the AU. As a consequence, the gravitational constant μ is equal to $4\pi^2$.

If $\mathcal{G}_1, \mathcal{G}_2$ and $\mathcal{G}_3$ are defined such that

$$\begin{aligned} \mathcal{G}_1 &= -(A-B)(3A+B), \\ \mathcal{G}_2 &= (A-B)^2, \\ \mathcal{G}_3 &= 4\pi\mu\rho_\odot - 2(B^2 - A^2) \end{aligned} \tag{2}$$

the equations of motion in Cartesian coordinates are:

$$\begin{aligned} \frac{d^2x}{dt^2} &= -\frac{\mu M_\odot}{r^3}x - \mathcal{G}_1 x' \cos(\Omega_0 t) + \mathcal{G}_2 y' \sin(\Omega_0 t), \\ \frac{d^2y}{dt^2} &= -\frac{\mu M_\odot}{r^3}y - \mathcal{G}_1 x' \sin(\Omega_0 t) - \mathcal{G}_2 y' \cos(\Omega_0 t), \\ \frac{d^2z}{dt^2} &= -\frac{\mu M_\odot}{r^3}z - \mathcal{G}_3 z, \end{aligned} \tag{3}$$

where x, y, z are the coordinates of the comet in the fixed frame [thus $x' = x \cos(\Omega_0 t) + y \sin(\Omega_0 t)$ and $y' = -x \sin(\Omega_0 t) + y \cos(\Omega_0 t)$]. Equations 3 will be referred to as the Cartesian model, and its results will be denoted by the subscript $\mathcal{C}$ when necessary. The RADAU integrator described by Everhart (1985) is used at the 15th-order, with LL$= 12$, to integrate Eq. 3. This integrator was chosen because it is fast, reliable and accurate compared to other non-symplectic integrators.

2.2 The Hamiltonian model with Delaunay elements

We use Delaunay's variables $L = \sqrt{\mu a}, G = \sqrt{\mu a(1-e^2)}, H = G \cos i, l = M, g = \omega$ and $h = \Omega$, where $a, e, i, M, \omega, \Omega$ are the cometary semi-major axis, eccentricity, inclination, mean anomaly, argument of perihelion and longitude of node (all the angles being measured in the fixed Galactic frame). When the Hamiltonian is averaged with respect to the mean anomaly over one period, L is conserved. Moreover, if the radial component of the Galactic tide is neglected, i.e., setting $\mathcal{G}_1 = \mathcal{G}_2 = 0$, the set of equations becomes:

$$\left\langle \frac{dG}{dt} \right\rangle = -\mathcal{G}_3 \frac{5L^2}{4\mu^2} \left(L^2 - G^2\right) \left(1 - \frac{H^2}{G^2}\right) \sin 2g, \tag{4a}$$

$$\left\langle \frac{dg}{dt} \right\rangle = \mathcal{G}_3 \frac{L^2 G}{2\mu^2} \left[1 - 5 \sin^2 g \left(1 - \frac{L^2 H^2}{G^4}\right)\right], \tag{4b}$$

where $\langle\,\rangle$ indicates that the equations are averaged with respect to the mean anomaly M (see Appendix A for the complete equations including the radial component).

2.3 The Hamiltonian model with Matese elements

Equations 4a and 4b become singular when $e = 1$. Therefore, in Paper II for $e \geq 0.99$, we replaced the Hamiltonian model with Delaunay elements by Lagrangian equations of motion in the variables: $a, H_b = -\sqrt{\mu a(1-e^2)} \cos \alpha, H = \sqrt{\mu a(1-e^2)} \cos i, b$, and l. Here b and l are the latitude and longitude of perihelion of the comet, and α is the angle between the orbital plane and the plane orthogonal to the Galactic plane and passing through the perihelion and the Galactic poles, measured from the south Galactic pole to the cometary velocity (see Fig. 1).

In the present paper, for $e \geq 0.99$, we instead adopt the Hamiltonian formalism, using the variables: $L, \Theta = H_b, H, M, \theta = b$ and $\lambda = l$. This set of elements will be referred to as Matese elements since it first appeared in Matese and Whitman (1992).

Similar elements have been used elsewhere in the literature in order to remove the singularity at $e = 1$: Tremaine (2000) for elliptic collision orbits, and Tommei (2006) for hyperbolic collision orbits. The Matese elements are slightly different from those

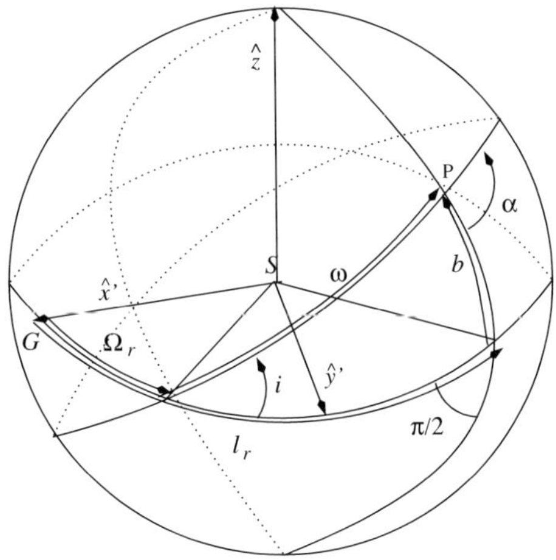

Fig. 1 Location of the angles $i, \omega, \Omega_r, b, l$ and α in the rotating frame centred on the Sun S. G indicates the direction of the Galactic centre, and P the perihelion of the cometary orbit. At any time t, λ and λ_r, and Ω and Ω_r are related by: $\lambda = \lambda_r + \Omega_0 t$ and $\Omega = \Omega_r + \Omega_0 t$. Note that $\lambda_r = lr$

used in the quoted papers, but the procedure to define them is similar. The proof of the canonicity of the Matese elements is given in Appendix B.

The Hamiltonian $\mathcal{H}$ of the motion of a comet under the action of the Galactic tide and the Sun is given by:

$$\mathcal{H} = -\frac{\mu}{2a} + \mathcal{G}_1 \frac{x'^2}{2} + \mathcal{G}_2 \frac{y'^2}{2} + \mathcal{G}_3 \frac{z^2}{2}. \tag{5}$$

Substituting x', y' and z by the Matese elements in Eq. 5, and averaging with respect to the mean anomaly, gives:

$$\begin{aligned} \langle \mathcal{H} \rangle \quad = \quad & -\frac{\mu^2}{2L^2} + \frac{L^2}{4\mu^2 \cos^2 \theta} \Big\{ \Big(-4\Theta^2 \cos^2 \theta - 4H^2 + 5 \cos^2 \theta L^2 \Big) \\ & \times \Big(\mathcal{G}_1 \cos^2 \lambda_r + \mathcal{G}_2 \sin^2 \lambda_r \Big) \cos^2 \theta \\ & + \mathcal{G}_1 (\Theta \cos \theta \sin \theta \cos \lambda_r + H \sin \lambda_r)^2 \\ & + \mathcal{G}_2 (\Theta \cos \theta \sin \theta \sin \lambda_r + H \cos \lambda_r)^2 \\ & + \mathcal{G}_3 \Big[\Big(-4\Theta^2 \cos^2 \theta - 4H^2 + 5 \cos^2 \theta L^2 \Big) \sin^2 \theta + \Theta^2 \cos \theta^4 \Big] \Big\}. \end{aligned} \tag{6}$$

The Hamiltonian equations of motion have the form:

$$\begin{aligned} \left\langle \frac{\mathrm{d}L}{\mathrm{d}t} \right\rangle &= -\frac{\partial \langle \mathcal{H} \rangle}{\partial M} & \left\langle \frac{\mathrm{d}l}{\mathrm{d}t} \right\rangle &= \frac{\partial \langle \mathcal{H} \rangle}{\partial L}, \\ \left\langle \frac{\mathrm{d}\Theta}{\mathrm{d}t} \right\rangle &= -\frac{\partial \langle \mathcal{H} \rangle}{\partial \theta} & \left\langle \frac{\mathrm{d}\theta}{\mathrm{d}t} \right\rangle &= \frac{\partial \langle \mathcal{H} \rangle}{\partial \Theta}, \\ \left\langle \frac{\mathrm{d}H}{\mathrm{d}t} \right\rangle &= -\frac{\partial \langle \mathcal{H} \rangle}{\partial \lambda} & \left\langle \frac{\mathrm{d}\lambda}{\mathrm{d}t} \right\rangle &= \frac{\partial \langle \mathcal{H} \rangle}{\partial \Theta}, \end{aligned}$$

which gives, when the radial component is neglected (see Appendix C for the complete equations):

$$\left\langle \frac{\mathrm{d}\Theta}{\mathrm{d}t} \right\rangle = \frac{L^2 \mathcal{G}_3}{2\mu^2} \frac{\sin\theta}{\cos^3\theta} \left(5(\Theta^2 - L^2)\cos^4\theta + 4H^2 \right), \tag{7}$$

$$\left\langle \frac{\mathrm{d}\theta}{\mathrm{d}t} \right\rangle = \frac{L^2 \Theta \mathcal{G}_3}{2\mu^2} \left(1 - 5\sin^2\theta \right). \tag{8}$$

The singularity at $\cos\theta = 0$ is evident (as in Paper II), but these equations show that the singularity at $e = 1$ has indeed disappeared, and that the Taylor development of the solution of this system is well defined when $e = 1$. This simplifies the results of Paper II.

2.4 The mappings

In this paper, we make use of two mappings. These are built following the method of Paper I, for the two Hamiltonian models obtained in Sects. 2.2 and 2.3. If x is the solution for either of the Hamiltonians, its Taylor development is:

$$\mathrm{x}(T + \Delta T) = \mathrm{x}(T) + \sum_{n=1}^{+\infty} \left. \frac{\mathrm{d}^n \mathrm{x}}{\mathrm{d}t^n} \right|_T \frac{\Delta T^n}{n!}, \tag{9}$$

where the subscript T indicates that the quantities are computed at time T.

Taking ΔT equal to one orbital period of the comet, which is consistent with the averaging of the equations, this development provides us with an easy way to derive mappings of different orders using an algebraic manipulator.

2.5 The hybrid model

To construct the hybrid model, we first have to determine the domain of reliability of each Hamiltonian model. We use the same procedure as in Paper II: 400 000 sets of starting orbital elements are randomly chosen, as follows:

- the initial semi-major axis a_0 is in the range 3 000–100 000 AU, with a uniform distribution in $\log a_0$;
- the initial eccentricity e_0 is uniformly distributed within the range 0–0.9999;
- the initial inclination i_0 is such that $\cos i_0$ is uniformly distributed between -1 and 1;
- the initial argument of perihelion ω_0, longitude of the ascending node Ω_0, and mean anomaly M_0 for direct numerical integration, are uniformly distributed within the range 0–2π.

Using these sets of initial orbital elements, we integrated the equations of motion over one orbital period using the Cartesian model and one of the two Hamiltonian models. We define the relative error E as:

$$E = \left| \frac{q_{\mathcal{H}} - q_{\mathcal{C}}}{q_0} \right| \times 100, \tag{10}$$

where $q_{\mathcal{H}}$ and $q_{\mathcal{C}}$ denote the values of the perihelion distance at the end of the integration as computed by the Hamiltonian and Cartesian models, respectively, and q_0 is the initial value of the perihelion distance.

We divide the $(e_0 - \log a_0)$ plane into 60×70 cells. In each cell we compute the error E^*, such that 99% of the errors found in the cell are smaller than E^*. We then compute the best analytical fit:

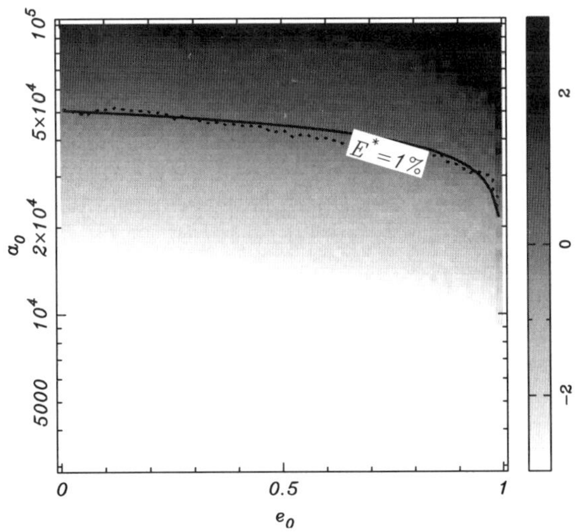

Fig. 2 Grey-scale representation of the error $\log(E^*)$ in the $(e_0 - \log a_0)$ plane. The *dotted curve* is the level curve of $E^* = 1$ (see the text for the definition of E^*), and the *full-drawn curve* is the best analytical fit given by $a_c = 57809(1-e)^{0.179}$ AU

$$a_c = a^* \times (1-e)^{\eta} \tag{11}$$

to the level curve of $E^* = 1$. The value of a_c, for any given value e of the eccentricity, is the upper limit of the semi-major axis, below which the Hamiltonian models are considered reliable.

Figure 2 shows the results obtained for a Galactic tide without radial component, and a local Galactic disk density $\rho_\odot = 0.1 M_\odot\,\mathrm{pc}^{-3}$. For this model, the parameters in Eq. 11 are found to be $a^* = 57809$ AU and $\eta = 0.179$.

For each Galactic model, i.e., for each set of parameters A, B and $\rho_\odot$, we construct the hybrid model as follows. If $a < a_c$ and $e > 0.99$, we adopt the mapping of order 3 obtained from the Hamiltonian model using Matese elements; if $a < a_c$ and $e < 0.99$, we adopt the mapping of order 3 obtained from the Hamiltonian model using Delaunay elements (Paper II). If $a > a_c$, we use the Cartesian model.

In Paper II, it was shown that the inclusion of the radial component of the tide does not change the value of a^* and η. Thus a direct scaling could be used when changing the value of $\mathcal{G}_3$ or of the local density $\rho_\odot$—see Eq. 2.

However, since the reliability of the hybrid model relies on the parameters a^* and η, it is important to determine their correct values. Thus the above experiment was repeated for the different values of $\rho_\odot$ considered in this paper (see Sect. 4 for a discussion of the choice of these values). Table 1 shows the values obtained for a^* and η. These will be used in the following, according to the value used for $\rho_\odot$.

Table 1 Values of a^* and η for different choices of $\rho_\odot$

$\rho_\odot$ ($M_\odot\,\mathrm{pc}^{-3}$)	a^*(AU)	η
$\rho_{min} = 0.07$	64 268	0.179
$\rho_{ref} = 0.1$	57 809	0.179
$\rho_{med} = 0.15$	50 609	0.177
$\rho_{max} = 0.26$	41 686	0.170

3 Long-term integrations

Let us now study the long-term effect of the Galactic tide on the transport of Oort cloud comets to the inner Solar System. The initial conditions are chosen such that:

- the initial semi-major axis a_0 is between 3 000 and 100 000 AU, with a uniform distribution in a_0^{-1};
- the initial eccentricity e_0 is chosen to yield $q_0 > 32$ AU, with a frequency function $f(e_0) \propto e_0$;
- the initial inclination i_0 is obtained from a uniform distribution of $\cos i_0$ between -1 and 1;
- the initial mean anomaly M_0, argument of perihelion ω_0 and longitude of the ascending node Ω_0, are uniformly distributed in the range 0–2π.

This choice of the initial conditions corresponds to a thermalized Oort cloud (Hills 1981), i.e., a cloud that has reached a steady state under stellar perturbations. Moreover, all the comets are initially outside the planetary region. The distribution of the semi-major axes is arbitrary; however, according to previous results (Duncan et al. 1987; Levison et al. 2001), our distribution generates an outer Oort cloud much more populated than the real outer Oort cloud appears to be. This will be useful, in the framework of the present study, in order to obtain better statistics for this part of the cloud.

We integrate a sample of 10^6 fictitious test comets over a time scale of 5 Gyr. The integration of a comet is stopped as soon as its perihelion distance $q \leq 15$ AU. Below this threshold, comets are most likely removed by planetary perturbations within one orbital period. This region will hereafter be referred to as the *target region*. In Sect. 3.1, the integrations take into account only the normal component of the tide, while the influence of the radial component is included in Sect. 3.2.

In both cases we adopt, as the value of the local density, $\rho_\odot = 0.10\, M_\odot\, \mathrm{pc}^{-3}$. Moreover, in Sect. 3.2, a flat Galactic rotation curve (i.e., the circular speed around the Galactic centre does not vary with the distance), with Oort constants $A = -B = 13\, \mathrm{km\, s}^{-1}\, \mathrm{kpc}^{-1}$, is assumed. The above values are among those commonly used in studies of Oort cloud comet dynamics (e.g., Levison et al. 2001; Papers I and II).

3.1 Long-term dynamics with only the vertical tide

For this first experiment, including only the normal (or "vertical") component of the tide, the reliability of the hybrid model is checked using two different values of a^* (see Sect. 2.5): the first one is the nominal value i.e. $a^* = 57\,809$ AU, given in Table 1, and the second, $a^* = 15\,000$ AU, is a control value. With the control value of a^*, numerical integrations in Cartesian coordinates are performed on a much wider domain than for the nominal value.

The results obtained with the two values of a^* differ by less than 0.2%, as regards the number of comets entering the target region over various time spans. The distributions of orbital elements of the comets entering the target region either overlap exactly, or show small differences owing to small number statistics. For example, in the last 500 Myr of the integration, eight comets have $q < 10$ AU for the nominal a^*, whereas there are six for the control value.

Rough estimates of the computational time for the integrations on a 2.8 GHz XEON computer are less than 48 hours for the nominal a^*, almost 15 days for the control value and more than 3 months if only Cartesian integrations are performed.

In Fig. 3, we plot the number of comets entering the target region versus time. The flux is seen to be three times smaller at the end of the integration than at the beginning.

To understand this result, it is useful to consider that, if (1) the radial component of the tide is neglected, and (2) the semi-major axis is small enough to justify the averaging of the Hamiltonian with respect to the mean anomaly, then the system is essentially integrable (see Matese and Whitman 1989, 1992; Breiter and Ratajczak 2005). We call this situation the *integrable approximation*.[2] This will be a key factor for explaining most results of our study.

In this case, the evolution of the perihelion distance is strictly periodic and, for given values of $L = \sqrt{\mu a}$ and $H = \sqrt{\mu a(1-e^2)} \cos i, q$ depends only on ω. The minimum reached by q over one cycle is then constant. Hence, the orbital domain from which comets may enter the target region is well defined and represents only a small part of the Oort cloud. This domain will become depleted as time increases, since our integration stops when a comet reaches the target region. Therefore, the flux of comets entering that region must decrease with time.

Let us now consider the distributions of $q, \cos i, \omega$ and $1/a$ for the comets entering the target region at the beginning and at the end of our integration. In Fig. 4, these distributions are plotted for the first and the last 500 Myr (left and right columns, respectively). The $1/a$ distribution of the comets entering the target region between 500 Myr and 1 Gyr has also been plotted in order to help us discuss the variations of the flux at the beginning of the integration.

Moreover, we have divided the Oort cloud into four regions, according to the semi-major axis of the comets: (1) CI, the *inner Oort cloud*, corresponding to $a < 10^4$ AU; (2) CII, the *central inner Oort cloud*, corresponding to $10^4 \leqslant a < 3 \times 10^4$ AU; (3) CIII, the *central outer Oort cloud*, corresponding to $3 \times 10^4 \leqslant a < 5 \times 10^4$ AU; and (4) CIV, the *outer Oort cloud*, corresponding to $a \geqslant 5 \times 10^4$ AU. For each distribution the bins are shaded according to the proportion of comets coming from each of these

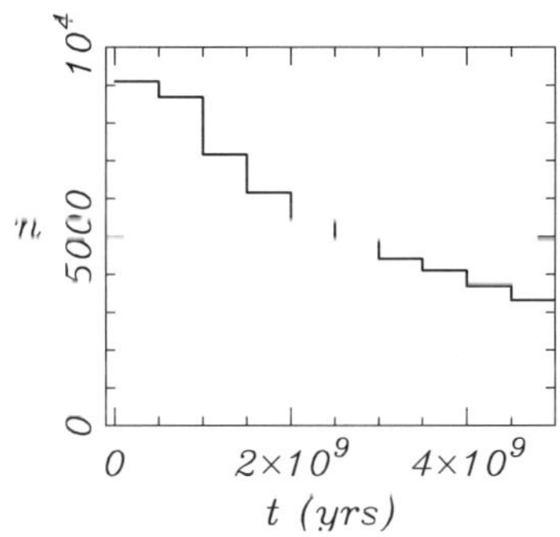

Fig. 3 Number of comets, per 500 Myr, entering the target region versus time

[2] This configuration is not the only case of integrability. For instance, the case $\mathcal{G}_1 = \mathcal{G}_2$ leads to an integrable system as well, but this situation does not correspond to our Galaxy. Thus we will not discuss it further. (see Elipe and Ferrer 1994 for a general study on the integrability of systems of this kind.)

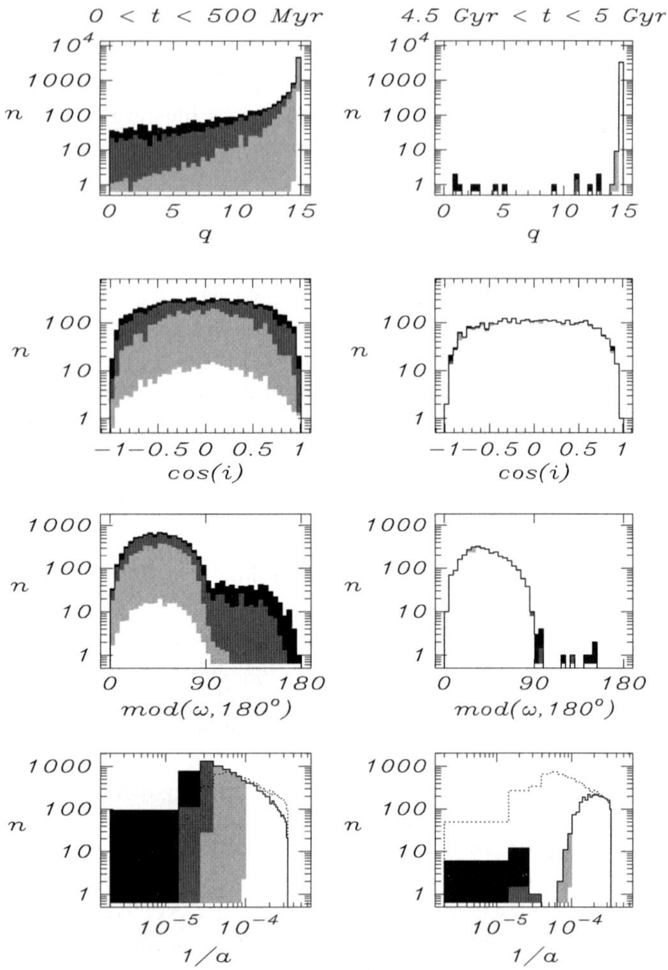

Fig. 4 Distributions of orbital elements for comets entering the target region. The left column corresponds to the flux in the first 500 Myr and the right column to the flux in the last 500 Myr. Each bin is shaded with respect to the relative population in the inner (*white*), central inner (*light grey*), central outer (*dark grey*) and outer Oort cloud (*black*). Two-shade bins in the distribution of $1/a$ are due to the fact that the bin borders do not coincide with the borders of the cloud regions. The *dotted curves* in both distributions of $1/a$ correspond to the $1/a$ distribution of the flux between 500 Myr and 1 Gyr

regions: white for the inner cloud, light grey for the central inner cloud, dark grey for the central outer cloud and black for the outer cloud. In the following, whenever we mention these regions as sources of comets for the target region, we refer to the corresponding parts where transfer into the target region is possible.

3.1.1 Perihelion distance distribution in the target region

Concerning the distribution of q, we note that, during the first 500 Myr, the target region is mostly fed by comets coming from the two central regions (CII and CIII) of the Oort cloud. The central inner cloud contributes more than 50% of the comets with $q \approx 10$ AU, but less than 30% of those with $q < 5$ AU. For the latter q-range the central outer cloud is the main contributor, with more than 60% of the comets.

The contribution of the outer cloud is more or less constant and its low value is mainly due to the fact that this region is initially less populated. The fact that the inner cloud produces only comets with $14 < q \leq 15$ AU is an artifact due to our choice of perihelion distance threshold. Indeed, for the comets in the inner cloud the Galactic tide is not strong enough to move the cometary perihelion distance far inside the target region during one orbital period.

At the end of the integrations, the situation is completely different. Indeed, while the peak between 14 and 15 AU is still present and is mainly due to the inner cloud ($\approx 99\%$ of the comets with $q > 14.6$ AU come from CI), the rest of the distribution has nearly disappeared. Only 56 comets come from the other three regions of the cloud, with six comets having a perihelion distance smaller than 5 AU coming from the outer Oort cloud.

3.1.2 *Inverse semi-major axis distribution in the target region*

The $1/a$ distributions of Fig. 4 show that, in agreement with the q distributions, in the last 500 Myr almost no comet comes from the two central regions. A marginal flux from the outer cloud is still present, whereas the flux from the inner cloud remains almost the same as in the beginning.

This behaviour can be explained using the integrable approximation. In this case q librates with a period whose lower limit is given by (see Matese and Whitman 1992):

$$P_q^{\min} = \frac{1}{4\sqrt{5\mu a^3}\rho_{\odot}}. \tag{12}$$

The period of the q-cycles is hence longer for smaller semi-major axis. Indeed we find that, after 4.5 Gyr, as a increases, the central regions become more and more depleted, and even completely empty for $a > 15\,500$ AU. For this threshold, using Eq. 12, the minimum period of the q-cycles is $P_q^{\min} = 804$ Myr. This is much shorter than the total integration time, so that all comets will have performed at least one complete cycle, thus emptying completely these regions of the Oort cloud.

When the integrable approximation is not valid, although the quantity $H = \sqrt{\mu a(1-e^2)}\cos i$ is still conserved ($\mathrm{d}H/\mathrm{d}t = 0$ since the radial component is neglected, i.e. $\mathcal{G}_1 = \mathcal{G}_2 = 0$; see Appendix A), the quantity $L = \sqrt{\mu a}$ is not constant because the averaging of the Hamiltonian is not valid. Consequently, the minimum of q over one quasi-period is no longer constant. Hence, a marginal flux of comets from the outer cloud into the target region survives for the whole time span. This occurs for comets with $a > 25\,400$ AU, "peaking" around 60 000 AU. This flux of comets, even if marginal, is of great importance as regards observable comets, since all the comets with a final perihelion distance $q < 5$ AU arrive from there.

Another interesting feature concerns the $1/a$ distribution of comets in the target region coming from the inner cloud. In Fig. 4, looking at the dotted curve plotted in the $1/a$ distribution frames, we note that (1) the flux coming from CI increases during the first Gyr, and (2) this flux then remains constant until the end of the integration for comets with $a < 4700$ AU (this value corresponds to the peak of the $1/a$ distribution at the end of the integration). Point (2) is easily explained considering the value of the minimum period of the q-cycles, which is equal to $P_q^{\min} = 4.850$ Gyr when $a = 4700$ AU. Consequently, for smaller semi-major axes the comets did not have enough time to perform a complete cycle, which implies that the flux remains constant.

Concerning the increase of the flux at the beginning of the integration, let us consider Eqs. 4a and 4b, which give the variations of $G = \sqrt{\mu a(1-e^2)}$ and $g = \omega$ with respect to time. In Eq. 4b, we can neglect the term $\frac{L^2H^2}{G^4} = \frac{\cos^2 i}{1-e^2}$ for comets that have a small minimum perihelion distance $q_{\min}$, whenever q is not too close to $q_{\min}$. Indeed, since $\sqrt{1-e^2}\cos i$ is conserved, we have:

$$\frac{\cos^2 i}{1-e^2} \leq \frac{1-e_{\max}^2}{(1-e^2)^2} \leq \frac{1}{1-e_{\max}^2}, \tag{13}$$

where $e_{\max}$ is the maximum value that e may reach. When $e_{\max}$ is close to 1, the right-hand member may grow very large. However, since there is a vertical asymptote at 1, the middle member will quickly decrease with e. For instance, if $q_{\min} = 10$ AU and $a = 10^4$ AU, then $L^2H^2/G^4 = 1$ for $q = 230$ AU and is smaller than 0.05 when $q = 1\,000$ AU. Using the approximation $L^2H^2/G^4 \approx 0$ in Eq. 4b, we easily see that the variation of ω vanishes when $\mathrm{mod}(\omega, 180°) \simeq \pm\arcsin(\sqrt{1/5}) \simeq \pm 26.6°$. Since comets entering the target region have suffered large perihelion distance perturbations, it follows that $\mathrm{mod}(\omega, 180°)$ is likely librating about 90° (Matese and Whitman, 1989, 1992; Breiter et al. 1996). Thus the value of 26.6° corresponds to a minimum of $\mathrm{mod}(\omega, 180°)$ during its libration cycle. This value is also known to be a barrier that the Galactic latitude of the cometary perihelion cannot cross in the integrable approximation (Matese and Whitman 1989; Breiter et al. 1996).[3]

Considering Eq. 4a, which gives the variation of $G = \sqrt{\mu a(1-e^2)}$ (i.e., of q, since $L = \sqrt{\mu a}$ is constant when the equations are averaged over the mean anomaly, see Sect. 2.2), as a function of the argument of perihelion, we note that the maximum rate

[3] This result may be deduced from Eq. 8. This equation, and the Hamiltonian formalism, give the relation

$$\frac{\partial^2 \langle \mathcal{H} \rangle}{\partial \Theta^2} = \frac{1}{\Theta}\frac{\partial \langle \mathcal{H} \rangle}{\partial \Theta}.$$

Using the formula

$$\frac{\mathrm{d}^n f(\theta,\Theta)}{\mathrm{d}t^n} = \sum_{k=0}^{n} C_k^n \frac{\partial^k}{\partial \theta^k}\frac{\partial^{n-k}}{\partial \Theta^{n-k}} f(\theta,\Theta) \frac{\mathrm{d}^k \theta}{\mathrm{d}t^k}\frac{\mathrm{d}^{n-k}\Theta}{\mathrm{d}t^{n-k}}$$

with $f(\theta,\Theta) = \frac{\mathrm{d}\theta}{\mathrm{d}t}$, we may easily show that

$$\frac{\mathrm{d}^n \theta}{\mathrm{d}t^n} = 0 \quad \forall n > 1,$$

if

$$\frac{\mathrm{d}\theta}{\mathrm{d}t} = 0.$$

The latter equation is verified for $\theta \simeq \pm 26.6°$, showing that this value corresponds to a stationary solution. Thus, 26.6° becomes a barrier for θ.

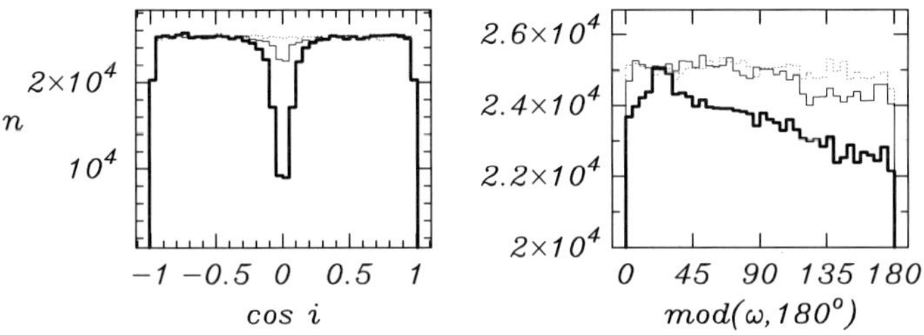

Fig. 5 Distribution of the cosine of the Galactic inclination and argument of perihelion of comets in the Oort cloud at time $t = 0$ (*thin dotted line*), $t = 5 \times 10^8$ years (*thin full line*), and $t = 5 \times 10^9$ years (*thick full line*)

of decrease of q occurs when $\text{mod}(\omega, 180°) = 45°$, just after $\text{mod}(\omega, 180°)$ has reached its minimum value of 26.6°. Thus, as time increases, the values of $\text{mod}(\omega, 180°)$ accumulate around 26.6°, producing a reservoir of comets with q decreasing rapidly, which results in an increased flux into the target region.

This scenario is confirmed by our numerical data. Indeed, the flux of comets with $a < 10^4$ AU toward the target region increases by an amount of 1 568 comets between the first 500 Myr and the successive 500 Myr time intervals, while the number of comets in the Oort cloud with $a < 10^4$ AU and with a value of $\text{mod}(\omega, 180°)$ in the range [13°, 45°] increases by an amount of 1 334 comets between $t = 0$ and $t = 500$ Myr (see Fig. 5 and later comments).

3.1.3 *Distribution of* mod(ω, 180°) *in the target region*

The distribution of the argument of perihelion can be explained using Eq. 4a. We note that q decreases when $\text{mod}(\omega, 180°) < 90°$ and increases otherwise. Since all the comets considered in our sample had their perihelion distance decreasing during the orbital revolution preceding their entrance into the target region, the value of $\text{mod}(\omega, 180°)$ was in the range 0–90° during that revolution.

Thus, if the variation of ω over one orbital period is not too large, it will still be in almost the same range when the comet enters the target region. Since the variations of ω over one orbital period are $\propto a^3$ (see Eq. 4b), the values of $\text{mod}(\omega, 180°)$ will be less and less confined to [0, 90°] as a increases. This is seen clearly in the distribution at the beginning of the integrations (the number of comets in the target region with $\text{mod}(\omega, 180°) > 90°$ is 1, 37, 419 and 176 for CI–CIV, respectively), and also at the end of the integration, the corresponding numbers of comets being 0, 0, 2 and 11.

At the end of the integration the $\text{mod}(\omega, 180°)$ distribution reaches its maximum for $27° < \text{mod}(\omega, 180°) < 36°$. This is again a signature of the accumulation of $\text{mod}(\omega, 180°)$ toward 26.6° just before the onset of the strongest tidal perturbations on q.

3.1.4 *Distribution of* cos *i in the target region*

The main characteristic of the cos i distributions is the rapid fall-off toward $|\cos i| = 1$, and a broad maximum for $\cos i \approx 0$. No strong selection effect is observed related to the source regions, and the main difference between the distributions at the beginning and the end of the integration, apart from the already observed fact that the flux from the central cloud regions nearly disappears, is that the distribution at the end has become somewhat flattened. Indeed, at the beginning of the integration, the number of comets coming from CI increases by a factor $\simeq 3$ from $|\cos i| = 0.75$ to $|\cos i| = 0$, while at the end of the integration, the corresponding increase is just a factor $\simeq 1.5$.

We have already mentioned that the comets entering the target region are likely to have a librating perihelion argument. It has also been shown (see, e.g., Matese and Whitman 1989) that this libration requires the Galactic perihelion latitude b to be greater than $b^* = 26.6°$ in absolute value. However, using the relation $\cos i = \cos b \cos \alpha$, where α is defined in Fig. 1, one has $|\cos i| < \cos b < \cos b^* = 0.89$. This explains the fall-off of the $\cos i$ distributions when $|\cos i| > 0.8$, especially at the end of the integration.

The slight preference of comets with $\cos i \approx 0$ is explained by the fact that the tide is more efficient in reducing the perihelion distance, when the orbital plane is orthogonal to the Galactic plane, i.e., when $\cos i \approx 0$. More precisely, in the integrable approximation, the maximal value reached by the eccentricity is given by Breiter and Ratajczak (2005):

$$e_{\max}^2 = \frac{4 - 5\delta^2 - \beta + \sqrt{(4 - 5\delta^2 - \beta)^2 + \beta}}{8}, \tag{14}$$

where $\beta = e^2(1 - 5 \sin^2 i \sin^2 \omega)$, and $\delta = \sqrt{1 - e^2} \cos i$, are both constants of motions (β is related to the Hamiltonian and δ to $H = \sqrt{\mu a(1 - e^2)} \cos i$, as detailed by Breiter and Ratajczak 2005). It is easily found that $e_{\max}^2 \to 1$ when $|\delta| \to 0$. Thus, small values of $|\delta|$ induce a small minimum perihelion distance $q_{\min}$. Hence, for a given semi-major axis and perihelion distance, the smallest $q_{\min}$ are obtained for the smallest values of $|\cos i|$.

In short, whatever the perihelion distance, the tide will be more efficient in sending comets into the target region for small $|\cos i|$. However, the distribution is not strongly peaked at $\cos i = 0$ since, due to the conservation of β, $|\cos i|$ increases when q decreases.

The flatness of the $\cos i$ distribution of comets entering the target region in the last 500 Myr is explained by the fact that the Oort cloud gets depleted for $\cos i \approx 0$. Due to the quasi-integrability of the dynamics, this domain is not replenished from other parts of the cloud, so that the flux of comets with $\cos i \approx 0$ is reduced more quickly than for higher values of $|\cos i|$. The effect is small, because at the end of the integration the comets mainly come from CI and, as already mentioned, this region of the cloud has not been strongly depleted.

3.1.5 *Distribution of* $\cos i$ *and* $\mathrm{mod}(\omega, 180°)$ *in the cloud*

The distributions of $\cos i$ and $\mathrm{mod}(\omega, 180°)$ of comets in the Oort cloud at times $t = 0, t = 500$ Myr, and $t = 5$ Gyr, are plotted in Fig. 5.

These distributions confirm some of the previously described results, namely:

- the gap that develops for $\cos i \approx 0$ shows that this range of Galactic inclinations is the only one that gets significantly depleted;
- the strong accumulation of comets, at $t = 5$ Gyr, with argument of perihelion between 22.5 and 27° is consistent with a minimum of the ω libration cycle at 26.6°;

- the depletion of the argument of perihelion outside the range of this accumulation is duc to thc dcplction of the Oort cloud itself;
- the stronger depletion of the interval $90° < \text{mod}\,(\omega, 360°) < 180°$ as time increases is due to the fact that it corresponds to increasing q (see Eq. 4a), and since comets entering the target region are not integrated any more, there are less and less comets with increasing perihelion distance as time passes.

3.2 The effect of the radial tide

In Paper I, it was shown that the radial component of the tide affects the transport of comets from the Oort cloud to the inner Solar System, as it allows cometary perihelia to reach variable minimum values. However, the general effects of the radial component on the distributions of orbital elements were not studied.

In this section, we set the values of the Oort constants A and B, which characterize the radial component of the tide (see Eq. 1), to $A = -B = 13\,\text{km}\,\text{kpc}^{-1}$, as was done in Papers I and II.

As found in those papers, the total flux of comets from the Oort cloud towards the target region is slightly increased with respect to the case without a radial tide (see Fig. 6): 9 435 instead of 9 109, and 3 668 instead of 3 323, for the comets entering in the first and last 500 Myr, respectively.

On the other hand, with respect to Fig. 4, the distributions of orbital elements shown in Fig. 7 for the model including the radial tide are rather different for the first 500 Myr and drastically different for the last 500 Myr. The main reason for these differences is that the integrable approximation fails for smaller semi-major axes—and thus the q-cycles are not constant anymore—when the radial component of the tide is included.

For the first 500 Myr, the largest differences with respect to Fig. 4 concern the distribution of mod(ω, 90°), for which values greater than 90° are obtained even for comets coming from the inner cloud (10 such comets instead of 1), and the distribution of cos i, where an increase of the number of comets with $|\cos i| \approx 1$ is observed. Indeed, during the first 500 Myr, 38 comets entering the target region had $|\cos i| > 0.96$ when the radial component was neglected, but this number increases to 90 when the radial component is included. This increase of the flux for comets whose orbital plane is close to the Galactic plane was already observed in Paper I.

However, it is at the end of the integration that the differences are the largest. In fact, 48 comets instead of 6 are now entering the target region with a perihelion distance $q < 5$ AU in the last 500 Myr. These comets come only from the central

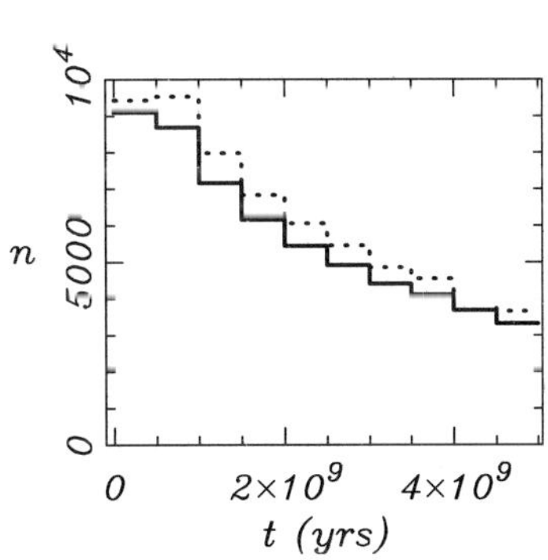

Fig. 6 Number of comets entering the target region during 500 Myr intervals versus time for the Galactic model with (*dotted curve*) and without (*full curve*) the radial component of the tide

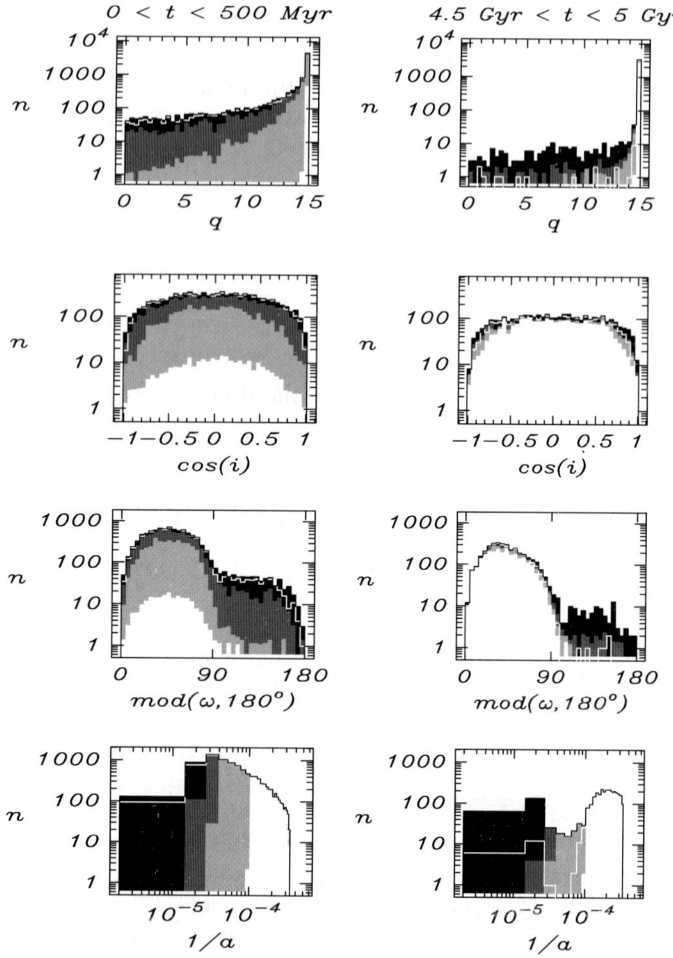

Fig. 7 The same as Fig. 4, but for a Galactic model with a radial tide and flat rotation curve. The *white lines* correspond to the distributions shown in Fig. 4

outer and outer regions of the cloud. Moreover, while the influx of comets from CI is almost unchanged (the increase is about 2%), comets coming from CII may now have a perihelion distance as small as 8.5 AU instead of 14 AU, which shows that the radial component allows comets to penetrate deeper inside the planetary region.

As a matter of fact, when the radial tide is included, the integrable approximation ceases to be valid for $a > 11\,200$ AU, while without the radial component the validity is lost only for $a > 25\,400$ AU. Hence, the minimum of the perihelion distance may now vary between different cycles for a wider range of semi-major axes. Thus it tends to reach lower values than in the integrable approximation.

A further illustration of this phenomenon is obtained by considering the $1/a$ distribution of comets entering the target region in the last 500 Myr. Indeed, a flux coming from the central cloud is now present with 122 and 53 comets coming from CII and CIII, respectively, while there were only 36 and 3 comets without the radial component (Fig. 4). The loss of validity of the integrable approximation is also the reason why the values of $\mathrm{mod}(\omega, 180°)$ are now less confined to $[0, 90°]$.

Furthermore, the distribution of cos i of comets entering the target region in the last 500 Myr confirms that it is the radial component that causes the increase of the flux of comets from CII and CIII. In fact, most of these comets have $|\cos i| > 0.61$, (94/122 and 34/53 for CII and CIII, respectively), implying that their orbital planes are close to the Galactic plane, i.e. the configuration where the radial component is the most efficient.

4 The influence of the Galactic parameters

4.1 The local density of the Galactic disk

In studies of Oort cloud dynamics, various sets of Galactic parameters have been used, due to the fact that these parameters are not very well constrained. This is the case for the local density of the Galactic disk, which enters into the calculation of the normal component of the Galactic tide. We recall that: $\mathcal{G}_3 = 4\pi\mu\rho_\odot - 2(B^2 - A^2)$, and it is easily verified that the first term dominates in the right-hand member.

Thus, the values of the local density $\rho_\odot$ of the Galactic disk used in the Oort cloud literature span from 0.08 (Matese and Lissauer, 2004) to 0.18 $M_\odot\,\text{pc}^{-3}$ (Matese and Whitman 1992), with two values being used more often: 0.1 $M_\odot\,\text{pc}^{-3}$ (Levison et al. 2001; Fouchard 2004; Fouchard et al. 2005) and 0.15 $M_\odot\,\text{pc}^{-3}$ (Wiegert and Tremaine 1999; Neslušan and Jakubík 2005).

Hipparcos results tend to favour a value of $\rho_\odot$ close to 0.1 $M_\odot\,\text{pc}^{-3}$ (Holmberg and Flynn 2000). However, García-Sánchez et al. (2001) have pointed out an incompleteness of the Hipparcos data that tends to lower the value of $\rho_\odot$. If one considers all the different values of $\rho_\odot$ computed in the last 20 years, one finds a straight average of 0.15 (Holmberg and Flynn 2000) with the extreme highest value of 0.26 $M_\odot\,\text{pc}^{-3}$ (Bahcall et al. 1992).

In the present work, to explore a wide range of possibilities, four values of $\rho_\odot$ will be used: $\rho_{\text{min}} = 0.07$, $\rho_{\text{ref}} = 0.1$ (the value already used in the previous section), $\rho_{\text{med}} = 0.15$ and $\rho_{\text{max}} = 0.26$ (all these values are expressed in $M_\odot\,\text{pc}^{-3}$). Let us start our analysis by disregarding the radial component.

Figure 8 shows the flux into the target region versus time for the four different values of $\rho_\odot$. At the beginning of the integration the values of the flux are very different (more than three times greater for ρ_{max} than for ρ_{min}), but at the end of the integration

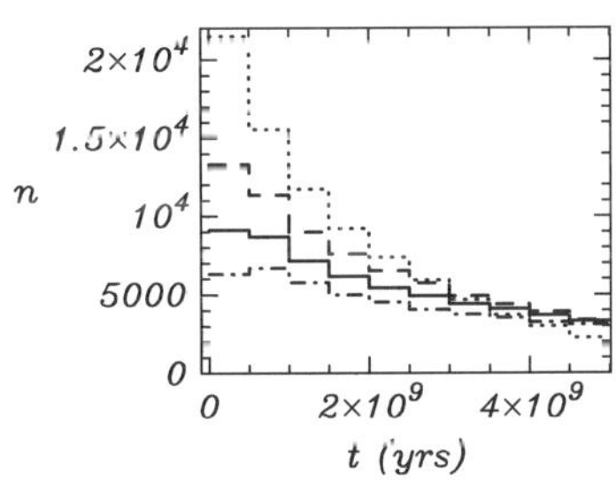

Fig. 8 Number of comets entering the target region during 500 Myr intervals versus time for $\rho_{\text{min}} = 0.07$ (*dot-dashed curve*), $\rho_{\text{ref}} = 0.1$ (*full curve*), $\rho_{\text{med}} = 0.15$ (*dashed curve*) and $\rho_{\text{max}} = 0.26\,M_\odot\,\text{pc}^{-3}$ (*dotted curve*). Only the vertical component of the tide is considered

all four fluxes are comparable, with the minimum flux obtained for the largest value of $\rho_\odot$.

This result may be explained as follows. On the one hand, using Eqs. 4a and 2 with $A = B = 0$ and making the approximation $G \propto \sqrt{q}$, we get for the amount of variation of q during one orbital revolution:

$$\Delta q \propto \sqrt{q}\rho_\odot. \tag{15}$$

On the other hand, the lower bound of the period of q variation is $P_q^{\min} \propto \rho_\odot^{-1}$ (Eq. 12). As a consequence, when $\rho_\odot$ increases, the perturbations of q obviously increase (inducing an increased flux towards the target region). However, $P_q^{\min}$ decreases and, when the integrable approximation is valid, the orbital domain from which comets may enter the target region gets depleted more quickly.

Figure 9 shows the distributions of orbital elements of comets entering the target region in the first 500 Myr, for the four values of $\rho_\odot$. The increase of the flux as $\rho_\odot$ increases is the main difference between the distributions. The shapes of the distributions are approximately the same, but the contributions of the four different regions of the cloud are quite different, which is most apparent on the q and $\mod(\omega, 180°)$ distributions.

Indeed, as the value of $\rho_\odot$ increases, the contribution of the central inner cloud is drastically increased for comets with $q < 5$ AU, with numbers of comets equal to 13, 82, 224 and 606 for $\rho_\odot = \rho_{\rm min}, \rho_{\rm ref}, \rho_{\rm med}$ and $\rho_{\rm max}$, respectively. This is a direct consequence of the increase of the typical q-perturbation with $\rho_\odot$, as given by Eq. 15. Indeed, a quick check of this equation may be performed by considering, in the q-distributions, the lowest bin containing comets from CI.

Using the lower bound of this bin for $\rho_{\rm min}$ and $\rho_{\rm med}$ (because these bins contain more than 100 comets), and the middle of the bin for $\rho_{\rm ref}$ and $\rho_{\rm max}$ (as they are much less populated, with less than 30 comets), we can compute the typical perturbations of q, finding $\Delta q = 0.37$, 0.56, 0.75 and 1.32 AU for $\rho_{\rm min}, \rho_{\rm ref}, \rho_{\rm med}$ and $\rho_{\rm max}$, respectively. Using Eq. 15, we find $\Delta q = 0.53$, 0.79 and 1.37 AU for $\rho_{\rm ref}, \rho_{\rm med}$ and $\rho_{\rm max}$, respectively, if normalization is done with the Δq found with $\rho_{\rm min}$. This confirms the validity of Eq. 15, when the integrable approximation is valid.

Concerning the distributions of $\mod(\omega, 180°)$, we note that the main difference is the increase of comets from CII with $\mod(\omega, 180°) > 90°$ as $\rho_\odot$ increases: the four values obtained are 15, 37, 119 and 505. This drastic increase is due to: (1) the fact that the perturbations of ω are proportional to $\rho_\odot$ (see Eq. 4b), so that the values of ω upon entry into the target region become less confined to the interval $[0°; 90°]$, and (2) the failure of the integrable approximation for smaller semi-major axes when $\rho_\odot$ increases.

This failure of the integrable approximation is more evident considering the distributions of orbital elements of comets entering the target region in the last 500 Myr, as shown in Fig. 10. Looking at the $1/a$ distributions, we note that some features occur for each value of $\rho_\odot$, i.e.: (1) CI is the main contributor to the flux, and (2) CIII and CIV contribute only marginally.

The fluxes coming from CI and CII decrease, as $\rho_\odot$ increases. This is explained by the behaviour of the lower limit of the q libration period, i.e., $P_q^{\min} \propto \rho_\odot^{-1}$. Thus CI and CII are emptied more quickly, and at smaller semi-major axes, with increasing density. Indeed, considering the $1/a$ distributions for increasing values of $\rho_\odot$, the gaps seen between the CI and CIII + CIV contributions terminate on the right side

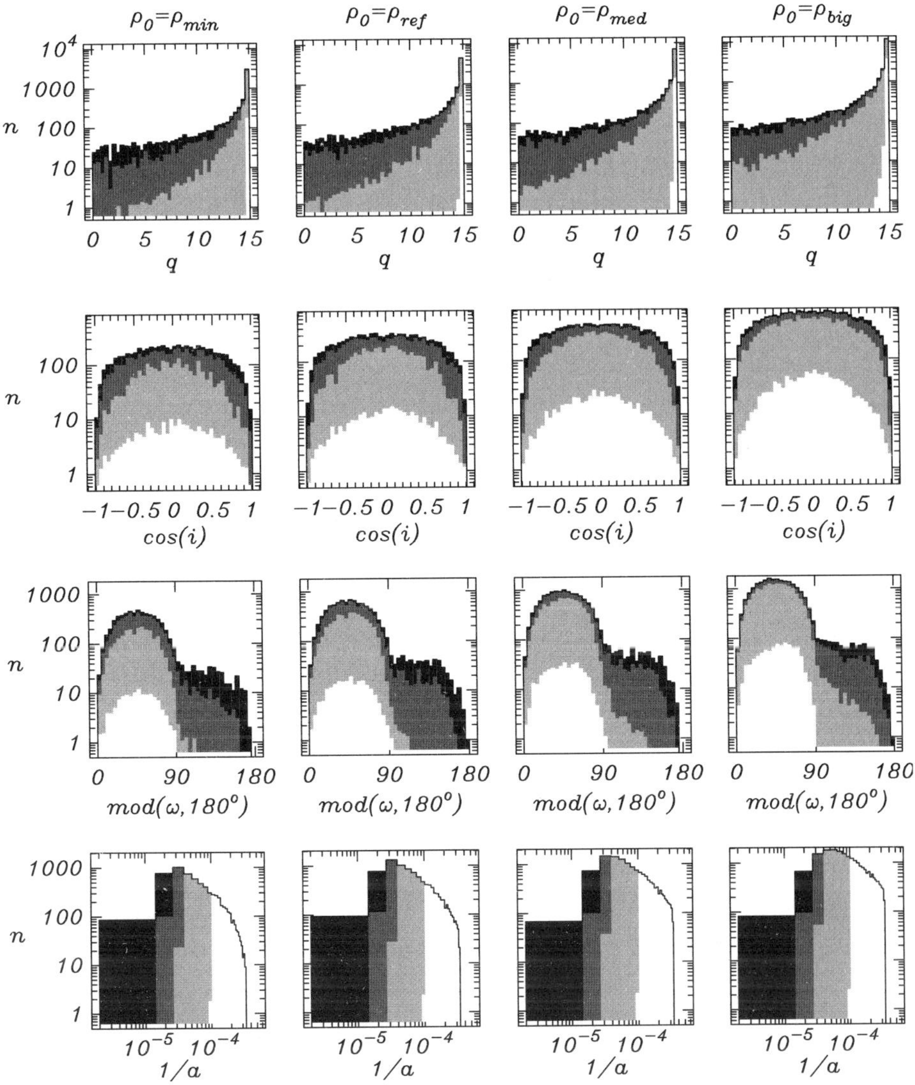

Fig. 9 The same as Fig. 4 but for models with different local density of the Galactic disk (shown above each column). Only the flux in the first 500 Myr is considered here

at decreasing values of a: 19 300, 15 500, 13 000 and 11 200 AU. These values correspond, respectively, to $P_q^{\min} = 830$, 804, 700 and 500 Myr. The decrease of $P_q^{\min}$, as $\rho_{\odot}$ increases, means that the values of P_q spread over a wider range with increasing density.

Moreover, as $\rho_{\odot}$ increases, the peak flux towards the target region occurs at decreasing values of a: $a = 6\,300$, 4 700, 4 050 and 3 150 AU for $\rho_{\min}$, ρ_{ref}, ρ_{med} and $\rho_{\max}$, respectively. These values correspond to $P_q^{\min} = 4.5$, 4.85, 4 and 3.5 Gyr, respectively. Since the real values of P_q are larger and scatter over a range that increases with $\rho_{\odot}$, the above locations of the peak fluxes can be understood as corresponding to the

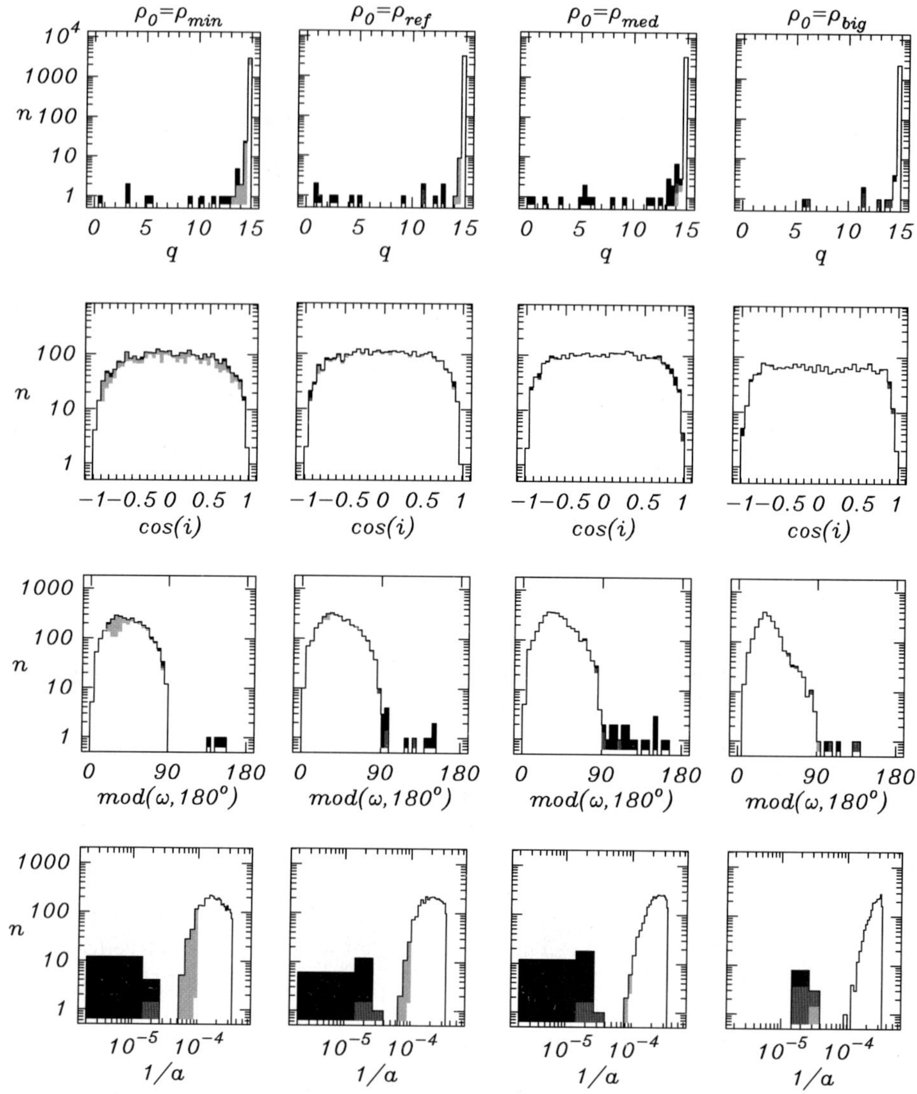

Fig. 10 The same as Fig. 9, but for comets entering the target region in the last 500 Myr

semi-major axes below which no depletion of the source region occurs during the time span of our integration.

As regards the distributions of mod(ω, 180°), we note that they all have a maximum for mod(ω, 180°) ranging in [27°, 36°]. Consequently, for comets in CI the integrable approximation is valid, as well as the accumulation of mod(ω, 180°) around 26.6°. One may note that, for $\rho_{\min}$, this property is also true for comets in CII, where a noticeable contribution is observed in the same range.

The shape of the mod(ω, 180°) distributions, which becomes sharper for larger $\rho_\odot$, may be explained considering that the perturbations of ω are proportional to $\rho_\odot$ (see Eq. 4b). Indeed, as soon as mod(ω, 180°) is far from 26.6°, it evolves more quickly for higher values of $\rho_\odot$.

Finally, the distributions of cos i become flatter when $\rho_\odot$ increases. In the case of $\rho_{\max}$, a slight decrease is even apparent around cos $i = 0$, which means that the reservoir is getting depleted even for CI. Indeed, it has already been mentioned that the domain where cos i is close to zero is the main contributor to the flux of comets into the target region.

4.1.1 The effect of the radial component

To see what happens when a radial component, corresponding to a flat rotation curve, with $A = -B = 13\,\mathrm{km\,s\,kpc^{-1}}$, is included into the four Galactic models, we will focus on the flux at the end of the integration, since this is where the differences between the models are the most evident.

Figure 11 shows the distributions of orbital elements of comets entering the target region in the last 500 Myr. Again, we note that the main effect of the radial component manifests itself as a failure of the integrable approximation, giving results similar to those obtained for ρ_{ref} in Sect. 3.2.

By comparing with Fig. 10, we find that: (1) the flux of comets towards small q increases drastically; (2) the values of mod$(\omega, 180°)$ are less confined to $[0°; 90°]$; (3) the flux coming from CII, CIII (and CIV for $\rho_\odot = \rho_{\max}$), which had almost disappeared when no radial component was considered, is now still present at the end of the integration.

However, the effects of the radial component show up for smaller a when $\rho_\odot$ is small rather than when it is large. In fact, the numbers of comets coming from the different cloud regions shown in Table 2 indicate that the radial component is more effective in supplying a flux from CII for small values of $\rho_\odot$, e.g. $\rho_{\min}$, while it becomes most effective for the flux from CIII for the largest density. For the flux coming from CIV, the integrable approximation has been completely broken, both because the values of a are too large and because the radial component is included.

However, two effects are still competing: (1) the efficiency of the tide in reducing q is greater for larger values of $\rho_\odot$, and (2) the cloud is depleted more quickly for larger $\rho_\odot$, thus reducing the flux at the end of the integration. These two opposing effects explain the comparable fluxes coming from CIV for the different values of $\rho_\odot$ used.

4.2 The Oort constants

The values of the Oort constants A and B that characterize the radial component of the tide (see Eq. 1) are not well known. However, the uncertainties are smaller than for $\rho_\odot$ (Mignard 2000; Olling and Dehnen 2003). Anyway, their values imply a radial component almost one order smaller than the normal component of the tide.

Basically, one has the following properties of the Oort constants. (1) $B - A = \Omega_0$, where Ω_0 is the Sun's angular velocity about the Galactic center, and (2) if $A = -B$, then the rotation curve in the solar neighbourhood is flat, i.e., the circular speed is constant with respect to the radius. If $A > -B$, it is decreasing, and if $A < -B$, it is increasing.

In studies of Oort cloud dynamics, in which the radial component of the tide has been considered, a flat rotation curve was usually assumed, like we did in the previous sections.

We will now investigate the effects of two further sets of Oort constants: one corresponding to a decreasing rotation curve, characterized by $A = 14.5$ and

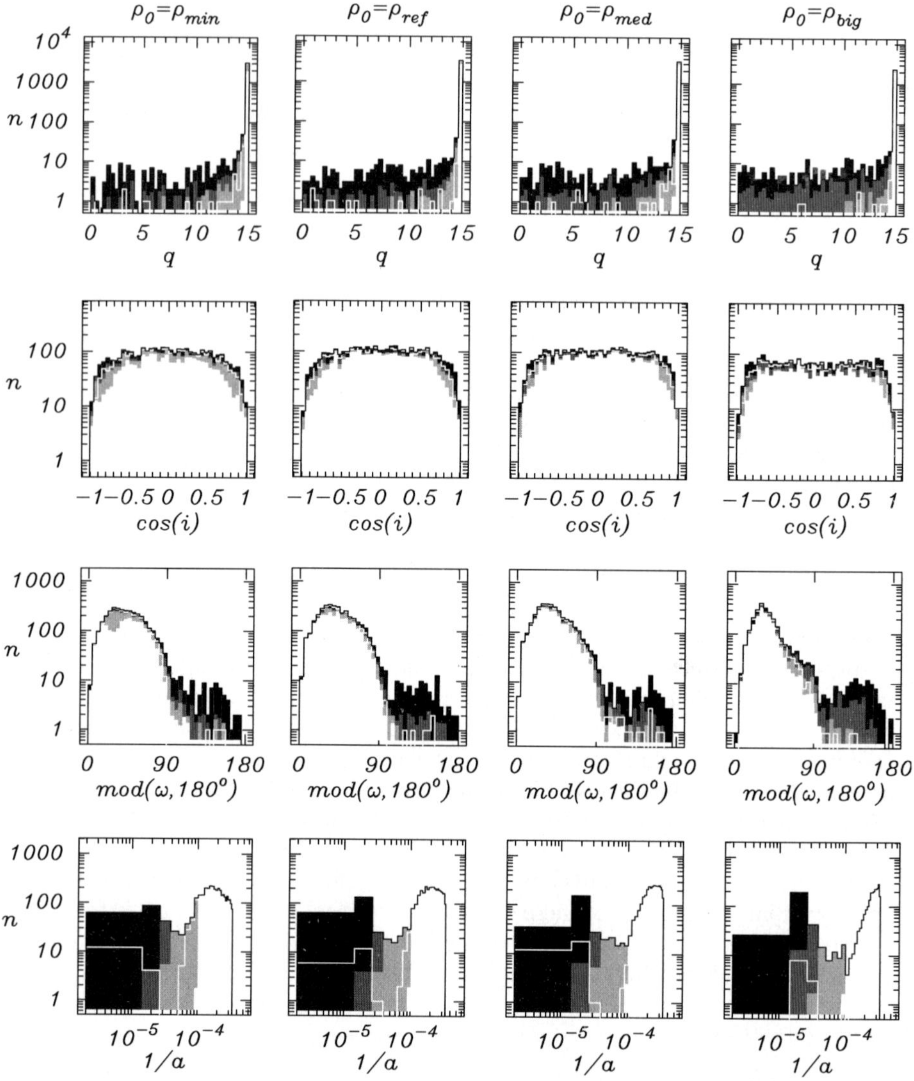

Fig. 11 The same as Fig. 10, but for models with different local densities of the Galactic disk (shown above each column) and with a radial component corresponding to a flat rotation curve. Only the flux in the last 500 Myr is shown

Table 2 Number of comets coming from different regions of the Oort cloud and entering the target region in the last 500 Myr for the four different values of $\rho_\odot$, using a model including the radial tide

$\rho_\odot$	ρ_{min}	ρ_{ref}	ρ_{med}	ρ_{max}
CII	264	125	99	65
CIII	51	53	71	133
CIV	119	164	123	136

$B = -12\,\mathrm{km\,s^{-1}\,kpc^{-1}}$, and one corresponding to an increasing rotation curve, characterized by $A = 11.3$ and $B = -13.9\,\mathrm{km\,s^{-1}\,kpc^{-1}}$. Like in Sect. 3, the local density of the Galactic disk is set equal to ρ_{ref}.

Figure 12 shows the distributions of the orbital elements of the comets entering the target region in the first 500 Myr (left) and the last 500 Myr (right). No clear differences are apparent between the two models, no matter which orbital element is considered. A similar conclusion can be drawn from the comparison of Figs. 12 and 7. The only

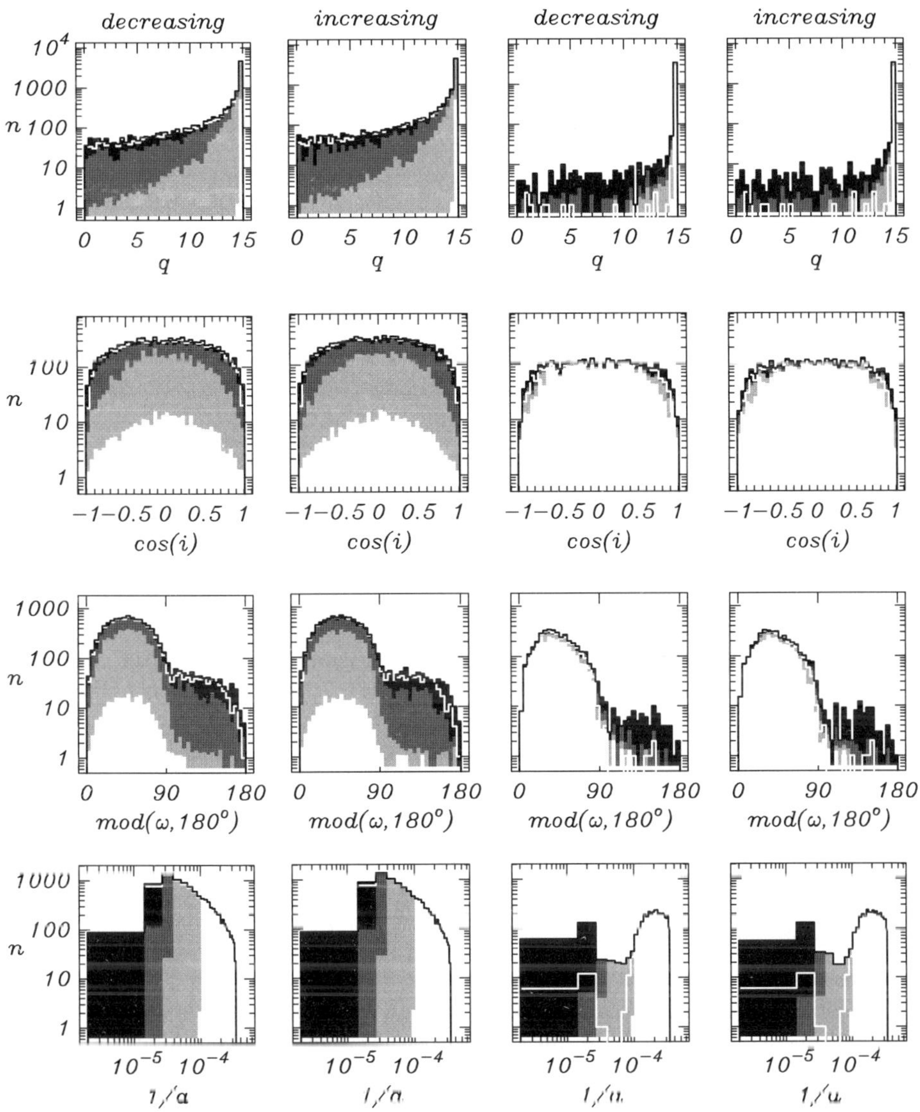

Fig. 12 The same as Fig. 7 for two different sets of Oort constants: $A = 14.5$ and $B = -12$ (columns labelled *decreasing*) and $A = 11.3$ and $B = -13.9$ (columns labelled *increasing*). The local density of the Galactic disk is ρ_{ref} in both cases. The two left columns correspond to the flux in the first 500 Myr, and the two right columns correspond to the last 500 Myr

differences appear, trivially, in cases of small number statistics. Hence, deviations from a flat rotation curve do not seem to affect the dynamics of the Oort cloud.

5 Conclusions

We have studied the long-term effects of the Galactic tide on Oort cloud comets using the "hybrid model" (see Papers I, II and Sect. 2). This allows to perform integrations of 10^6 comets over 5 Gyr in a reasonable amount of time (less than 48 hours on a 2.8 GHz XEON computer, instead of about 3 months when integrating the full equations of motion). The computations were performed, both considering only the normal component of the tide (model $\mathcal{A}$), and taking into account both the normal and radial tidal components (model $\mathcal{B}$).

We have found that the main factor governing the effects of the Galactic tide on the orbital evolution of comets is the validity (or invalidity) of the "integrable approximation", i.e., whether (or not) the dynamics is well described by an integrable model. In fact, the integrability of the dynamics implies that the orbital domain from which comets enter the target region is well defined and represents only a small part of the Oort cloud.

The main results of our numerical investigation are:

- For both models, $\mathcal{A}$ and $\mathcal{B}$, the flux of comets entering the target region ($q \leq 15$ AU) is three times greater at the beginning of the integration than at the end. This is due to the validity of the integrable approximation in the regions bounded by $a < 25\,400$ AU for model $\mathcal{A}$ and $a < 11\,200$ AU for model $\mathcal{B}$. As time passes, the part of these regions from which comets may enter the target region gets depleted, and the flux of comets entering the target region decreases.
- As found also in Paper I, the number of comets entering the target region is greater when the radial component is taken into account (model $\mathcal{B}$), than in the model where this component is neglected.
- For both models, the Galactic tide induces an accumulation of comets with argument of perihelion ω (modulo 180°) close to 26.6°. As a consequence, the flux of comets with $a < 10^4$ AU increases during the first 500 Myr.

On the other hand, the distributions of orbital elements for model $\mathcal{B}$ are very different from those of model $\mathcal{A}$:

- The minimum values of q are smaller when the radial component is included, and during the last 500 Myr of our integrations the number of comets entering the target region with $q < 5$ AU is about eight times larger than in model $\mathcal{A}$.
- The flux of comets coming from the central regions of the Oort cloud ($10^4 < a < 5 \times 10^4$ AU) is almost equal to zero when only the normal component of the tide is considered, while it is relatively large when the radial component is included.
- The flux of comets whose orbital planes are close to the Galactic plane ($|\cos i| \simeq 1$) is larger in model $\mathcal{B}$ than in model $\mathcal{A}$ (see Paper I).

We have also analysed the effects of the local density $\rho_\odot$ of the Galactic disk. When $\rho_\odot$ increases, the Oort cloud is depleted more rapidly (e.g., during the first Gyr the flux entering the target region is three times higher for $\rho_\odot = 0.26\,M_\odot\,\mathrm{pc}^{-3}$ than for $\rho_\odot = 0.07\,M\odot\,\mathrm{pc}^{-3}$), and the perihelion distances q reach lower values.

Finally, we have not found significant changes in the orbital evolution of comets with different sets of values for the Oort constants.

All the Galactic models used in the present study make use of the hypothesis that the Galactic potential is axisymmetric. Some studies have indicated that this might not be so (Olling and Dehnen 2003), even though the non-axisymmetry may be drastically reduced by shifting the location of the Galactic centre (Mignard 2000). In any case, the influence of the non-axisymmetry should be very small. Moreover, since the motion of the Sun around the Galactic centre is poorly known on very long-time scales, all our Galactic models adopt constant values of $\rho_\odot$, A and B.

The aim of the present paper is to analyse separately various aspects of the influence of the Galactic field, potentially affecting the dynamics of comets in the Oort cloud, in order to get a quantitative idea of their relative importance. The dynamics of comets in the real Oort cloud is much more complicated, since passing stars and planetary perturbations are present and have to be taken into account. These two mechanisms will be introduced in a forthcoming paper.

Acknowledgements M.F. is grateful to ESA for financial support at INAF-IASF. Part of this work has been performed using the computing facilities provided by the program "Simulations Interactives et Visualisation en Astronomie et Mécanique 2 (SIVAM2)". H.R. gratefully acknowledges the support of the Swedish National Space Board via Grant Nr. 64/02:3.

Appendix A: Complete Hamiltonian equations of the model using Delaunay's elements

$$
\begin{aligned}
\left\langle \frac{\mathrm{d}L}{\mathrm{d}t} \right\rangle &= 0, \\
\left\langle \frac{\mathrm{d}G}{\mathrm{d}t} \right\rangle &= -\frac{5L^2}{2\mu^2}\left(L^2 - G^2\right)\left\{ \cos g \sin g \Big[\mathcal{G}_3 \left(1 - \frac{H^2}{G^2}\right) \right. \\
&\quad + \left(\mathcal{G}_1 \sin^2 h_r + \mathcal{G}_2 \cos^2 h_r\right) \frac{H^2}{G^2} - \mathcal{G}_1 \cos^2 h_r \\
&\quad \left. - \mathcal{G}_2 \sin^2 h_r \Big] - (\mathcal{G}_1 - \mathcal{G}_2)\left(\cos^2 g - \sin^2 g\right) \cos h_r \sin h_r \frac{H}{G} \right\}, \\
\left\langle \frac{\mathrm{d}H}{\mathrm{d}t} \right\rangle &= -\frac{L^2}{2\mu^2}(\mathcal{G}_1 - \mathcal{G}_2)\left\{ 5(L^2 - G^2)\frac{H}{G} \cos g \sin g \right. \\
&\quad + \left(\cos^2 h_r - \sin^2 h_r\right) + \sin h_r \cos h_r \Big[G^2 - H^2 \\
&\quad \left. + 5(L^2 - G^2)\left(\cos^2 g - \sin^2 g \frac{H^2}{G^2}\right) \Big] \right\}, \\
\left\langle \frac{\mathrm{d}g}{\mathrm{d}t} \right\rangle &= \frac{L^2 G}{2\mu^2}\left\{ \mathcal{G}_3 \Big[1 - 5 \sin^2 g \left(1 - \frac{L^2 H^2}{G^4}\right) \Big] \right. \\
&\quad + \left(\mathcal{G}_1 \cos^2 h_r + \mathcal{G}_2 \sin^2 h_r\right)\left(1 - 5 \cos^2 g\right) \\
&\quad - 5\left(\mathcal{G}_1 \sin^2 h_r + \mathcal{G}_2 \cos^2 h_r\right) \frac{L^2 H^2}{G^4} \sin^2 g \\
&\quad \left. + 5(\mathcal{G}_1 - \mathcal{G}_2) \cos g \sin g \cos h_r \sin h_r \left(G^2 + L^2\right) \frac{H}{G^3} \right\},
\end{aligned}
$$

$$\left\langle \frac{\mathrm{d}h}{\mathrm{d}t} \right\rangle = \frac{L^2}{2G\mu^2}\Bigg\{ \Big(\mathcal{G}_1 \sin^2 h_r + \mathcal{G}_2 \cos^2 h_r - \mathcal{G}_3\Big) \times \Big[G^2 + 5(L^2 - G^2)\sin^2 g\Big]\frac{H}{G} - 5(\mathcal{G}_1 - \mathcal{G}_2) \times (L^2 - G^2)\cos g \sin g \cos h_r \sin h_r \Bigg\}.$$

Appendix B: Proof of the canonicity of the Matese elements

In order to show that the Matese elements $L = \sqrt{\mu a}, \Theta = \sqrt{\mu a(1-e^2)}\cos\alpha$, $H = \sqrt{\mu a(1-e^2)}\cos i, M, \theta = b$ and $\lambda = l$ (see Fig. 1 for the definition of α, l and b) are canonical, let us consider a canonical transformation from the Delaunay elements to new momenta and coordinates $(\mathbf{I}, \mathbf{w})$, defined by the mixed-variable generating function:

$$\mathcal{S} = Lw_1 + \left(w_3 - \frac{\pi}{2}\right)H + \frac{\pi}{2}G \pm \left[G \arccos\frac{G \sin w_2}{\sqrt{G^2 - H^2}} - H \arccos\frac{H \tan w_2}{\sqrt{G^2 - H^2}}\right], \tag{16}$$

where $-\frac{\pi}{2} \le w_2 \le \frac{\pi}{2}$ and $0 \le w_1, w_3 < 2\pi$. The arguments of the arccosines have absolute values less than 1 if $|\sin w_2| \le \sin i$, i.e., $-\min(i, \pi - i) \le w_2 \le \min(i, \pi - i)$.

We can now compute the values of I_1, I_2, I_3:

$$I_1 = \frac{\partial \mathcal{S}}{\partial w_1} = L \quad \text{and} \quad I_2 = \frac{\partial \mathcal{S}}{\partial w_2} = \mp\frac{\sqrt{G^2\cos^2 w_2 - H^2}}{\cos w_2},$$

where the upper sign holds when $\frac{3\pi}{2} > g > \frac{\pi}{2}$, and the lower sign when $-\frac{\pi}{2} < g < \frac{\pi}{2}$.

Using the relations $-\cos\alpha\cos\theta = \cos g \sin i$ and $\sin i \sin g = \sin\theta$ we obtain:

$$\cos\alpha = \frac{\cos g \sin i}{-\cos\theta} = \pm\frac{\sqrt{G^2\cos^2\theta - H^2}}{G\cos\theta}$$

with the same rule as before for the sign. Identifying with the above expression for I_2, we indeed find:

$$I_2 = -G\cos\alpha = \Theta.$$

Then, we easily find:

$$I_3 = \frac{\partial \mathcal{S}}{\partial w_3} = H \quad \text{and} \quad l = \frac{\partial \mathcal{S}}{\partial L}.$$

For g, we have:

$$g = \frac{\partial \mathcal{S}}{\partial G} = \frac{\pi}{2} \pm \frac{\partial}{\partial G}\left[G\cos^{-1}\frac{G\sin w_2}{\sqrt{G^2 - H^2}} - H\cos^{-1}\frac{H\tan w_2}{\sqrt{G^2 - H^2}}\right],$$

which yields:

$$\sin g = \frac{\sin w_2}{\sin i}. \tag{17}$$

For h:

$$h = \frac{\partial \mathcal{S}}{\partial H} = w_3 - \frac{\pi}{2} \pm \frac{\partial}{\partial H}\left[G \cos^{-1} \frac{G \sin w_2}{\sqrt{G^2 - H^2}} - H \cos^{-1} \frac{H \tan w_2}{\sqrt{G^2 - H^2}}\right]$$

and hence:

$$\sin(h - w_3) \tan i = -\tan w_2. \tag{18}$$

Finally, identifying Eqs. 17 and 18 with the relations $\sin g \sin i = \sin\theta$ and $\sin(h-\lambda)\tan i = -\tan\theta$, we obtain $w_2 = \theta$ and $w_3 = \lambda$.

Appendix C: Complete Hamiltonian equations of the model using Matese elements

$$\left\langle \frac{\mathrm{d}L}{\mathrm{d}t} \right\rangle = 0,$$

$$\begin{aligned}
\left\langle \frac{\mathrm{d}\Theta}{\mathrm{d}t} \right\rangle = \frac{L^2}{2\mu^2}\Big[& \cos\theta \sin\theta(-4\Theta^2 + 5L^2)(\mathcal{G}_1 \cos^2\lambda_r + \mathcal{G}_2 \sin^2\lambda_r) \\
& -\mathcal{G}_1\left(\Theta \cos\theta \cos\lambda_r + H \sin\lambda_r \frac{\sin\theta}{\cos^2\theta}\right)\left(\Theta \sin\theta \cos\lambda_r + H \sin\lambda_r \frac{1}{\cos\theta}\right) \\
& -\mathcal{G}_2\left(\Theta \cos\theta \cos\lambda_r - H \cos\lambda_r \frac{\sin\theta}{\cos^2\theta}\right)\left(\Theta \sin\theta \sin\lambda_r - H \cos\lambda_r \frac{1}{\cos\theta}\right) \\
& +\mathcal{G}_3 \frac{\sin\theta}{\cos^3 b}\left(5(\Theta^2 - L^2)\cos^4\theta + 4H^2\right)\Big],
\end{aligned}$$

$$\begin{aligned}
\left\langle \frac{\mathrm{d}H}{\mathrm{d}t} \right\rangle = \frac{L^2(\mathcal{G}_1 - \mathcal{G}_2)}{2\mu^2 \cos^2\theta}\Big[& (-4\Theta^2 \cos^2\theta - 4H^2 + 5L^2 \cos^2\theta)\cos^2\theta \cos\lambda_r \sin^2\lambda_r \\
& + (\Theta \cos\theta \sin\theta \sin\lambda_r - H \cos\lambda_r)(\Theta \cos\theta \sin\theta \cos\lambda_r + H \sin\lambda_r)\Big],
\end{aligned}$$

$$\begin{aligned}
\left\langle \frac{\mathrm{d}\theta}{\mathrm{d}t} \right\rangle = \frac{L^2}{2\mu^2 \cos^2\theta}\Big\{ & -4\Theta \cos^4\theta\left(\mathcal{G}_1 \cos^2\lambda_r + \mathcal{G}_2 \sin^2\lambda_r\right) \\
& + \mathcal{G}_1 (\Theta \cos\theta \sin\theta \cos\lambda_r + H \sin\lambda_r)\cos\theta \sin\theta \cos\lambda_r \\
& + \mathcal{G}_2 (\Theta \cos\theta \sin\theta \sin\lambda_r - H \cos\lambda_r)\cos\theta \sin\theta \sin\lambda_r \\
& + \mathcal{G}_3 \Theta \cos^2\theta(1 - 5\sin^2\theta)\Big\},
\end{aligned}$$

$$\begin{aligned}
\left\langle \frac{\mathrm{d}\lambda}{\mathrm{d}t} \right\rangle = \frac{L^2}{2\mu^2 \cos^2\theta}\Big\{ & -4H \cos^2\theta\left(\mathcal{G}_1 \cos^2\lambda_r + \mathcal{G}_2 \sin^2\lambda_r\right) \\
& + \mathcal{G}_1 \sin\lambda_r (\Theta \cos\theta \sin\theta \cos\lambda_r + H \sin\lambda_r) \\
& - \mathcal{G}_2 \cos\lambda_r (\Theta \cos\theta \sin\theta \sin\lambda_r - H \cos\lambda_r) - 4\mathcal{G}_3 H \sin^2\theta\Big\}.
\end{aligned}$$

References

Bahcall, J.N., Flynn, C., Gould, A.: Local dark matter from a carefully selected sample. ApJ **389**, 234–250 (1992)

Bailey, M.E.: The structure and evolution of the solar system comet cloud. MNRAS **204**, 603–633 (1983)

Brasser, R.: Some properties of a two-body system under the influence of the Galactic tidal field. MNRAS **324**, 1109–1116 (2001)

Breiter, S., Ratajczak, R.: Vectorial elements for the Galactic disc tide effects in cometary motion. MNRAS, **364**, 1222–1228 (2005)

Breiter, S., Dybczyński, P.A., Elipe, A.: The action of the galactic disk on the Oort cloud comets. Qualitative study. A & A **315**, 618–624 (1996)

Duncan, M., Quinn, T., Tremaine, S.: The formation and extent of the solar system comet cloud. AJ **94**, 1330–1338 (1987)

Dybczyński, P.A.: Simulating observable comets II. Simultaneous stellar and galactic action. A & A **441**, 783–790 (2005)

Elipe, A., Ferrer, S.: Reductions, relative equilibria, and bifurcations in the generalized van der Waals potential: relation to the integrable cases. Phys. Rev. Lett. **72**, 985–988 (1994)

Everhart, E.: An efficient integrator that uses Gauss-Radau spacings. In: Carusi, A., Valsecchi, G.B. (eds.) Proc. IAU Colloq. 83, Dynamics of Comets: Their Origin and Evolution, p. 185, Reidel, Dordrecht (1985)

Fouchard, M.: New fast models of the galactic tide. MNRAS **349**, 347–356 (2004)

Fouchard, M., Froeschlé, C., Matese, J.J., Valsecchi, G.B.: Comparison between different models of the galactic tidal effects on cometary orbits. Celest. Mech. Dyn. Astron. **93**, 231–264 (2005)

García-Sánchez, J., Weissman, P.R., Preston, R.A., Jones, D.L., Lestrade, J.-F., Latham, D.W., Stefanik, R.P., Paredes, J.M.: Stellar encounters with the solar system. A & A **379**, 634–659 (2001)

Hills, J.G.: Comet showers and the steady-state infall of comets from the Oort cloud, Astron. J. **86**, 1730–1740 (1981)

Heisler, J.: Monte Carlo simulations of the Oort comet cloud. Icarus **75**, 104–121 (1990)

Heisler, J., Tremaine, S.: The influence of the galactic tidal field on the Oort comet cloud, Icarus **65**, 13–26 (1986)

Holmberg, J., Flynn, C.: The local density of matter mapped by Hipparcos. MNRAS **313**, 209–216 (2000)

Levison, H., Dones, L., Duncan, M.J.: The origin of Halley-type comets: probing the inner Oort cloud. Astron. J. **121**, 2253–2267 (2001)

Matese, J.J., Lissauer, J.J.: Characteristics and frequency of weak stellar impulses of the Oort cloud. Icarus **157**, 228–240 (2002)

Matese, J.J., Lissauer, J.J.: Perihelion evolution of observed new comets implies the dominance of the galactic tide in making Oort cloud comets discernable. Icarus **170**, 508–513 (2004)

Matese, J.J., Whitman, P.G.: The galactic disk tidal field and the nonrandom distribution of observed Oort cloud comets. Icarus **82**, 389–401 (1989)

Matese, J.J., Whitman, P.G.: A model of the galactic tidal interaction with the Oort comet cloud, Celest. Mech. Dyn. Astron. **54**, 13–35 (1992)

Mignard, F.: Local galactic kinematics from Hipparcos proper motions. A & A **354**, 522–536 (2000)

Neslušan, L., Jakubík, M.: Some characteristics of the outer Oort cloud as inferred from observations of new comets. A & A **437**, 1093–1108 (2005)

Olling, R.P., Dehnen, W.: The Oort constants measured from proper motions. ApJ **599**, 275–296 (2003)

Olling, R.P., Merrifield, M.R.: Refining the Oort and Galactic constants. MNRAS **297**, 943–952 (1998)

Oort, J.H.: The structure of the cloud of comets surrounding the Solar System and a hypothesis concerning its origin. Bull. Astron. Inst. Neth. **11**, 91–110 (1950)

Tommei, G: Canonical elements for Öpik theory. Celest. Mech. Dyn. Astron., **94**, 173–195 (2006)

Tremaine, S.: Canonical elements for collision orbits. Celest. Mech. Dyn. Astron. **79**, 231–233 (2000)

Wiegert, P., Tremaine, S.: The evolution of long-period comets. Icarus **137**, 84–121 (1999)

Celestial Mechanics and Dynamical Astronomy (2006) 95:327–344
DOI 10.1007/s10569-006-9022-0

ORIGINAL ARTICLE

Dependence on the observational time intervals and domain of convergence of orbital determination methods

Alessandra Celletti · Gabriella Pinzari

Received: 15 November 2005 / Revised: 21 March 2006 /
Accepted: 13 April 2006 / Published online: 15 August 2006

Abstract In the framework of the orbital determination methods, we study some properties related to the algorithms developed by Gauss, Laplace and Mossotti. In particular, we investigate the dependence of such methods upon the size of the intervals between successive observations, encompassing also the case of two nearby observations performed within the same night. Moreover we study the convergence of Gauss algorithm by computing the maximal eigenvalue of the jacobian matrix associated to the Gauss map. Applications to asteroids and Kuiper belt objects are considered.

Keywords Orbital determination · Gauss method · Laplace method · Mossotti method

1 Introduction

The determination of the orbital motion of a celestial body can be obtained through the celebrated methods of Gauss or Laplace, once a certain number (at least 3) of astronomical observations are available (see Poincaré 1906; Moulton 1914; Plummer 1918; Herrick 1937 for discussions on Laplace and Gauss methods). An alternative technique was developed by Mossotti in the 19th century. The three methods (Gauss, Laplace and Mossotti) have been extensively reviewed and compared in Celletti and Pinzari (2005). In this work, we want to explore the dependence of the three techniques upon the observational time intervals. It is relevant to quote the recent results obtained

A. Celletti
Dipartimento di Matematica, Università di Roma Tor Vergata, Via della Ricerca Scientifica 1, I-00133 Roma, Italy
e-mail: celletti@mat.uniroma2.it

G. Pinzari (✉)
Dipartimento di Matematica, Università "Roma Tre", Largo S. L. Murialdo 1, I-00146 Roma, Italy
e-mail: pinzari@mat.uniroma3.it

in Milani et al. (2004, 2005), Milani and Knezevic (2005); in these works the authors investigate the problem of the orbit determination using two pairs of data, each of them composed by observations at very close times. As a consequence very few observational data are available to compute the orbit and efficient mathematical methods have been developed to sample the *admissible region* for the undetermined variables.

Let t_1, t_2, t_3 be the times of the three observations; having fixed the intermediate time t_2, we vary the time intervals $t_2 - t_1$ and $t_3 - t_2$, ranging from a few hours (whenever two observations are performed on the same night) to several days. Two sets of data are investigated: the first 10,000 numbered asteroids and 615 Kuiper belt objects. While in the first case Gauss method provides the best results, the orbital determination of Kuiper belt objects seems to privilege Laplace method, being Mossotti's technique intermediate in all cases. For the selected samples of data, the recovery of the orbits of the asteroidal belt improves as the time intervals decrease, while it improves within the Kuiper belt objects whenever the time intervals increase. A statistic of the successful results in terms of the elliptic elements (semi-major axis, eccentricity and inclination) is also performed. In the second part of the paper, we concentrate on Gauss algorithm to investigate the stability domain of such method, by looking at the eigenvalues of the jacobian matrix associated to the Gauss map. We provide a numerical investigation performed on asteroids and Kuiper belt objects. We also develop an analytical estimate of the first-order computation of the largest eigenvalue; we prove a proposition ensuring the convergence of Gauss method, which is related to the contractive character of the Gauss map, at least for small values of the observational times.

2 Implementation of Gauss, Laplace and Mossotti methods

2.1 Basics of the methods

With reference to a heliocentric frame let us denote the unknown elements of the asteroid as follows: a is the semi-major axis, e is the eccentricity, i denotes the inclination, ω is the argument of perihelion, Ω is the longitude of the ascending node and M is the mean anomaly at a fixed epoch T. We assume that the ecliptic geocentric longitudes and latitudes, say λ_i and β_i, $i = 1, \dots, N$, are given through N observations at times t_i referred to the epoch T. Moreover, let $t \to \vec{a}(t)$ denote the Sun–Earth vector, $t \to \vec{r}(t)$ is the Sun–asteroid vector, while $t \to \rho(t)$ is the geocentric distance and $t \to \vec{b}(t)$ with $|\vec{b}(t)| = 1$ denotes the Earth–object direction.

We assume to perform three observations at times t_1, t_2, t_3. The time intervals $t_{ij} = t_j - t_i$, $i, j = 1, 2, 3$, are regarded as small quantities of order ε; for some positive constants γ_{12}, γ_{23}, with $\gamma_{12} + \gamma_{23} = 1$, we set

$$\varepsilon \equiv t_{13} \qquad t_{12} = \gamma_{12}\varepsilon \qquad t_{23} = \gamma_{23}\varepsilon. \tag{1}$$

Let $\vec{k}$ be the unit vector perpendicular to the plane of the orbit; the coplanarity condition of the vectors $\vec{r}_i = \vec{a}_i + \rho_i \vec{b}_i$, $i = 1, 2, 3$, reads as

$$n_{23}\vec{r}_1 - n_{13}\vec{r}_2 + n_{12}\vec{r}_3 = 0,$$

where $n_{ij} = \vec{r}_i \wedge \vec{r}_j \cdot \vec{k}$ is twice the oriented area of the triangle spanned by $\vec{r}_i$ and $\vec{r}_j$. If $\vec{b}_1, \vec{b}_2, \vec{b}_3$ are linearly independent, one can express ρ_i as linear functions (with coefficients of $O(\varepsilon^{-2})$) of the ratios n_{ik}/n_{lk} with $i \neq l \neq k$. The first goal of Gauss method is to find a good approximation of ρ_i, say up to terms of $O(\varepsilon)$. To this end, let

S_{ij} be the areas of the elliptic sectors spanned between t_i and t_j, and let $\eta_{ij} = n_{ij}/S_{ij}$, f_{ij} be half the angle between $\vec{r}_i$, $\vec{r}_j$. Denote by $z = (P, Q)$ a new set of quantities, called Gauss parameters, defined as

$$\begin{aligned} P &= \tfrac{n_{12}}{n_{23}} = \tfrac{\gamma_{12}}{\gamma_{23}} f(\eta_{12}, \eta_{23}), \\ Q &= 2r_2^3\left(\tfrac{n_{12}+n_{23}}{n_{13}} - 1\right) = \gamma_{12}\gamma_{23}\, \varepsilon^2 g\left(\eta_{12}, \eta_{23}, \tfrac{r_1}{r_2}, \tfrac{r_2}{r_3}, f_{12}, f_{23}\right), \end{aligned} \tag{2}$$

where f and g are suitable functions differing from one up to $O(\varepsilon^2)$, $O(\varepsilon)$, respectively (see Celletti and Pinzari 2005). The quantities ρ_i can be expressed in terms of P, Q as

$$\rho_2 = G_2(P, Q, \rho_2), \qquad \rho_1 = G_1(P, Q, \rho_2), \qquad \rho_3 = G_3(P, Q, \rho_2)$$

for suitable functions G_i, $i = 1, 2, 3$ (see Appendix B for explicit expressions of the G_i). In particular $\rho_2 = \rho_2(P, Q)$ is a solution of an implicit equation, from which we derive $\rho_1 = \rho_1(P, Q)$, $\rho_3 = \rho_3(P, Q)$. Finally, setting

$$P_0 = \gamma_{12}/\gamma_{23}, \qquad Q_0 = \gamma_{12}\gamma_{23}\, \varepsilon^2 \tag{3}$$

one finds that $G_i(P, Q, \rho_2) = G_i(P_0, Q_0, \rho_2) + O(\varepsilon)$, namely $\rho_i = \rho_{i,0} + O(\varepsilon)$, where $\rho_{i,0} = \rho_i(P_0, Q_0)$.

Gauss algorithm is inductively based on the following steps:

(1) start from $z_0 = (P_0, Q_0)$;
(2) given $z_n = (P_n, Q_n)$, compute $\rho_{2,n} = \rho_2(P_n, Q_n)$ trying to solve the implicit equation $\rho_{2,n} = G_2(P_n, Q_n, \rho_{2,n})$ and let, for $i = 1, 3$, $\rho_{i,n} = \rho_i(P_n, Q_n)$. The three vectors $\vec{r}_{i,n} = \vec{a}_i + \rho_{i,n}\vec{b}_i$, $i = 1, 2, 3$ are shown to be coplanar;
(3) if the endpoints of $\vec{r}_{1,n}, \vec{r}_{2,n}, \vec{r}_{3,n}$ are not on a straight line, there exists a unique conic $\mathcal{C}_n$ through $\vec{r}_{1,n}, \vec{r}_{2,n}, \vec{r}_{3,n}$; compute the quantities $\eta_{ij,n}$, $f_{ij,n}$ $r_{i,n}$ on $\mathcal{C}_n$;
(4) determine the new parameters $z_{n+1} = (P_{n+1}, Q_{n+1})$ through (2), where the r.h.s. are computed with $\eta_{ij,n}$, $f_{ij,n}$, $r_{i,n}$ replacing η_{ij}, f_{ij}, r_i. Such procedure defines the Gauss map $\mathcal{F}(\mathcal{C}, t_2, \gamma_{12}, \gamma_{23}, \varepsilon) \equiv \mathcal{F}_G$ as $z_{n+1} = \mathcal{F}_G(z_n)$;
(5) look for a fixed point of the Gauss map, motivated by the fact that a conic section $\mathcal{C}$ (on which a Keplerian motion takes place) is a solution of Gauss problem if and only if it corresponds to a fixed point of $\mathcal{F}_G$.

We can finally summarize Gauss method (Gauss 1809; see also Gallavotti 1980) with the following

Theorem 2.1 *Let $\mathcal{C}, t_2, \gamma_{12}, \gamma_{23}, \varepsilon$ be such that $\vec{b}_1, \vec{b}_2, \vec{b}_3$ are linearly independent, and $\partial_\rho G_2(P, Q, \rho)|_{\rho_2} \neq 1$, where $z = (P, Q)$ is the fixed point of $\mathcal{F}_G$, defined in (2). Let D be the domain of definition of $\mathcal{F}_G$, $U \subset D$ a neighbourhood of z, V a neighbourhood of ρ_2, $\rho : z' = (P', Q') \in U \to \rho(P', Q') \in V$ be the smooth solution of $\rho = G_2(P', Q', \rho)$ such that $\rho(P, Q) = \rho_2$. If $z_0 \in U$, the associated conic section $\mathcal{C}_0$ verifies: $\mathcal{C} - \mathcal{C}_0 = O(\varepsilon)$. Finally, if $z_n \subset U$, the associated conic section $\mathcal{C}_n$ verifies: $\mathcal{C} - \mathcal{C}_n = O(\varepsilon^{n+1})$.*

A different approach is provided by Laplace method, whose aim is to find an approximation of the position $\vec{r}$ and the velocity $\vec{v}$, so to determine the unknown orbit. Let $r = r(\rho) = |\vec{a} + \rho\vec{b}|$ be the heliocentric distance; using the equations of motion, one gets an implicit equation in the unknowns $\dot{\lambda}, \dot{\beta}, \ddot{\lambda}, \ddot{\beta}$:

$$\rho = \frac{d_1}{d}\left(\frac{1}{r^3} - \frac{1}{a^3}\right) \equiv L(d_1/d, \rho). \tag{4}$$

Moreover, one finds that $\dot{\rho} = \frac{d_2}{d}\left(\frac{1}{r^3} - \frac{1}{a^3}\right)$, with $d = d(\lambda, \beta, \dot{\lambda}, \dot{\beta}, \ddot{\lambda}, \ddot{\beta})$, $d_1 = d_1(\lambda, \beta, \dot{\lambda}, \dot{\beta})$, $d_2 = d_2(\lambda, \beta, \ddot{\lambda}, \ddot{\beta})$ (see Celletti and Pinzari 2005, for the explicit expressions of d, d_1, d_2). Given the N observations $(\lambda_1, \beta_1), (\lambda_2, \beta_2), \ldots, (\lambda_N, \beta_N)$, Laplace method (Laplace 1780) consists in replacing $\dot{\lambda}, \ddot{\lambda}$ (equivalently $\dot{\beta}, \ddot{\beta}$) by the derivatives of some interpolating polynomials of degree $N-1$ obtained through the observed data $(t_1, \lambda_1), (t_2, \lambda_2), \ldots, (t_N, \lambda_N)$ (equivalently $(t_1, \beta_1), (t_2, \beta_2), \ldots, (t_N, \beta_N)$).

An alternative technique was developed by Mossotti (1942) and it is based on the following procedure. Writing the coplanarity condition among $\vec{r}(t), \vec{r}_2, \vec{v}_2$ as

$$\vec{r}(t) = T(t)\vec{r}_2 + V(t)\vec{v}_2 \tag{5}$$

and developing the equation of motion $\ddot{\vec{r}} = -\vec{r}/r^3$ in Taylor series with initial data $\vec{r}(t_2) = \vec{r}_2, \dot{\vec{r}}(t_2) = \vec{v}_2$, one obtains

$$T(t) = 1 - \frac{(t-t_2)^2}{2r_2^3}h(t), \qquad V(t) = (t-t_2)k(t),$$

where $h(t)$ and $k(t)$ are suitable functions; if h_i and k_i denote their values at times t_i, one can show that h_i and k_i differ from one up to $O(\varepsilon)$. Using (2.5) computed at t_1 and t_3, one can express ρ_2 and $\vec{v}_2$ as

$$\begin{aligned} \rho_2 &= M(h_1, h_3, k_1, k_3, \rho_2) = M(1,1,1,1,\rho_2) + O(\varepsilon), \\ \vec{v}_2 &= \vec{N}(h_1, h_3, k_1, k_3, \rho_2) = \vec{N}(1,1,1,1,\rho_2) + O(\varepsilon) \end{aligned}$$

for suitable (vector) functions $M, \vec{N}$. In conclusion, it turns out that ρ_2 is a solution of an implicit equation, which can be solved in analogy to Gauss method.

2.2 Iteration of the methods

A major advantage of Gauss method with respect to the others is that it provides an iterative procedure to find better approximations of the solution. On the contrary, the methods of Laplace (implemented over three observations) and Mossotti were originally limited to the first-order approximation. However, an iterative scheme can be implemented along the following lines.

Let us consider first the method of Laplace. Let $R(t)$ denote the remainder function of order 3 of the series expansion of $\lambda(t)$ around t_2, namely $\lambda(t) = P(t) + R(t)$, with $P(t) = \lambda_2 + \dot{\lambda}(t_2)(t-t_2) + \frac{\ddot{\lambda}(t_2)}{2}(t-t_2)^2$ (obviously $R(t_2) = 0$). In other words, $\dot{\lambda}_2 \equiv \dot{\lambda}(t_2)$, $\ddot{\lambda}_2 \equiv \ddot{\lambda}(t_2)$ are the derivatives of the interpolating polynomial $t \to P(t)$ of degree 2 through $\lambda_1 - R_1, \lambda_2, \lambda_3 - R_3$ (here, $R_i = R(t_i)$), at times t_1, t_2, t_3. Similarly for $\dot{\beta}(t_2), \ddot{\beta}(t_2)$, where the remainder functions are denoted as S_1, S_3. When $\dot{\lambda}_2$, $\dot{\beta}_2$, $\ddot{\lambda}_2$, $\ddot{\beta}_2$ are expressed as functions of R_1, R_3, S_1, S_3, Eq. 4, with $t = t_2$, takes the form (without changing the symbol for L) $\rho_2 = L(R_1, R_3, S_1, S_3, \rho_2)$; the first approximation ($N = 3$) of Laplace corresponds to take $R_i = S_i = 0$ ($i = 1, 3$). We are therefore led to define a sequence of remainder functions $R_{i,n}, S_{i,n}$ as follows:

(1) Start with $R_{1,0} = R_{3,0} = 0$ ($S_{1,0} = S_{3,0} = 0$).
(2) Given $R_{1,n}$, $R_{3,n}$ ($S_{1,n}, S_{3,n}$), let $\dot{\lambda}_n$, $\ddot{\lambda}_n$ ($\dot{\beta}_n, \ddot{\beta}_n$) be defined as the derivatives of the interpolating polynomial $t \to P_n(t)$ ($t \to Q_n(t)$) of degree 2 through $\lambda_1 - R_{1,n}$, λ_2, $\lambda_3 - R_{3,n}$ ($\beta_1 - S_{1,n}, \beta_2, \beta_3 - S_{3,n}$) at times t_1, t_2, t_3, respectively. Let $d_n = d(\lambda_2, \beta_2, \dot{\lambda}_n, \dot{\beta}_n, \ddot{\lambda}_n, \ddot{\beta}_n), d_{1,n} = d_1(\lambda_2, \beta_2, \dot{\lambda}_n, \dot{\beta}_n), d_{2,n} = d_2(\lambda_2, \beta_2, \ddot{\lambda}_n, \ddot{\beta}_n)$. If $d_n \neq 0$, compute the position $\vec{r}_{2,n}$ and the velocity $\vec{v}_{2,n}$. Let C_n, be the conic

describing a Keplerian motion with initial data $\vec{r}_{2,n}, \vec{v}_{2,n}$ (whenever the latter vectors are not parallel), and let $t \to \lambda_n(t), t \to \beta_n(t)$ be the motion of the angles.

(3) Define $R_{i,n+1}$, $S_{i,n+1}$ as the remainder functions of order 3 of the Taylor expansion of $t \to \lambda_n(t), t \to \beta_n(t)$ around $t = t_2$. Introduce the Laplace map $\mathcal{F}_{\rm L}$ as

$$(R_{1,n+1}, R_{3,n+1}, S_{1,n+1}, S_{3,n+1}) = \mathcal{F}_{\rm L}(R_{1,n}, R_{3,n}, S_{1,n}, S_{3,n}).$$

Like for Gauss, all fixed points of $\mathcal{F}_{\rm L}$ provide a solution of the problem, while the nth iteration of $\mathcal{F}_{\rm L}$ gives an approximation of the unknown or bit up to terms of order $O(\varepsilon^n)$, provided that $d \neq 0, \partial_\rho L(R_1, R_3, S_1, S_3, \rho)|_{\rho_2} \neq 1$ and $(R_{1,n}, R_{3,n}, S_{1,n}, S_{3,n})$ belongs to a suitable neighbourhood of (R_1, R_3, S_1, S_3). Let us now present an iterative scheme for the method developed by Mossotti. Define the sequence $h_{i,n}, k_{i,n} (i = 1, 3)$ as follows:

(1) Start with $h_{i,0} = k_{i,0} = 1$.
(2) Given $h_{i,n}$, $k_{i,n}$, let $\vec{r}_{2,n}$, $\vec{v}_{2,n}$ be the vectors obtained replacing h_i, k_i with $h_{i,n}$, $k_{i,n}$. If $\vec{r}_{2,n}, \vec{v}_{2,n}$ are not parallel, let $\mathcal{C}_n$ be the corresponding conic. Finally, let $\vec{r}_{1,n}, \vec{r}_{3,n}$ denote the positions of the same body at times t_1, t_3, respectively.
(3) Define $h_{i,n+1}$, $k_{i,n+1}$ by means of the relations

$$T_{i,n+1} = 1 - \frac{(t_i - t_2)^2}{2r_{2,n}^3} h_{i,n+1}, \quad V_{i,n+1} = (t_i - t_2)k_{i,n+1}, \quad i = 1, 3,$$

where $T_{i,n+1}, V_{i,n+1}$ are the coefficients of the linear relations providing $\vec{r}_{1,n}, \vec{r}_{3,n}$ as a combination of $\vec{r}_{2,n}, \vec{v}_{2,n}$ in analogy to (5). Let the Mossotti map $\mathcal{F}_{\rm M}$ be defined as

$$(h_{1,n+1}, h_{3,n+1}, k_{1,n+1}, k_{3,n+1}) = \mathcal{F}_{\rm M}(h_{1,n}, h_{3,n}, k_{1,n}, k_{3,n}).$$

As for the previous methods, all fixed points of $\mathcal{F}_{\rm M}$ define a solution of the problem, and the nth iteration of $\mathcal{F}_{\rm M}$ provides an approximation of the unknown orbit up to terms of order $O(\varepsilon^n)$, whenever $\partial_\rho M(h_1, h_3, k_1, k_3, \rho)|_{\rho_2} \neq 1, \vec{b}_1 \wedge \vec{b}_2 \cdot \vec{b}_3 \neq 0$ and for $(h_{1,n}, h_{3,n}, k_{1,n}, k_{3,n})$ in a suitable neighbourhood of (h_1, h_3, k_1, k_3).

3 Dependence on the times of observations

In order to study the dependence on the intervals among the times of observations, we consider two samples given by the first 10,000 numbered asteroids and by 615 Kuiper belt objects.[1] We apply Gauss, Mossotti and Laplace methods for different time intervals t_{12} and t_{23}, where the central time t_2 is the real observational time as provided by the astronomical data (see footnote 1). Starting from the elements $(a, e, i, \omega, \Omega, M)$ at the epoch t_2, and given the time intervals t_{12} and t_{23}, we compute the geocentric longitude and latitude at times t_1, t_2, t_3 by means of the coordinates of the object and that of the Earth (see Appendix A). Finally, we apply Gauss, Mossotti and Laplace methods, iterating the procedure as described in the previous section until convergence is reached. In order to be sure that a given method converges in a significant range around the given time t_2 (and not only for the specific time t_2), we

[1] The astronomical data of the asteroids can be found on the web site “Astdys” at http://hamilton.dm.unipi.it/cgi–bin/astdys/astibo; the astronomical data of the Kuiper belt objects can be found at the ephemerides page by D. Jewitt at http://www.ifa.hawaii.edu/faculty/jewitt/kb.html.

proceed as follows. Define $t_{ij}^n \equiv t_{ij}+n/2$, where $n=0,\pm1,\pm2$; if the method converges for the above time lapses t_{12}^n and t_{23}^n ($n=0,\pm1,\pm2$), then we say that the method is successful, otherwise we decide that the method fails.

We consider several choices of the time intervals t_{ij} from 3 to 90 days. Moreover, to cover the case of two observations performed within the same night, we selected t_{12} of the order of some hours and t_{23} ranging from 5 to 30 days. The results are summarized in Table 1, where the first percentage refers to the asteroids, while the second number of each method refers to Kuiper belt objects. Concerning the main belt, one concludes that Gauss method provides the best result, while Mossotti is more successful than Laplace; the opposite conclusion holds for the Kuiper belt objects. For equal time intervals $t_{12}=t_{23}$ (i.e. the first 8 lines of Table 1), the number of successful cases within the asteroidal belt increases as the time interval decreases, while (again) the opposite conclusion can be drawn for the Kuiper belt objects. As discussed in the following section, one might expect that whenever the time interval ε among the observations is sufficiently small (say $\varepsilon<\bar\varepsilon$), Gauss method (as well as the other techniques) converges. Of course $\bar\varepsilon$ depends on $\mathcal{C},\gamma_{12},\gamma_{23}$ (and t_2), implying that smaller is ε, greater is the number of converging orbits for fixed values of γ_{12},γ_{23}. On the other hand, the dependence of $\bar\varepsilon$ on γ_{12},γ_{23} implies that t_{12},t_{23} cannot be chosen *too* small, otherwise $\mathcal{C}$ (as well as its approximants $\mathcal{C}_n$) is badly determined. The latter effect is particularly relevant when the semi-major axis is large as it happens for the Kuiper belt (notice that the mean anomalies between two observations differ by $M_{ij}=t_{ij}a^{-3/2}$ and that the difference v_{ij} between the true anomalies, and henceforth between the t_{ij}, goes to zero with M_{ij}).

In order to see the distribution of the previous results as functions of the semi-major axis, eccentricity and inclination, we compute the percentages of successful results of the first 10,000 numbered asteroids by considering four different regions in a,e,i, each one being composed by 2,500 objects. The results are provided in Table 2 for the time intervals $t_{12}=1^h$ and $t_{23}=5^d$ and in Table 3 for $t_{12}=t_{23}=10^d$. In particular, Table 3

Table 1 Percentage of successful results for Gauss, Mossotti and Laplace methods; the first number refers to the asteroids (e.g. 99.86, first line of Gauss method), while the second to Kuiper's objects (e.g. 79.67, same line)

t_{12}	t_{23}	Gauss	Mossotti	Laplace
3^d	3^d	99.86/79.67	99.55/92.03	99.00/93.33
5^d	5^d	99.87/93.33	99.45/93.98	98.90/93.98
10^d	10^d	99.78/93.98	99.23/94.30	98.73/94.63
15^d	15^d	99.58/94.47	99.27/94.47	98.54/94.63
30^d	30^d	99.45/94.63	99.36/94.47	98.17/94.63
60^d	60^d	98.77/94.63	98.41/94.63	96.00/94.63
90^d	90^d	96.80/94.63	96.73/94.63	94.32/94.63
10^d	30^d	99.60/94.63	99.45/94.63	98.01/94.63
5^d	10^d	99.82/94.47	99.56/94.63	98.63/94.63
1^h	5^d	99.77/7.32	99.72/54.79	98.82/93.17
5^h	5^d	99.87/17.40	99.77/78.53	98.86/93.66
1^h	10^d	99.80/17.40	99.66/79.84	98.60/94.31
5^h	10^d	99.81/53.17	99.67/88.62	98.55/94.30
1^h	30^d	99.68/63.25	99.62/90.24	97.59/94.63
5^h	30^d	99.70/83.85	99.64/92.84	97.61/94.63

Table 2 Percentage of successful results for Gauss, Mossotti and Laplace methods in terms of semimajor axis a (in AU), eccentricity e, inclination i (in degrees)

	Gauss	Mossotti	Laplace
$0 \le a < 2.341$	99.56	99.04	97.36
$2.341 \le a < 2.6144$	99.96	99.96	98.48
$2.6144 \le a < 3.0053$	99.80	99.96	99.52
$3.0053 \le a < 100$	99.76	99.92	99.92
$0 \le e < 0.094$	99.68	99.92	99.60
$0.094 \le e < 0.140244$	99.92	99.92	99.56
$0.140244 \le e < 0.187321$	99.84	99.64	98.52
$0.187321 \le e < 1$	99.64	99.40	97.60
$0 \le i < 3.2185$	99.72	99.76	98.68
$3.2185 \le i < 6.0218$	99.84	99.56	98.36
$6.0218 \le i < 10.918$	99.72	99.80	99.08
$10.918 \le i < 360$	99.80	99.76	99.16

Each parameter region is composed by 2,500 objects belonging to the first 10,000 numbered asteroids. The time intervals are $t_{12} = 1^h$ and $t_{23} = 5^d$

Table 3 Percentage of successful results for Gauss, Mossotti and Laplace methods in terms of semimajor axis a (in AU), eccentricity e, inclination i (in degrees)

	Gauss	Mossotti	Laplace
$0 \le a < 2.341$	99.56	97.28	96.88
$2.341 \le a < 2.6144$	99.80	99.68	98.40
$2.6144 \le a < 3.0053$	99.88	99.96	99.72
$3.0053 \le a < 100$	99.88	100	99.92
$0 \le e < 0.094$	99.92	99.80	99.56
$0.094 \le e < 0.140244$	99.84	99.84	99.48
$0.140244 \le e < 0.187321$	99.80	99.40	98.92
$0.187321 \le e < 1$	99.56	97.88	96.96
$0 \le i < 3.2185$	99.84	99.60	99.08
$3.2185 \le i < 6.0218$	99.76	98.68	97.92
$6.0218 \le i < 10.918$	99.76	99.56	98.84
$10.918 \le i < 360$	99.76	99.08	99.08

Each parameter region is composed by 2,500 objects belonging to the first 10,000 numbered asteroids. The time intervals are $t_{12} = 10^d$ and $t_{23} = 10^d$

shows that the success of all methods (slightly) grows if the semi-major axis increases, though a more reliable test should be performed over sample data with equally spaced values of the semi-major axes. On the other hand, all methods seem to be independent on the value of the inclination, while only Laplace method is affected by the value of the eccentricity, performing better for lower eccentricities.

4 Convergence of Gauss algorithm: computation of the eigenvalues of the Jacobian matrix

In the framework of Theorem 2.1, we investigate whether $\mathcal{F}_G : z' = (\zeta_1, \zeta_2) \rightarrow \mathcal{F}_G(z') \equiv (\mathcal{F}_G^1(z'), \mathcal{F}_G^2(z'))$ can be indefinitely iterated from the initial point z_0 and, eventually, if the nth iterate $z_n = \mathcal{F}_G^n(z_0)$ tends to its fixed point $z = z(\mathcal{C}, t_2, \gamma_{12}, \gamma_{23}, \varepsilon)$. Let $W \subset U$ be a closed convex neighbourhood of z; by Lagrange's theorem, if $z_1, z_2 \in W$, there exists z_1^*, z_2^* belonging to the interval (z_1, z_2), such that

$\mathcal{F}_G(z_1)-\mathcal{F}_G(z_2)=\partial\mathcal{F}_G(z_1^*,z_2^*)(z_1-z_2)$, where $\partial\mathcal{F}_G(z_1^*,z_2^*)$ has entries $\partial_{\zeta_j}\mathcal{F}^i_G(z_i^*)$, for $z'=(\zeta_1,\zeta_2)$. Let us assume that the complex eigenvalues $\lambda_1(x,y),\lambda_2(x,y)$ of $\partial\mathcal{F}_G(x,y)$ verify, for $x,y\in W$

$$\lambda_1(x,y)\neq\lambda_2(x,y), \qquad |\lambda_i(x,y)|\leq\theta<1. \tag{6}$$

For $z_1\neq z_2\in W$, let $\vec{v}_i^*\in\mathbf{C}^2$ denote the eigenvector corresponding to $\lambda_i^*\equiv\lambda_i(z_1^*,z_2^*)$; we define $d(z_1,z_2)\equiv|\alpha_1|+|\alpha_2|$, where $\alpha_1,\alpha_2\in\mathbf{C}$ are such that $z_1-z_2=\alpha_1\vec{v}_1^*+\alpha_2\vec{v}_2^*$. Otherwise, for $z_1=z_2$ we set $d(z_1,z_2)=0$. With this choice of the metric, $\mathcal{F}_G$ becomes a contraction on W, being $\mathcal{F}_G(z_1)-\mathcal{F}_G(z_2)=\lambda_1^*\alpha_1\vec{v}_1^*+\lambda_2^*\alpha_2\vec{v}_2^*$. On the other hand, one can conclude by continuity that setting $x=y=z(\mathcal{C},t_2,\gamma_{12},\gamma_{23},\varepsilon)$, if $\lambda_i(z,z)\equiv\lambda_i(z)$ verify

$$\lambda_1(z)\neq\lambda_2(z) \tag{7}$$

and

$$\mu(\mathcal{C},t_2,\gamma_{12},\gamma_{23},\varepsilon)\equiv\mathrm{Max}_{i=1,2}|\lambda_i(z)|<1 \tag{8}$$

then, there exists a suitable closed convex set W containing z where (6) holds, namely, $\mathcal{F}_G$ is a contraction. As a consequence, its unique fixed point z in W can be obtained as the limit $z=\lim_{n\to\infty}z_n$, starting from *any* $z_0\in W$. We will see (Proposition 4.1 below) that, under slightly stronger assumptions than in Theorem 2.1 (see (9) and (10) below), condition (8) is always satisfied, provided ε is small enough. The assumptions we make are the following:

(1) The vectors $\vec{b}_2=\vec{b}(t_2),\dot{\vec{b}}_2=\dot{\vec{b}}(t_2),\ddot{\vec{b}}_2=\ddot{\vec{b}}(t_2)$ are linearly independent:

$$\vec{b}_2\wedge\dot{\vec{b}}_2\cdot\ddot{\vec{b}}_2\neq0; \tag{9}$$

(2) setting $\vec{a}_2=\vec{a}(t_2)$, one has

$$\mathcal{D}\equiv3\frac{\vec{b}_2\wedge\dot{\vec{b}}_2\cdot\vec{a}_2}{\vec{b}_2\wedge\dot{\vec{b}}_2\cdot\ddot{\vec{b}}_2}\frac{\rho_2+\vec{a}_2\cdot\vec{b}_2}{r_2^5}\neq1. \tag{10}$$

Remark 4.1 The independence of the $\vec{b}_i$'s is required by Gauss algorithm. Indeed, for $\varepsilon<1$ let us expand in Taylor series as

$$\vec{b}_1=\vec{b}_2-\dot{\vec{b}}_2\,\gamma_{12}\,\varepsilon+\ddot{\vec{b}}_2\frac{\gamma_{12}^2}{2}\varepsilon^2+o(\varepsilon^3)$$

$$\vec{b}_3=\vec{b}_2+\dot{\vec{b}}_2\,\gamma_{23}\varepsilon+\ddot{\vec{b}}_2\frac{\gamma_{23}^2}{2}\varepsilon^2+o(\varepsilon^3); \tag{11}$$

then, by (9) for ε small one finds that $|\vec{b}_1\wedge\vec{b}_2\cdot\vec{b}_3|=|\frac{1}{2}\vec{b}_2\wedge\dot{\vec{b}}_2\cdot\ddot{\vec{b}}_2\,\gamma_{12}\gamma_{23}\,\varepsilon^3+o(\varepsilon^4)|>0$. With a similar argument, one finds that condition (10) implies that for ε small $\partial_{\rho_2}G_2(P,Q,\rho_2)\neq1$ allowing to solve Gauss equation.

Proposition 4.1 *For any* $\mathcal{C},t_2$ *such that conditions* (9) *and* (10) *are satisfied, one has* $\mu(\mathcal{C},t_2,\gamma_{12},\gamma_{23},\varepsilon)\to0$ *as* $\varepsilon\to0$.

The proof is given in Appendix B.

In order to prove the contractive character of $\mathcal{F}_G$ for $0<\varepsilon<\bar{\varepsilon}$ for a suitable $\bar{\varepsilon}$ (and, consequently, the convergence of Gauss algorithm for $0<\varepsilon<\bar{\varepsilon}$, at least if $\bar{\varepsilon}$ is so small that the initial point $z_0 = (P_0, Q_0)$, defined in (3), belongs to W), we still need the assumption (7). In this context, we provide in Appendix B a sufficient condition (Corollary B.1), based on the computation of $\lambda_1(z), \lambda_2(z)$ at the first order in ε.

4.1 Eigenvalues of $\partial\mathcal{F}_G(z)$

Motivated by the previous discussion and by the fact that the explicit computation of μ is extremely long, we determine numerically the elements of the jacobian matrix $\partial\mathcal{F}_G(z)$, which yield the eigenvalues $|\lambda_1(z)|, |\lambda_2(z)|$. We let t_{12}, t_{23} vary, while t_2 is fixed equal to a given epoch (MJD 53450 for the asteroids, while it changes for Kuiper belt objects according to the astronomical data of footnote 1). More precisely, for each $\mathcal{C}$ (with related set of elements $(a, e, i, \omega, \Omega, M)$ at time t_2) and for each choice of t_{12}, t_{23}, we compute the three vectors $\vec{r}_1, \vec{r}_2, \vec{r}_3$. Together with the three Sun–Earth vectors $\vec{a}_1, \vec{a}_2, \vec{a}_3$, we obtain the Earth–object directions $\vec{b}_1, \vec{b}_2, \vec{b}_3$, which provide the Gauss map $\mathcal{F}_G$ and its fixed point z.

The Jacobian $\partial\mathcal{F}_G(z)$ is computed through a polynomial interpolation. Let us consider, for example, the computation of the first element $\partial_P P'(P, Q)$ (for the other derivatives, the computation is quite similar), where $\mathcal{F}_G = (P', Q')$. Having fixed Q, we choose an odd number (say, $2n+1$) of points $P_i = P + \mathrm{i}h$, $i = -n, \ldots, n$, equally spaced and symmetrically distributed around P with constant step-size h, such that $2nh = 0.1$. Denoting by F_i the value of P' at $z_i = (P_i, Q)$, we approximate $\partial_P P'(P, Q)$ with the quantity $\sum_{|i|\le n,\, i\ne 0} \frac{(-1)^{i+1}}{ih} \frac{(n!)^2}{(n-i)!(n+i)!} F_i$. The overall number of nodes is such that the difference between the values of the derivatives is smaller than 0.001 as n increases to $n+1$. The computational details are provided in Appendix C.

4.2 Eigenvalues of asteroids and Kuiper belt objects

We compute the eigenvalues of the jacobian matrix of the Gauss map, following the algorithm outlined in the previous sections. Over a sample of 100 asteroids of the main belt we found 20 objects with at least one eigenvalue with modulus greater than one. Typically the graph of the maximum modulus of the eigenvalues versus the time intervals t_{12} or t_{23} is provided in Fig. 1 (left panel), where t_{12} and t_{23} are taken between

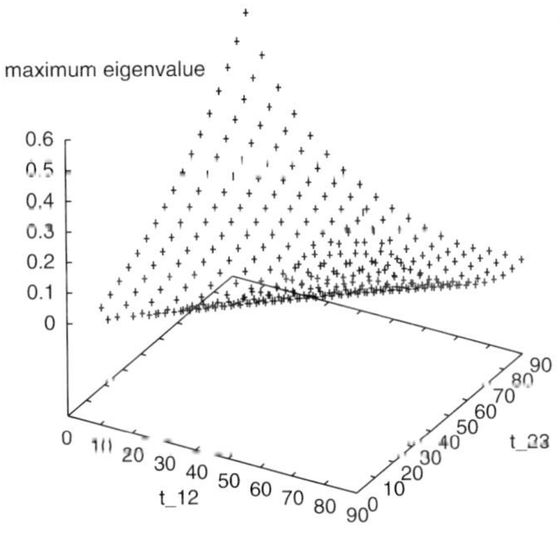

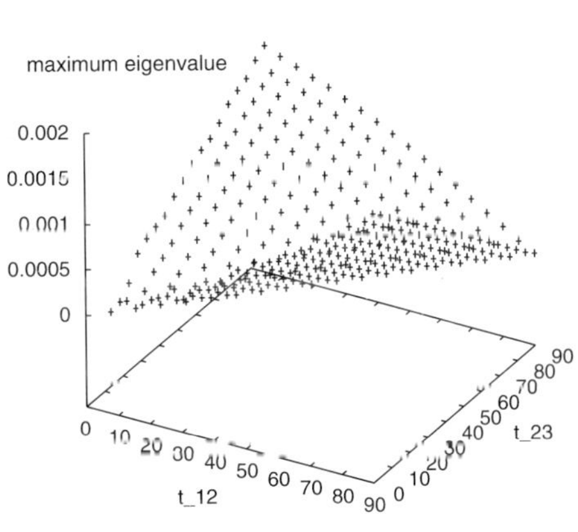

Fig. 1 Maximum modulus of the eigenvalues of the jacobian matrix versus the time intervals t_{12} and t_{23}. Left: asteroid number 8; Right: Kuiper belt object number 12

0 and 90 days with a time-step equal to 5 days. This example refers to the asteroid number 8, whose elements are $a = 0.2012$ AU, $e = 0.1563$, $i = 5.8869°$, $\omega = 284.9649°$, $\Omega = 111.0326°$, $M = 81.1258°$ at epoch MJD 53450. A similar procedure was adopted for the 615 objects of the Kuiper belt; however, contrary to the main belt objects we have not found any sample showing an eigenvalue greater than one. A typical picture of the first eigenvalue of a Kuiper belt object is provided in Fig. 1 (right panel), which corresponds to the Kuiper belt object number 12, whose elements are $a = 42.3035$ AU, $e = 0.2174$, $i = 14.0299°$, $\omega = 236.5808°$, $\Omega = 56.2982°$, $M = 336.4332°$ at epoch MJD 53400.5. We remark that in both cases the graph of $|\lambda_1|$ versus t_{12}, t_{23} is roughly symmetric with respect to the line $t_{12} = t_{23}$, where the eigenvalue approximately attains its minimum; in this situation the contractive character of the Gauss map is stronger. This remarks confirms indeed that Gauss method gives better results whenever equal observational time intervals are considered. In this case Gauss (1809) noticed that at the first iteration his method gives errors of $O(\varepsilon^2)$, instead than $O(\varepsilon)$ as obtained using different time intervals; furthermore, as remarked also by Poincaré (1906), the errors after n iterations become of order $O(\varepsilon^{2n})$ using equal time intervals, while they are $O(\varepsilon^n)$ at different time intervals.

Acknowledgements We are deeply grateful to Giovanni Federico Gronchi and to Andrea Milani for useful discussions and suggestions.

Appendix

A Computation of the longitude and latitude from the elliptic elements

We derive the ecliptic geocentric longitude and latitude from the elliptic elements, without taking into account topocentric corrections or aberrational effects. We restrict to consider $e < 1$. Let a, e, i, ω, Ω, M be the elliptic elements at a fixed reference epoch $T = 0$; let t_1, t_2, t_3 be the times of observations with $t_{12} = t_2 - t_1$, $t_{23} = t_3 - t_2$. The mean anomaly at time t_2 is given by $M_2 \equiv M(t_2) = M + nt_2$, where $n = ka^{-3/2}$ is the mean motion with $k = 0.985608°$/day. Similarly one has $M_1 = M_2 - nt_{12}, M_3 = M_2 + nt_{23}$. The eccentric anomalies ξ_1, ξ_2, ξ_3 at t_1, t_2, t_3 are obtained solving Kepler's equation $\xi_i - e\sin\xi_i = M_i (i = 1,2,3)$. Let $\vec{s} = (x,y,z)$ be the coordinates of the asteroid in the orbital frame with the x-axis coinciding with the perihelion line, i.e. $x = a(\cos\xi - e), y = a(1-e^2)^{1/2}\sin\xi, z = 0$. Replacing ξ with ξ_1, ξ_2, ξ_3, one obtains the position vectors $\vec{s}_1$, $\vec{s}_2$, $\vec{s}_3$, which must be transformed in the ecliptic frame by means of the following three rotations:

(a) a rotation of angle ω around the z-axis;
(b) a rotation of angle i around the x-axis;
(c) a rotation of angle Ω around the z-axis.

Let the resulting vectors in the ecliptic frame be denoted as $\vec{s}_i^{(e)} (i = 1,2,3)$; with a similar procedure one obtains the Earth's coordinates $\vec{a}_i^{(e)} (i = 1,2,3)$. Defining the generic geocentric vectors as $\vec{R} \equiv \vec{s}^{(e)} - \vec{a}^{(e)} \equiv (X,Y,Z)$, the longitude of $\vec{R}$ is given by the expression $\lambda = \tan^{-1}(Y/X)$ if $X > 0$ and $\lambda = \tan^{-1}(Y/X) + \pi$ if $X < 0$, while the latitude is given by $\beta = \sin^{-1}(Z/(X^2+Y^2+Z^2)^{1/2})$.

B Proof of Proposition 4.1

In this appendix, we give a proof of Proposition 4.1 as a byproduct of Proposition B.1 below. Moreover (see Corollary B.1), we provide a sufficient condition to ensure that $\mathcal{F}_G$ is a contraction for ε small. Let $\bar{\mathcal{C}}$ be a conic, and let $\bar{z} = (\bar{P}, \bar{Q})$ be its Gauss parameters.[2] We recall that we keep t_2 fixed, while t_1, t_3 are varied; let ε be the time interval between the first and the third observation and, as in (1), let $t_1 = t_2 - \gamma_{12}\varepsilon$, $t_3 = t_2 + \gamma_{23}\varepsilon$. Denote by $\mathcal{F}_G : z = (P, Q) \to z' = (P', Q')$ the Gauss map, defined in a suitable neighbourhood of $\bar{z}$. We want to compute the eigenvalues of the jacobian matrix of $\mathcal{F}_G$, which we denote as $\mathcal{J} = \mathcal{J}(\bar{\mathcal{C}}, t_2, \gamma_{12}, \gamma_{23}, \varepsilon) \equiv \{\mathcal{J}_{ij}\}_{i,j=1,2}$.

Proposition B.1 *Fix t_2 and $\bar{\mathcal{C}}$ such that conditions* (9) *and* (10) *are satisfied. Then, there exist $\hat{\mathcal{J}}_{11}, \hat{\mathcal{J}}_{12}, \hat{\mathcal{J}}_{21}, \hat{\mathcal{J}}_{22}$ depending on $\bar{\mathcal{C}}, t_2, \gamma_{12}, \gamma_{23}$, such that*

$$\mathcal{J}_{11} = \partial_P \; P'(\bar{P}, \bar{Q}) = \hat{\mathcal{J}}_{11}\varepsilon + o(\varepsilon^2) \qquad \mathcal{J}_{12} = \partial_Q \; P'(\bar{P}, \bar{Q}) = \hat{\mathcal{J}}_{12} + o(\varepsilon),$$
$$\mathcal{J}_{21} = \partial_P \; Q'(\bar{P}, \bar{Q}) = \hat{\mathcal{J}}_{21}\varepsilon^2 + o(\varepsilon^3) \qquad \mathcal{J}_{22} = \partial_Q \; Q'(\bar{P}, \bar{Q}) = \hat{\mathcal{J}}_{22}\varepsilon + o(\varepsilon^2).$$

Remark B.1 The eigenvalues $\lambda_1, \lambda_2 \in \mathbf{C}$ of $\mathcal{J}$ can be written as $\lambda_j = \hat{\iota}_j\varepsilon + o(\varepsilon^2)(j = 1,2)$ with $\hat{\iota}_j = \tau \pm \sqrt{\tau^2 - \delta}$, where $\tau = (\hat{\mathcal{J}}_{11} + \hat{\mathcal{J}}_{22})/2$ and $\delta = \hat{\mathcal{J}}_{11}\hat{\mathcal{J}}_2 - \hat{\mathcal{J}}_{21}\hat{\mathcal{J}}_{12}$ are the semi–trace and determinant of $\hat{\mathcal{J}} = \{\hat{\mathcal{J}}_{ij}\}_{i,j=1,2}$. Moreover, if $\bar{\mathcal{C}}, t_2, \gamma_{12}, \gamma_{23}$ are such that $\Delta \equiv \tau^2 - \delta \neq 0$, then, $\lambda_1(\bar{z}) \neq \lambda_2(\bar{z})$ for $\varepsilon > 0$ sufficiently small.

Corollary B.1 *Let $\bar{\mathcal{C}}, t_2$ verify* (9) *and* (10) *and let γ_{12}, γ_{23} be chosen such that $\Delta \neq 0$. Then, there exists $\bar{\varepsilon} > 0$ such that, if $0 < \varepsilon < \bar{\varepsilon}$, the mapping $\mathcal{F}_G : (W, d) \to \mathbf{R}^2$ is a contraction.*

Let us first recall the definition of the Gauss map, referring to Celletti and Pinzari (2005), for details. Let $\rho_2(P, Q)$ be the solution of Gauss equation:

$$\rho_2 = G_2(P, Q, \rho_2) \equiv -\vec{c}_2 \cdot \vec{a}_2 + \frac{\vec{c}_2 \cdot \vec{a}_1 + \vec{c}_2 \cdot \vec{a}_3 P}{P+1}\left(1 + \frac{Q}{2r_2^3}\right),$$

where $\vec{c}_i = \dfrac{\vec{b}_j \wedge \vec{b}_k}{\vec{b}_1 \wedge \vec{b}_2 \cdot \vec{b}_3}\varepsilon_{jki}$, $r_2 = |\vec{a}_2 + \rho_2\vec{b}_2|$ and $\varepsilon_{jki} = 1$ if $\{j,k,i\}$ is an even permutation of $\{1,2,3\}$, $\varepsilon_{jki} = -1$ otherwise. Let $\rho_1 \equiv \rho_1(P, Q), \rho_3 \equiv \rho_3(P, Q)$ be defined as

$$\rho_1 = -\vec{c}_1 \cdot \vec{a}_1 + \frac{P+1}{1 + \frac{Q}{2r_2(\rho_2(P,Q))^3}}\vec{c}_1 \cdot \vec{a}_2 - P\vec{c}_1 \cdot \vec{a}_3 \equiv G_1(P, Q, \rho_2(P, Q)),$$
$$\rho_3 = -\frac{1}{P}\vec{c}_3 \cdot \vec{a}_1 + \frac{P+1}{P\left(1 + \frac{Q}{2r_2(\rho_2(P,Q))^3}\right)}\vec{c}_3 \cdot \vec{a}_2 - \vec{c}_3 \cdot \vec{a}_3 = G_3(P, Q, \rho_2(P, Q)).$$

Let $\vec{r}_i(P, Q), i = 1,2,3$, be written as

$$\vec{r}_i - \vec{r}_i(P, Q) = \vec{a}_i + \rho_i(P, Q)\vec{b}_i. \tag{12}$$

It can be shown (Celletti and Pinzari 2005) that $\vec{r}_1, \vec{r}_2, \vec{r}_3$ are coplanar and define a unique conic $\mathcal{C} = \mathcal{C}(P, Q)$ with a focus in their common origin. We also recall that the eccentricity $e = e(P, Q)$ of $\mathcal{C}$ and the argument of perihelion $g = g(P, Q)$ are given by

[2] Barred quantities will refer to $\bar{\mathcal{C}}$.

$$e = \frac{\sqrt{A^2+B^2}}{|n_{12}+n_{23}-n_{13}|},$$
$$\cos g = \frac{B}{\sqrt{A^2+B^2}} s, \qquad \sin g = -\frac{A}{\sqrt{A^2+B^2}} s, \tag{13}$$

where $n_{ij} = n_{ij}(P,Q)$ is the oriented area of the triangle formed by $\vec{r}_i, \vec{r}_j$, $s \equiv \mathrm{sgn}(n_{12}+n_{23}-n_{13})$ and, denoting by $v_{ij} \equiv 2f_{ij}$ the angle formed by $\vec{r}_i, \vec{r}_j$, one has

$$A \equiv r_2(r_3-r_1) + r_1(r_2-r_3)\cos v_{12} + r_3(r_1-r_2)\cos v_{23},$$
$$B \equiv -r_1(r_2-r_3)\sin v_{12} + r_3(r_1-r_2)\sin v_{23}. \tag{14}$$

Moreover, let P, Q be expressed by

$$P = \frac{n_{12}}{n_{23}}, \qquad Q = 2r_2^3\left(\frac{n_{12}+n_{23}}{n_{13}} - 1\right) \tag{15}$$

and let $\eta_{ij} = \eta_{ij}(P,Q)$ denote the ratio of the area of the triangle formed by $\vec{r}_i$ and $\vec{r}_j$ with the corresponding conic sector. The *Gauss map* $\mathcal{F}_G$ is finally defined by $z' = \mathcal{F}_G(z)$, where $z' = (P', Q')$ takes the form

$$P' = \frac{\gamma_{12}}{\gamma_{23}}\frac{\eta_{12}}{\eta_{23}}, \qquad Q' = \varepsilon^2\gamma_{12}\gamma_{23}\frac{r_2^2}{r_1 r_3}\frac{\eta_{12}\eta_{23}}{\cos f_{12}\cos f_{23}\cos f_{13}}. \tag{16}$$

The proof of Proposition B.1 is obtained through some technical lemmas, which provide estimates of the derivatives of $\eta_{ij}, r_i/r_j, f_{ij}$, appearing in (16) (assumptions (9) and (10) are assumed throughout all this appendix).

Lemma B.1 *There exist two constants $\mathcal{R}_P, \mathcal{R}_Q$ depending on $\bar{C}, t_2, \gamma_{12}, \gamma_{23}$, such that for $i = 1,2,3$ one has*

$$\partial_P \varepsilon\rho_i(\bar{P},\bar{Q}) = \mathcal{R}_P + o(\varepsilon), \qquad \partial_Q \varepsilon^2\rho_i(\bar{P},\bar{Q}) = \mathcal{R}_Q + o(\varepsilon). \tag{17}$$

Proof Using (11), (9) and $\vec{a}_3 - \vec{a}_1 = \dot{\vec{a}}_2\varepsilon + o(\varepsilon^2)$, denoting for short $\vec{B} \equiv -2\frac{\vec{b}_2 \wedge \dot{\vec{b}}_2}{\vec{b}_2 \wedge \dot{\vec{b}}_2 \cdot \ddot{\vec{b}}_2 \gamma_{12}\gamma_{23}\varepsilon^2}$ and recalling that $\vec{c}_2 = \frac{\vec{b}_3 \wedge \vec{b}_1}{\vec{b}_1 \wedge \vec{b}_2 \cdot \vec{b}_3}$, one has

$$\vec{c}_2 = \vec{B} + o(\varepsilon^{-1}), \qquad \vec{c}_2 \cdot (\vec{a}_3 - \vec{a}_1) = \vec{B} \cdot \dot{\vec{a}}_2\varepsilon + o(1). \tag{18}$$

The implicit function theorem shows that $\rho_2(P,Q)$ is a smooth function of (P,Q), such that

$$\partial_P\rho_2(\bar{P},\bar{Q}) = \frac{\partial_P G_2(\bar{P},\bar{Q},\bar{\rho}_2)}{1-\partial G_2(\bar{P},\bar{Q},\bar{\rho}_2)} = \frac{\frac{\vec{c}_2\cdot(\vec{a}_3-\vec{a}_1)}{(P+1)^2}}{1+3\vec{c}_2\cdot\vec{a}_3\frac{\bar{Q}}{2\bar{r}_2^5}(\bar{\rho}_2+\vec{a}_2\cdot\vec{b}_2)}$$
$$+3\frac{\frac{[\vec{c}_2\cdot(\vec{a}_3-\vec{a}_1)]^2}{(P+1)^3}\frac{\bar{Q}}{2\bar{r}_2^5}(\bar{\rho}_2+\vec{a}_2\cdot\vec{b}_2)}{\left[1+3\vec{c}_2\cdot\vec{a}_3\frac{\bar{Q}}{2\bar{r}_2^5}(\bar{\rho}_2+\vec{a}_2\cdot\vec{b}_2)\right]^2} + o(\varepsilon). \tag{19}$$

An explicit expression up to $o(1)$ is obtained using (18), (19) and the estimates for $\bar{P},\bar{Q}$ given by $\bar{P}=\frac{\gamma_{12}}{\gamma_{23}}+o(\varepsilon^2),\bar{Q}=\gamma_{12}\gamma_{23}\varepsilon^2+o(\varepsilon^3)$:

$$\partial_P\rho_2(\bar{P},\bar{Q})=-\frac{2}{1-\mathcal{D}}\frac{\vec{b}_2\wedge\dot{\vec{b}}_2\cdot\vec{a}_2}{\vec{b}_2\wedge\dot{\vec{b}}_2\cdot\ddot{\vec{b}}_2}\frac{\gamma_{23}}{\gamma_{12}\varepsilon}+o(1)\equiv\frac{\mathcal{R}_P}{\varepsilon}+o(1).$$

Similar computations allow to conclude that $\partial_P\rho_1(\bar{P},\bar{Q})\equiv\frac{\mathcal{R}_P}{\varepsilon}+o(1)$ and that $\partial_P\rho_3(\bar{P},\bar{Q})\frac{\mathcal{R}_P}{\varepsilon}+o(1)$. Concerning the derivative with respect to Q, one finds that

$$\partial_Q\rho_2(\bar{P},\bar{Q})=\frac{\partial_Q G_2(\bar{P},\bar{Q},\bar{\rho}_2)}{1-\partial_{\rho_2}G_2(\bar{P},\bar{Q},\bar{\rho}_2)}$$

$$=\frac{\left[\vec{c}_2\cdot\vec{a}_3-\frac{\vec{c}_2\cdot(\vec{a}_3-\vec{a}_1)}{\bar{P}+1}\right]\frac{1}{2\bar{r}_2^3}}{1+3\left[\vec{c}_2\cdot\vec{a}_3-\frac{\vec{c}_2\cdot(\vec{a}_3-\vec{a}_1)}{\bar{P}+1}\right]\frac{\bar{Q}}{2\bar{r}_2^5}(\bar{\rho}_2+\vec{a}_2\cdot\vec{b}_2)}$$

and one easily finds that $\mathcal{R}_Q$ in (17) takes the expression $\mathcal{R}_Q\equiv\frac{1}{2(1-\mathcal{D})}\frac{\varepsilon^2\vec{B}\cdot\vec{a}_2}{\bar{r}_2^3}$.

As a corollary of the previous lemma we have the following result.

Lemma B.2 *For any $i\neq j$, there exist constants $\mathcal{R}_P^{*i,j},\mathcal{R}_Q^{*i,j}$ depending on $\bar{\mathcal{C}},t_2,\gamma_{12},\gamma_{23}$, such that $\partial_P(r_i-r_j)=\mathcal{R}_P^{*i,j}+o(\varepsilon)$, $\partial_Q\,\varepsilon(r_i-r_j)=\mathcal{R}_Q^{*i,j}+o(\varepsilon)$ (a similar expression is valid also for r_i/r_j).*

Next we have the following

Lemma B.3 *There exist two constants $\mathcal{N}_P,\mathcal{N}_Q$ depending on $\bar{\mathcal{C}},t_2,\gamma_{12},\gamma_{23}$, such that*

$$\partial_P\,n_{13}(\bar{P},\bar{Q})=\mathcal{N}_P+o(\varepsilon),\qquad\partial_Q\,\varepsilon n_{13}(\bar{P},\bar{Q})=\mathcal{N}_Q+o(\varepsilon).\tag{20}$$

Proof Let $\vec{k}(P,Q)=\frac{\vec{r}_1(P,Q)\wedge\vec{r}_3(P,Q)}{|\vec{r}_1(P,Q)\wedge\vec{r}_3(P,Q)|}$ be a unit vector normal to the plane formed by $\vec{r}_1,\vec{r}_2,\vec{r}_3$. Then, $n_{13}(P,Q)=\vec{r}_1\wedge\vec{r}_3\cdot\vec{k}$ and

$$\partial_P n_{13}(\bar{P},\bar{Q})=(\partial_P\vec{r}_1\wedge\vec{\bar{r}}_3\cdot\vec{\bar{k}}+\vec{\bar{r}}_1\wedge\partial_P\vec{\bar{r}}_3\cdot\vec{\bar{k}}+\vec{\bar{r}}_1\wedge\vec{\bar{r}}_3\cdot\partial_P\vec{k})\,|_{(\bar{P},\bar{Q})}.$$

Last term is zero, since $\partial_P\vec{k}(\bar{P},\bar{Q})$ is perpendicular to $\vec{\bar{k}}$ and therefore, it is linearly dependent with $\vec{\bar{r}}_1,\vec{\bar{r}}_3$. For the remaining terms, using (12) we have

$$(\partial_P\vec{r}_1\wedge\vec{\bar{r}}_3\cdot\vec{k})\,|_{(\bar{P},\bar{Q})}=(\vec{b}_1\wedge\vec{a}_3\cdot\vec{\bar{k}}+\bar{\rho}_3\vec{b}_1\wedge\vec{b}_3\cdot\vec{\bar{k}})\,\partial_P\rho_1(\bar{P},\bar{Q}),$$
$$(\vec{\bar{r}}_1\wedge\partial_P\vec{\bar{r}}_3\cdot\vec{\bar{k}})\,|_{(\bar{P},\bar{Q})}=(\vec{a}_1\wedge\vec{b}_3\cdot\vec{k}+\bar{\rho}_1\vec{b}_1\wedge\vec{b}_3\cdot\vec{\bar{k}})\,\partial_P\rho_3(\bar{P},\bar{Q}).$$

By (11), the two terms in parenthesis are both equalto $\bar{\rho}_2\vec{b}_2\wedge\dot{\vec{b}}_2\cdot\vec{\bar{k}}\varepsilon$ up to $o(\varepsilon^2)$, while for the first term we remark that $\vec{b}_1\wedge\vec{a}_3+\vec{a}_1\wedge\vec{b}_3=(\vec{a}_2\wedge\dot{\vec{b}}_2-\dot{\vec{a}}_2\wedge\vec{b}_2)\varepsilon+o(\varepsilon^2)$. Casting together the previous formulae and using Lemma B.1, we conclude that

$$\partial_P n_{13}(\bar{P},\bar{Q})=(\vec{a}_2\wedge\dot{\vec{b}}_2\cdot\vec{\bar{k}}-\dot{\vec{a}}_2\wedge\vec{b}_2\cdot\vec{\bar{k}}+2\bar{\rho}_2\vec{b}_2\wedge\dot{\vec{b}}_2\cdot\vec{\bar{k}})\,\mathcal{R}_P+o(\varepsilon),$$

which can be written as $\partial_P n_{13}(\bar{P},\bar{Q})\equiv\mathcal{N}_{P+o}(\varepsilon)$ for a suitable constant $\mathcal{N}_P$. In a similar way one obtains the second of (20).

Remark B.2 Similar results hold for $\vec{\partial} n_{12}(\bar{P},\bar{Q}), \bar{\partial} n_{23}(\bar{P},\bar{Q})$. More precisely, for $(i,j) = (1,2),(2,3)$, one has $\partial_P n_{ij} = \mathcal{N}_P\, \gamma_{ij} + o(\varepsilon), \partial_Q \varepsilon n_{ij} = \mathcal{N}_Q \gamma_{ij} + o(\varepsilon)$. As a consequence of Lemmas B.1, B.3 and of the previous remark, a similar estimate holds for v_{ij}, being $\sin v_{ij} = n_{ij}/(r_i r_j)$.

Lemma B.4 *There exist two constants* $\mathcal{S}_P, \mathcal{S}_Q$ *depending on* $\bar{\mathcal{C}}, t_2, \gamma_{12}, \gamma_{23}$*, such that*

$$\partial_P(\sin v_{12} + \sin v_{23} - \sin v_{13})(\bar{P},\bar{Q}) = \mathcal{S}_P\, \varepsilon^2 + o(\varepsilon^4),$$
$$\partial_Q(\sin v_{12} + \sin v_{23} - \sin v_{13})(\bar{P},\bar{Q}) = \mathcal{S}_Q\, \varepsilon + o(\varepsilon^3).$$

Next step is to evaluate the derivatives of the eccentricity $e(P,Q)$ of $\mathcal{C}(P,Q)$.

Lemma B.5 *There exist two constants* $\mathcal{E}_P, \mathcal{E}_Q$ *depending on* $\bar{\mathcal{C}}, t_2, \gamma_{12}, \gamma_{23}$*, such that* $\partial_P\, \varepsilon e(\bar{P},\bar{Q}) = \mathcal{E}_P + o(\varepsilon)$, $\partial_Q\, \varepsilon^2 e(\bar{P},\bar{Q}) = \mathcal{E}_Q + o(\varepsilon)$.

Proof From (13), we obtain ($\vec{\partial} \equiv (\partial_P, \partial_Q)$):

$$\vec{\partial} e(\bar{P},\bar{Q}) = \frac{\vec{\partial}\sqrt{A^2+B^2}(\bar{P},\bar{Q})}{|\bar{n}_{12}+\bar{n}_{23}-\bar{n}_{13}|} - \bar{e}\frac{\vec{\partial}(n_{12}+n_{23}-n_{13})(\bar{P},\bar{Q})}{\bar{n}_{12}+\bar{n}_{23}-\bar{n}_{13}};$$

therefore, we can take $\mathcal{E}_P = \mathcal{E}_P^1 - \bar{e}\mathcal{E}_P^2, \mathcal{E}_Q = \mathcal{E}_Q^1 - \bar{e}\mathcal{E}_Q^2$, where $\mathcal{E}_P^i, \mathcal{E}_Q^i$ are such that

$$\varepsilon\frac{\partial_P\sqrt{A^2+B^2}(\bar{P},\bar{Q})}{|\bar{n}_{12}+\bar{n}_{23}-\bar{n}_{13}|} = \mathcal{E}_P^1 + o(\varepsilon),$$
$$\varepsilon^2\frac{\partial_Q\sqrt{A^2+B^2}(\bar{P},\bar{Q})}{|\bar{n}_{12}+\bar{n}_{23}-\bar{n}_{13}|} = \mathcal{E}_Q^1 + o(\varepsilon) \tag{21}$$

and

$$\varepsilon\frac{\partial_P(n_{12}+n_{23}-n_{13})(\bar{P},\bar{Q})}{(\bar{n}_{12}+\bar{n}_{23}-\bar{n}_{13})} = \mathcal{E}_P^2 + o(\varepsilon),$$
$$\varepsilon^2\frac{\partial_Q(n_{12}+n_{23}-n_{13})(\bar{P},\bar{Q})}{(\bar{n}_{12}+\bar{n}_{23}-\bar{n}_{13})} = \mathcal{E}_Q^2 + o(\varepsilon). \tag{22}$$

To prove (21) we proceed as follows. From the second of (15) with $(P,Q) = (\bar{P},\bar{Q})$, one has

$$|\bar{n}_{12}+\bar{n}_{23}-\bar{n}_{13}| = \bar{n}_{13}\frac{\bar{Q}}{2\bar{r}_2^3} = \sqrt{\bar{p}}\frac{\gamma_{12}\gamma_{23}}{2\bar{r}_2^3}\varepsilon^3 + o(\varepsilon^4), \tag{23}$$

where $\bar{p}$ is the parameter of $\bar{\mathcal{C}}$ and $\bar{n}_{13} = \sqrt{\bar{p}}\,\varepsilon + o(\varepsilon^2)$. Using (13) one has

$$\vec{\partial}\sqrt{A^2+B^2}(\bar{P},\bar{Q}) = \hat{\bar{g}}\cdot\vec{\partial}R^{\perp}, \tag{24}$$

where $R^{\perp} = (sB, -sA)$. Therefore, we need to evaluate $\vec{\partial}A(P,Q), \partial B(P,Q)$. To this end, rewrite (14) as

$$A = -r_1(r_2-r_3)(1-\cos v_{12}) - r_3(r_1-r_2)(1-\cos v_{23})$$
$$B = r_1 r_3(\sin v_{12} + \sin v_{23} - \sin v_{13}) - (n_{12}+n_{23}-n_{13}), \tag{25}$$

where we used $n_{ij} = r_i r_j \sin v_{ij}$. From (25) one has

$$\vec{\partial}A(\bar{P},\bar{Q}) = \sum_{i=1}^{3}\vec{A}_i, \qquad \vec{\partial}B(\bar{P},\bar{Q}) = \sum_{i=1}^{3}\vec{B}_i,$$

where

$$\vec{A}_1 = -\vec{\partial} r_1(\bar{P},\bar{Q})(\bar{r}_2 - \bar{r}_3)(1 - \cos\bar{v}_{12}) - \vec{\partial} r_3(\bar{P},\bar{Q})(\bar{r}_1 - \bar{r}_2)(1 - \cos\bar{v}_{23}),$$
$$\vec{A}_2 = -\bar{r}_1\vec{\partial}(r_2 - r_3)(\bar{P},\bar{Q})(1 - \cos\bar{v}_{12}) - \bar{r}_3\vec{\partial}(r_1 - r_2)(\bar{P},\bar{Q})(1 - \cos\bar{v}_{23}),$$
$$\vec{A}_3 = -\bar{r}_1(\bar{r}_2 - \bar{r}_3)\vec{\partial}(1 - \cos v_{12})(\bar{P},\bar{Q}) - \bar{r}_3(\bar{r}_1 - \bar{r}_2)\vec{\partial}(1 - \cos v_{23})(\bar{P},\bar{Q}),$$
$$\vec{B}_1 = \vec{\partial} r_1\bar{r}_3(\sin\bar{v}_{12} + \sin\bar{v}_{23} - \sin\bar{v}_{13}) + \bar{r}_1\vec{\partial} r_3(\sin\bar{v}_{12} + \sin\bar{v}_{23} - \sin\bar{v}_{13}),$$
$$\vec{B}_2 = \bar{r}_1\bar{r}_3\vec{\partial}(\sin v_{12} + \sin v_{23} - \sin v_{13}),$$
$$\vec{B}_3 = -\vec{\partial}(n_{12} + n_{23} - n_{13}).$$

Using Taylor formula for $\bar{r}_1, \bar{r}_3, \bar{v}_1, \bar{v}_3$ and recalling Lemmas B.1, B.2, B.4, we find that for suitable constants $\mathcal{A}_P, \mathcal{A}_Q, \mathcal{B}_P, \mathcal{B}_Q$, one has

$$\begin{aligned} \partial_P A(\bar{P},\bar{Q}) &= \mathcal{A}_P\,\varepsilon^2 + o(\varepsilon^3), \quad \partial_Q A(\bar{P},\bar{Q}) = \mathcal{A}_Q\,\varepsilon + o(\varepsilon^2),\\ \partial_P B(\bar{P},\bar{Q}) &= \mathcal{B}_P\,\varepsilon^2 + o(\varepsilon^3), \quad \partial_Q B(\bar{P},\bar{Q}) = \mathcal{B}_Q\,\varepsilon + o(\varepsilon^2). \end{aligned} \tag{26}$$

The proof of (21) is obtained casting together (26), (24) and (23). The proof of (22) is quite similar: using (15) we have

$$\frac{\vec{\partial}\,[n_{12} + n_{23} - n_{13}]\,(\bar{P},\bar{Q})}{\bar{n}_{12} + \bar{n}_{23} - \bar{n}_{13}} = \frac{\vec{\partial} n_{13}(\bar{P},\bar{Q})}{\bar{n}_{13}} + \frac{\vec{\partial} Q(\bar{P},\bar{Q})}{Q} - 3\frac{\vec{\partial} r_2(\bar{P},\bar{Q})}{\bar{r}_2}.$$

Therefore, by Lemmas B.1, B.3, we obtain (22).

We remark that (26) allows to evaluate the derivatives of the true anomaly $v_2 = -g$; indeed, taking the gradient of $\tan v_2 = A/B$ (see (13)), one has:

$$\begin{aligned} \vec{\partial} v_2(\bar{P},\bar{Q}) &= \cos^2\bar{v}_2\left[\frac{\vec{\partial} A(\bar{P},\bar{Q})}{\bar{B}} - \frac{\bar{A}}{\bar{B}^2}\vec{\partial} B(\bar{P},\bar{Q})\right]\\ &= \bar{s}\frac{\cos\bar{g}}{\sqrt{\bar{A}^2 + \bar{B}^2}}\left[\vec{\partial} A(\bar{P},\bar{Q}) + \tan\bar{g}\,\vec{\partial} B(\bar{P},\bar{Q})\right], \end{aligned}$$

where $\sqrt{\bar{A}^2 + \bar{B}^2} = \bar{e}|\bar{n}_{12} + \bar{n}_{23} - \bar{n}_{13}| = \frac{\bar{e}\sqrt{\bar{p}}\gamma_{12}\gamma_{23}}{2\bar{r}_2^3}\varepsilon^3 + o(\varepsilon^4)$ (see (13), (23)). Therefore, we obtain the following.

Lemma B.6 *There exist two constants $\mathcal{N}_P^2, \mathcal{N}_Q^2$ depending on $\bar{\mathcal{C}}, t_2, \gamma_{12}, \gamma_{23}$ such that*

$$\varepsilon\,\partial_P v_2(\bar{P},\bar{Q}) = \mathcal{N}_P^2 + o(\varepsilon), \qquad \varepsilon^2\partial_Q v_2(\bar{P},\bar{Q}) = \mathcal{N}_Q^2 + o(\varepsilon).$$

Finally we are able to compute the lowest orders of the quantities $\eta_{ij} = \eta_{ij}(P,Q) = n_{ij}/S_{ij}$ appearing in the definition of $\mathcal{F}_G$ (see (16)). For simplicity we assume to deal with an elliptic trajectory, i.e. $\bar{e} < 1$, though the results can be extended to any value of the eccentricity. Let $z = (P,Q)$ vary in a small neighbourhood of $\bar{z} = (\bar{P},\bar{Q})$. If $\xi_i = \xi_i(P,Q)$ denotes the eccentric anomaly and if $M_i = M_i(P,Q) = \xi_i - e\sin\xi_i$ is the mean anomaly, the quantity η_{23} can be expressed as

$$\eta_{23} = \frac{\sin(\xi_3 - \xi_2) - e(\sin\xi_3 - \sin\xi_2)}{M_{23}} = 1 - \frac{\xi_{23} - \sin\xi_{23}}{M_{23}},$$

where $\xi_{ij} = \xi_j - \xi_i$, $M_{ij} = M_j - M_i$. Therefore, we have

$$\begin{aligned}\vec{\partial}\eta_{23}(\bar{P},\bar{Q}) &= -\frac{\vec{\partial}(\xi_{23} - \sin\xi_{23})}{\bar{M}_{23}} + \frac{(\bar{\xi}_{23} - \sin\bar{\xi}_{23})\vec{\partial}M_{23}}{\bar{M}_{23}^2} \\ &= -\left(\frac{\bar{\xi}_{23}^2}{2\bar{M}_{23}} + o\left(\bar{\xi}_{23}^4/\bar{M}_{23}\right)\right)\vec{\partial}\xi_{23}(\bar{P},\bar{Q}) \\ &\quad + \left(\frac{\bar{\xi}_{23}^3}{6\bar{M}_{23}^2} + o\left(\bar{\xi}_{23}^5/\bar{M}_{23}^2\right)\right)\vec{\partial}M_{23}(\bar{P},\bar{Q}) \\ &= [\varepsilon\mathcal{E}_1 + o(\varepsilon^3)]\vec{\partial}\xi_{23}(\bar{P},\bar{Q}) + [\varepsilon\mathcal{E}_2 + o(\varepsilon^3)]\vec{\partial}M_{23}(\bar{P},\bar{Q}), \qquad (27)\end{aligned}$$

where we used $\bar{M}_{ij} = \gamma_{ij}\bar{a}^{-3/2}\varepsilon, \bar{\xi}_{ij} = \bar{M}_{ij}/(1 - \bar{e}\cos\bar{\xi}_i) + o(\varepsilon^2)$, with $\mathcal{E}_1, \mathcal{E}_2$ being two suitable constants. We proceed to compute $\vec{\partial}\xi_{23}(\bar{P},\bar{Q})$, $\vec{\partial}M_{23}(\bar{P},\bar{Q})$. Using the classical relations

$$\xi_i = 2\tan^{-1}\left(f(e)\tan\frac{v_i}{2}\right), \qquad f(e) \equiv \sqrt{\frac{1-e}{1+e}}$$

and recalling Lemmas B.5 and B.6, one finds that

$$\begin{aligned}&\partial_P\xi_{23}(\bar{P},\bar{Q}) = \mathcal{X}_P^{23} + o(\varepsilon), \qquad \partial_{Q}\varepsilon\xi_{23}(\bar{P},\bar{Q}) = \mathcal{X}_Q^{23} + o(\varepsilon), \\ &\partial_P M_{23}(\bar{P},\bar{Q}) = \mathcal{M}_P^{23} + o(\varepsilon), \qquad \partial_{Q}\varepsilon M_{23}(\bar{P},\bar{Q}) = \mathcal{M}_Q^{23} + o(\varepsilon) \qquad (28)\end{aligned}$$

for some quantities $\mathcal{X}_Q^{23}, \mathcal{X}_Q^{23}, \mathcal{M}_P^{23}, \mathcal{M}_Q^{23}$ depending only on $\bar{C}, \gamma_{12}, \gamma_{23}$. Inserting (28) in (27), we obtain the following

Lemma B.7 *Let $i \neq j \in \{1,2,3\}$. There exist two constants $\varepsilon_P^{ij}, \varepsilon_Q^{ij}$ depending on $\bar{C}, t_2, \gamma_{12}, \gamma_{23}$, such that*

$$\partial_P\eta_{ij}(\bar{P},\bar{Q}) = \mathcal{E}_P^{ij}\varepsilon + o(\varepsilon^2), \qquad \partial_Q\eta_{ij}(\bar{P},\bar{Q}) = \mathcal{E}_Q^{ij} + o(\varepsilon).$$

We are finally ready to complete the
Proof of proposition B.1 From the definition of the Gauss map (16), one has

$$\begin{aligned}\vec{\partial}P'(\bar{P},\bar{Q}) &= \frac{\bar{\eta}_{12}}{\bar{\eta}_{23}}\left[\frac{\vec{\partial}\eta_{12}(\bar{P},\bar{Q})}{\bar{\eta}_{12}} - \frac{\vec{\partial}\eta_{23}(\bar{P},\bar{Q})}{\bar{\eta}_{23}}\right] \\ \vec{\partial}Q'(\bar{P},\bar{Q}) &= \gamma_{12}\gamma_{23}\bar{q}\left[\varepsilon^2\frac{\vec{\partial}r_2/r_1(\bar{P},\bar{Q})}{\bar{r}_2/\bar{r}_1} + \varepsilon^2\frac{\vec{\partial}r_2/r_3(\bar{P},\bar{Q})}{\bar{r}_2/\bar{r}_3}\right] \\ &\quad + \gamma_{12}\gamma_{23}\bar{q}\left[\varepsilon^2\frac{\vec{\partial}\eta_{12}(\bar{P},\bar{Q})}{\bar{\eta}_{12}} + \varepsilon^2\frac{\vec{\partial}\eta_{23}(\bar{P},\bar{Q})}{\bar{\eta}_{23}}\right] - \gamma_{12}\gamma_{23}\bar{q} \\ &\quad \times\left[\varepsilon^2\tan\bar{f}_{12}\vec{\partial}f_{12}(\bar{P},\bar{Q}) + \varepsilon^2\tan\bar{f}_{23}\vec{\partial}f_{23}(\bar{P},\bar{Q}) + \varepsilon^2\tan\bar{f}_{13}\vec{\partial}f_{13}(\bar{P},\bar{Q})\right],\end{aligned}$$

where $\bar{q} \equiv \frac{\bar{Q}}{\gamma_{12}\gamma_{23}\varepsilon^2} = \frac{\bar{r}_2^2}{\bar{r}_1\bar{r}_3}\bar{\eta}_{12}\bar{\eta}_{23}\frac{1}{\cos\bar{f}_{12}\cos\bar{f}_{23}\cos\bar{f}_{13}}$. For $i \neq j$, let $\bar{\eta}_{ij} = 1 + o(\varepsilon^2)$, $\bar{r}_i/\bar{r}_j = 1 + o(\varepsilon)$, $\cos\bar{f}_{ij} = 1 + o(\varepsilon^2)$; therefore, $\bar{q} = 1 + o(\varepsilon)$ and using Lemma B.7 to evaluate $\vec{\partial}\eta_{ij}(\bar{P},\bar{Q})$, Lemma B.2 to evaluate $\vec{\partial}[r_i/r_j](\bar{P},\bar{Q})$ and the Remark B.2 to evaluate $\vec{\partial}f_{ij}(\bar{P},\bar{Q}) = \vec{\partial}v_{ij}(\bar{P},\bar{Q})/2$, we find the result of Proposition B.1.

C Computation of the derivatives by polynomial interpolation

Suppose we want to compute the derivative at some point $\bar{x}$ of the function $x \to f(x)$, using a polynomial interpolation. Let $x_i = \bar{x} + ih, i = -n, \ldots, n$ be the nodes around $\bar{x}$ and let $y_i = f(x_i)$; we define the interpolating Laplace polynomial $\mathcal{P}_n$ of degree $2n$ as

$$\mathcal{P}_n(x) = \sum_{i=-n}^{n} \frac{\prod_{j\neq i}(x - x_j)}{\prod_{j\neq i}(x_i - x_j)} y_i.$$

After the change of variable $s = (x - \bar{x})/h$, one obtains

$$\mathcal{P}_n(\bar{x} + sh) = \sum_{i=-n}^{n} \frac{\Pi_{j\neq i}(s - j)}{\Pi_{j\neq i}(i - j)} y_i \equiv \mathcal{Q}_n(s).$$

The derivative $\mathrm{d}f(\bar{x})/\mathrm{d}x$ is approximated by $\mathrm{d}\mathcal{P}_n(\bar{x})/\mathrm{d}x = h^{-1}\mathrm{d}\mathcal{Q}_n(0)/\mathrm{d}s$. Let us consider first the term with $i = 0$:

$$\frac{\Pi_{j\neq 0}(s - j)}{\Pi_{j\neq 0}(-j)} y_0 = \frac{(s-n)(s-n+1)\cdots(s-1)(s+1)\cdots(s+n-1)(s+n)}{(-1)^n (j!)^2} y_0.$$

This term is an even function of s, so that its derivative at $s = 0$ is zero. On the other hand, deriving (through Leibnitz rule) with respect to s the remaining terms of the sum, for any $i \neq 0$ one has:

$$\frac{(s-n)\cdots[s-(i+1)][s-(i-1)]\cdots(s)\cdots(s+n)}{(i-n)\cdots(-1)(1)\cdots(i+n)} y_i$$
$$= \frac{(s-n)\cdots[s-(i+1)][s-(i-1)]\cdots(s)\cdots(s+n)}{(-1)^{n-i}(n-i)!(n+i)!} y_i;$$

computing these terms at $s = 0$, the only one which survives is given by

$$\frac{(-1)^{i+1}}{i} \frac{(n!)^2}{(n-i)!(n+i)!} y_i \quad (i \neq 0).$$

Finally, one concludes that

$$\frac{\mathrm{d}\mathcal{P}_n}{\mathrm{d}x}(\bar{x}) = \frac{1}{h}\frac{\mathrm{d}\mathcal{Q}_n}{\mathrm{d}s}(0) = \sum_{|i|\leq n,\, i\neq 0} \frac{(-1)^{i+1}}{ih} \frac{(n!)^2}{(n-i)!(n+i)!} y_i.$$

References

Celletti, A., Pinzari, G.: Four classical methods for determining planetary elliptic elements: a comparison. Celest. Mech. Dyn. Astr. **93**(1), 1–52 (2005)

Gallavotti, G.: Meccanica Elementare. P. Boringhieri (ed.), Torino, 2nd edn. pp. 498–516 (1980)

Gauss, C.F.: Theory of the Motion of the Heavenly Bodies Moving about the Sun in Conic Sections; Dover Publication. New York (1963)

Herrick, S. Jr.: On the Laplacian and Gaussian orbit methods. Astr. Soc. Pac. **49**(287), 17–23 (1937)

Laplace, P.S.: Memoires de l'Académie Royale des Sciences de Paris. Coll. Works **10**, 93–146 (1780)

Milani, A., Gronchi, G.F., De' Michieli Vitturi, M., Knezevic, Z.: Orbit determination with very short arcs. I admissible regions. Celest. Mech. Dyn. Astr. **90**(1–2), 57–85 (2004)

Milani, A., Gronchi, G.F., Knezevic, Z., Sansaturio, M.E., Arratia, O.: Orbit determination with very short arcs. Icarus **179**(2), 350–374 (2005)

Milani, A., Knezevic, Z.: From astrometry to celestial mechanics: orbit determination with very short arcs. Celest. Mech. Dyn. Astr. **92**(1–3), 1–18 (2005)

Moulton, F.R.: Memoir on the theory of determining orbits. Astrono. J. iss. 661-662-663, **28**, 103–124 (1914)
Mossotti, O.F.: Sopra la Determinazione delle Orbite dei Corpi Celesti per Mezzo di Tre Osservazioni, Scritti. Pisa, Domus Galileana, original version: Memoria Postuma (1942)
Plummer, H.C.: An introductory treatise on dynamical astronomy. Cambridge University Press, Cambridge; Dover Publication, New York (1960)
Poincare, H.: Sur la détermination des orbites par la méthode de Laplace. Bull. Astr. **23**, 161–187 (1906)

Celestial Mechanics and Dynamical Astronomy (2006) 95:345–356
DOI 10.1007/s10569-006-9028-7

REVIEW ARTICLE

Collision risk against space debris in Earth orbits

A. Rossi · G. B. Valsecchi

Received: 14 December 2005 / Revised: 27 March 2006 /
Accepted: 4 May 2006 / Published online: 15 August 2006

Abstract Öpik's formulae for the probability of collision are applied to the analysis of the collision risk against space debris in Low-Earth Orbit (LEO) and Medium Earth Orbit. The simple analytical formulation of Öpik's theory makes it applicable to complex dynamical systems, such as the interaction of the ISS with the whole debris population in LEO The effect of a fragmentation within a multiplane constellation can also be addressed. The analysis of the evolution of the collision risk in Earth orbit shows the need of effective mitigation measures to limit the growth of the collision risk and of the fragmentation debris in the next century.

Keywords Space debris · Impact risk · Öpik theory

1 Introduction

Currently about 10,000 objects with diameter larger than about 10 cm are orbiting the Earth. Most of these objects are cataloged by the United States Air Force Space Command (AFSPC) in the Two-Line Element (TLE) catalog. The limiting size of the objects included in the catalog (due to limitations in sensors power and in observation and data processing procedures) is about 5–10 cm below a few thousands kilometers of altitude and about 0.5–1 m in higher orbits (up to the geostationary ones). The orbits of the TLE catalog objects are maintained thanks to the observations performed by the Space Surveillance Network (SSN).

The SSN is composed by 25 sensors, both radars and optical sensors. To be included in the catalog the orbit of an object has to be known with an accuracy that allows to predict its next appearance over the network of sensors that is used to

A. Rossi (✉)
ISTI–CNR, Via Moruzzi 1, I-56124 Pisa, Italy
e-mail: alessandro.rossi@isti.cnr.it

G. B. Valsecchi
IASF–INAF, Roma, Italy

maintain the catalog itself. About 99% of the mass in orbit is due to the large objects included in the catalog. In addition to these larger objects, about 350,000 particles larger than 1 cm and 3×10^8 objects larger than 1 mm are polluting the near Earth environment.

To highlight the danger posed by these smaller particles, it should be kept in mind that the average impact velocity in Low-Earth Orbit (LEO) is about 10 km/s. This means that a ~1 mm particle is able, e.g., to penetrate the Reinforced Carbon–Carbon (RCC) tiles on the Space Shuttle wings leading edges; this event may cause a loss of the Shuttle during reentry. A particle around 5 mm is able to directly penetrate the Shuttle cabin.

Only 6% of all the objects currently present in space are active satellites. All the rest are space debris, composed by spent upper stages, inactive satellites, fragments from in-orbit explosions and collisions, mission-related objects (such as sensor covers), drops of liquid metal used as refrigerant by nuclear reactors carried on-board the soviet reconnaissance satellites (RORSATs), tiny dipoles used for telecommunication tests (West Ford Needles), and particles from Solid Rocket Motors (SRM) exhausts (ranging from micrometer sized dust to centimeter sized slag).

Figure 1 shows the density of objects for three different size regimes as a function of altitude and highlight the three main zones of accumulation in space: the region of the LEO (below about 2,000 km), the Medium Earth Orbits (MEO, between 2,000 km and about 36,000 km), and the High-Earth Orbits (HEO, above 36,000 km).

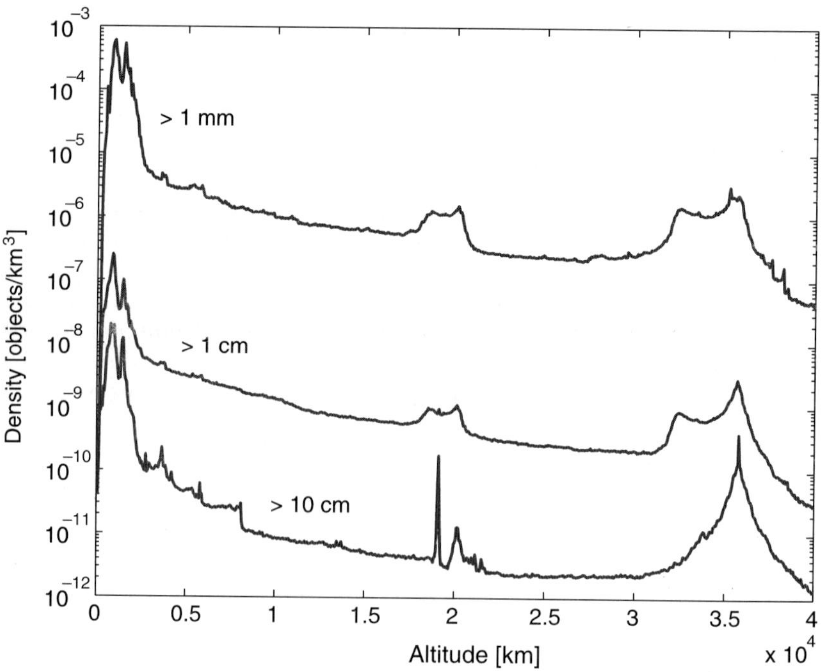

Fig. 1 Density of objects as a function of altitude for three different size thresholds: objects with diameter larger than 1 mm, 1 cm, and 10 cm

2 Known in-orbit collisions

The overcrowding of the space around the Earth makes collisions a serious threat. Currently hundreds of close approaches (i.e., passes within less than 1 km) between cataloged objects occur on a daily basis and indeed a few accidental collisions have already been recorded in LEO. Although still very unlikely from a statistical point of view, collisions are going to become the most important source of debris in a not too distant future.

On July 24, 1996, a mere 40 years after the first object ever entered the space above the planet, the first accidental collision between an operational satellite and a piece of debris was recorded: the French micro-satellite *Cerise* was hit, at the relative velocity of 14.77 km/s, by a fragment, of about 10 cm^2, coming from the 1986 explosion of an Ariane 1 rocket upper stage (Alby et al. 1997). The impact happened in a polar circular LEO ($h \simeq 700$ km). All of a sudden, collisions with space debris became a reality.

On January 17, 2005 a 31-year-old U.S. rocket body was hit by a fragment from the third stage of a Chinese CZ-4 launch vehicle, which had exploded in March 2000. The impact happened in a circular retrograde LEO ($h \simeq 885$ km) above the south polar region. The rocket body was about 1 m^2 and the radar cross-section of the fragment was 0.06 m^2. Three debris were cataloged from the impact.

Then recently, while analyzing historical tracking data, it was realized that another collision in LEO happened many years ago. In December 1991, a Russian non-functional navigation satellite (Cosmos 1934) was hit by a piece of debris from a similar spacecraft, Cosmos 926. The objects were in similar circular LEOs ($h \simeq 980$ km) with $i = 83°$. Two debris were cataloged a few weeks after the event.

As can be seen from Fig. 2 the collisions happened in high-density regions of LEO. Moreover all the orbits have large inclinations, which means that the spacecraft were repeatedly crossing the high-density regions at high latitude.

3 Collision risk evaluation

The number of collisions between two orbiting particles expected during a time interval Δt can be formally expressed as $P_i\ (R + r)^2\ \Delta t$, where r and R are the projectile and target radius, respectively (both bodies are assumed to be spherical), and P_i is the so-called *intrinsic collision probability per unit of time*, a quantity depending only on the two sets of orbital elements; P_i may be interpreted as the collision rate between two bodies for which $(r+R) = 1$ m. Of course, $P_i = 0$ if the two orbits cannot intersect each other for geometrical reasons; this occurs when the apocenter distance of the inner orbit is smaller than the pericenter distance of the outer one.

A very simple, approximate estimate of P_i can be obtained by applying a particle-in-a-box (PIAB) model. For the LEO environment, we have an average relative velocity $V \approx 10$ km/s and a volume $W \approx 10^{12}$ km^3. Therefore, the PIAB model yields a mean intrinsic collision probability of $\approx 10^{-9}/m^2/$year.

Improving on the simplistic approach of the PIAB model, in a similar way as done for the Asteroid Belt, using Wetherill's algorithm and an *nCUBE* 2 multi-computer (Wetherill 1967), Rossi and Farinella (1992) calculated the mutual collision probabilities and impact velocities for a set of 2,700 objects with perigee < 1600 km. The average intrinsic collision probability P_i was found to be $1.105 \pm 0.812 \times 10^{-9}/m^2/$year and the average impact velocity $V = 9.65 \pm 0.88$ km/s. While the average value of P_i

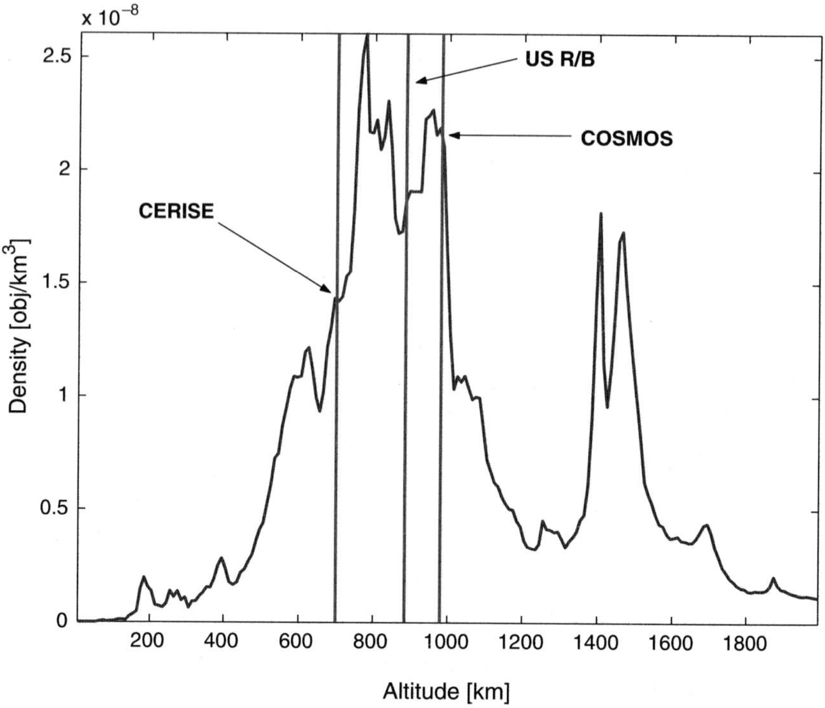

Fig. 2 Spatial density of objects larger than 10 cm in LEO. The *vertical lines* mark the altitude of the known collision events

is remarkably close to the one calculated with the PIAB method, from Fig. 3 it can be noted how the actual collision rate can grow by up to a factor 3 (depending on the chosen orbital elements).

Wetherill's algorithm is actually an extension of the method developed by Öpik (1976) to evaluate collision probabilities between small Solar system bodies and the planets. As a matter of fact the original Öpik method can be shown to be particularly well suited to analyze the collision risk in LEO.

Öpik's analytical expressions relate in a simple way the semi-major axis a, eccentricity e, and inclination i of the projectile orbit to the magnitude and direction of the relative velocity vector at impact $\vec{U}$, in a reference frame that is well suited to describe the impact risk for a target in a circular LEO.

The great advantage of Öpik's approach is that the probability and the geometry of the impacts on a given target can be expressed by means of simple analytical relations. In fact, given a projectile and a target on a circular orbit whose radius is used as unit of length, the intrinsic collision probability per unit time in Öpik's theory is simply given by:

$$P = \frac{U}{2\pi^2 a^{1.5} |U_x| \sin I} \tag{1}$$

(we have assumed $\sqrt{\mathrm{GM}_\oplus} = 1$), where $U = |\vec{U}|$ is the projectile's velocity relative to the target (in units of the target's geocentric velocity):

$$U = \sqrt{3 - \frac{1}{a} - 2\sqrt{a(1-e^2)}\cos I}\,, \tag{2}$$

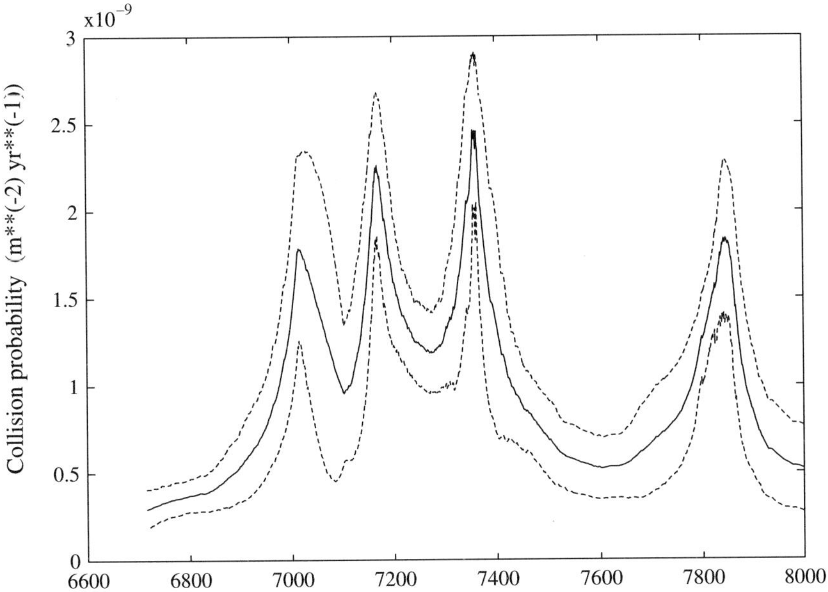

Fig. 3 Running-box mean of the collision probability (m²/year) with $\pm 1\sigma$ variation, as a function of the semi-major axis, for the objects with semi-major axis < 8,000 km

U_x is the x-component of $\vec{U}$ (in a frame centered on the target, with the y-axis pointing in the direction of the target's instantaneous motion and the x-axis pointing away from the Earth) and I is the inclination of the orbit of the projectile relative to that of the target (Valsecchi et al. 2000). By means of the components of the vector $\vec{U}$, given by:

$$U_x = \pm\sqrt{2 - \frac{1}{a} - a(1 - e^2)},$$

$$U_y = \sqrt{a(1 - e^2)}\cos I - 1,$$

$$U_z = \pm\sqrt{a(1 - e^2)}\sin I.$$

we can introduce two angles, θ and ϕ, such that:

$$U_x = U\sin\theta\sin\phi, \qquad U_y = U\cos\theta, \qquad U_z = U\sin\theta\cos\phi.$$

Therefore:

$$\cos\theta = \frac{U_y}{U}, \qquad \tan\phi = \frac{U_x}{U_z}. \tag{3}$$

The angle θ plays the rôle of the colatitude (angle between the vector U and the y-axis), and ϕ that of the longitude, measured from the U–y plane to the y–z plane; θ and ϕ define the direction of arrival of the projectile, in the frame co-moving with the target.

By means of Eqs. 1–3, we have completely characterized the risk of collision between an object in a circular LEO and a population of projectiles in any crossing orbit, having at hand the velocity, the direction and the probability of collision. Thanks to its simple analytical formulation, Öpik's theory is particularly well suited to analyze the collision risk of selected targets with large populations of projectiles over long-time spans. The risk of collision against the whole population of debris or against a given cloud of particles following a fragmentation event in space can also be visualized in a compact and informative way, as shown in Valsecchi et al. (2000, 2003).

3.1 Collision risk on the international space station

A particularly sensitive target in LEO, due to its large dimensions and to the presence of humans on board, is of course the International Space Station (ISS). The USSPACECOM routinely predicts all the conjunctions between any catalogued object and a $40 \times 80 \times 80\,\text{km}^3$ box about the ISS position. If the collision probability between any of these objects and the ISS is stably larger than 10^{-4} (after a few consecutive orbit determination runs of the projectile and the target), an avoidance maneuver is performed, whenever possible. Currently the ISS is performing about two avoidance maneuvers per year (Foster 2001; M. Matney, private communication). Since the ISS is shielded, at best, against debris smaller than 1 cm all the debris that are within these two thresholds (i.e., between the shielded and the tracked population) are a potential threat to its safety. Note that some ISS modules are not even shielded up to the 1 cm threshold and are therefore more susceptible to damage.

To visualize the impact risk on the ISS it is convenient to use a Hammer–Aitoff (equal-area) projection of the celestial sphere. Figure 4 shows such a projection. The map is centered on the instantaneous direction of motion of the ISS, and rotating around the Earth, so that the relative longitude of the latter, marked with a small circle in the diagram, is $\lambda - \lambda_\oplus = -90°$, and its latitude is $B = 0°$. The center, at

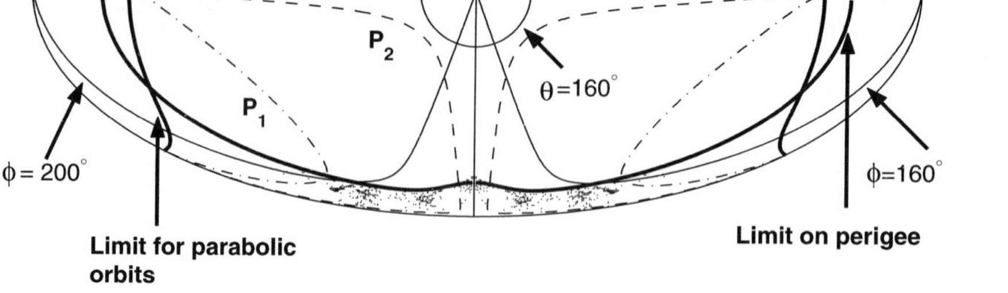

Fig. 4 Distribution of the potential impactor on the ISS, having impact velocities between 4 and 5 km/s. The equal-probability curves correspond to impact probabilities of $P_2 = 10^{-9}/\text{m}^2/\text{year}$ (*dashed line*) and $P_1 = 10^{-10}/\text{m}^2/\text{year}$ (*dash–dotted line*)

$\lambda - \lambda_{\oplus} = 0°$, $B = 0°$, coincides with the instantaneous direction of motion of the ISS. The circle around the center denotes $\theta = 160°$, and the four curves emanating from the center denote (from upper right, clockwise) $\phi = 20°$, $\phi = 160°$, $\phi = 200°$, $\phi = 340°$. Note that $\theta = 180°$ represents the direction of an object that has a head-on collision with the target, whereas the circle $\theta = 160°$ represents a cone including the objects that impact the target from a direction of 160° from the motion of the target. Of course the user can choose to show on the plot the curves for any desired value of θ and ϕ. Here we just choose a set of reasonable values to exemplify the method. As pointed out by the arrows in the plot, the contours corresponding to parabolic orbits and to orbits having perigee at about 150 km of altitude are drawn (thick solid lines, identified by the respective arrows). Orbits bound to the Earth and not prone to immediate atmospheric reentry must occupy the region delimited by these two curves.

The symmetry of the four quadrants in the diagram in Fig. 4 is related to the fact that for each set of a, e, i of the projectile orbit, there are four directions from which it can approach the ISS, for appropriate values of ω. These possibilities correspond to collisions taking place at the ascending or at the descending node of the orbit (relative to the ISS orbit), in the pre-perigee or post-perigee part of the orbit of the projectile. Correspondingly, one can easily verify that, for each set of a, e, i, there is a single set of U, θ, but four values of ϕ, each associated to one of the possibilities listed before.

Exploiting the analytical relations given by Öpik, it is possible to draw contours of equal impact probability in the projection of the celestial sphere given in Fig. 4.

In particular, the collision probability can be written in terms of θ and ϕ. Therefore, given a probability value $P = P(\theta, \phi)$, four lines can be drawn, in the four quadrants of the sky plane, such that the projectiles lying on these lines have the given probability P of impacting against the target. These are the dashed and dash-dotted curved lines in Fig. 4. The particles falling inside these lines have a probability of impact higher than P, while the opposite holds for projectiles falling outside the lines.

In a plot such as Fig. 4, the population of potential impactors, with impact velocities U lying in an interval $U_i < U \leq U_i + \Delta U$, with $\Delta U = 1$ km/s, are shown ($U_i = 4$ km/s in Fig. 4). As it can be noted, the projectiles come from close to the border, i.e. from the back of the ISS. The impacting directions are also not far from the local horizontal plane (defined by $\sin\phi = 0$, i.e., the central vertical line in Fig. 4), within less than 30° of that plane. All the crossing objects have impact probabilities $> P_1$; only those closer to the ISS orbital plane have impact probabilities $> P_2$.

The orbits and the nature of the projectiles populating Fig. 4 can be inferred from the population model used in the simulations (in our case, the European Space Agency space debris reference model MASTER 99). Most of the projectiles are in moderate to high-eccentricity orbits. About half of them have semi-major axis $a \sim 13{,}000$ km (up to about 20,000 km) and their perigee close to the ISS orbital altitude. Their inclination ranges from about 20 to about 38°. More than 95% of these particles are SRM related debris, the rest being fragmentation debris.

By plotting the crossing population at different velocity (i.e., energy) levels, the whole picture of the collision risk for the ISS can be drawn (Valsecchi and Rossi 2002). The general conclusions of the collision risk analysis for the ISS have been:

- the potential impactors tend to come from directions close to the $\sin\phi = 0$ plane, the deviation from this plane being a decreasing function of the impact velocity U;
- most of the objects crossing the ISS orbit have impact velocities between about 9 and 12 km/s, with impact probability in excess of $P_1 = 10^{-10}/\text{m}^2/\text{year}$;

- the SRM-related particles dominate the potential impactors for nearly all the velocity range; only at very high velocities, above 13 km/s, the impactor population is dominated by fragments.

Coupling Öpik's method with a model simulating the long-term evolution of the whole debris population (see Sect. 4), a simulation of the evolution of the flux of debris on the ISS for the next 20 years has been performed. In this case the impactor population is again given by the original MASTER population, plus the result of the space activities (launches, explosion, collisions, SRM firings) in these two decades. The results show that the SRM particles are the major contributors to the flux below an impact energy level of 10^7 J. Since these particles have low masses and lower impact velocities they do not contribute for higher impact energies. The flux shows periodic variations (a factor 2–3) that are are mainly related to the ISS altitude variations (the ISS is flying at a variable altitude corresponding to a nearly constant atmospheric density). In this respect, the most critical part of the mission is the central one when the ISS will orbit in the altitude band around 450 km, with fluxes up to a few parts in $10^{-6}/\mathrm{m}^2/\mathrm{year}$, for impact energies in excess of 10^6 J.

3.2 Collision risk on satellite constellations

Another set of particularly sensible targets in space are the multi-plane satellite constellations. The complex interaction between the dynamics of a debris cloud generated by the accidental fragmentation of a constellation spacecraft and the overall dynamics of the satellites in the constellation planes can be effectively studied by Öpik's method.

This interaction is due to the interplay between the collision debris orbital evolution, under the effect of the initial impulse after the break-up and of the geopotential (and the air drag in LEO), and the global precessional motion of the planes of the constellation, under the effect of the Earth's J_2. The result is a dangerous collective behavior that can strongly enhance the collision risk within the constellation, with respect to the background flux normally affecting the constellation orbit.

We studied different Walker-like LEO constellations (Rossi et al. 1999; Rossi and Valsecchi 2001). The largest such constellation currently orbiting is IRIDIUM. It consists of 66 LEO satellites (plus six spares) orbiting in six different orbital planes at an altitude of about 780 km and with an inclination of 86.4° to the Earth's equator.

In the simulations, an Iridium satellite is broken up by a 1-kg projectile at a relative speed of 10 km/s. The resulting fragment swarm (some 8,000 bodies heavier than 1 g) are evolved under Earth oblateness and air drag perturbations. Then, the cumulative impact probability on the other constellation satellites, as a function of time, is calculated by means of the method described in Sect. 3.

We have found that, in the impact energy range between about 10^7 and 10^8 J, corresponding to disruptive projectiles, the collision probability due to the fragmentation debris stays higher than the background level from the general orbiting population for several years. This means that, after the initial break-up, there will be a probability of the order of 10% that a second one will follow within 5 years, and that eventually this may trigger a collisional chain-reaction effect (at the constellation altitude) with a characteristic time scale of about one century, much less than the current estimates with the general debris population (some 300–500 yr) (Cordelli et al. 1998).

The same kind of study has been performed for the MEO constellations used for global navigation, namely GPS, GLONASS, and the forthcoming European

GALILEO (Rossi et al. 2004). The densities in the MEO region are about two orders of magnitude lower than in LEO (there are less objects and more space). Nonetheless the estimates call for about 60,000 objects with diameter >1 cm possibly crossing the Navigation Satellites Orbits.

To minimize the impact risk in the region, at the End-of-Life (EOL) the GPS spacecrafts are moved to a disposal region, about 500 km above the operational orbit. Recently it has been shown that the disposal orbits are potentially unstable with a long-term increase of the eccentricity (Chao and Gick 2004). The cause of this long-term eccentricity growth is a resonance condition resulting from third body and geopotential perturbations. This eccentricity growth may represent an hazard for the long-term disposal of the spent satellites and upper stages since they can re-enter the operative zone, if the disposal orbit is not stable.

As a matter of fact the operational satellites can perform avoidance maneuvers if the projectile is large enough to be tracked from ground. The problem can arise from the accumulation, in the graveyard zones above the constellations, of non-operational spent spacecraft, unable to perform avoidance maneuvers. If a fragmentation would ever happen in the orbits of the navigation constellation the resulting cloud of debris would be very difficult to track from the ground sensors, due to the large distance, and would therefore represent a serious hazard for all the spacecraft in the region.

Similarly to what has been done for the LEO constellations, the collision risk for the navigation constellations, following a fragmentation of a spacecraft, can be studied. An extension of the theory presented in Sect. 3 has to be devised to apply Öpik's theory to the MEO region, arising from the fact that the precession rates of the perigee argument ($\dot{\omega}$) is about two orders of magnitude smaller at the GPS altitude than in LEO ($\dot{\omega}_{\mathrm{GPS}} \simeq -0.02°$/day). Since a basic assumption of Öpik's theory is that the argument of perigee ω of the projectile orbit, evaluated using as reference plane the orbital plane of the target, is randomly distributed between 0 and 2π, this slower evolution prevents the direct application of our original method to MEOs. In Rossi et al. (2004) a simplified approximation was presented allowing a preliminary study of the problem.

In general terms, the slower dynamics (precession of the angular arguments of an orbit) of the MEO region prevent the appearance of the strong global effects observed for the LEO constellations. In all the simulations it has been observed that the flux following a generic fragmentation is by far larger than the low-background flux in MEO. The values still account for low risks in terms of impacts per year (of the order of 10^{-3}–10^{-4} over a decade). However, the very sensible applications of the navigation constellation call for a high level of reliability that could be seriously endangered by such prolonged levels of debris fluxes. The results also show a strong interrelation of the three constellations. That is in some cases the space distribution of the three systems is such that a larger flux is experienced if the fragmentation event happens on a different constellation. In these cases particles coming from one constellation get a ΔV, from the fragmentation, that modify their orbit and allows them to reach the other constellations altitude. This is particularly true when considering the GPS and the GLONASS constellations. These results stress the need to consider harmonized mitigation policies for the three constellations. That is, the navigation constellation orbital regime should be viewed a whole region to be protected against possible future debris growth. This problem is being analyzed and discussed at the international level in the Inter-Agency Debris Coordination Committee (IADC), the main international committee established to study and face the space debris issue.

The IADC (see the web page: http://www.iadc-online.org) periodically gathers the representatives of all the main space agencies to share the results of the researches on the different aspects of the field. Although the IADC has no power to legislate or enforce any rule, its role is of paramount importance in issuing recommendations that can be passed to higher political levels within each country. The IADC is also in contact with the United Nations Committee on the Peaceful Uses of Outer Space (COPUOS).

4 Long-term evolution of the collision risk

Having established the tools to evaluate and manage the current collision risk in Earth orbit, another important issue is the forecast of the collision risk in the next decades. This evaluation is especially important to device effective strategies to reduce the growth of the risk through appropriate mitigation measures.

Mathematical models and large numerical codes have been developed to simulate the interplay of all the physical processes involved in the evolution of the debris population, modeling the known source mechanisms, injecting objects in space, and sink mechanisms, removing objects from space (Rossi et al. 1994, 1998).

In a typical simulation campaign, a standard scenario of the future evolution, assuming the "business-as-usual" continuation of space activities is compared with a number of other scenarios where different physical models and mitigation measures are used. The reference scenario is characterized by a launch activity deduced from the traffic observed over the last 5 years, adjusted by taking into account the phasing out of obsolete launchers and the introduction of new rocket families. A total of about 60 launches per year is foreseen on average. Mission-related objects are released according to the current practices and no de-orbiting or re-orbiting of spacecraft and upper stages is performed at the EOL. The explosion statistic is based over the last 5 years, with an average of 2.4 explosions/year. It is assumed that the progressive introduction, over the coming 5–25 years, of explosion prevention measures on the systems currently in use, will lead to no more explosions after 2030. As far as the production of SRM slag is concerned, a minimum use of SRM is envisaged, based on current and planned practices (the transition to new launchers and larger commercial spacecraft is reducing considerably the reliance on such propulsion systems).

In Fig. 5, the number of catastrophic collisions expected in the next 100 years is shown, for the reference scenario, by the thick solid lines ($\pm 1\sigma$ variations). In this scenario, about 60 catastrophic collisions should be expected, with a clearly more than linear growth in the cumulative number of events. Looking at the number of debris in the next 100 years, it could be noted that the objects larger than 1 cm will double, displaying a more than linear pace too, and that the collision debris will become the dominant source of particles. Only the adoption of effective mitigation measures will allow a significant reduction of future collisions. The dashed curve in Fig. 5 shows (along with the 1σ boundaries) the expected number of collision in a scenario where a number of mitigation measures are undertaken:

- no mission related object is released after the year 2020;
- starting from the year 2010 the re-orbiting of spacecraft at EOL is performed. The GEO satellites are re-orbited, at EOL, about 300 km above the GEO altitude. The spacecraft in LEO are moved either to super-LEO storage zones (above 2,000 km)

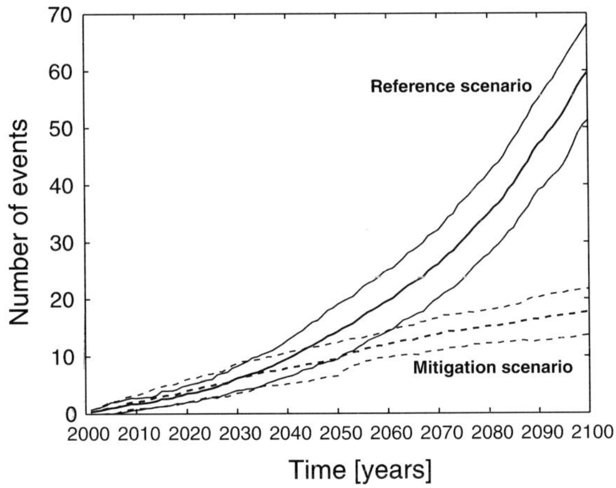

Fig. 5 Cumulative number of catastrophic collisions in the two different scenarios (*thick solid and dashed lines*) $\pm 1\sigma$ variations (*thin solid and dashed lines*), coming from the MonteCarlo simulation process (see text for details)

or they are maneuvered to orbits with a residual lifetime (under the effect of the air drag) of 25 years;

– starting from the year 2010, all upper stages with perigee height $h_p \geq 1{,}400$ km are left where they are, while those with $h_p < 1{,}400$ km, or in high-eccentricity orbits crossing the LEO region, are immediately de-orbited at EOL.

The EOL mitigation measures obtain the important result of stabilizing, and then progressively reducing, the number of intact objects in LEO. This means that the target area is reduced, thus reducing the collision activity. In fact, in the mitigation scenario, the number of expected collisions drops to about 20, showing a linear growth in the next 100 years. Note that different residual spacecraft lifetimes have been simulated (from 25 to 75 years). The results indicate that even a residual life of 75 years in LEO is able to reach this important goal, averting the long-term onset of an exponential growth, but of course there is a temporal delay and a correspondingly higher maximum population with respect to the scenarios where 25 and 50 years are considered.

These results clearly show the paramount importance of the mitigation measures, that are currently being discussed at the international level, to minimize the collision risk in Earth orbit and to safeguard the space environment for the future generations.

5 Conclusions

The analysis of the collision risk for targets in circular orbits in LEO and MEO can be effectively performed by the Öpik's theory. The collision risk on the ISS can be fully characterized by this method. The complex interaction between the multiplane constellations and a large cloud of debris can also be analyzed, both in LEO and in MEO. The analysis of the evolution of the collision risk in Earth orbit for the next century showed again the need of effective mitigation measures, to reduce the non-linear growth of the number of catastrophic collisions and the related number of fragmentation debris.

Acknowledgements The authors wish to thank the two anonymous referees for their useful comments that helped to improve the paper.

References

Alby, F., Lansard, E., Michal, T.: Collision of Cerise with space debris. In: Proceedings of the Second European Conference on Space Debris. Proceedings ESA SP-393, pp. 589–594. Noordwijk, Netherlands (1997)

Chao, C.C., Gick, R.A.: Long-term evolution of navigation satellite orbits: GPS/GLONASS/GALILEO. Adv. Space Res. **34**, 1221–1226 (2004)

Cordelli, A., Farinella, P., Rossi, A.: The influence of the fragmentation threshold on the long-term evolution of the orbital debris population. Planet. Space. Sci. **46**, 691–699 (1998)

Foster, J.L., Jr.: The analytic basis for debris avoidance operations for the International Space Station. In: Proceedings of the Third European Conference on Space Debris. Proceedings ESA SP-473, pp. 441–445. Noordwijk, Netherlands (2001)

Öpik, E.J.: Interplanetary Encounters. Elsevier, New York, USA (1976)

Rossi, A., Farinella, P.: Collision rates and impact velocities for bodies in low Earth orbit. ESA J. **16**, 339–348 (1992)

Rossi, A., Cordelli, A., Farinella, P., Anselmo, L.: Collisional evolution of the Earth's orbital debris cloud. J. Geophys. Res. **99**(E11), 23195–23210 (1994)

Rossi, A., Anselmo, L., Cordelli, A., Farinella, P., Pardini, C.: Modelling the evolution of the space debris population. Planet. Space Sci. **46**, 1583–1596 (1998)

Rossi, A., Valsecchi, G.B., Farinella, P.: Risk of collision for constellation satellites, Nature **399**, 743–744 (1999)

Rossi, A., Valsecchi, G.B.: Self generated debris hazard for satellite constellations. In: Second International Workshop on : Satellite Constellations and Formation flying. Haifa, Israel. February 19–20, 2001

Rossi, A., Valsecchi, G.B., Perozzi, E.: Risk of collision for the navigation constellations: the case of the forthcoming Galileo. J. Astron. Sci. **52**, 455–474 (2004)

Valsecchi, G.B., Rossi, A., Farinella, P.: Visualizing impact probabilities of space debris. Space Debris, **1**(2), 143–158 (2000)

Valsecchi, G.B., Rossi, A.: Analysis of the space debris impact risk on the International Space Station. Celest. Mech. Dyn. Astron **83**, 63–76 (2002)

Valsecchi G.B., Farinella, P., Rossi, A.: Collision risk: a new method for assessing and visualizing it. Acta Astron. **53**, 203–217 (2003)

Wetherill, G.W.: Collisions in the asteroid belt. J. Geophys. Res. **72**, 2429–2444 (1967)

Celestial Mechanics and Dynamical Astronomy (2006) 95:357–370
DOI 10.1007/s10569-006-9008-y

ORIGINAL ARTICLE

On target for Venus – set oriented computation of energy efficient low thrust trajectories

Michael Dellnitz · Oliver Junge · Marcus Post · Bianca Thiere

Received: 2 November 2005 / Revised: 29 January 2006 /
Accepted: 24 February 2006 / Published online: 15 August 2006

Abstract Recently new techniques for the design of energy efficient trajectories for space missions have been proposed that are based on the circular restricted three body problem as the underlying mathematical model. These techniques exploit the structure and geometry of certain invariant sets and associated invariant manifolds in phase space to systematically construct energy efficient flight paths. In this paper, we extend this model in order to account for a continuously applied control force on the spacecraft as realized by certain low thrust propulsion systems. We show how the techniques for the trajectory design can be suitably augmented and compute approximations to trajectories for a mission to Venus.

Keywords Set oriented numerics · Dynamical system · Earth venus transfer · Three body problem · Low thrust trajectory · Invariant manifold · Reachable set · Space mission design

1 Introduction

A new paradigm for the construction of energy efficient trajectories for spacecraft is currently emerging. It heavily bases on concepts and techniques from the theory and numerical treatment of dynamical systems. The basic strategy is the following: instead of a two body problem, as in more classical approaches, one considers a restricted three body problem as the mathematical model for the motion of the spacecraft. This enables one to exploit the intricate structure and geometry of certain invariant sets and their stable and unstable manifolds—which are not present in two body problems—as candidate regions for energy efficient trajectories. For example, this approach has recently been used in the design of the trajectory for the *Genesis discovery mission* (Lo et al., 2001).

M. Dellnitz · M. Post (✉) · B. Thiere
Faculty of Computer Science, Electrical Engineering and Mathematics, University of Paderborn, 33095 Paderborn, Germany
e-mail: mpost@math.upb.de

O. Junge
Center for Mathematical Sciences, Munich University of Technology, 85747 Garching, Germany
e-mail: junge@ma.tum.de

Building on this basic concept, techniques have been proposed that synthesize partial orbits from different three body problems into a single one, yielding energy efficient trajectories with eventually very complicated itineraries (Koon et al. 2000b, 2002). In (Koon et al. 2002), a *petit grand tour* between the moons of Jupiter has been constructed by this approach. The idea of the technique is as follows: one computes the intersection of parts of the stable resp. unstable manifold of two specific periodic orbits in the vicinity of two moons, respectively, with a suitably chosen surface. After a transformation of these two curves into a common coordinate system one identifies points on them that lie close to each other—ideally one searches for intersection points. Typically, however, these two curves will not intersect in the chosen surface, so a certain (impulsive) maneuver of the spacecraft will be necessary in order to transit from the part of the trajectory on the unstable manifold to the one on the stable manifold. In a final step this "patched 3-body approximation" to a trajectory is used as an initial guess for standard local solvers using the full n-body dynamics of the solar system as the underlying model.

The approach of patching 3-body problems as sketched above is tailored for spacecraft with impulsive thrust engines. Recently however, interest has grown in continuously thrusting engines that exert small forces on the spacecraft only. For these, the usual restricted three body problem is not an adequate model, since one needs to incorporate the control forces.

In this paper, we propose an extension of the patched 3-body approach to the case of a continuously controlled spacecraft. Roughly speaking, the stable and unstable manifolds are replaced by certain (forward and backward) reachable sets in phase space. Using set oriented numerical tools we can efficiently compute the corresponding sets of intersection of these two reachable sets.

The paper is organized as follows: in Sect. 2, we briefly review the planar circular restricted three body problem that serves as a starting point for the discussion. Section 3 contains a sketch of the patched 3-body technique. In Sect. 4, we present the augmented three body model that incorporates a continuously acting control acceleration. The description of the generalized patching approach is given in Sect. 5, with comments on the implementation following in Sect. 6. In Sect. 7, we apply the procedure in order to compute a trajectory for a mission to Venus.

2 The planar circular restricted three body problem

As alluded to in the Introduction, typically the full n-body problem is too complicated for a detailed investigation of its dynamics. The classical patched conics approach breaks this model into several two-body problems whose solutions can easily be written down analytically. However, it turns out (McGehee 1969; Koon et al. 2000a), that it is worthwhile to hazard the consequences of considering a more complicated model, the (*planar*) *circular restricted three body problem* (*PCR3BP*).

Let us briefly recall the basics of this model—for a more detailed exposition see, e.g. (Szebehely 1967; Abraham and Marsden 1978; Meyer and Hall 1992). The PCR3BP models the motion of a particle of very small mass in the gravitational field of two heavy bodies (like, e.g. the Sun and the Earth). These two *primaries* move in a plane counterclockwise on circles about their common center of mass with the same constant angular velocity. One assumes that the third body moves in the same plane and does not influence the motion of the primaries while it is only influenced by the gravitational forces of the primaries.

In a normalized rotating coordinate system the origin is the center of mass and the two primaries are fixed on the x-axis at $(-\mu, 0)$ and $(1-\mu, 0)$, respectively, where $\mu = m_1/(m_1+$

m_2) and m_1 and m_2 are the masses of the primaries. In this paper we are considering the two systems Sun–Earth–Spacecraft and Sun–Venus–Spacecraft with μ-values of

$$\mu_{SE} = 3.04041307864 \times 10^{-6} \quad \text{and} \quad \mu_{SV} = 2.44770642702 \times 10^{-6},$$

respectively.

The equations of motion for the spacecraft with position (x_1, x_2) in rotating coordinates are given by

$$\ddot{x}_1 - 2\dot{x}_2 = \Omega_{x_1}(x_1, x_2), \qquad \ddot{x}_2 + 2\dot{x}_1 = \Omega_{x_2}(x_1, x_2) \tag{1}$$

with

$$\Omega(x_1, x_2) = \frac{x_1^2 + x_2^2}{2} + \frac{1-\mu}{r_1} + \frac{\mu}{r_2} + \frac{\mu(1-\mu)}{2}$$

and

$$r_1 = \sqrt{(x_1 + \mu)^2 + x_2^2}, \qquad r_2 = \sqrt{(x_1 - 1 + \mu)^2 + x_2^2}.$$

$\Omega_{x_1}, \Omega_{x_2}$ are the partial derivatives of Ω with respect to the variables x_1, x_2.

Equation 1 have a first integral, the *Jacobi integral*, given by

$$C(x_1, x_2, \dot{x}_1, \dot{x}_2) = -(\dot{x}_1^2 + \dot{x}_2^2) + 2\Omega(x_1, x_2). \tag{2}$$

The system posesses five equilibrium points (the *Lagrange points*): the collinear points L_1, L_2, and L_3 on the x-axis and the equilateral points L_4 and L_5. The three-dimensional manifolds of constant C-values are invariant under the flow of (1), their projection onto position-space, the *Hill's region*, determines the allowed region for the motion of the spacecraft (cf. Fig. 1a).

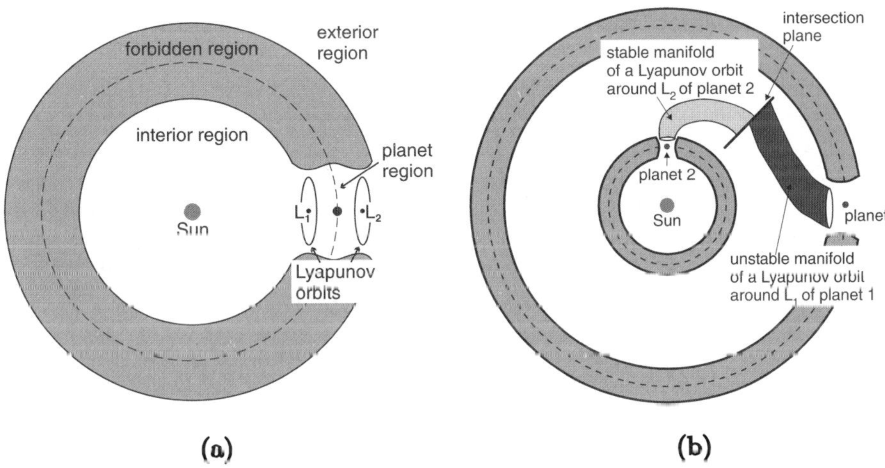

Fig. 1 (**a**) Projection of an energy surface onto position space (schematic) for a value of the Jacobi integral for which the spacecraft is able to transit between the exterior and the interior region. (**b**) Sketch of the "patched 3-body approach" (cf. Koon et al. 2000b, 2002). The idea is to travel within certain invariant manifold "tubes" possibly including an impulsive maneuver at the intersection plane

3 Coupling 3-body problems

The idea of patching 3-body problems essentially relies on two key observations:

1. For suitable energy values (i.e. values of the Jacobi integral (2)) there exist periodic solutions, the *Lyapunov orbits* (cf. Fig. 1a), of (1) in the vicinity of the equilibrium points L_1 and L_2 that are unstable in both time directions. Their unstable resp. stable manifolds W^u resp. W^s are (topologically) cylinders that locally partition the three-dimensional energy surface into two sets: (1) transit orbits, that pass between the *interior region* and the *planet region* in the case of an L_1-Lyapunov orbit or between the *exterior region* and the *planet region* in the case of L_2, and (2) non-transit orbits that stay in the exterior or interior region (McGehee 1969; Koon et al. 2000a).
2. By "embedding" one PCR3BP into a second one, parts of the unstable manifold of a Lyapunov orbit in one system may come close to the stable manifold of a Lyapunov orbit in the other system (where, for a moment, it may help to imagine that the two systems do not move relative to each other) (cf. Fig. 1b). It may thus be possible for a spacecraft to "bridge the gap" between two pieces of trajectories in the vicinity of these manifolds by exerting an impulsive maneuver (Koon et al. 2000b, 2002).

One way to detect a close approach of two such invariant manifolds is to reduce the dimensionality of the problem. One computes the intersection of the two manifolds with a suitable intersection plane (cf. Fig. 1b) and determines points of close approach in this surface—for example by inspecting projections onto 2D-coordinate planes. This approach has in fact been used for a systematic construction of trajectories that follow prescribed itineraries around and between the Jovian moons (Koon et al. 2002).

4 A controlled three body problem

In current mission concepts, like for the ESA interplanetary mission *BepiColombo* to Mercury and the current *Smart I* mission, ion propulsion systems are being investigated that continuously exert a small force on the spacecraft ("low-thrust propulsion"). The planar circular restricted 3-body problem (1) does not model this continuous thrusting capability and the model needs to be enhanced by a suitably defined control term. Here, we will restrict our considerations to the special case of a control force whose direction is defined by the spacecraft's velocity, since it is necessary that the acceleration and velocity vectors are parallel for the force to have a maximum impact onto the kinetic energy of the spacecraft. This is due to the fact that the time derivative of kinetic energy solely depends on the dot product of the spacecraft's velocity and its acceleration if the mass is assumed to be constant (see Gerthsen and Vogel 1993).

The control term which is to be included into the model is therefore parametrized by a single real value u, determining the magnitude of the control acceleration. We do not take into account that the mass of the spacecraft changes during its flight. If one takes this into account the model (Eq. 3) does not change because it describes a spacecraft of neglegible mass. The only effect would be that the same acceleration u can be achieved by less driving force if the mass decreases over time. This would allow to employ a higher upper bound for u at a later time which means that we maintain a conservative estimate for the upper bound of u.

The velocity vector of the spacecraft has to be viewed with respect to the inertial coordinate system and not the rotating one. In view of this, one is lead to the following control

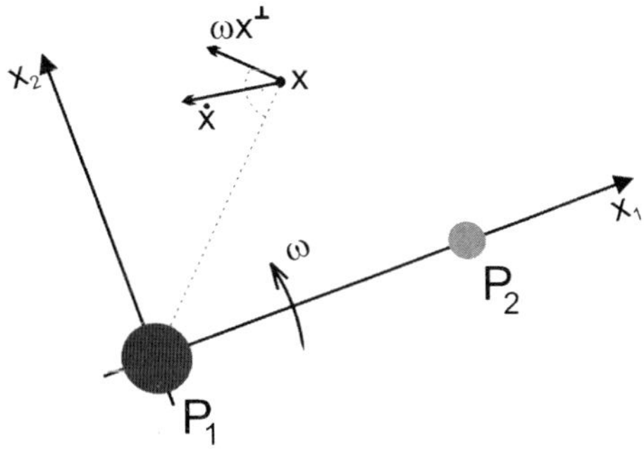

Fig. 2 The velocity of the spacecraft with respect to the inertial frame is given by $\dot{x} + \omega x^{\perp}$

system, modeling the motion of the spacecraft under the influence of its low thrust propulsion system in rotating coordinates (cf. Fig. 2):

$$\ddot{x} + 2\dot{x}^{\perp} = \nabla\Omega(x) + u\frac{\dot{x} + \omega x^{\perp}}{\|\dot{x} + \omega x^{\perp}\|}. \tag{3}$$

Here, $u = u(t) \in [u_{\min}, u_{\max}] \subset R$ denotes the magnitude of the control force, $x = (x_1, x_2)$, $x^{\perp} = (-x_2, x_1)$ and ω is the common angular velocity of the primaries.

In a mission to Venus the spacecraft will get closer to the Sun, meaning that part of its potential energy with respect to the Sun will be transformed into kinetic energy. Because the total 2BP energy (sun–particle) is higher for the Earth orbit than for the Venus orbit the 2BP energy has to be reduced on the way to Venus. As a consequence, the spacecraft's kinetic energy has to be reduced during its flight such that the 2BP energy matches the one of Venus. Thus, in our concrete application the control values u will actually be negative.

5 Coupling controlled 3-body problems

Obviously, every solution of (1) is also a solution of (3) for the control function $u \equiv 0$. We are going to exploit this fact in order to generalize the patched 3-body approach as described in Sect. 3 to the case of controlled 3-body problems. We are still going to use the L_1- and L_2- Lyapunov orbits as "gateways" for the transition between the interior, the planet and the exterior regions. However, instead of computing the relevant invariant manifolds of these periodic orbits, we compute certain *reachable sets* (see, e.g. Colonius and Kliemann 2000), i.e., sets in phase space that can be accessed by the spacecraft when employing a certain control function.

5.1 Reachable sets

We denote by $\phi(t, z, u)$ the solution of the control system (3) for a given initial point $z = (x, \dot{x})$ in the phase space at $t_0 = 0$ and a given admissible control function $u \in \mathcal{U} = \{u\colon \mathbb{R} \to [u_{\min}, u_{\max}], u \text{ admissible}\}$. Here $u_{\min}, u_{\max} \in \mathbb{R}$ are predetermined bounds on the magnitude of the control force, and the attribute "admissible" alludes to the fact that only a certain subset of functions is allowed. Both the bounds and the set of admissible control functions will be determined by the design of the thrusters. For example, the set of admissible control functions could be the set of piecewise constant functions, where the minimal length of an interval on which the function is constant is determined by how fast the magnitude of the accelerating force can be changed within the thrusters.

For a set S in phase space Z (S being an element of the power set $\mathbb{P}(Z)$) and a given function $\tau\colon S \times \mathcal{U} \mapsto \mathbb{R}$, we call $\mathcal{R}\colon \mathbb{P}(Z) \times (S \times \mathcal{U} \mapsto \mathbb{R}) \mapsto \mathbb{P}(Z)$ with

$$\mathcal{R}(S, \tau) = \{\phi(\tau(z, u), z, u) | u \in \mathcal{U}, z \in S\}$$

the set which is (τ)-*reachable from* S. Later on, we will choose $\tau(x, u)$ in such a way that the reachable sets are contained in the intersection plane.

5.2 Patched controlled 3-body systems

The idea is, roughly speaking, to mimic the patched 3-body approach while replacing the invariant manifolds of the Lyapunov orbits by certain reachable sets. We describe the approach by considering a mission from an outer planet (e.g. Earth) to an inner planet (e.g. Venus).

For two suitable sets $\mathcal{O}_1$ and $\mathcal{O}_2$ (in the vicinity of an L_1-Lyapunov orbit of Earth and an L_2-Lyapunov orbit of Venus, respectively) one computes associated reachable sets $\mathcal{R}(\mathcal{O}_1, \tau_1) \subset \Sigma_1$ and $\mathcal{R}(\mathcal{O}_2, \tau_2) \subset \Sigma_2$ within suitably chosen intersection planes Σ_1 and Σ_2 in each system, respectively. After a transformation of one of these reachable sets into the other rotating system, the intersection of them is determined. We will describe efficient methods that allow to compute an outer covering of this intersection in Sect. 6.

Abstractly, the procedure can be summarized as follows:

1. Identify suitable sets $\mathcal{O}_1$ and $\mathcal{O}_2$ in the phase space of the two 3-body problems, respectively. They should be chosen such that all points in $\mathcal{O}_1$ belong to trajectories that transit from the planet region into the interior region of Earth and those in $\mathcal{O}_2$ transit from the exterior region into the planet region of Venus. Furthermore, in each of the two 3-body problems, choose an intersection plane $\Sigma_i = \{\theta = \theta_i\}$, where $(r, \theta) = (r(x), \theta(x))$ are polar coordinates for the position x of the spacecraft and θ_i is a suitable angle (see also step 3).
2. For points $z_1 \in \mathcal{O}_1$ and $z_2 \in \mathcal{O}_2$ and an admissible control function u, let

$$\tau_1(z_1, u) = \inf\{t > 0 | \phi(t, z_1, u) \in \Sigma_1\} \quad \text{and}$$
$$\tau_2(z_2, u) = \sup\{t < 0 | \phi(t, z_2, u) \in \Sigma_2\}.$$

   For $i = 1, 2$, compute

$$\mathcal{R}(\mathcal{O}_i, \tau_i) = \{\phi(\tau_i(z, u), z, u) | z \in \mathcal{O}_i, u \in \mathcal{U}\} \subset \Sigma_i. \tag{4}$$

3. In order to transform one of the reachable sets $\mathcal{R}(\mathcal{O}_1, \tau_1)$ or $\mathcal{R}(\mathcal{O}_2, \tau_2)$ into the other rotating frame, let $\theta(t)$ be the phase angle between the two planets as seen in the rotating frame of the inner planet. We need to choose a time t_0 such that $\theta(t_0) = \theta_1 - \theta_2$. One can consider t_0 to be the time when the spacecraft arrives at the intersection plane. Using t_0 to transform $\mathcal{R}(\mathcal{O}_1, \tau_1)$ into the rotating frame of the inner planet, we obtain the set $\hat{\mathcal{R}}(\mathcal{O}_1, \tau_1) \subset \Sigma_2$. Note that here we exploit the fact that both systems are autonomous.
4. Compute the intersection

$$\hat{\mathcal{R}}(\mathcal{O}_1, \tau_1) \cap \mathcal{R}(\mathcal{O}_2, \tau_2) \subset \Sigma_2 \tag{5}$$

   (see Sect. 6). If this intersection turns out to be empty, typically one needs to increase the range $[u_{\min}, u_{\max}]$ of the control functions or to choose the section angles θ_1, θ_2 differently. By construction, for each point $z \in \hat{\mathcal{R}}(\mathcal{O}_1, \tau_1) \cap \mathcal{R}(\mathcal{O}_2, \tau_2)$, there exist admissible control functions u_1 and u_2 and times $t_1 = -\tau_1(\tilde{z}, u_1)$, $t_2 = -\tau_2(z, u_2)$, such that $\phi(t_1, \tilde{z}, u_1) \in \mathcal{O}_1$ and $\phi(t_2, z, u_2) \in \mathcal{O}_2$, where $\tilde{z}$ are the coordinates of z with respect to the rotating frame of the outer planet at the phase angle $\theta(t_0)$ between the two planets.

Thus, by construction of the sets $\mathcal{O}_1$ and $\mathcal{O}_2$ we have found a controlled trajectory that transits from the outer planet region into the inner planet region.

6 Implementation

6.1 Computing the reachable sets

For the purpose of this paper we restrict ourselves to constant control functions and choose a grid of control values in $[u_{\min}, u_{\max}]$. Although very simple, already this leads to quite satisfactory results. For each of the values on this grid, we numerically integrate the control system (3) by an embedded Runge–Kutta scheme with adaptive stepsize control as implemented in the code DOP853 by Hairer et al. (1993). After each integration step, we check whether the computed trajectory has crossed the intersection plane under consideration and, if this is the case, we start Newton's method in order to obtain a point in the intersection plane.We store that point, together with the corresponding control value.

6.2 Computing the intersection

In step 4 of the algorithm we need to compute the intersection $\hat{\mathcal{R}}(\mathcal{O}_1, \tau_1) \cap \mathcal{R}(\mathcal{O}_2, \tau_2) \subset \Sigma_2$ of the two reachable sets in a common section. We use a *set oriented approach* in order to compute this intersection. Roughly speaking, we construct coverings of $\hat{\mathcal{R}}(\mathcal{O}_1, \tau_1)$ and $\mathcal{R}(\mathcal{O}_2, \tau_2)$ by collections of subsets of Σ_2 and identify those subsets that belong to both coverings. This approach has been used before in the context of the detection of connecting orbits in parameter dependent ordinary differential equations (see Dellnitz et al. 2001).

More precisely, let $\mathcal{P} = \{P_1, \ldots, P_p\}$ be a finite partition of some relevant bounded part of Σ_2 determined by the region between the two planets under consideration. We compute

$$\mathcal{P}_1 = \{P \in \mathcal{P} | \hat{\mathcal{R}}(\mathcal{O}_1, \tau_1) \cap P \neq \emptyset\} \quad \text{and}$$
$$\mathcal{P}_2 = \{P \in \mathcal{P} | \mathcal{R}(\mathcal{O}_2, \tau_2) \cap P \neq \emptyset\}$$

and finally

$$\mathcal{T}_{1,2} = \mathcal{P}_1 \cap \mathcal{P}_2.$$

By construction, the set $\cup_{p \in \mathcal{T}_{1,2}} P$ contains the intersection $\hat{\mathcal{R}}(\mathcal{O}_1, \tau_1) \cap \mathcal{R}(\mathcal{O}_2, \tau_2)$.

6.3 Data structure

We now briefly comment on the data structures that we used in the implementation for a detailed description see Dellnitz and Hohmann 1996, 1997; Dellnitz and Junge 1999.

The elements (P_k above) of the coverings are generalized rectangles (*boxes*) of appropriate dimension. These are stored in a binary tree which implicitely defines the geometry of each box, i.e., only the geometry of the root box has to be stored explicitely. When computing $\mathcal{P}_1$ and $\mathcal{P}_2$, in each set $P \in \mathcal{P}_1$ or $P \in \mathcal{P}_2$ we furthermore store the magnitude of the applied control force and the minimal $\Delta V = \int u\, dt$ that is necessary to reach the set P from either $\mathcal{O}_1$ or $\mathcal{O}_2$. Additionally, we store two flags in each box which describe to which reachable set the box belongs to. This allows us to efficiently determine the intersection by just evaluating these flags. What is more, whenever the intersection $\mathcal{T}_{1,2}$ consists of more than one box, the stored values enable us to choose trajectories with a minimal ΔV (with respect to the chosen control range, intersection planes and gateway sets).

7 On target for Venus

We apply our method for the construction of continuously controlled interplanetary trajectories to the design of a mission to Venus. In 2005, the European Space Agency launched *VenusExpress*, a mission to Venus that sends a *MarsExpress*-like spacecraft into an elliptical orbit around Venus via a sequence of impulsive thrust maneuvers. The transfer time from Earth is around 150 days, while the required ΔV amounts to roughly 1,500 m/s (ESA 2001; Fabrega et al. 2003). The interplanetary low-thrust orbit that we are going to construct in this section corresponds to a flight time of roughly 1.4 years, applying a ΔV of approximately 3,300 m/s. Since typical low-thrust propulsion systems (as in the ESA mission *Smart I* and the planned cornerstone mission *BepiColombo* for example) have a specific impulse which is approximately one order of magnitude larger than the one of chemical engines, these figures amount to a dramatic decrease in the amount of on-board fuel: at the expense of roughly the three-fold flight time the weight of the fuel can be reduced to at least 1/3 of what is used for *VenusExpress*.

7.1 Computational details

We are now going to comment on the specific details of the computation for the Earth–Venus transfer trajectory (cf. Sect. 5). The step with the highest computational effort is the second one because there many trajectories have to be computed. The other steps are computationally less expensive. Note, however, that the first step may require some a priori studies or knowledge. Finally, the phase space resolution (size of the boxes) for storing the different reachable sets may be limited by the memory of the computer.

1. For the construction of the "gateway set" $\mathcal{O}_1$ we consider the L_1-Lyapunov orbit $\mathcal{L}_1$ associated with the value $C_1 = 3.0005$ of the Jacobi integral in the Sun–Earth PCR3BP. This value results from experimenting with several different values and eventually bears further optimization potential. We compute the intersection A_1 of its interior local unstable manifold (i.e. the piece of its local unstable manifold that extends into the interior region) with the section $\Gamma = \{x_1 = 0.98\}$ in the given energy surface $\{C = C_1\}$. The manifold is computed by integrating points in the unstable eigenspace of the Lyapunov orbit. Let $\bar{A}_1$ denote the points that are enclosed by the closed curve A_1 in this two-dimensional surface. We set

$$\mathcal{O}_1 = \mathcal{L}_1 \cup (\bar{A}_1 \backslash A_1).$$

   Analogously, we compute A_2, $\bar{A}_2$ and $\mathcal{O}_2$ in the Sun–Venus system, using again a value of $C_2 = 3.0005$ for the Jacobi integral. As intersection planes we choose $\Sigma_1 = \Sigma_2 = \{\theta = \pi/4\}$—since by experimenting this turned out to yield a good compromise between transfer time and ΔV.
2. We have been using constant control functions only, employing 800 mN as an upper bound for the maximal thrust and using a grid size of 1 mN. The bound is in accordance with the capabilities of the thrusters that are planned to be used in connection with the *BepiColombo* mission. Here, we assumed a mass of 4,000 kg for the spacecraft.
3. Figure 3 shows coverings of the sets $\hat{\mathcal{R}}(\mathcal{O}_1, \tau_1)$ and $\mathcal{R}(\mathcal{O}_2, \tau_2)$, as well as a covering of their intersection $\mathcal{T}_{1,2}$, projected onto the $(x_1, \dot{x}_1)$-plane. According to the minimum ΔV we have chosen a point $p \in \mathcal{T}_{1,2}$ (or more precisely a box) the corresponding trajectory of which is shown in Fig. 4—in the inertial frame as well as in both rotating frames. It requires a (constant) control force of $u_1 = -651$ mN in the first phase (i.e. while

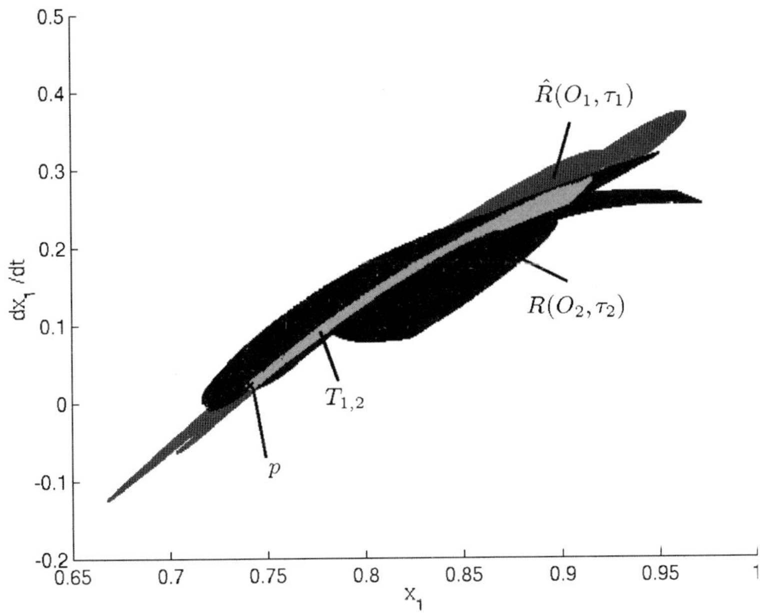

Fig. 3 Intersection $\mathcal{T}_{1,2}$ (*light grey*, 9387 boxes) of two reachable sets in a common intersection plane. $\hat{\mathcal{R}}(\mathcal{O}_1, \tau_1)$ (*dark grey*, 121075 boxes): reachable set of the gateway set of Earth, $\mathcal{R}(\mathcal{O}_2, \tau_2)$ (*black*, 171579 boxes): reachable set of the gateway set of Venus. The figure shows a projection of the covering in 3-space onto the $(x_1, \dot{x}_1)$-plane (normalized units). The computation of these sets took 2.8 h on a 3.2 GHz Xeon processor

travelling from $\mathcal{O}_1$ to Σ_1) and of $u_2 = -96$ mN in the second phase. The corresponding flight times are $|\tau_1| = 0.51$ and $|\tau_2| = 0.92$ years, amounting to a total ΔV of approximately 3,300 m/s. Please, note that the flight time is influenced by the choice of θ. Note that there still exists a discontinuity in the computed trajectory when switching from the first to the second phase. This is due to the fact that the two pieces of the trajectory are only forced to end in the same box in the intersection plane. However, the radii of the boxes are rather small, namely roughly 10,000 km in position space and $\approx$35 m/s in the velocity coordinates. Therefore, we expect the computed trajectory to be a very good initial guess for a standard local solver for a suitably formulated optimal control problem (like, e.g., a collocation or multiple shooting approach, see Deuflhard et al. 1976; von Stryk 1993; Deuflhard and Bornemann 2002; Stoer and Bulirsch 2002).

7.2 Linking in the planets

So far, our construction comprised the computation of pieces of controlled trajectories linking the two gateway sets $\mathcal{O}_1$ and $\mathcal{O}_2$ in the neighborhood of the Sun–Earth L_1 and the Sun–Venus L_2 Lagrange points. While missions like the *Genesis discovery mission* (Lo et al. 2001) have shown that one might reach these sets at the expense of very little fuel, it would be interesting to get at least a rough estimate on the flight time and the corresponding ΔV for the transfers between the planets and the gateway sets in our case. In particular, it might be worthwhile to find a compromise between flight time and ΔV.

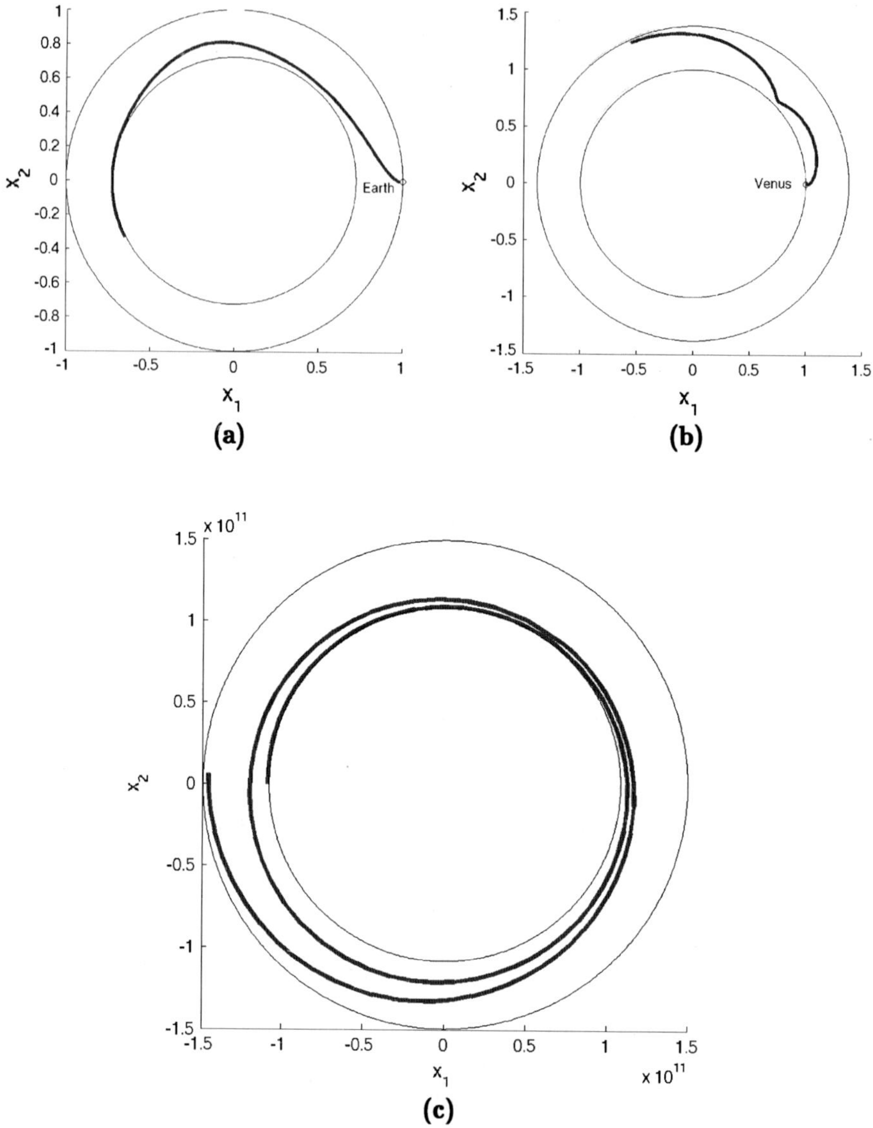

Fig. 4 Approximation of an interplanetary trajectory: joining the gateway sets $\mathcal{O}_1$ (near the Sun–Earth L_1) and $\mathcal{O}_2$ (near the Sun–Venus L_2). (a) Rotating frame of Earth (normalized units), (b) Rotating frame of Venus (normalized units) and (c) Inertial frame (coordinates in m)

To this end we computed extensions of the trajectories between the gateway sets into regions around Earth and Venus, respectively. An idea similar to this approach can be found in Gómez et al. (2001).

7.3 From Earth to $\mathcal{O}_1$

The starting point $x_1 \in \mathcal{O}_1$ of the interplanetary patched trajectory computed in the previous section is contained in $\bar{A}_1 \setminus A_1$, i.e., lies "within" the local unstable manifold tube

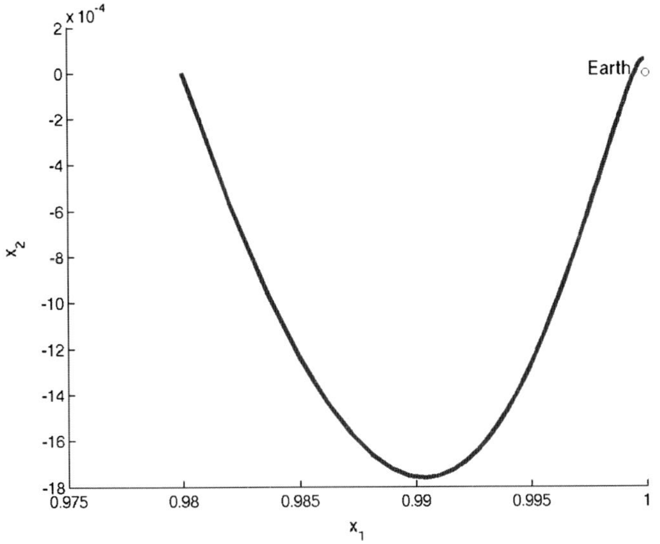

(a)

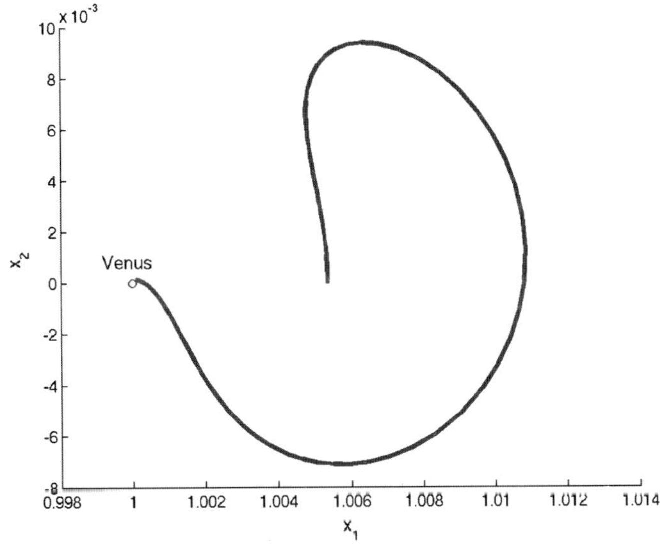

(b)

Fig. 5 Controlled transit between the gateway sets $\mathcal{O}_1$ and $\mathcal{O}_2$ and the planets (projection onto configuration space, normalized units in the respective Sun-planet rotating frame). (**a**) From Earth to the gateway set $\mathcal{O}_1$, control force -700 mN, flight time 0.1 years, $\Delta V = 630$ m/s and (**b**) From the gateway set $\mathcal{O}_2$ to Venus, control force -9 mN, flight time 0.27 years, $\Delta V = 20$ m/s

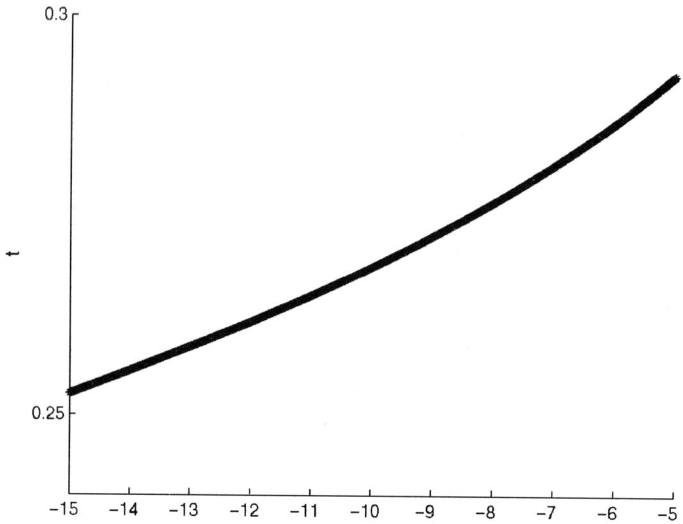

(a)

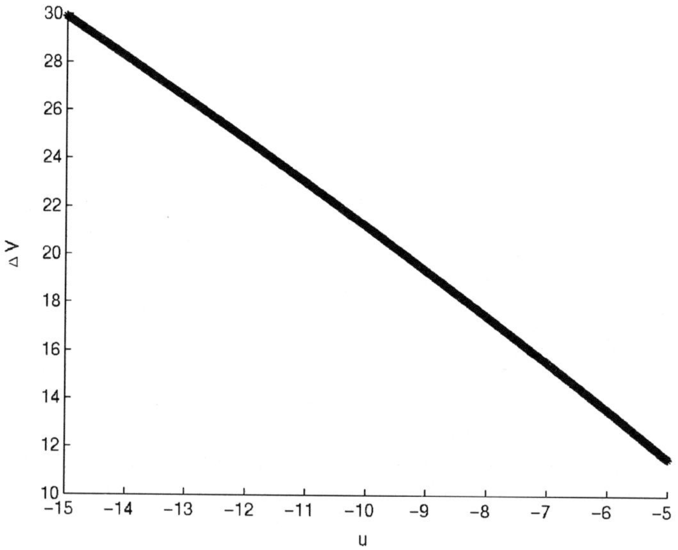

(b)

Fig. 6 Transfer time t (years) and ΔV (m/s) in dependence of the applied control force u (mN) for the transit from the gateway set $\mathcal{O}_2$ into a neighborhood of Venus

of $\mathcal{L}_1$ which extends into the interior region. Using x_1 as initial value and the prescribed range $[-800, 0]$ mN of control forces, we computed the associated trajectories of the controlled Sun–Earth PCR3BP backward in time until they crossed the section $\Sigma_E = \{x_1 = 1 - \mu - 10^{-4}\}$. It turns out that for control values between -800 and -650 mN these trajectories approach Earth up to a distance of approximately 15,000 km which we considered close enough for our purposes. Figure 5a shows the trajectory for $u_1 = -700\,\text{mN}$, requiring a ΔV of 630 m/s and a flight time of slightly more than 0.1 years.

7.4 From $\mathcal{O}_2$ to Venus

The endpoint $x_2 \in \mathcal{O}_2$ of the interplanetary patched trajectory actually lies on the stable manifold (i.e. on the part that locally extends into the exterior region) of the relevant Sun–Venus L_2-Lyapunov orbit $\mathcal{L}_2$. The transfer from $\mathcal{L}_2$ to Venus is almost free, however, one can slightly decrease the transfer time by employing a small control. Figure 5b shows one possible trajectory from $\mathcal{L}_2$ into a 10,000 km neighborhood of Venus, employing a control force of $-9\,\text{mN}$, yielding a ΔV of approximately 20 m/s and a transfer time of 0.27 years. In this case, we used a cross section $\Sigma_V = \{x_1 = 1-\mu+10^{-4}\}$. Additionally Fig. 6 shows how the transfer time and the required ΔV for this piece of the trajectory depend on the applied control force.

7.5 The complete journey

Choosing $u_1 = -700\,\text{mN}$ for the transfer from Earth to the gateway set $\mathcal{O}_1$ and $u_2 = -9\,\text{mN}$ for the trajectory from $\mathcal{O}_2$ to Venus, we finally end up with a flight time of roughly 1.8 years and a corresponding ΔV of slightly less than 4,000 m/s for the complete journey from Earth to Venus. Again, note that these are rough estimates and that the trajectory that we constructed should be viewed as an initial guess for a local solver that uses a more detailed model of the solar system.

8 Conclusion

This paper advocates a new approach to the construction of interplanetary low-thrust trajectories. It is based on a recently developed technique for the design of energy efficient trajectories that exploits the structure of the stable and unstable manifolds of certain periodic orbits in the vicinity of the L_1 and L_2 Lagrange points in the circular restricted three body problem. We incorporated a continuously applied, typically small force into the model and showed how one can generalize the concept of invariant manifolds to this context by employing the notion of reachable sets. In combination with set oriented numerical techniques for the efficient computation of the intersection of two suitable reachable sets we constructed an approximate low-thrust trajectory from Earth to Venus that uses a ΔV of approximately 4,000 m/s, while requiring a flight time of roughly 1.8 years.

In comparison to *VenusExpress* which requires a ΔV of roughly 1,500 m/s and a flight time of around 150 days (ESA, 2001, Fabrega et al. 2003) the interplanetary low-thrust orbit that we constructed requires only about 1/3 of the on board fuel mass because of the higher specific impulse of low-thrust engines.

In general, our technique has two particular advantages: the first one is that no potentially risky swing-by maneuvers are required. The second one is that the approach inherently provides an increased flexibility with respect to the launch date of the mission. This is due to fact that in principle a spacecraft can stay arbitrarily long near some periodic orbit in the vicinity of the L_1 or L_2 Lagrange point such that a required launch date can be met.

Acknowledgements We thank Shane Ross and Albert Seifried for helpful discussions on the contents of this paper. We also gratefully acknowledge support by Marc Steckling (EADS Astrium GmbH). Finally, we are much obliged to one of the referees for a very careful review of this paper. The research is (partly) supported by the DFG Research Training Group GK- 693 of the Paderborn Institute for Scientific Computation (PaSCo).

References

Abraham, R., Marsden, J.: Foundations of mechanics, 2nd edn Addison-Wesley, Reading, MA (1978)

Colonius, F., Kliemann, W.: The dynamics of control. Systems & Control: Foundations & Applications. Birkhäuser Boston Inc., Boston, MA. With an appendix by Lars Grüne (2000)

Dellnitz, M., Hohmann, A.: The computation of unstable manifolds using subdivision and continuation. In: Broer, H., van Gils, S., Hoveijn, I., Takens, F. (ed.) Nonlinear dynamical systems and chaos, pp. 449–459, Birkhäuser, *PNLDE* 19, (1996)

Dellnitz, M., Hohmann, A.: A subdivision algorithm for the computation of unstable manifolds and global attractors. Numerische Mathematik, **75**, 293–317 (1997)

Dellnitz, M., Junge, O.: On the approximation of complicated dynamical behavior. SIAM J. Numer. Anal. **36**(2), 491–515 (1999)

Dellnitz, M., Junge, O., Thiere, B.: The numerical detection of connecting orbits. Discrete Contin. Dyn. Syst. Ser. B, **1**(1), 125–135 (2001)

Deuflhard, P., Bornemann, F.: Scientific computing with ordinary differential equations, Texts in Applied Mathematics, Vol. 42. Springer-Verlag, New York (2002)

Deuflhard, P., Pesch, H.-J., Rentrop, P.: A modified continuation method for the numerical solution of nonlinear two-point boundary value problems by shooting techniques. Numer. Math. **26**(3), 327–343 (1976)

ESA: Venus Express mission definition report. European Space Agency ESA-SCI, **6**, 36–40 (2001)

Fabrega, J., Schirmann, T., Schmidt, R., McCoy, D.: Venus Express: the first european mission to Venus. Int. Astronautical Congress, IAC-03-Q.2.06, 1–11 (2003)

Gerthsen, C., Vogel, H.: *Physik.* Springer, Berlin (1993)

Gómez, G., Jorba, À., Simó, C., Masdemont, J.: Dynamics and mission design near libration points. Vol. III, World Scientific Monograph Series in Mathematics, Vol. 4 World Scientific Publishing Co. Inc., River Edge, NJ (2001)

Hairer, E., Nørsett, S.P., Wanner, G.: Solving ordinary differential equations. I. Non-stiff problems. Series in Computational Mathematics, Vol.8, 2nd edn. Springer-Verlag, Berlin, (1993)

Koon, W., Lo, M., Marsden, J., Ross, S.: Heteroclinic connections between periodic orbits and resonance transitions in celestial mechanics. Chaos, **10**, 427–469 (2000a)

Koon, W., Lo, M., Marsden, J., Ross, S.: Shoot the moon. In: AAS/AIAA Astrodynamics specialist conference, Florida, **105**, 1017–1030 (2000b)

Koon, W., Lo, M., Marsden, J., Ross, S.: Constructing a low energy transfer between jovian moons. Contemp. Math. **292**, 129–145 (2002)

Lo, M., Williams, B., Bollman, W., Han, D., Hahn, Y., Bell, J., Hirst, E., Corwin, R., Hong, P., Howell, K., Barden, B., Wilson, R.: Genesis mission design. J. Astronautical Sci. **49**, 169–184 (2001)

McGehee, R.: Some homoclinic orbits for the restricted 3-body problem. PhD thesis, University of Wisconsin, (1969)

Meyer, K., Hall, R.: Hamiltonian mechanics and the n-body problem. Applied Mathematical Sciences, Springer-Verlag, Berlin (1992)

Stoer, J., Bulirsch, R.: Introduction to numerical analysis, Vol. 12. Springer-Verlag, New York (2002)

Szebehely, V.: Theory of orbits—the restricted problem of three bodies. Academic Press, New York (1967)

von Stryk, O.: Numerical solution of optimal control problems by direct collocation. In: Bulirsch, R., Miele, A., Stoer, J., Well, K.-H. (ed.) Optimal control—calculus of variations, optimal control theory and numerical methods, Internat. Ser. Numer. Math., pp. Vol. 111, 129–143. Birkhäuser, Basel (1993)

Celestial Mechanics and Dynamical Astronomy (2006) 95:371–390
DOI 10.1007/s10569-006-9020-2

ORIGINAL ARTICLE

Weak stability boundary trajectories for the deployment of lunar spacecraft constellations

Christian Circi · Paolo Teofilatto

Received: 7 November 2005 / Revised: 17 March 2006 /
Accepted: 1 April 2006 / Published online: 15 August 2006

Abstract Suitable lunar constellation coverage can be obtained by separating the satellites in inclinations and node angles. It is shown in the paper that a relevant saving of velocity variation ΔV can be achieved using weak stability boundary trajectories. The weakly stable dynamics of such transfers allows the separation of the satellites from the nominal orbit to the required orbit planes with a small amount of ΔV. This paper also shows that only one different set of orbital parameters at Moon can be reached with the same ΔV manoeuvre starting from a nominal trajectory and ending at a fixed periselenium altitude. In fact, such a feature is proved to be common to other simpler dynamical systems, such as the two- and three-body problems.

Keywords Restricted three- and four-body problem · Weak stability lunar transfer · Lunar constellation

1 Introduction

Lunar constellation missions are foreseen in the future to study the lunar environment at various latitudes and longitudes particularly on the side facing away from the Earth. Observations of the lunar polar region are also of particular interest in view of the development of a future lunar base (Bussey and Spudis, 2005). Then it is of interest to consider a lunar constellation with polar orbits to observe the polar regions and inclined orbits for scientific uses. Global coverage can be obtained by varying the orbit inclination or the right ascension of ascending node (Ω). In this paper, a combined variation of two parameters is proposed to achieve global coverage. For example, global coverage can be achieved with three spacecraft is equally spaced on the polar orbit $i_1 = 90°$ and node angle $\Omega_1 = 0°$, and three spacecrafts equally spaced on an orbit with

C. Circi (✉)· P. Teofilatto
Department of Aerospace and Astronautics Engineering, School of Aerospace Engineering, University of Rome, Via Eudossiana 16, Rome, 00184, Italy
e-mail: christian.circi@uniroma1.it

$i_2 = 55°$, $\Omega_2 = 140°$. The orbits of all six spacecrafts are circular with radius 3,738 km. However, such a high altitude (2,000 km) could be incompatible with the payload characteristics. If a low altitude is required, the satellites can be set in orbits of radius $R = 1{,}938$ km (Graziani, 2005). The coverage will no longer be global and continuous, but with seven satellites in polar orbit and five satellites at inclination of 55° continuous local coverage, in the polar and 55° latitude band, can be ensured with a swath equal to about 1,600 km. In any case, an orbital control strategy must be considered, as the higher harmonics of the lunar gravitational field give a significant variation of the satellite eccentricity (D'Avanzo et al., 1997) so the satellites of the constellation must have orbital control engines and enough propellant mass for orbit maintenance. Then it is of particular importance to reduce as much as possible the on-board propellant mass needed for the lunar transfer and for the final orbit injection. A convenient option is the use of the so-called weak stability boundary (WSB) lunar transfers. These transfers use low-energy orbits found by Belbruno (Belbruno, 1990) and already tested in a Japanese mission (Belbruno and Miller, 1990; Kawaguchi et al. 1995). Figure 1 shows an example of a WSB lunar transfer: in the origin is the Earth, and the circular orbit is the orbit followed by the Moon during the spacecraft transfer orbit.

Note that the spacecraft goes farther than 10^6 km from the Earth to take advantage of the Sun's gravitational effect; if the Sun's gravitational field is neglected, one would obtain the elliptical orbit shown in Fig. 1. In fact, the Sun increases the perigee of

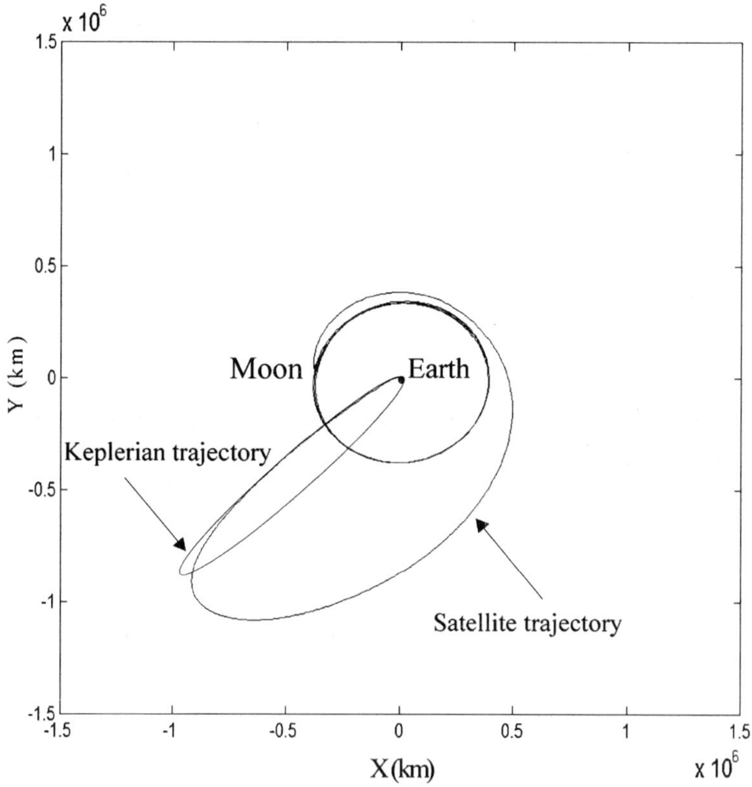

Fig. 1 A weak stability boundary (*WSB*) lunar transfer

the orbit until it reaches the moon's orbit radius, and it provides the spacecraft with just the right energy to allow a lunar ballistic capture. Therefore, the first ΔV given by the launcher at Earth departure, is bigger than the first impulse necessary for Hohmann-like transfer, but due to the ballistic capture, the braking ΔV at Moon, given by the satellite engine, is by far smaller, and thus about 20% of the on-board propellant mass can be saved (Belbruno, 2004). Some physical and analytical insights into the WSB dynamics can also be found in Kawaguchi et al. (1995), Koon et al. (2000, 2001), Circi and Teofilatto (2001), Koon et al. (2001).

Then, to save as much propellant as possible, a nominal WSB trajectory is used as a transfer trajectory to the moon and the satellites are spread on the planes of the constellation by small impulsive burns performed in different points of the nominal trajectory. It seems convenient to separate the satellites in a region of low gravitational field in the Earth–Moon–Sun system. It is expected that in such a region a small impulse produces rather different orbital parameters at Moon, so a substantial saving can be achieved with respect to a constellation deployment performed at Moon arrival. In fact the "weak stability" character of the nominal trajectory is such that small impulsive separation manoeuvres, corresponding to a variation of velocity of 20 cm/s (the separation velocity, which can be achieved by a small spring separation device), given at certain points of the nominal trajectory allow the formation of these 12 satellites in the two lunar constellation planes with a ΔV saving with respect to a classical constellation deployment strategy. Therefore, this strategy is proved to be effective for the kind of constellation considered (two planes). However, for a more general constellation design different strategies must be pursued. In fact, it is proved in the paper that the values of the final orbit parameter at the Moon are frozen to just two sets of values if impulses of the same intensity are performed at different points of the nominal trajectory, while keeping the periselenium altitude h_p fixed at a nominal value. This sort of regular behaviour seems to be surprising in the setting of "weakly stable" dynamics; however, it is shown that this behaviour is a common feature of the restricted two-, three- and four-body problems. In particular the restricted (bicircular) four-body problem can be considered a good approximation of the dynamics involved in a WSB dynamic trajectory.

This paper is organized as follows: in Sect. 2, the proposed strategy for the lunar constellation deployment is described and the ΔV saving with respect to a traditional manoeuvre is determined. In Sect. 3, it is shown that the orbital parameters at the Moon, obtained by small perturbation of a nominal WSB trajectory, are frozen to just two sets of values, if the periselenium distance is kept fixed. In Sects. 4 and 5, this property is proved to be common to other simpler dynamical systems, such as the Keplerian problem (Sect. 4) and the restricted three- and four-body problems (Sect. 5).

2 Lunar constellation deployment by WSB trajectories

The basic idea is to take advantage of the WSB dynamics to deploy a constellation of satellites at circular orbits of radius R saving as much propellant as possible. A feasible strategy is to consider a spacecraft on a WSB trajectory having the periselenium distance equal to R. For example, a constellation of satellites of commune radius $R = 1{,}938$ km is considered, then nominal WSB trajectory with periselenium altitude equal to 200 km polar orbit at Moon is chosen. The total ΔV needed to achieve a circular lunar orbit at 200 km of altitude, from a circular Earth orbit at 200 km of altitude, using WSB transfers is 3.83 km/s ($\Delta V_1 = 3.19$ km/s, $\Delta V_2 = 0.64$ km/s). On the other hand, an ordinary Hohmann-like transfer having the same final parameters at Moon requires 3.98 km/s ($\Delta V_1 = 3.13$ km/s, $\Delta V_2 = 0.85$ km/s). It is important to note that the second impulse, at the moon pericentre, is given by the satellite, while the first one is given by the last stage of the launcher. For the second ΔV, the difference between WSB and classical transfer is about 200 m/s and this saving directly influences the satellite propellant mass. Then a substantial saving (about 20% of the on-board propellant mass) is achieved in reaching the polar lunar orbit of the satellite constellation. A numerical experiment is now executed to explore which lunar trajectories are attainable and what ΔV saving can be obtained in the constellation deployment performing small impulses along the nominal WSB transfer. For instance, let the effect of six impulses of 20 cm/s given along both the directions ($\pm\hat{c}_1, \pm\hat{c}_2, \pm\hat{c}_3$) of each of the three inertial axes be considered. These impulses are simultaneously performed at a point in the nominal trajectory. In the present example, such a point corresponds to 44 days of transfer time. In Fig. 2, the point of the nominal trajectory where the six impulses have been simultaneously performed is shown (separation time). Even if the relative ΔV between the six trajectories is rather small, their relative distance increases with time and the six trajectories have different approaches to the Moon.

In fact, three of the six trajectories miss the lunar capture: these are the trajectories generated by the impulses given along the $\hat{c}_1, -\hat{c}_2, -\hat{c}_3$ directions. The other three trajectories, generated by impulses given along the $-\hat{c}_1, \hat{c}_2$, and $\hat{c}_3$, directions, are captured and the parameters are shown in Table 1 together with their nominal trajectory parameters. Note that different periselenium altitudes and angles, inclination and nodal angles are achieved at Moon.

In the hypothesis of a constellation of circular orbits of commune radius R, it is convenient to select only those impulses generating trajectories with periselenium height h_p close to the nominal trajectory periselenium altitude $h_\mathrm{p} = 200$ km (so $R = 200 + 1{,}738$ km is equal to the circular orbit radius required by the constella-

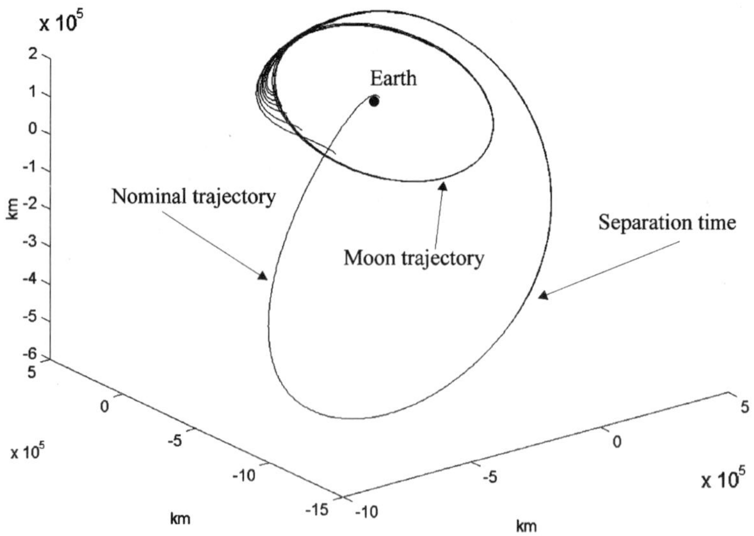

Fig. 2 The reference weak stability boundary (*WSB*) lunar transfer

Table 1 Orbital parameters at moon pericentre

Trajectory	h_p (km)	Transfer time (days)	a (km)	e	i (deg)	Ω (deg)	ω (deg)	ϑ (deg)
Nominal	203	91.92	32,909	0.941	89.99	0.033	120	−8.33
$-\hat{c}_1$	589	90.85	37,158	0.937	55.36	143.15	177.42	5.49
$\hat{c}_2$	14,327	90.45	47,813	0.663	52.79	167.08	50.62	12.13
$-\hat{c}_3$	2,293	90.72	38,542	0.895	55.02	152.28	72.25	−5.23

tion design). In fact, only orbits with $h_p < 1{,}000$ km will be considered: one can see that just one of these six trajectories satisfies such a requirement. This trajectory has $h_p = 589$ km, inclination $i = 55°$ and $\Omega = 143°$, and so a variation of inclination and Ω is achieved with a small variation of velocity ΔV. Note that a global lunar coverage can be achieved with seven satellites on the nominal lunar orbit (that has parameters $i_1 = 90°, \Omega_1 = 0°$), and five satellites on the varied orbit ($i_1 = 55°, \Omega_1 = 143°$) and all circular orbits at $R = 1{,}938$ km.

Note that to reduce the periselenium altitude from 598 km to the nominal altitude (200 km), one needs only 5 m/s. Using a traditional strategy the same coverage properties can be obtained by varying, for example, the inclination at aposelenium of the nominal trajectory. Then a variation of velocity 122 m/s is needed for every satellite to be set in the non-nominal plane. Therefore, according to this constellation design, five satellites increase the total required ΔV by about 600 m/s.

3 Separation manoeuvre on WSB lunar transfer

This constellation deployment has been realized with an impulse given in direction $-\hat{c}_1$ performed after 44 days of transfer time. To determine if alternative lunar satellite configurations can be achieved changing the separation time, the nominal trajectory is divided into 416 points, corresponding to the steps of the numerical integration. Moreover, to investigate the effect of the different possibilities, 18 impulsive burns, rather than six, are given at every point of the trajectory. The 18 burns are given in both the directions of each of the inertial reference system axes ($\pm\hat{c}_1, \pm\hat{c}_2, \pm\hat{c}_3$), and in the diagonal directions ($\pm\frac{\sqrt{2}}{2}\hat{c}_i \pm \frac{\sqrt{2}}{2}\hat{c}_j$), $i, j = 1, 3, i < j$, and they have been enumerated according to Table 2.

The burns have all the same intensity equal to $\Delta V = 20$ cm/s. In Fig. 3, sequences of burns performed at different points of the nominal trajectory are shown: the burns

Table 2 Burns direction

1	2	3	4	5	6
$\hat{c}_1$	$-\hat{c}_1$	$\hat{c}_2$	$-\hat{c}_2$	$\hat{c}_3$	$-\hat{c}_3$
7	8	9	10	11	12
$\frac{\sqrt{2}}{2}\hat{c}_1 + \frac{\sqrt{2}}{2}\hat{c}_2$	$\frac{\sqrt{2}}{2}\hat{c}_1 - \frac{\sqrt{2}}{2}\hat{c}_2$	$-\frac{\sqrt{2}}{2}\hat{c}_1 + \frac{\sqrt{2}}{2}\hat{c}_2$	$-\frac{\sqrt{2}}{2}\hat{c}_1 - \frac{\sqrt{2}}{2}\hat{c}_2$	$\frac{\sqrt{2}}{2}\hat{c}_2 + \frac{\sqrt{2}}{2}\hat{c}_3$	$\frac{\sqrt{2}}{2}\hat{c}_2 - \frac{\sqrt{2}}{2}\hat{c}_3$
13	14	15	16	17	18
$-\frac{\sqrt{2}}{2}\hat{c}_2 + \frac{\sqrt{2}}{2}\hat{c}_3$	$-\frac{\sqrt{2}}{2}\hat{c}_2 - \frac{\sqrt{2}}{2}\hat{c}_3$	$\frac{\sqrt{2}}{2}\hat{c}_1 + \frac{\sqrt{2}}{2}\hat{c}_3$	$-\frac{\sqrt{2}}{2}\hat{c}_1 + \frac{\sqrt{2}}{2}\hat{c}_3$	$\frac{\sqrt{2}}{2}\hat{c}_1 - \frac{\sqrt{2}}{2}\hat{c}_3$	$-\frac{\sqrt{2}}{2}\hat{c}_1 - \frac{\sqrt{2}}{2}\hat{c}_3$

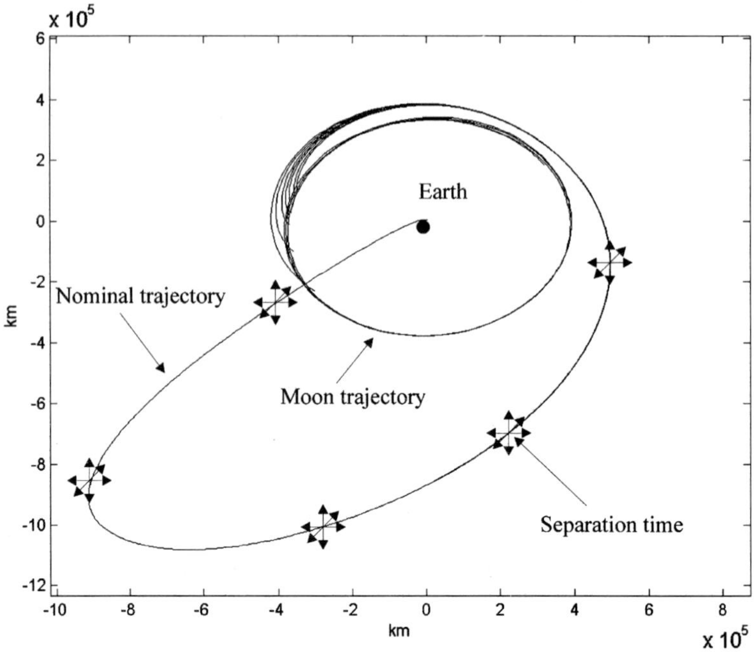

Fig. 3 Sequence of burns on the nominal trajectory at different times

are depicted by a cluster of arrows. Only those burns that produce a periselenium altitude h_p close to the nominal value are taken into account, and in Fig. 4 the final lunar parameters of the trajectories separated from the nominal WSB trajectory at different times are shown.

Note that a close-to-nominal altitude is achieved by only a few of the 18 trajectories generated at each separation time. Moreover, these trajectories have parameters at the Moon close to one of the two sets of parameters:

$$\begin{aligned} &i = 90°,\ \Omega = 0°,\ \omega = 120° \quad \text{(nominal WSB trajectory)} \\ &i = 55°,\ \Omega = 140°,\ \omega = 170° \quad \text{(varied WSB trajectory)} \end{aligned} \tag{1}$$

Figures 5–7 determine which of the 18 burns and at which separation time, lead to lunar capture with close-to-nominal altitude: in fact, the numbers over the circled points refer to the burn direction. For instance, in the range of separation time from 0 to 10 days, the successful burns are initially in direction 8, then in direction 17 and finally in direction 15. To give more physical insight into these results, Fig. 5 shows the angle between the burn direction and the velocity direction $\hat{v}$ of the nominal trajectory at burn time. Figure 6 shows the angle between the burn direction and the normal to the trajectory plane $\hat{h}$, and Fig. 7 shows the angle between the burn direction and the normal to the velocity direction in the trajectory plane $\hat{l}$.

For instance, let us consider in Fig. 5 the burn direction 2 occurring between the separation time 40 days and the separation time 60 days. The angle between the burn and the velocity direction varies approximately from 150° to 180°. Figure 6 shows that the burn direction 2 is in the trajectory plane (since the angle of the burn direction

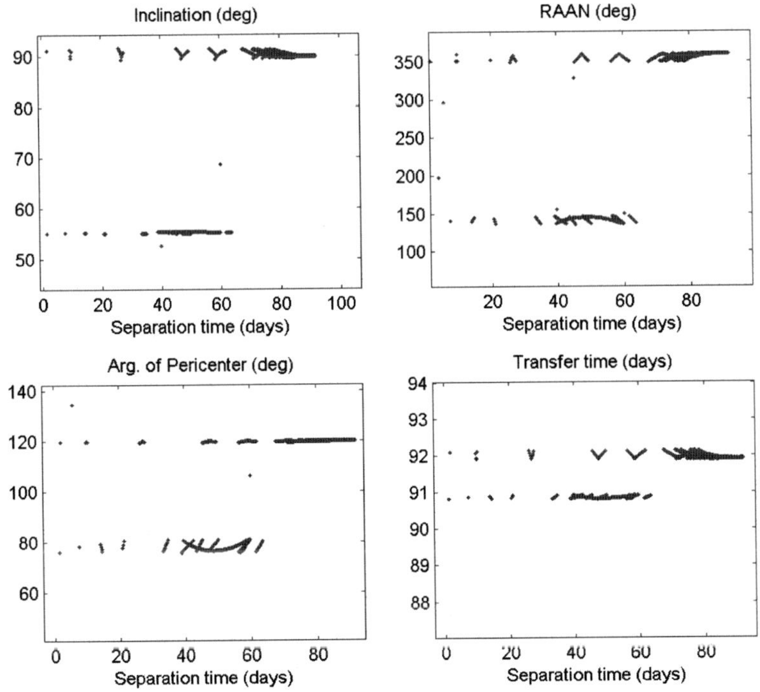

Fig. 4 Orbital parameters at Moon for different separation times from the nominal trajectory

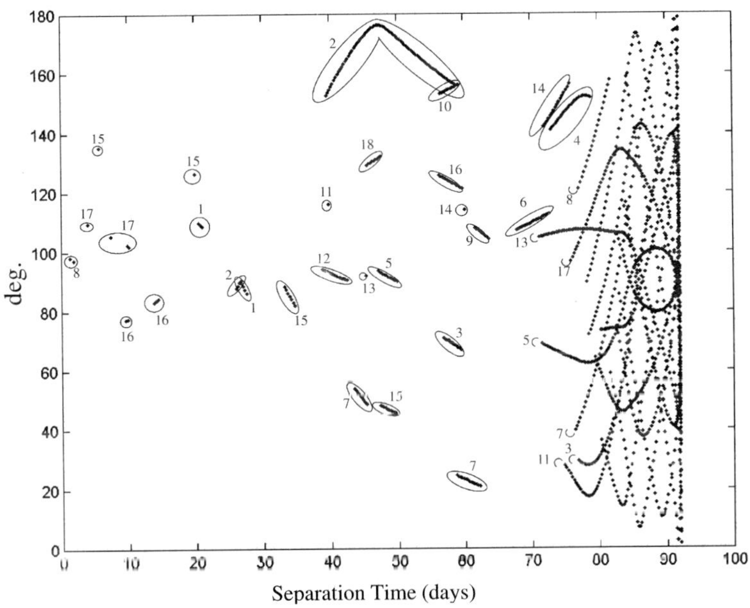

Fig. 5 Successful burn directions and their angle with respect to velocity direction

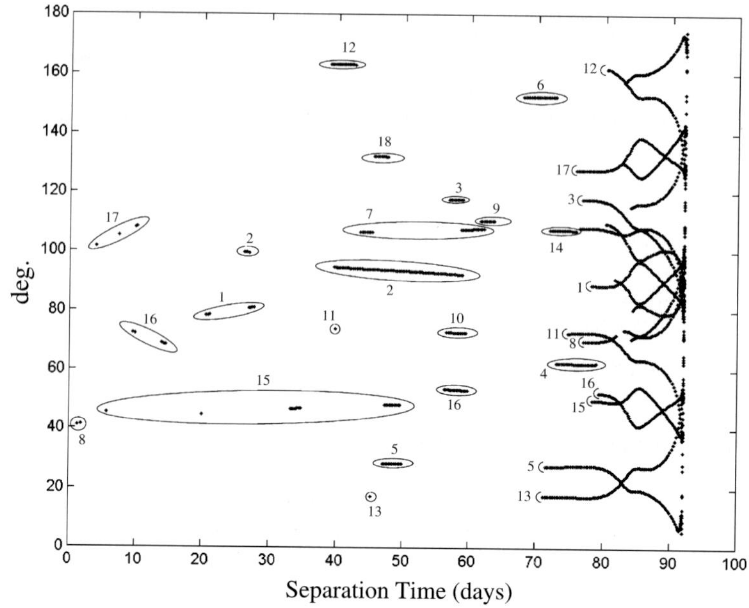

Fig. 6 Successful burn directions and their angle with respect to $\hat{h}$

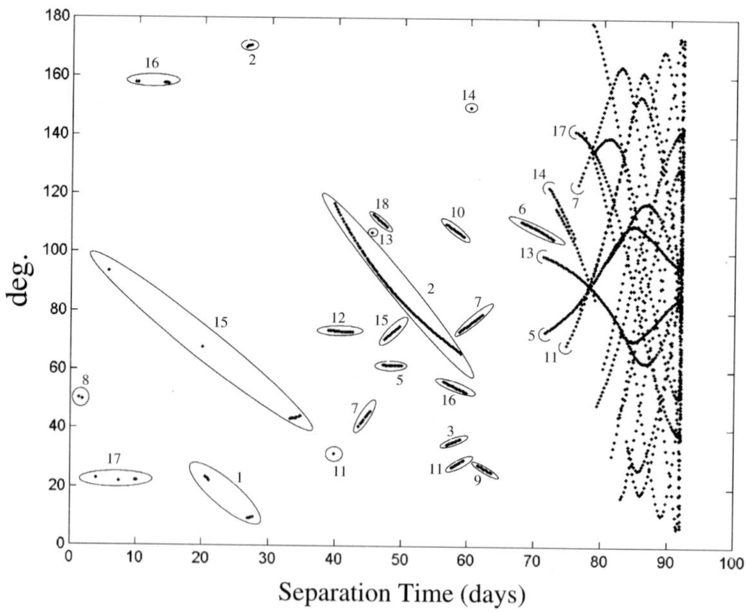

Fig. 7 Successful burn directions and their angle with respect to $\hat{l}$

angle with the normal to the trajectory plane is almost 90°) and then Fig. 7 shows the complementary of the angle of Fig. 5. This successful burn strategy is reported in Fig. 8 (braking impulsive burns). Note that in the same range of separation time (from 40 to 60 days) the successful burn direction 12 is parallel to $-\hat{h}$. Other possibilities can be determined by inspection of Figs. 5–7.

It turns out that several satellites can be set into the varied orbit (1) performing small impulsive separation burns. Moreover, such a manoeuvre can be performed at several different points of the nominal trajectory. To summarize, the proposed lunar constellation deployment seems to be rather convenient since:

- There is large freedom in the satellite separation time.
- The strategy is effective in terms of ΔV.

4 Families of trajectories in the Keplerian problem

The separation manoeuvre determined in the previous section is convenient for satellite constellation consisting of two planes. If the constellation design requires other orbital planes one could think of applying the same strategy, performing different impulses at different points of the nominal trajectory. However, Fig. 4 shows that

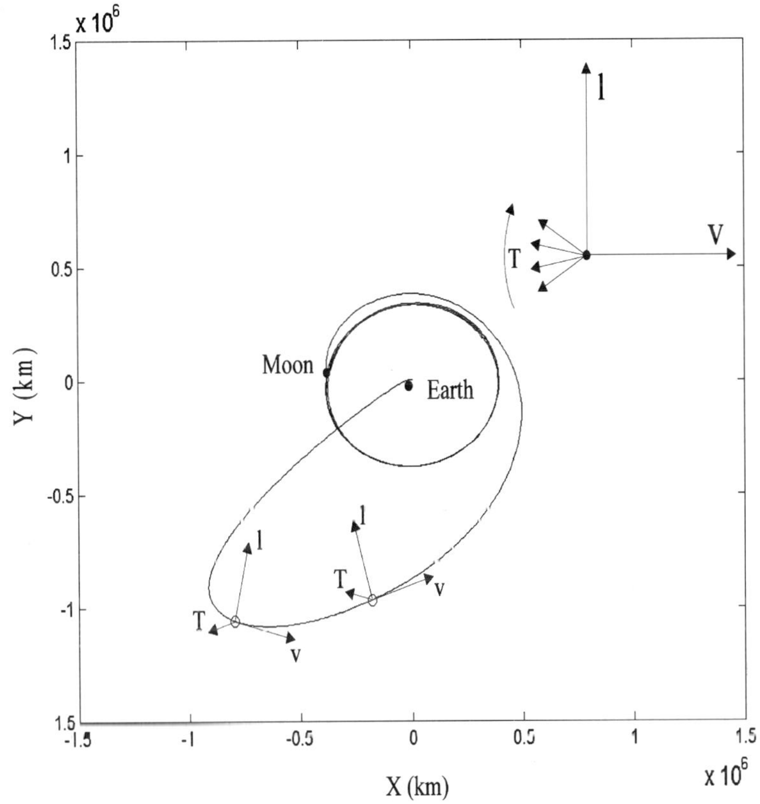

Fig. 8 Burn direction 2 at separation time from 40 to 60 days

it is not possible to vary the final orbit parameters i and Ω performing impulses of the same intensity at different points of the nominal trajectory, while maintaining h_p close to the nominal value. It seems a bit surprising to see such a result in the setting of "weakly stable" dynamics. In fact, these results show that there exists just two families of trajectories generated by small ΔV performed on a nominal WSB lunar transfer: (1) close-to-nominal trajectories (nominal family), (2) trajectories having Moon parameters similar to (1) (varied family). These two families have the following features:

(1) All the trajectories in both the families have almost the same periselenium altitude.
(2) The parameters i, Ω, ω of the nominal family are different with respect to the parameters of the varied family.

In this section, an attempt is made to understand such dynamical behaviour. Results will be obtained in simpler dynamical systems; in particular the Keplerian case will be analysed first. Families of trajectories passing through the same initial point $\mathbf{r}_0$ and having the same pericentre distance $\mathbf{r}_1$ will be determined. Of course, the orbits of these families will have different semimajor axes a and eccentricities e; the values of the perigee argument ω will be characterized. To define Keplerian orbits connecting point $\mathbf{r}_0$ to pericentre $\mathbf{r}_1$, let ϕ be the transfer angle between the two vectors (Fig. 9). Then the Keplerian conic from $\mathbf{r}_0$ to $\mathbf{r}_1$ has parameters (Battin, 1987):

$$e = \frac{1 - \frac{\mathbf{r}_0}{\mathbf{r}_1}}{1 - \frac{\mathbf{r}_0}{\mathbf{r}_1}\cos(\phi)}, \qquad a = \frac{\mathbf{r}_1}{1 - e} \tag{2}$$

and the velocity V_0 at $\mathbf{r}_0$ of a direct orbit has radial and tangential velocity:

$$\begin{aligned} V_r &= -\sqrt{\frac{\mu}{\mathbf{r}_0}}\frac{e\sin(\phi)}{\sqrt{1+e\cos(\phi)}} \\ V_\vartheta &= -\sqrt{\frac{\mu}{\mathbf{r}_0}}\sqrt{1+e\cos(\phi)} \end{aligned} \tag{3}$$

Varying the transfer angle ϕ in (3) one gets the velocities at $\mathbf{r}_0$ to reach pericentre $\mathbf{r}_1$. The velocities (3) draw a hyperbola in the (V_r, V_ϑ) plane and the equal energy curves are just circles of radius $V_0 = \sqrt{V_r{}^2 + V_\vartheta{}^2}$. Figure 10 shows the hyperbola and some equal energy curves for $\mathbf{r}_0 = 10^6$ km and $\mathbf{r}_1 = 384{,}400$ km (approximately the Earth–Moon distance).

The branch of hyperbola with radial velocity $V_r > 0$ corresponds to orbits where $\mathbf{r}_0$ is after the pericentre; in the other branch ($V_r < 0$) the spacecraft at $\mathbf{r}_0$ is before the pericentre. The point $V_r = 0$ corresponds to $\mathbf{r}_0$ at apogee ($\phi = \pi$) and defines the minimum energy (Hohmann) transfer from $\mathbf{r}_0$ to $\mathbf{r}_1$. Note that elliptic transfer is possible only for $\mathbf{r}_0$ in the apogee region. The equal energy curves intersect the velocity locus hyperbola at two points: such intersection points are close only if the orbit in $\mathbf{r}_0$ is close to the minimum energy orbit transfer. In such a case with a small ΔV, it is possible to switch from one branch of the velocity locus to the other. Of course, the burn direction that produces such a ΔV must be given in an appropriate direction. Namely if (f_r, f_ϑ) are the radial and tangential components of the burn performed at the true anomaly ϑ of an orbit of parameters $(a, e, \omega, i, \Omega)$, then the variation of these parameters can be found according to the planetary Lagrange equations; in particular,

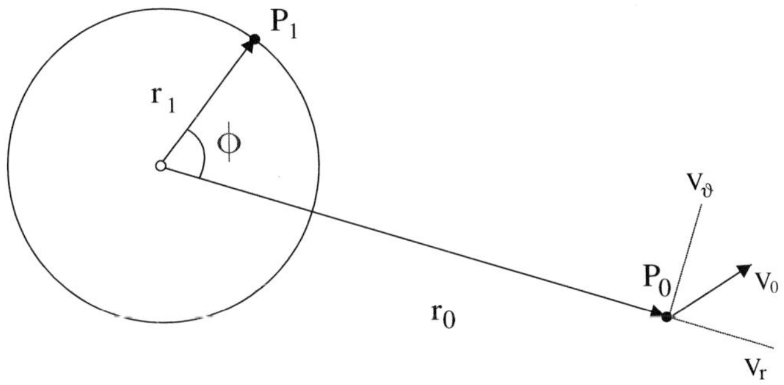

Fig. 9 Transfer angle from $\mathbf{r}_0$ to $\mathbf{r}_1$

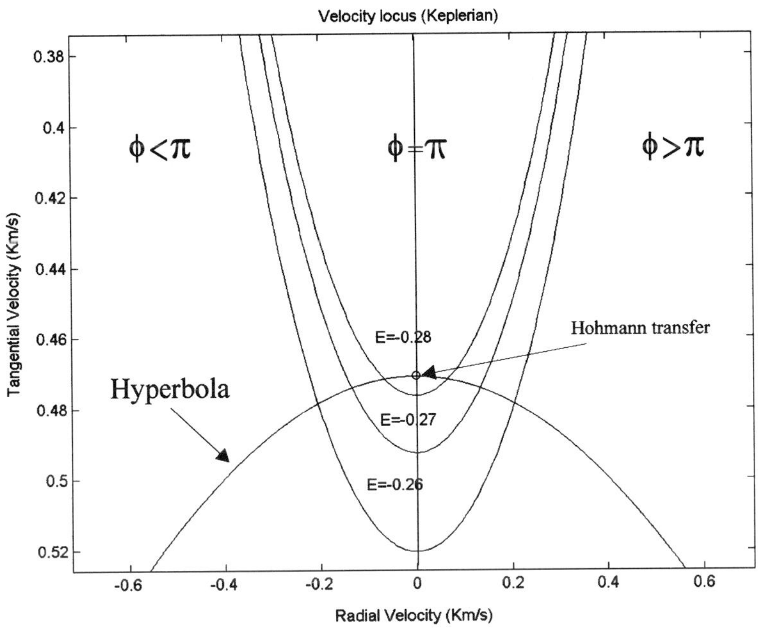

Fig. 10 Velocity locus for the Keplerian orbits

$$
\begin{aligned}
\dot{a} &= \frac{2a^2}{h}\left[ef_r\sin(\vartheta)+(1+e\cos(\vartheta))f_\vartheta\right] \\
\dot{e} &= \frac{r}{h}\left[\left(e+\cos(\vartheta)+\frac{h^2}{\mu r}\cos(\vartheta)\right)f_\vartheta+\frac{h^2}{\mu r}f_r\sin(\vartheta)\right] \\
\dot{\omega} &= \frac{r}{he}\left[\sin(\vartheta)(2+e\cos(\vartheta))f_\vartheta-\cos(\vartheta)(1+e\cos(\vartheta))f_r\right]
\end{aligned}
\tag{4}
$$

If the burn must be such that the pericentre distance $r_p = a(1-e)$ is constant, one has the equation $\dot{r}_p = 0$, which leads to

$$\frac{\dot{a}(1-e)}{a} = \dot{e} \tag{5}$$

This relation leads to the (f_r, f_ϑ) components of the burn:

$$\frac{f_r}{f_\vartheta} = \frac{\dfrac{e+\cos(\vartheta)}{1+e\cos(\vartheta)} + \cos(\vartheta) - \dfrac{2(1+e\cos(\vartheta))}{1+e}}{\sin(\vartheta)\left(\dfrac{2e}{1+e} - 1\right)} \tag{6}$$

That is, to maintain a given pericentre distance, a burn given at point $\mathbf{r}_0$ of true anomaly ϑ and in a nominal orbit of semi-major a and eccentricity e, must have the direction defined by the angle σ taken from the radial direction $\hat{r}_0$, where $\sigma = a\tan(\frac{f_r}{f_\vartheta})$ and the ratio f_r/f_ϑ is given by (6). Then, one has the two relations:

$$\Delta V^2 = \Delta f_r^2 + \Delta f_\vartheta^2 \tag{7}$$

$$\Delta f_r = \pm \sqrt{\frac{\Delta V^2}{1+\dfrac{1}{\left(\dfrac{\Delta f_r}{\Delta f_\vartheta}\right)^2}}} \tag{8}$$

where $\dfrac{\Delta f_r}{\Delta f_\vartheta} = \dfrac{f_r}{f_\vartheta} = \Delta f_{r\vartheta}$ is equal to (6). Then, according to (4), the variation $\Delta\omega$ of perigee argument of the nominal orbit is equal to:

$$\Delta\omega = \pm\frac{h}{\mu e}\frac{\sqrt{\frac{\Delta V^2}{1+\frac{1}{(\Delta f_{r\vartheta})^2}}}}{(1+e\cos(\vartheta))} \left[\frac{\sin(\vartheta)^2(2+e\cos(\vartheta))\left(\dfrac{2e}{1+e}-1\right)}{\dfrac{e\cos(\vartheta)}{1+e\cos(\vartheta)} + \cos(\vartheta) - \dfrac{2(1+e\cos(\vartheta))}{1+e}} - \cos(\vartheta)(1+e\cos(\vartheta))\right], \tag{9}$$

where ϑ is the true anomaly of the point $\mathbf{r}_0(\vartheta)$ of the nominal orbit where the burn is performed. The plot of the variation of pericentre argument $\Delta\omega$ as point $\mathbf{r}_0$ varies on the nominal orbit (i.e., as the true anomaly ϑ, or the corresponding time, varies) is in Fig. 11 where the variation of velocity produced by the impulsive burn is $\Delta V = 20\,\text{cm/s}$. Note that in the Keplerian model $\Delta\omega$ is proportional to ΔV and that only two solutions exist.

In this figure, a continuous variation of argument of pericentre is shown, different from the behaviour of ω in Fig. 4. However, it turns out that close to pericentre there is an almost zero variation $\Delta\omega$; that is, the argument of pericentre stays close to the nominal value, as can be seen for the WSB case of Fig. 4 (the nominal family). Moreover, at the apocentre region (corresponding to about 40 days), the rate of variation of ω decreases; that is, the burns about the apogee region produce almost the same pericentre arguments, as happens in Fig. 4 (the varied family).

5 Families of trajectories in the restricted three- and four-body problem

In the restricted three-body problem setting, the definition of orbits connecting point $\mathbf{r}_0$ to a fixed pericentre distance has already been discussed in Hadjidemetriou (1968), where $\mathbf{r}_0$ is in the inner Lagrangian point L_1 and a complete classification of

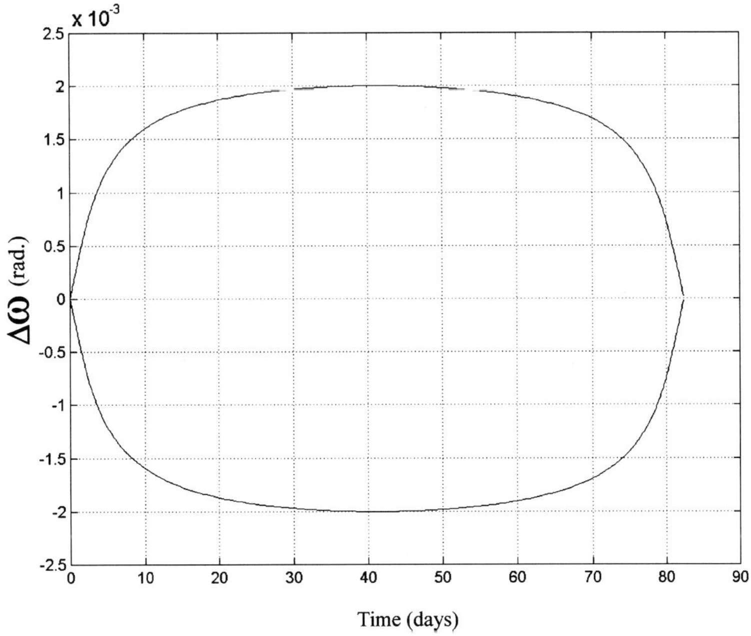

Fig. 11 Argument of pericentre variation for different values of burn point $\mathbf{r}_0$

two- and three-dimensional orbits according to their pericentre distance from both primaries is given. Note that in the same reference an analytic formula for the velocities V_0 needed to reach a pericentre distance $\mathbf{r}_1$ from a satellite in position $\mathbf{r}_0$ was given for the two-body problem defined in a reference frame rotating with unitary angular velocity ω_1:

$$\mathbf{r}_1^2\left[V_0^2 + 2V_0 \cdot \omega_1 \times \mathbf{r}_0 + (\omega_1 \times \mathbf{r}_0)^2 - 2\right] + 2\mathbf{r}_1$$
$$-\frac{1}{4}\left[\mathbf{r}_0^2 + \frac{2}{r_0} - 2 + 2V_0 \cdot \omega_1 \times \mathbf{r}_0 + (\omega_1 \times \mathbf{r}_0)^2\right]^2 = 0 \qquad (10)$$

(for $\mathbf{r}_0$ in L_1, one recognizes (23) of Hadjidemetriou (1968)). A numerical search is developed here in the restricted three-body problem related to the Earth–Moon system with $\mathbf{r}_0 = 10^6$ km (made dimensionless with respect to the average Earth–Moon distance EM $= 384,400$ km) having an angle of $-40°$ with respect to the Earth–Moon axis. The final point $\mathbf{r}_1$ is any point with periselenium of distance $R = 1,938$ km from the Moon centre of mass (dimensionless with respect to Earth–Moon). The reference frame used is the usual frame rotating with the Earth–Moon system (Fig. 12), with coordinate position (x, y) dimensionless with respect to the average Earth–Moon distance EM $= 384,400$ km, and dimensionless time $\tau = t/\omega_{EM}$ (where ω_{EM} is the angular velocity of the Earth–Moon system), so the component velocities (u, v) are dimensionless with respect to the average lunar velocity.

Figure 13 gives the horizontal and vertical velocities (u_0, v_0) at $\mathbf{r}_0$ in order to reach periselenium distance $\mathbf{r}_1$. The hyperbola of the Keplerian case (Fig. 10) shrinks into a curve with two branches that are close to each other. Note that with a small ΔV, it is possible to shift from one branch to the other.

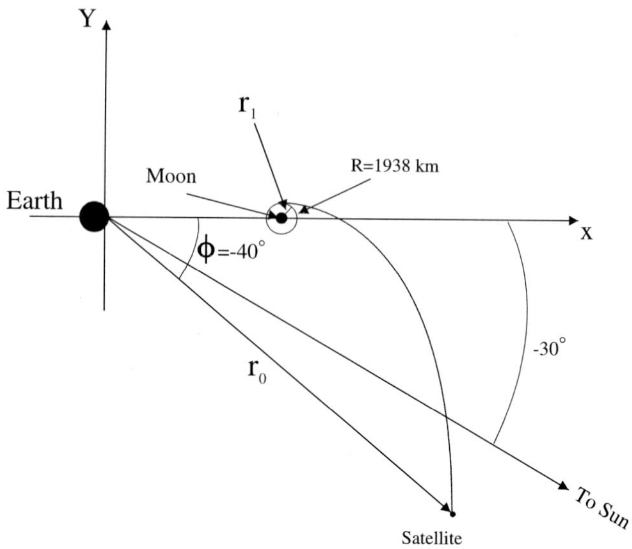

Fig. 12 Rotating Earth–Moon reference frame

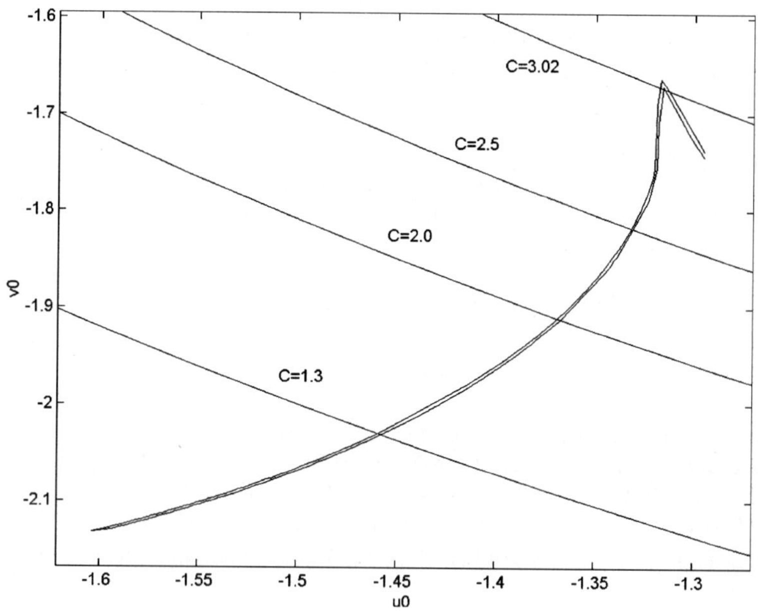

Fig. 13 Velocity locus (three-body problem, dimensionless velocities u_0, v_0)

In the velocity locus of Fig. 13, the equal Jacobi constant curves are shown. These curves are circles centred on the origin: note that the minimum energy transfer from $\mathbf{r}_0$ to the Moon, has Jacobi constant close to the value of the external Earth–Moon L_2 lagrangian point, so the upper region of the locus represents lunar transfer orbits with rather low energy. The transfer orbits of each of the two branches of the velocity locus have almost equal periselenium arguments ω; however, the value ω of the upper branch is different from the ω value of the lower branch, the variation being $\Delta\omega \simeq 100°$ (Fig. 14).

It follows that with a small variation of velocity ΔV it is possible to have a lunar transfer with the same periselenium distance, but rather different periselenium arguments. Different from the Keplerian case, such a possibility holds true for rather different energy values (Jacobi constant values) of the transfer orbit. The same analysis has been made taking into account the Sun's gravitational field in the so-called bicircular restricted four-body problem (Simò et al. 1995; Kazuyuki 2004). That is, the Earth–Moon system is rotating in circular orbit about the Sun and the Moon is rotating around the Earth in a circular orbit in the same plane. The reference system has its origin in the Earth–Moon system. It is rotating with angular velocity ω_{EM}, and position and time are dimensionless as in the Earth–Moon-restricted three-body problem. The same initial and final positions of the three-body problem are considered; that is, $\mathbf{r}_0 = 10^6$ km (made dimensionless with respect to the average Earth–Moon distance, EM), with angle of $-40°$ with respect to the Earth–Moon axis, the final point $\mathbf{r}_1$ is a periselenium of distance $R = 1{,}938$ km from the Moon (dimensionless with respect to EM) and the Sun is initially at angle $-30°$ from the Earth–Moon axis (Fig. 12). Because of such a configuration, the Sun's gravitational field will brake the spacecraft during the final lunar approach to allow ballistic capture (Circi and Teofilatto, 2001).

A two-dimensional search on the velocity components (u_0, v_0) is carried out in order to reach the Moon from $\mathbf{r}_0$ with periselenium distance $R = 1{,}938$ km using the following equations of the bicircular problem (with $z_0 = w_0 = 0$):

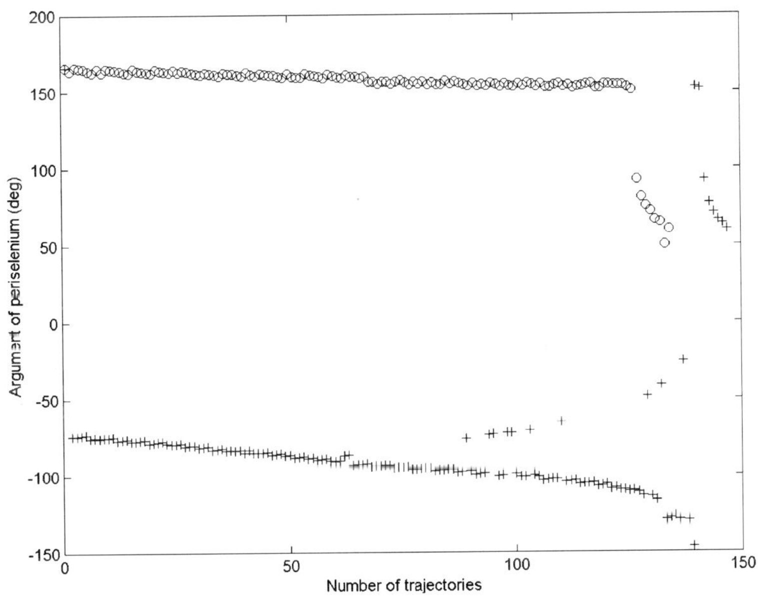

Fig. 14 Periselenium argument of the orbits in the velocity locus mentioned in text

$$
\begin{aligned}
\dot{x} &= u \\
\dot{y} &= v \\
\dot{z} &= w \\
\dot{u} &= 2v + x - \tfrac{1-\mu}{r_{\mathrm{E}}^3}(x+\mu) - \tfrac{\mu}{r_{\mathrm{M}}^3}(x-1+\mu) - \tfrac{m_S}{r_{\mathrm{S}}^3}(x-\rho\cos(\theta)) - \tfrac{m_S}{\rho^2}\cos(\theta) \\
\dot{v} &= -2u + y - \tfrac{1-\mu}{r_{\mathrm{E}}^3}y - \tfrac{\mu}{r_{\mathrm{M}}^3}y - \tfrac{m_S}{r_{\mathrm{S}}^3}(y-\rho\sin(\theta)) - \tfrac{m_S}{\rho^2}\sin(\theta) \\
\dot{w} &= -\tfrac{1-\mu}{r_{\mathrm{E}}^3}z - \tfrac{\mu}{r_{\mathrm{M}}^3}z - \tfrac{m_S}{r_{\mathrm{S}}^3}z \\
\dot{\theta} &= \omega_S \tau
\end{aligned}
\tag{11}
$$

where (Fig. 15)

$$
\begin{aligned}
r_{\mathrm{E}} &= \sqrt{(x+\mu)^2 + y^2 + z^2} \\
r_{\mathrm{M}} &= \sqrt{(x-1+\mu)^2 + y^2 + z^2} \\
r_{\mathrm{s}} &= \sqrt{(x-\rho\cos(\theta))^2 + (y-\rho\sin(\theta))^2 + z^2} \\
\rho &= 389.2, \mu = 0.01215, m_{\mathrm{S}} = 3.298 \cdot 10^5
\end{aligned}
\tag{12}
$$

Figure 16 shows the four-body problem velocity locus; this looks rather similar to the three-body velocity locus of Fig. 13.

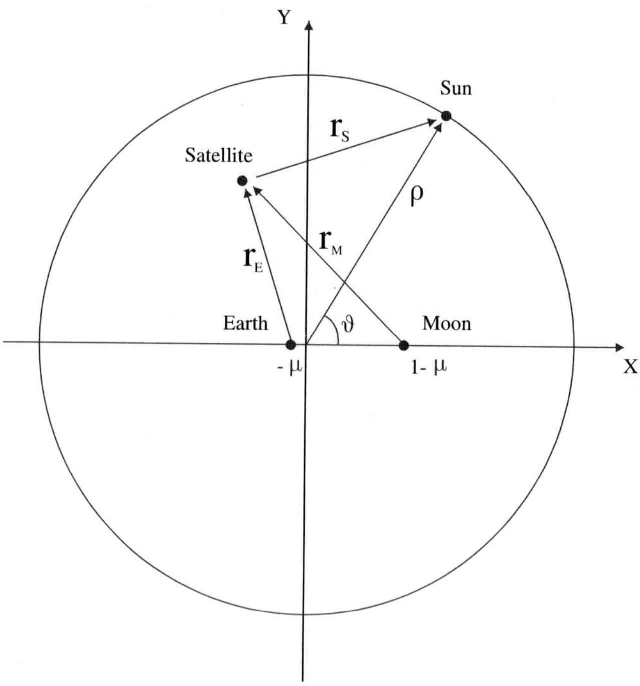

Fig. 15 Reference system in a bicircular model

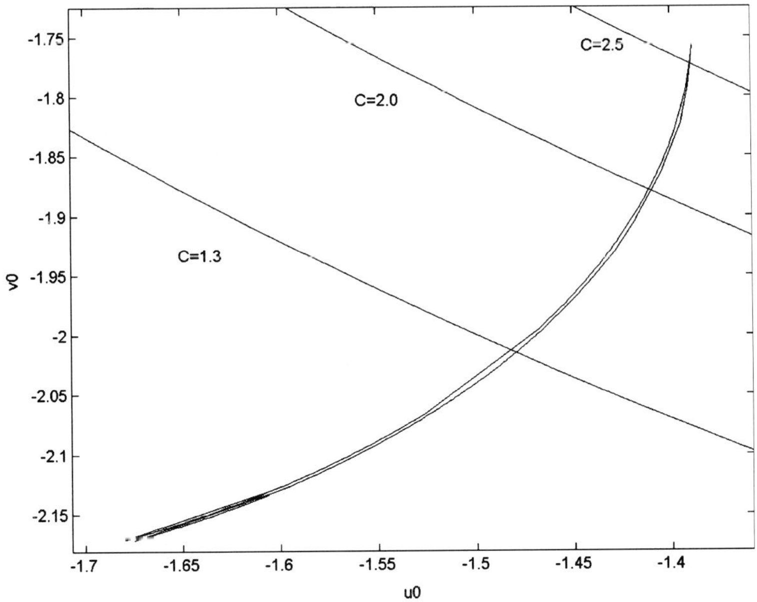

Fig. 16 Velocity locus (restricted four-body problem, dimensionless velocities u_0, v_0)

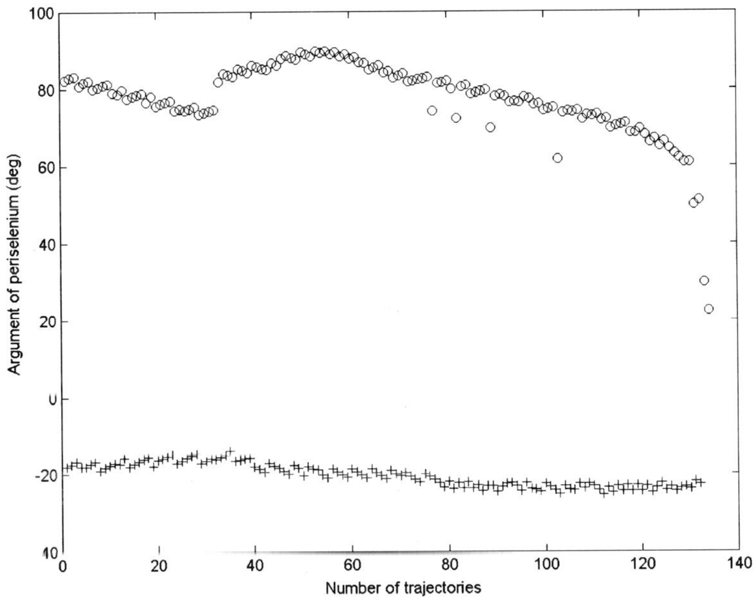

Fig. 17 Periselenium argument of the orbits in the velocity locus mentioned in text

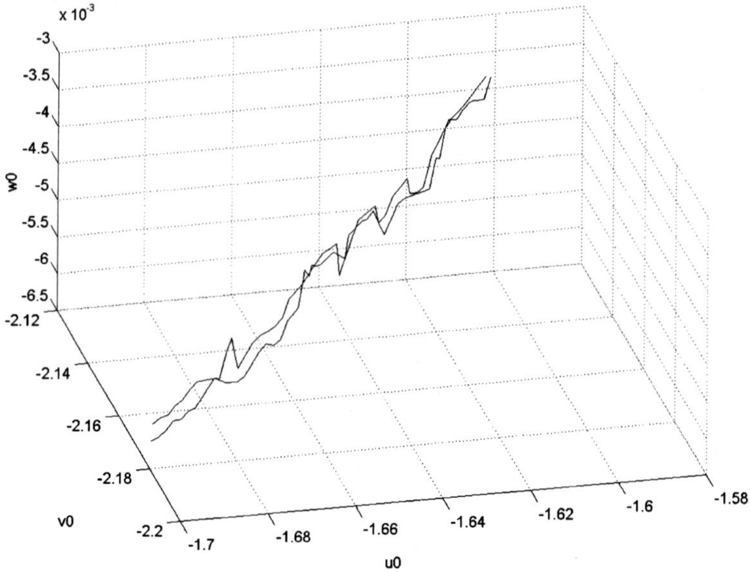

Fig. 18 Three-dimensional velocity locus (restricted four-body problem, dimensionless velocities)

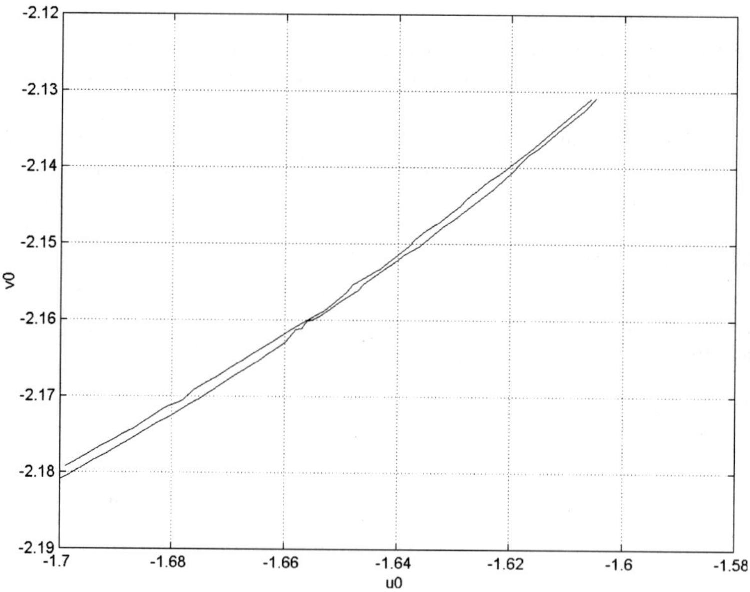

Fig. 19 Velocity locus: planar (u_0, v_0) projection of Fig. 18

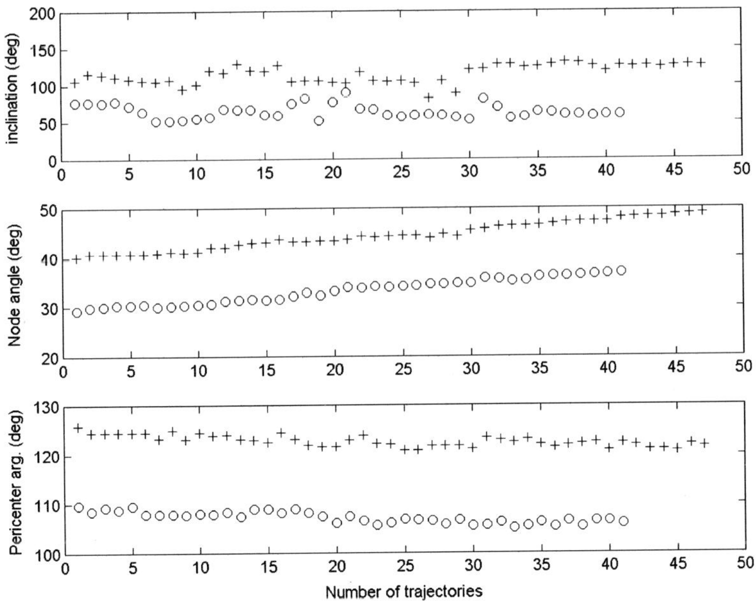

Fig. 20 Orbital parameters of the two families of Fig. 18

In Fig. 16, two families of trajectories can be again identified: in fact, as in the three-body problem case, two branches in the velocity locus are apparent. In Fig. 17, the periselenium argument of trajectories belonging to the two families is shown. Note that there is a difference of about 100° between the ω of the two families, as occurs in the three-body problem (Fig. 14).

Finally, a search on the three-dimensional components $u_0, v_0, w_0 \neq 0$ of the velocity $\mathbf{V}_0$ at $\mathbf{r}_0$ needed to reach the periselenium altitude r_1 is performed. The three-dimensional velocity locus is in Fig. 18. The planar u_0, v_0 projection of Fig. 18 is in Fig. 19, and it turns out to be qualitatively similar to the planar result of Fig. 16.

In this three-dimensional case, different parameters can be compared, such as inclination i, node angle Ω and the argument of perigee ω. Figure 20 shows that the orbital parameters in each of the two branches are similar, whereas the two branches have different orbital parameters.

6 Conclusions

A strategy for a lunar satellite constellation deployment is proposed, using a WSB lunar transfer and small perturbations from such a trajectory. The strategy is proved to be feasible and economical if the constellation design requires just two different orbital planes and for every satellite in a varied orbit about 120 m/s of ΔV are saved. The orbital parameters at the Moon obtained by small perturbation of a nominal WSB trajectory are frozen to a unique set of values, if the periselenium distance is kept fixed. Such a property resembles what happens in simpler dynamical systems. In fact, in the Keplerian problem families of trajectories having the same pericentre distance define a hyperbola in the velocity locus. The variation in perigee argument is as small as is the intensity of the burn and it varies in a continuous way with the point where the burn is performed. However, it is possible to note that if the burn is performed around the pericentre region, the pericentre argument ω of the varied trajectory is almost equal to the pericentre argument of the nominal trajectory. If the burn is about the apocentre region, the angle ω of the varied trajectory is different from the nominal one and all the varied trajectories separated from the nominal trajectory at the apocenter region have almost the same ω. These features are amplified in the restricted three- and four-body problem. It is shown that trajectories having the same pericentre distance define two families in the velocity locus. The variation in perigee argument is large if one shifts from one family to the other, and this can be done with a small variation of velocity. Considering a three-dimensional case, it is possible to compare other orbital parameters, such as inclination and node angles. Again, each family is defined by common values of orbital parameters that vary from one family to the other.

References

Battin, R.H.: An introduction to the mathematics and methods of astrodynamics. AIAA Education Series, New York (1987)

Belbruno, E.: Examples of nonlinear dynamics of ballistic capture in the Earth-Moon system. AIAA Paper no. 2896 (1990)

Belbruno, E., Miller, J.: A ballistic lunar capture trajectory for the Japanese spacecraft Hiten. JPL Report IOM 312/90.4–1317 (1990)

Belbruno, E.: Capture dynamics and chaotic motions in celestial mechanics, pp. 144–148. Princeton University Press, Princeton (2004)

Bussey, B., Spudis, P.: The lunar poles: mapping the future lunar base. Moon Base: a Challenge for Humanity, International Conference, Venice Workshop, 26–27 May (2005)

Circi, C., Teofilatto, P.: On the dynamics of weak stability boundary lunar transfers. Celest. Mech. Dyn. Astr. **79**(1), 41–72 (2001)

D'Avanzo, P., Teofilatto, P., Ulivieri, C.: Long term effects on lunar orbiter. Acta Astronautica **40**(1), 13–20 (1997)

Graziani, F.: LUNISAT: an university satellite to the Moon. Paper presented in CELMEC IV, San Martino al Cimino, Italy, 12 September (2005)

Hadjidemetriou, J.: A classification of orbits through L_1 in the restricted three body problem. The Astronomical J. **73**(2), 104–110 (1968)

Kawaguchi, J., Yamakawa, H., Uesugi, T., Matsuo, H.: On making use of lunar and solar gravity assist for lunar A and planet B missions. Acta Astronautica, **35**, 633–642 (1995)

Kazuyuki, Y.: Sun-perturbed earth-to-moon transfers with low energy and moderate flight time. Celest. Mech. Dyn. Astr. **90**(3–4), 197–212 (2004)

Koon, W., Lo, M., Marsden, J., Ross, S.: Dynamical system, the three- body problem and space mission design. International Conference on Differential Equations, Berlin (2000)

Koon, W., Lo, M., Marsden, J., Ross, S.: Low energy transfer to the moon. Celest. Mech. Dyn. Astr. **81**(1), 63–73 (2001)

Simò, C., Gomez, G., Jorba, A., Masdemont, J.: The bicircular model near the triangular libration point of the RTBP. In: Roy, A., Steves, B.A. (eds.) From Newton to Chaos, pp. 343–370. Plenum Press, New York (1995)

Celestial Mechanics and Dynamical Astronomy (2006) 95:391–405
DOI 10.1007/s10569-006-9024-y

ORIGINAL ARTICLE

Aero-gravity assist maneuvers: controlled dynamics modeling and optimization

R. Armellin · M. Lavagna · A. Ercoli-Finzi

Received: 7 November 2005 / Revised: 9 April 2006 /
Accepted: 14 April 2006 / Published online: 15 August 2006

Abstract The aero-gravity assist maneuver is here proposed as a tool to improve the efficiency of the gravity assist as, thanks to the interaction with the planetary atmospheres, the angular deviation of the velocity vector can be definitely increased. Even though the drag reduces the spacecraft velocity, the overall Δv gain could be remarkable whenever a high lift-to-drag vehicle is supposed to fly. Earlier studies offer simplified approaches according to both the dynamics modeling and the atmospheric trajectory constraints. In this paper a 3D dynamical model is adopted and a more realistic L/D performance for the hypersonic vehicle is assumed. Some relevant aspects related to the multidisciplinary design have been considered such as heating rates and structural loads bounding. Comparisons between in and out of plane maneuvering have been performed by assuming, as control variables, either the angle of attack or the bank angle, respectively. The optimal control problem has been solved by selecting a direct method approach. The dynamics has been transcribed into a set of non-linear constraints and the arising non-linear programming problem has been solved with a sequential quadratic programming solver. To gain the global optimum convergence the initial guess has been supplied by solving the same problem by a direct shooting technique and a genetic optimizer.

Keywords Aero-gravity assist maneuver · Optimal control problem · Multiple shooting technique

1 Introduction

Planetary gravity assist is a consolidated maneuver to limit the propellant mass required for interplanetary missions. That maneuver allows rotating the spacecraft relative velocity thanks to the momentum exchange occurring between the planet and

R. Armellin (✉) · M. Lavagna · A. Ercoli-Finzi
Aerospace Engineering Department, Politecnico di Milano,
Via La Masa 34, 20156 Milano, Italy
e-mail: armellin@aero.polimi.it

the probe. Therefore, as a result, the outcoming heliocentric velocity vector can be controlled, both in magnitude and direction. The momentum exchange depends on the eccentricity of the hyperbola described by the spacecraft in the planetary reference frame, which is a function of the spacecraft incoming velocity, the pericenter radius and the planetary mass. Pericenter radius is, obviously, lower bounded by planetary radius and the spacecraft incoming velocity is generally high for high energy missions; as a consequence only massive planets allow great maneuver effectiveness. Unfortunately only few planets in the solar system satisfy this requirement.

The aero-gravity assist (AGA) has been first proposed by McRonald and Randolph (1992) as a solution for the gravity assist (GA) efficiency increase whenever planets with atmosphere are selected for the swingby. During an AGA maneuver the spacecraft enters the planet atmosphere and exploits the aerodynamic force to augment the centripetal force provided by the planet allowing an increase in the overall velocity bending, i.e., the momentum exchange. Earlier works focused on simulating a possible atmospheric sling trajectory, adopted a simple analytical model, and assumed a circular atmospheric path with no regard for optimality (Lewis and McRonald 1992; Bonfiglio et al. 2000). Lohar et al. (1994) and Johnson (2002) presented a more realistic mathematical model, but still 2D, and obtained an optimal trajectory to maximize spacecraft heliocentric velocity. Others studies assume an extended mathematical model: heating constraints are included (Lohar et al. 1995, 1997) and a 3D trajectory is considered to exploit the maneuver for heliocentric plane change (Lohar et al. 1996); however, in the latter case, the constant altitude approximation is still preserved to limit the optimization convergence issues.

A step forward is given in this work as the optimization problem is posed with no altitude constraints on a 3D dynamics model. Either the angle of attack or the bank angle profile is here selected to control the atmospheric trajectory. The objective function could be selected either to maximize or minimize the spacecraft heliocentric velocity and different constraints can be imposed to limit the heating rate and the structural load experienced during the atmospheric path; the heliocentric plane change amount can be freely settled too.

To assure the AGA maneuver effectiveness, the drag losses must be bounded: therefore, the vehicle aerodynamics set must offer a high L/D in hypersonic flying conditions. Such vehicles, called waveriders, have been extensively studied by Anderson, Lewis, et al. for both Earth and planetary applications. The authors extensively refer to these works for the vehicle aerodynamic properties estimate.

The paper is organized as follows: first the dynamics, the aerodynamics and the heating rate models applied are described; secondly the optimization problem is formalized; finally simulations concerning Mars aero-gravity assist are discussed.

2 The selected models

2.1 Equations of motion

The vehicle dynamics is written in a local non-inertial reference frame x, y, z attached to the vehicle with x axis aligned with the local vertical, y pointing toward east and z to complete the right handed triad. In addition an inertial planetocentric ecliptic reference frame X, Y, Z is assumed: in particular the in plane X axis is collinear to the vernal equinox direction. The relative orientation between the two frames is defined

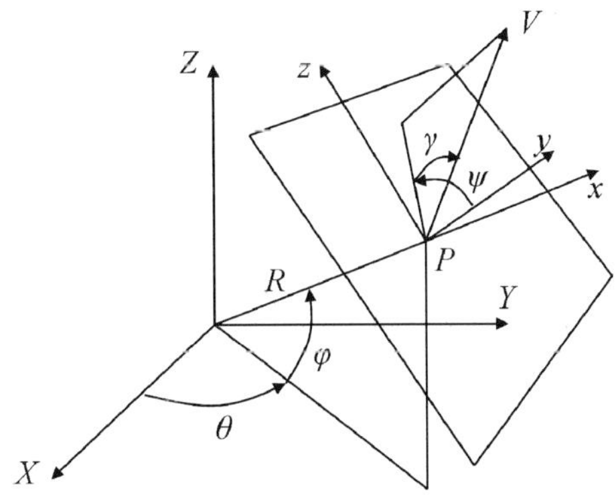

Fig. 1 Local reference frame and state vector variables $\mathbf{x} = \{R, \theta, \varphi, V, \gamma, \psi\}$ definition

by the longitude θ and the latitude ϕ, whereas the flight path angle γ and the heading angle ψ are introduced to identify the vehicle velocity vector in the local reference frame. For the sake of clearness these quantities are highlighted in Fig. 1. The motion of the vehicle center of mass is described by the following set of ordinary differential equations (ODEs)

$$\begin{cases} \dot{R} &= V \sin\gamma \\ \dot{\theta} &= \dfrac{V\cos\gamma\cos\psi}{R\cos\varphi} \\ \dot{\varphi} &= \dfrac{V\cos\gamma\sin\psi}{R} \\ \dot{V} &= \dfrac{D}{m} - G\sin\gamma \\ V\dot{\gamma} &= \dfrac{L\cos\sigma}{m} - G\cos\gamma + \dfrac{V^2\cos\gamma}{R} \\ V\dot{\psi} &= \dfrac{L\sin\sigma}{m cos\gamma} - \dfrac{V^2\tan\varphi\cos\gamma\cos\psi}{R}, \end{cases} \tag{1}$$

where L and D are, respectively, lift and drag force and σ is the bank angle, defined as the angle between the lift vector and the plane described by the local vertical and the velocity vector. The first three equations represent the kinematic relations whereas the latter three represent the dynamics. To avoid the objective function Hessian and the constraints Jacobian ill-conditioning and to improve the optimization convergence, Eq.(1) have been scaled by

$$V_c = \sqrt{\mu/R_a}, \quad l = R_a, \quad \text{and } m = m_0,$$

where μ is the planetary gravity constant, R_a is the planetary sensible atmosphere radius and m_0 is the vehicle mass.

2.2 Waverider aerodynamic model

As already mentioned, AGA maneuvers ask for a hypersonic high lift to drag ratio, offered by the waverider shaped vehicles, first introduced by Nonweiler in 1959. Those vehicles' design is aimed in both keeping the shockwave attached to the leading edge and preserving the high pressure behind the shock along the bottom surface to obtain the so-called compression lift effect. High L/D ratios are, hence, obtained. The waverider shapes have been, usually, obtained starting from known flowfields generated behind either wedge, cone or power law shaped bodies (Corda and Anderson 1988; Starkey and Lewis 1999). Since the late 80s, design optimization of waverider shape has been the subject of several research projects. The Maryland University Hypersonics Team produced a great amount of work about this topic focusing attention on viscous (Bowcutt et al. 1987), gas rarefaction (Anderson et al. 1991a), off-design conditions (Starkey and Lewis 2000), and blunted nose effects on L/D performances (Gillum and Lewis 1997). Furthermore waveriders optimized for Cytherean and Martian atmosphere have been carried out; the obtained solutions, optimal according to aerodynamics and shape, turned out to be definitely similar to the Earth atmosphere optimization scenario (Anderson et al. 1991b).

In this work the aerodynamic performances of the vehicle are described by means of a drag–polar curve, as suggested in Gillum and Lewis (1997), expressed by

$$C_\mathrm{d} = C_{\mathrm{d}_0} + k|C_\mathrm{l}|^n \tag{2}$$

where C_l and C_d are the lift and the drag coefficient, respectively, C_{d_0} is the drag at zero lift, and k and n two constant coefficients. An optimized vehicle shape is here assumed with $(L/D)^* = (L/D)_{\max} = 3.7$ according to Anderson et al. 1991b; an aerodynamic performance degradation is considered: performance degradation is caused by the leading edge bluntness aimed to contain the heating rate, and by the Reynolds numbers experienced during the atmospheric passage simulated in this work, which are lower than the optimal design flight conditions. Moreover, a $C_\mathrm{l}^* = 0.0848$ (the value of the lift coefficient that maximizes L/D) has been extrapolated from data given in Anderson et al. (1991a and 1991b) and $n = 1.75$ is posted. It is worth noting that, due to atmosphere rarefaction, a substantial decrease in $(L/D)^*$ occurs during the atmospheric descent and breakaway phases. The rarefaction effects can be ignored within those two aforementioned phases, because of their very short duration and almost null lift action. The remaining coefficients describing the waverider aerodynamic behavior are computed as follows. The optimal lift coefficient formulation can be derived from the efficiency definition $L/D = \frac{C_\mathrm{l}}{C_{\mathrm{d}_0}+k|C_\mathrm{l}|^n}$ connected with the stationary condition $\frac{\partial L/D}{\partial C_\mathrm{l}} = 0$. The resulting system

$$\begin{cases} C_\mathrm{l}^* & = \left(\dfrac{C_{\mathrm{d}_0}}{(n-1)k}\right)^{\frac{1}{n}} \\ (L/D)^* & = \dfrac{1}{n}\left(\dfrac{1}{k}\right)^{\frac{1}{n}}\left(\dfrac{n-1}{C_{\mathrm{d}_0}}\right)^{\frac{n-1}{n}} \end{cases}, \tag{3}$$

can be solved to find the values $k = 0.9811$ and $C_{\mathrm{d}_0} = 0.0098$, thus leading to the drag polar profile shown in Fig. 2a. In accordance with the former computations the aerodynamic forces can be written as

$$\begin{cases} L = \frac{1}{2}\rho V^2 S \lambda C_\mathrm{l}^* \\ D = \frac{1}{2}\rho V^2 S C_\mathrm{d} \end{cases}, \tag{4}$$

where $\lambda = C_\mathrm{l}/C_\mathrm{l}^*$. For the ongoing simulation a more realistic value of 280 kg/m^2 for the coefficient $m/(SC_\mathrm{l}^*)$ is assumed, which is much greater than the value of 50 kg/m^2 adopted by McRonald and Randolph (1992). This more realistic value translates into

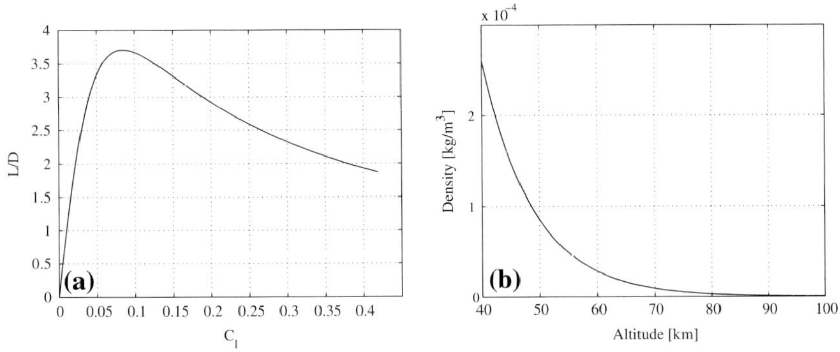

Fig. 2 Waverider drag polar curve $(L/D)^* = 3.7$ (**a**) and Mars density–altitude profile (**b**)

a smaller aerodynamic surface for a given spacecraft mass, therefore, once the atmospheric incoming conditions have been fixed, a higher density value is required to generate the aerodynamic force to accomplish the AGA, emphasizing heating rate and drag losses problems. To evaluate the aerodynamic forces an atmospheric model of the planet involved in the maneuver must be addressed. In this paper AGA maneuvers on Mars are analyzed and a classical exponential density profile is adopted, based on the interpolation of data collected by past missions in a range altitude of 40–100 km. The resulting density profile is shown in Fig. 2b.

2.3 Heating rate models

The heating rate experienced by the vehicle during its atmospheric path represents one of the major issues connected with the AGA maneuver design. Indeed the high vehicle velocity together with the fluid viscosity cause fluid high temperature, turning out into both high convective and radiative heating rates. The amount of those fluxes is here evaluated thanks to semi-analytical formulae reported by Anderson (1989) and Tauber and Sutton (1991) and already applied by Lohar et al. in previous works (1995, 1997). More specifically the convective heating rate comes from

$$q_{\mathrm{c}} = \frac{k\rho^{\frac{1}{2}} V^3}{(r_{\mathrm{n}})^{\frac{1}{2}}}, \tag{5}$$

where r_{n} is the vehicle nose radius. The simulations assumed a Martian atmosphere composed by 85% CO_2 and 15% N_2, entailing a $k = 1.35 \times 10^{-8}$.

The radiative heat flux is estimated by

$$q_{\mathrm{r}} = F r_{\mathrm{n}}^{a} \rho^{p} f(v), \tag{6}$$

where $f(v)$ are tabulated data depending upon the vehicle velocity and the atmosphere chemical composition, and the remaining quantities values are

$$F = 4.736 \times 10^4, \quad a = 1.072 \times 10^6\, V^{-1.88}\, \rho^{-0.325} \quad \text{and} \quad p = 1.22.$$

Through out the simulation a value of $r_{\mathrm{n}} = 0.3\,\mathrm{m}$, compatible with aerodynamic performances, has been considered.

3 Trajectory optimization

3.1 The optimal control problem: generalities

In this section a brief description of an optimal control problem is given (Betts 1998). Typically the dynamics of a system is described by a set of first order ODEs

$$\dot{\mathbf{x}} = \mathbf{f}(\mathbf{x}(t), \mathbf{u}(t), t), \tag{7}$$

where $\mathbf{x} = \{x_1, \ldots, x_n\}^T$ and $\mathbf{u} = \{u_1, \ldots, u_m\}^T$ with $(m \leq n)$ are state and control vectors, respectively. Initial conditions at time t_0 are defined by

$$\boldsymbol{\psi}_{0l} \leq \boldsymbol{\psi}(\mathbf{x}(t_0), \mathbf{u}(t_0), t_0) \leq \boldsymbol{\psi}_{0u}, \tag{8}$$

and terminal conditions at the final time t_f are defined by

$$\boldsymbol{\psi}_{\mathrm{f}l} \leq \boldsymbol{\psi}(\mathbf{x}(t_\mathrm{f}), \mathbf{u}(t_\mathrm{f}), t_\mathrm{f}) \leq \boldsymbol{\psi}_{\mathrm{f}u}. \tag{9}$$

In addition, the solution might satisfy algebraic path constraints of the form

$$\mathbf{g}_l \leq \mathbf{g}(\mathbf{x}(t), \mathbf{u}(t), t) \leq \mathbf{g}_u, \tag{10}$$

as well as simple bounds on the state variables

$$\boldsymbol{x}_l \leq \boldsymbol{x}(t) \leq \boldsymbol{x}_u, \tag{11}$$

and control variables

$$\mathbf{u}_l \leq \mathbf{u}(t) \leq \mathbf{u}_u. \tag{12}$$

Note that upper and lower bounds must be equal in the above expressions in order to obtain equality constraints. The optimal control problem is to determine the control profiles $\mathbf{u}(t)$ that minimize the objective function, written here in Mayer form,

$$J = \varphi(\mathbf{x}_\mathrm{f}, t_\mathrm{f}), \tag{13}$$

with the respect of the dynamics (7) and constraints (8–12).

3.2 Problem statement

The problem this paper refers to is now posted as an optimal control problem. The system dynamics (7) is represented by Eq.(1), according to the AGA maneuver, and $\mathbf{x} = \{R, \theta, \varphi, V, \gamma, \psi\}^T$ and $\mathbf{u} = \{\lambda, \sigma\}^T$ stay for the state and control vectors, respectively. The objective function is defined as

$$J = \mp \|\mathbf{V}_\mathrm{p} + \mathbf{V}_\infty^+(\boldsymbol{x}_\mathrm{f})\| = \mp \|\mathbf{V}_\mathrm{s}^+\|, \tag{14}$$

where $\mathbf{V}_\mathrm{p}$ and $\mathbf{V}_\infty^+(\boldsymbol{x}_\mathrm{f})$ are the planetary velocity vector and the spacecraft velocity going out from the planet sphere of influence. In the objective function the minus sign applies to maximization of the heliocentric velocity $\mathbf{V}_\mathrm{s}^+$, and plus sign applies to minimization of $\mathbf{V}_\mathrm{s}^+$. The initial conditions can be expressed as

$$\begin{cases} R(t_0) & = & R_\mathrm{a} \\ \mathbf{V}_\infty^-(\boldsymbol{x}_0) & = & \mathbf{V}_\infty^-, \end{cases} \tag{15}$$

which represents a set of one scalar and one vectorial constraints. The scalar constraint assures that the AGA maneuver begins at the planet sensible atmosphere interface, fixed, for Mars, at 100 km of altitude. The vectorial constraint states that the incoming velocity at the planet sphere of influence corresponding to the initial state $\boldsymbol{x}_0$, i.e., $\mathbf{V}_\infty^-(\boldsymbol{x}_0)$, must be equal to the spacecraft incoming velocity $\mathbf{V}_\infty^-$, imposed by the heliocentric trajectory analysis. Note that four constraints are imposed on the six state variables at t_0: the optimizer exploits the two remaining degrees of freedom to define the incoming hyperbola plane.

The final condition limits to

$$R(t_\mathrm{f}) = R_\mathrm{a}, \tag{16}$$

which assures that the AGA maneuver ends at the planet sensible atmosphere. Note that $\mathbf{V}_\infty^+(\boldsymbol{x}_\mathrm{f})$ and $\mathbf{V}_\infty^-(\boldsymbol{x}_0)$ dependence is expressed using a state transition matrix to avoid the numerical integration of the dynamics, in order to lower computational time and increase accuracy. The state vector bounds are

$$\begin{cases} R_\mathrm{s} & \le R(t) & \le R_\mathrm{a} \\ -\pi & \le \theta(t) & \le \pi \\ -\frac{\pi}{2} & \le \varphi(t) & \le \frac{\pi}{2} \\ V_\mathrm{e} & \le V(t) & \le V_{R_\mathrm{a}}, \\ -\frac{\pi}{2} & \le \gamma(t) & \le \frac{\pi}{2} \\ -\pi & \le \psi(t) & \le \pi \end{cases} \tag{17}$$

where R_s is the safe maneuver radius corresponding to an altitude of 40 km, V_e is Mars escape velocity and V_{R_a} is the velocity magnitude at Mars sensible atmosphere. The control vector is limited by

$$\begin{cases} -5 \le \lambda(t) \le 5 \\ -\pi \le \sigma(t) \le \pi. \end{cases} \tag{18}$$

Further constraints can be handled by establishing upper limits for the heating rates, the structural loads, and imposing a desired heliocentric orbital plane change α:

$$q = q_\mathrm{c}(\mathbf{x}(t), \mathbf{u}(t)) + q_\mathrm{r}(\mathbf{x}(t), \mathbf{u}(t)) \le q_{\max}, \tag{19}$$

$$n(\mathbf{x}(t), \mathbf{u}(t)) \le n_{\max}, \tag{20}$$

$$\arctan \frac{V_{\mathrm{s}\perp}^+}{V_{\mathrm{s}//}^+} = \alpha, \tag{21}$$

where $V_{\mathrm{s}\perp}^+$ and $V_{\mathrm{s}//}^+$ are, respectively, the ecliptic out of plane and the ecliptic in plane velocity components.

3.3 Solution technique

The problem stated in the previous section is here solved with a multiple shooting technique (Enright and Conway 1992). The time span $\{t_0, t_\mathrm{f}\}$ is split into $N-1$ equispaced intervals, where N is for the knots number. A cubic spline, defined within each former interval of time, models the control profile $\mathbf{u}(t)$; M points are kept to represent the cubic shape; therefore each of the $N-1$ time interval contains $M-1$ subintervals. On a generic knot i the variable vector is represented by

$$\mathbf{X}_i = \{\mathbf{x}_i, \tilde{\boldsymbol{\lambda}}_i \tilde{\boldsymbol{\sigma}}_i\} = \{R_i, \ldots, \psi_i, \lambda_i^1, \ldots, \lambda_i^{M-1}, \sigma_i^1, \ldots, \sigma_i^{M-1}\}^T, \tag{22}$$

with the exception for the last knot where

$$\mathbf{X}_N = \{R_N, \ldots, \psi_N, \lambda_N^1, \sigma_N^1\}^T. \tag{23}$$

The overall variables vector, remarking that the final time of the maneuver is unknown, is given by:

$$\mathbf{X} = \{t_\mathrm{f}\ \mathbf{X}_1, \ldots, \mathbf{X}_N\}^T. \tag{24}$$

The M points which define the cubic spline for the control vector, inside the i-th interval, are

$$\boldsymbol{\lambda}_\mathbf{i} = \left\{\lambda_i^1, \ldots, \lambda_i^{M-1}, \lambda_{i+1}^1\right\}^T \quad \text{and} \quad \boldsymbol{\sigma}_\mathbf{i} = \left\{\sigma_i^1 \ldots \sigma_i^{M-1} \sigma_{i+1}^1\right\}^T, \tag{25}$$

with $i = 1, \ldots, N-1$. Note that for each control variable, within the i-th interval, the M-th point defining the cubic spline is recalled by the solution offered for the first point of the cubic spline of the subsequent interval. The control profiles continuity is so implicitly satisfied and a $2(N-2)$ reduction on both the optimization variable vector and the equality constraint set is gained. In every time interval the interpolation process links each subset of knots values $\boldsymbol{\lambda}_i$ and $\boldsymbol{\sigma}_i$ to the corresponding time function $\lambda_i(t)$ and $\sigma_i(t)$. A 8th-order fixed-step Runge–Kutta scheme, using an absolute and relative tolerance of 10^{-9} (Stoer and Bulirsch 1993) inside each time interval $\{t_i, t_{i+1}\}$, is then applied to solve for the dynamics integration. As a consequence the computed values $\mathbf{x}_{i+1}^\mathrm{c}$ corresponding to every $\mathbf{x}_i$ are carried out and the dynamic equations are transcribed into $6(N-1)$ algebraic constraints

$$\mathbf{x}_i^\mathrm{c} - \mathbf{x}_i = 0, \tag{26}$$

for $i = 2, \ldots, N$. The control bounds (18) are rewritten as

$$\begin{cases} -5 \leq \boldsymbol{\lambda}_i \leq 5 \\ -\pi \leq \boldsymbol{\sigma}_i \leq \pi. \end{cases} \tag{27}$$

Furthermore to avoid the control vector elements first derivative discontinuities at interval interfaces the following constraints have been added

$$\begin{cases} \dot{\lambda}_{i|t_{i+1}} - \dot{\lambda}_{i+1|t_{i+1}} = 0 \\ \dot{\sigma}_{i|t_{i+1}} - \dot{\sigma}_{i+1|t_{i+1}} = 0 \end{cases}, \tag{28}$$

with $i = 1, \ldots, N-2$. The constraints on the initial and final states assume the following form

$$\begin{cases} R_1 = R_\mathrm{a} \\ R_N = R_\mathrm{a} \\ \mathbf{V}_\infty^-(\mathbf{x}_1) = \mathbf{V}_\infty^-, \end{cases} \tag{29}$$

while the constraints on the heating rate and the structural load already posted in (18) and (19) are formalized as follows

$$\max_i(q_\mathrm{c}(\mathbf{x}_i) + q_\mathrm{r}(\mathbf{x}_i)) \leq q_{\max}, \tag{30}$$

$$\max_i(n(\mathbf{x}_i)) \leq n_{\max}. \tag{31}$$

Finally the objective function (14) translates into

$$J = \mp\|\mathbf{V}_\mathrm{p} + \mathbf{V}_\infty^+(\mathbf{x}_N)\| = \mp\|\mathbf{V}_\mathrm{s}^+\|. \tag{32}$$

The optimal control problem is so transcribed into discrete form and solved using an SQP solver. The required initial estimate for first guess solution in terms of $\mathbf{x}_i, \boldsymbol{\lambda}_i$, and $\boldsymbol{\sigma}_i$ is satisfied by solving the same problem using a single shooting method with a simpler representation of the controls and a stochastic optimizer based on the genetic algorithms, to enhance convergence to the global minimum.

4 Simulations

4.1 Planar maneuver

In this section the control vector is limited to the lift coefficient ratio λ, keeping null the bank angle along the entire path through the atmosphere; as a consequence, a planar maneuver is studied. As expected, the optimizer selects the plane spanned by the $\mathbf{V}_\infty^-$ and $\mathbf{V}_\mathrm{p}$ vectors as the maneuver plane to maximize the momentum exchange between the vehicle and the planet. In the first two simulations the spacecraft heliocentric velocity maximization is addressed: more specifically, the former refers to an unconstrained heating rate scenario (AGA_{nhc}), whereas in the latter a constraint of $q_{\max} = 250\ \mathrm{W/cm}^2$ is considered (AGA_{hc}). The heating rate limit is comparable with the value considered for the thermal protection design of past missions to Mars, such as the Deep Space 2 mission (http: //www.nmp.nasa.gov/ds2/mission/events.html).

As far as the unconstrained maneuver is considered, the vehicle flies upside-down at nearly constant altitude, as Fig. 3 illustrates. A maximum L/D is selected to minimize the drag losses, with the exception of the descent and breakaway phases, as clearly reported in Fig. 4a. The overall velocity bending is such that the $\mathbf{V}_\mathrm{s}^+$ is maximized and, as pointed out by Fig. 5, this does not translate into alignment of the planet and spacecraft velocities. A larger bending angle demand requires the vehicle to pass through the atmosphere longer increasing the velocity loss due to drag. As a result, a trade-off between the bending angle and the change in the velocity magnitude exists. To better underline this point the analysis of simulations with the $\mathbf{V}_\infty^+$ and $\mathbf{V}_\mathrm{p}$ alignment enforced has been carried out. The lower performance of the maneuver is highlighted by the $\mathrm{AGA}_{//\mathbf{V}_\mathrm{p}}$ column in Table 1.

Whenever the heating rate constraint is of concern the identified guidance focuses on higher lift coefficient values in response to a higher altitude needed to satisfy the imposed constraint. The dotted line in Fig. 3a clearly depicts the effects of the heating

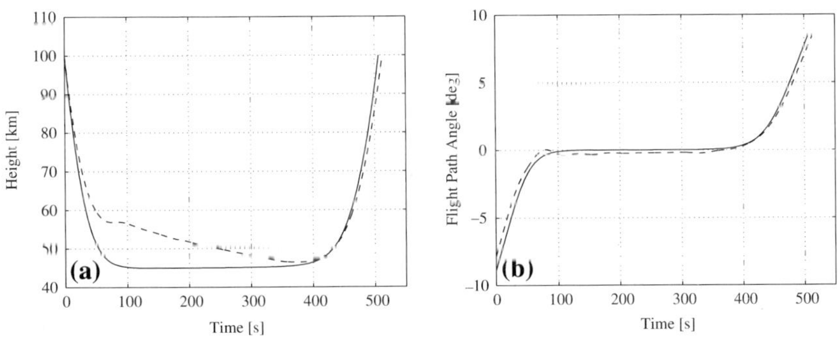

Fig. 3 2D maneuver altitude (**a**) and flight path angle (**b**): — AGA_{nhc}, - - AGA_{hc}

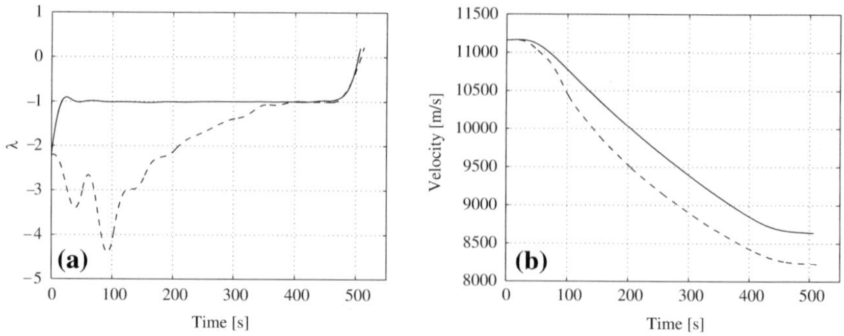

Fig. 4 2D maneuver λ profiles (**a**) and velocity vectors magnitude (**b**):— AGA_{nhc}, - - AGA_{hc}

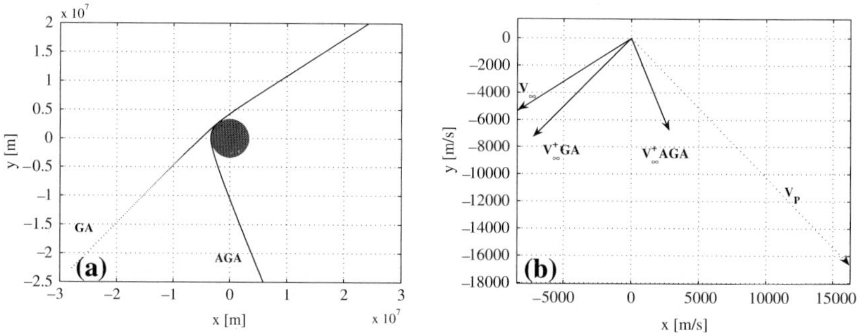

Fig. 5 2D maneuver comparison: trajectories (**a**) and velocity vectors (**b**) in the planetary reference frame

Table 1 Heliocentric plane change maneuvers comparison. Velocities are given in m/s, heating rates in W/cm^2, and angles in deg. Test case: $V_\infty^- = 10,000\,\mathrm{m/s}$

	V_S^+ max				V_S^+ min		
	AGA_{nhc}	AGA_{hc}	$AGA_{//V_p}$	GA	AGA_{nhc}	AGA_{hc}	GA
V_S^-	23378	23378	23378	23378	23378	23378	23378
V_∞^+	7078.5	6569.8	6291.1	10,000	7593.0	7064.1	10,000
V_S^+	29992	29506	29600	25454	15933	16474	21207
q_{max}	539.10	250	510.52	–	542.1	250	–
δ^*	80.05	80.08	101.96	12.54	66.71	65.74	12.54

*Overall velocity bending angle

rate constraint on the height history. A significant increase according to the drag losses occurs as a side effect, because of the L/D lowering (see Fig. 4b); the maximum L/D condition is restored as soon as the velocity magnitude has sufficiently decreased to respect the imposed heating rate bound. Heating rates comparison between constrained and unconstrained trajectories is given in Fig. 6; the values are reported for the discretization knots only.

To complete the 2D simulation framework, a maneuver to minimize the heliocentric velocity is performed as it is required for mission to inner planets. Even in this case, although less intuitive, the selected guidance answers a drag loss minimization

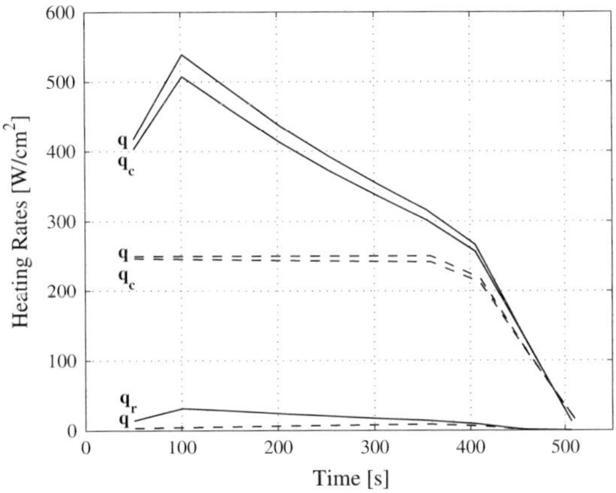

Fig. 6 2D maneuver heating rates: q total heating rate, q_c convective heating rate, q_r radiative heating rate. — AGA_{nhc}, - - - AGA_{hc}

strategy: most of the path is flown at maximum L/D, and the velocity minimization is gained by steering the spacecraft planetocentric velocity in the opposite direction of the planetary one.

Simulation results are summarized in Table 1: a comparison with a GA maneuver performed with a pericenter at 100 km of altitude is also offered. Even though the atmosphere interaction lowers the sphere of influence outgoing velocity, the AGA maneuver, regardless of the heating rate constraint, is greatly effective in better exploiting the chances offered by the planetary encounter to control the vehicle momentum variation. Indeed an increase of at least 15.9% in the V_s^+ is gained for the heliocentric velocity maximization case and a reduction of 22.3% for the minimization case.

For this and the following simulations a 4g constraint on the structural load, widely accepted in the literature, is imposed. However, in the runs this constraint is everywhere inactive as the maximum value reached is of 3.61g. As it is never the limiting constraint, more detailed investigation of this parameter will not be performed.

4.2 3D maneuver

This section deals with the V_s^+ maximization by controlling both the lift coefficient ratio and the bank angle. The previous section highlighted that, whenever no banking is allowed, the descent and breakaway phases must be performed through the angle of attack modulation only. The zero bank constraint removal leads to a more efficient maneuver, as far as the $(L/D)^*$ condition can be preserved along the entire atmospheric path. The vehicle is supposed to be trimmed at a constant, unknown angle of attack, whereas the bank angle profile is approximated by cubic splines: the former assumption allows containing the search space dimension whenever a 3D maneuver optimization scenario is dealt with. As expected, the detected optimal trajectory is characterized by the vehicle flying at $\lambda = 1$ for all atmospheric path long; the bank angle modulation is in charge of the descent and breakaway phases: it steers the lift vector out of the local vertical plane as highlighted by the solid line in Fig. 7a. It is worth noting that, apart from these two phases, the lift and the nadir directions are almost aligned and a nearly planar, constant altitude maneuver is flown, as shown in Fig. 8. The 3D maneuver turns out to be more efficient than the 2D; however,

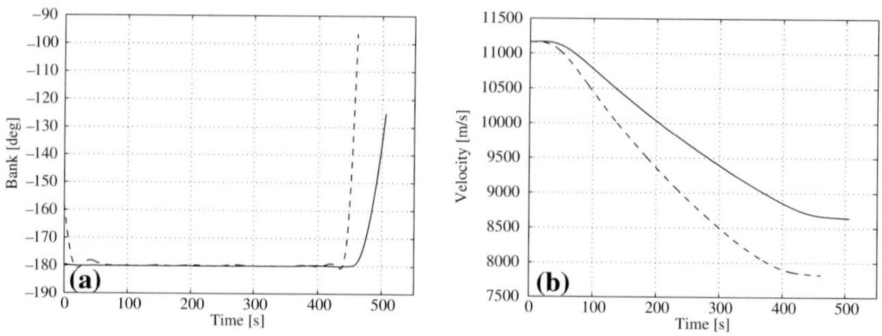

Fig. 7 3D maneuver bank angle profile (**a**) and velocity magnitude (**b**):— AGA_{nhc}, - - AGA_{hc}

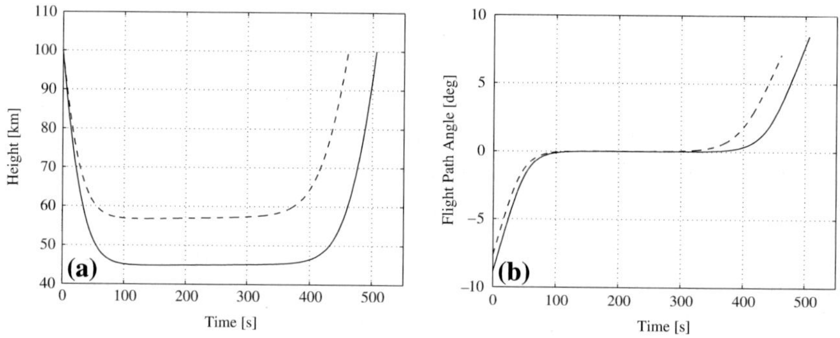

Fig. 8 3D maneuver altitude (**a**) and flight path angle (**b**):— AGA_{nhc}, - - AGA_{hc}

comparison of the first columns of Tables 1, 2 reveals that only a 1.7 m/s drag loss reduction is offered whenever the bank steering is allowed, which translates into a negligible difference on the V_s^+.

Let us now analyze the 3D heating rate constrained trajectories. The not-trimmed 3D maneuver scenario has been analyzed with both the controls represented by cubic splines: the 3D optimal control solution is almost undistinguishable from the 2D case as the null bank angle profile all over the flight path makes the lift vector aligned with nadir vector, and the angle of attack modulates to satisfy the heating rate constraint; whenever a high $m/(SC_l^*)$ vehicle is of concern coupling pitching and banking control could be an useless complication. On that basis a vehicle trimmed at a fixed angle of attack has been considered for the constrained maneuver too. This approach obviously leads to a sub-optimal maneuver in term of Δv gain with respect to the 2D maneuver settings. As a counterpoint it could be appealing for thermal load minimization and vehicle attitude control, due to the typical waveriders slender configuration.

Table 2 3D maneuvers comparison, test case: $V_\infty^- = 10{,}000$ m/s

	AGA_{nhc}	AGA_{hc}	GA
V_s^- [m/s]	23378	23378	23378
V_∞^+ [m/s]	7080.2	6059.1	10000
V_s^+ [m/s]	29992	28809	25453
q_{max} [W/cm^2]	541	250	–
δ^* [deg]	80.00	74.23	12.54

*Overall velocity bending angle

Figures 7, 8 offer, according to the dashed lines, the related simulation results. It is worth noting that the optimization process selects a value of $\lambda = 3.7274$ that leads to an equilibrium altitude of nearly 58 km and the satisfaction of heating constraint. Moreover, as the flight occurs at a constant altitude higher than the 2D maneuver height, a lower thermal load is experienced by the vehicle. The numerical results discussed in this section are resumed in Table 2 and compared with a GA maneuver with a pericenter at 100 km of altitude.

4.3 Heliocentric plane change maneuver

The effectiveness of AGA maneuvers whenever heliocentric plane changes are needed is considered. Firstly AGA planar maneuvers, both heating rate unconstrained and constrained, are analyzed to obtain helicentric orbital inclinations of 6, 10 and 14 deg, respectively, together with V_s^+ maximization and no constraints on the incoming planetary maneuver plane. Obviously in this case the maneuver is not confined in the plane defined by $\mathbf{V}_\infty^-$ and $\mathbf{V}_p$, as in Sect. 4.1, resulting in a lower energetic gain. As the orbital plane change demand is increased, the AGA plane maneuver gets farther from the one spanned by $\mathbf{V}_\infty^-$ and $\mathbf{V}_p$ vectors and the overall bending angle lowers, as indicated by Table 3. The effectiveness of these maneuvers clearly appears whenever compared with a GA characterized by the same $\mathbf{V}_\infty^-$ vector. Indeed a GA allows for a maximum inclination change of 5.1 deg together with a much lower value of $\mathbf{V}_s^+$, as shown in Table 3. As a further example a maneuver asking for a 6 deg plane change with the initial plane constrained to be defined by $\mathbf{V}_\infty^-$ and $\mathbf{V}_p$ is performed. To handle this constraint, modulation of both controls is needed: Fig. 9 reports both the optimal lift coefficient ratio and bank angle histories. The analyzed maneuver, with a bending angle of 72.23 deg, allows acquiring a $\mathbf{V}_s^+ = 28,540$ m/s, definitely greater than the heliocentric velocity obtainable with a simple GA maneuver that can offer a 5.1 deg heliocentric plane change only. It is worth noting that the more the velocity vector decreases, and consequently the centrifugal force, the more the banking effectiveness increases, as the lift surplus can be used to accomplish the plane change.

5 Conclusions

The paper analyzed the AGA maneuvers effectiveness in detail. Some constraints adopted in previous works, such as constant altitude approximation, have been dropped and a more realistic drag–polar curve has been applied together with a more

Table 3 Heliocentric plane change maneuvers comparison. Velocities are given in m/s, heating rates in W/cm^2, and angles in deg. Test case: $V_\infty^- = 10,000$ m/s

	$AGA_{2D,nhc}$			$AGA_{2D,hc}$			GA
inc	6	10	14	6	10	14	5.1
V_s^-	23378	23378	23378	23378	23378	23378	23378
V_∞^+	7220.8	7325.4	7450.6	6694.5	6786.0	7064.1	10000
V_s^+	29154	27921	25846	28578	27245	24729	23393
q_{max}	541.87	543.30	543.06	250	250	250	–
δ*	76.29	73.56	70.33	75.87	72.95	69.57	12.54

*Overall velocity bending angle

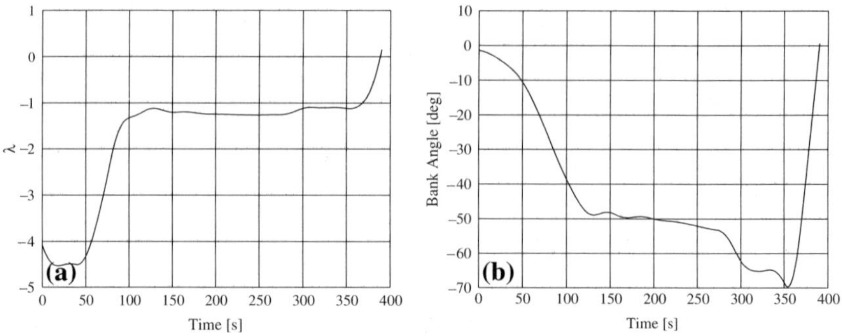

Fig. 9 $AGA_{3D,\,nhc}$ λ (**a**) and σ (**b**) profiles

credible $m/(SC_1^*)$ value. AGA maneuvers have been demonstrated to be an effective tool to augment the momentum exchange a GA can supply, allowing heliocentric velocity maximization, minimization and heliocentric plane changes.

Simulations show that the angle of attack control is more effective than the banking modulation as only small improvements are acquired according to drag losses reduction. The strategy of trimming the vehicle at a fixed angle of attack leads to sub-optimal solutions from the Δv point of view but it is still appealing if the vehicle attitude control and thermal load minimization are of concern.

The multiple shooting algorithm has been demonstrated to be a simple, flexible and effective approach to trajectory optimization within a frame of several constraint typologies and genetic algorithm has shown its capability to provide for good initial guesses.

Even though heat constrained trajectories have been carried out, heat fluxes still represent a great technological challenge, due to high vehicle atmospheric velocities coupled with the vehicle sharp leading edges aerodynamic requirement. The role played by uncertainties on dynamic, aerodynamic, and environment modeling must be addressed to better understand AGA maneuver feasibility.

Acknowledgement The authors would like to acknowledge the funding support of Carlo Gavazzi Space SpA.

References

Anderson, J.D.: Hypersonic and High Temperature Gas Dynamics. McGraw-Hill, New York (1989)

Anderson, J.D., Ferguson, F., Lewis, M.J.: Hypersonic waverider for high altitude applications. AIAA Paper 91–0530 (1991a)

Anderson, J.D., Lewis, M.J., Kothari, A.P., Corda, S.: Hypersonic waveriders for planetary atmospheres. J. Spacecraft **28**, 401–410 (1991b)

Betts, J.T.: Survey of numerical methods for trajectory optimization. J. Guid. Contr. Dyn. **21**, 193–207 (1998)

Bonfiglio, E.P., Longusky, J.M., Vinh, N.X.: Automated design of aerogravity-assist trajectories. J. Spacecraft Rockets **34**, 768–775 (2000)

Bowcutt, K.G., Anderson, J.D., Capriotti, D.: Viscous optimized hypersonic waveriders. AIAA Paper 87–0272 (1987)

Corda, S., Anderson, J.D.: Viscous optimized hypersonic waveriders designed from axisimmetric flow fields. AIAA Paper 88–0369 (1988)

Enright, P.J., Conway, B.A.: Discrete approximation to optimal trajectories using direct transcription and nonlinear programming. J. Guid. Contr. Dyn. **15**, 994–1001 (1992)

Gillum, M.J., Lewis, M.J.: Experimental results on a Mach 14 waverider with blunt leading edges. J. Aircraft **34**, 296–303(1997)

Johnson, W.R.: Optimization of Atmospheric Flythrough for Aero-gravity Assist Trajectories. AAE508 Report (2002)

Lewis, M.J., McRonald, A.D.: Design of hypersonic waveriders for aeroassisted interplanetary trajectories. J. Spacecraft Rockets **29**, 653–660 (1992)

Lohar, F.A., Mateescu, D., Misra A.K.: Optimal atmospheric trajectory for aero-gravity assist. Acta astronautica **32**, 89–96 (1994)

Lohar, F.A., Misra, A.K., Mateescu, D.: Optimal atmospheric trajectory for aerogravity assist with heat constraint. J. Guid. Contr. Dyn. **18**, 723–730 (1995)

Lohar, F.A., Mateescu, D., Misra, A.K.: Optimal atmospheric trajectory with heliocentric plane change. Acta astronautica **38**, 445–456 (1996)

Lohar, F.A., Misra, A.K., Mateescu, D.: Mars–jupiter aerogravity assist trajectories for high-energy transfer. J. Spacecraft Rockets **34**, 16–21(1997)

McRonald, A.D., Randolph J.E.: Hypersonic maneuvering for augmenting planetary gravity assist. J. Spacecraft Rockets **29**, 216–222(1992)

Starkey, R.P., Lewis, M.J.: Simple analytical model for parametric studies of hypersonic waveriders. J. Spacecraft Rockets **36**, 516–523 (1999)

Starkey R.P., Lewis, M.J.: Analytical off-design lift-to-drag-ratio analysis for hypersonic waveriders. J. Spacecraft Rockets **37**, 684–691 (2000)

Stoer, J. Bulirsch, R.: Introduction to Numerical Analysis. Springer, New York (1993)

Tauber, M.E., Sutton, K.: Stagnation-point radiative heating relations for earth and mars entries. J. Spacecraft Rockets **28**, 40–42 (1991)

Celestial Mechanics and Dynamical Astronomy (2006) 95:407–423
DOI 10.1007/s10569-006-9029-6

ORIGINAL ARTICLE

Analytical theory of a lunar artificial satellite with third body perturbations

Bernard De Saedeleer

Received: 15 November 2005 / Revised: 23 March 2006 /
Accepted: 7 May 2006 / Published online: 15 August 2006

Abstract We present here the first numerical results of our analytical theory of an artificial satellite of the Moon. The perturbation method used is the Lie Transform for averaging the Hamiltonian of the problem, in canonical variables: short-period terms (linked to l, the mean anomaly) are eliminated first. We achieved a quite complete averaged model with the main four perturbations, which are: the synchronous rotation of the Moon (rate $n_☾$), the oblateness J_2 of the Moon, the triaxiality C_{22} of the Moon ($C_{22} \approx J_2/10$) and the major third body effect of the Earth (ELP2000). The solution is developed in powers of small factors linked to these perturbations up to second-order; the initial perturbations being sorted ($n_☾$ is first-order while the others are second-order). The results are obtained in a closed form, without any series developments in eccentricity nor inclination, so the solution apply for a wide range of values. Numerical integrations are performed in order to validate our analytical theory. The effect of each perturbation is presented progressively and separately as far as possible, in order to achieve a better understanding of the underlying mechanisms. We also highlight the important fact that it is necessary to adapt the initial conditions from averaged to osculating values in order to validate our averaged model dedicated to mission analysis purposes.

Keywords Lunar artificial satellite · Third body · Lie · Hamiltonian · C_{22} · Earth

1 Introduction

We reached a corner stone in the development of our analytical theory of a lunar artificial satellite. For the first time, we achieved a complete averaged model with the main four perturbations, which are: the synchronous rotation of the Moon (rate $n_☾$),

B. De Saedeleer (✉)
Département de Mathématique, University of Namur, Rempart de la Vierge 8,
B-5000 Namur, Belgium
e-mail: Bernard.DeSaedeleer@fundp.ac.be

the oblateness J_2 of the Moon, the triaxiality C_{22} of the Moon ($C_{22} \approx J_2/10$) and the major third body effect of the Earth (ELP2000). Our goal is to build an averaged model for mission analysis purposes, and not to make any orbit determination.

In some previous paper (De Saedeleer 2004), we developed the perturbations in J_2 and C_{22}, and averaged them up to order J_2^2, C_{22}^2 and $J_2 \times C_{22}$. In another one (De Saedeleer and Henrard 2005), we detailed the development of the third body (Earth) perturbation by making use of the lunar theory ELP2000 (Chapront-Touzé and Chapront 1991).

Now, in this paper, we present our latest new results: the averaging of that third body perturbation and hence the building of a quite complete averaged model. Moreover, we present also here the first numerical integrations which come along with that averaged model, and which validate our analytical theory.

The perturbation method used is the Lie Transform for averaging the Hamiltonian of the problem, in canonical variables: short-period terms (linked to l, the mean anomaly) are eliminated first. The solution is developed in powers of small factors linked to these perturbations. The initial perturbations are sorted in such a way that $n_{☾}$ is first-order while the others are second-order. The averaging process is done up to second-order, which then means that the first-order effect of the perturbations is in fact captured.

Of course, the determination of the motion of a lunar satellite has already drawn some attention in the past (Oesterwinter 1970; Milani and Knežević 1995; Steichen, 1998a, b). So, we could extensively cross-check some of our results with the literature (see Sect. 3), but we also have gone a step further in the understanding of the dynamics.

It turns out that the problem of the lunar orbiter is quite interesting because its dynamics is different from the one of an artificial satellite of the Earth, by at least two aspects: the C_{22} lunar gravity term is only 1/10 of the J_2 term and the third body effect of the Earth on the lunar satellite is much larger than the effect of the Moon on a terrestrial satellite. So we have to account at least for these larger perturbations.

Our goal is not to go to very high order in J_2, nor to add many harmonics, while it could be done easily in principle, for example by addressing the complete zonal problem (De Saedeleer 2005); we rather want to highlight the main parameters affecting the dynamics, hence we deliberately choose to restrict the study to the aforementioned four main perturbations.

The structure of this paper is as follows. The geometry, variables and perturbations are described in Sect. 2; the averaged Hamiltonian is given in Sect. 3; the numerical integrations are introduced in Sect. 4; the effect of J_2 is adressed in Sect. 5; the additional effect of C_{22} and of ($n_{☾}$+ the Earth) is discussed in Sects. 6 and 7, respectively; the adaptation of the initial conditions from averaged to osculating values is discussed in Sect. 8 (with a detailed example given in Appendix); we then conclude in Sect. 9.

2 Geometry, variables and perturbations

We use here the canonical method of the Lie Transform (Deprit 1969). In order to keep the Hamiltonian formalism, it is required to work in canonical variables; we choose the classical Delaunay variables (l, g, h, L, G, H) defined as:

$$l = u - e\sin u, \qquad L = \sqrt{\mu a},$$
$$g = \omega, \qquad G = \sqrt{\mu a(1-e^2)}, \tag{1}$$
$$h = \Omega, \qquad H = \sqrt{\mu a(1-e^2)}\cos I,$$

where $(a, e, I, \omega, \Omega)$ are the keplerian elements, $\mu = GM_{☾}$, l and u are the mean and eccentric anomaly, respectively. In these variables, the unperturbed potential is simply written $-\mu^2/(2L^2)$.

Now we have to write all the perturbations in these variables and in an inertial frame; that is to say with respect to a constant direction in space. The inertial frame (x, y, z) is chosen so that its origin is taken at the center of the Moon and so that the (x, y) plane is the lunar equatorial plane (see Fig. 1).

In order to be able to use the expressions of the spherical harmonics for the potential, we first have to define spherical coordinates (r, λ', ϕ), so that the longitude of the satellite λ' starts from the x axis in the equatorial plane, the latitude ϕ being defined as the deviation from the (x, y) plane. Within that inertial frame, the perturbative potentials in J_2 and C_{22} may be written $(V_{20} - V_{22})$, with:

$$V_{20} = \frac{\mu}{r}\left(\frac{R}{r}\right)^2 J_2 P_{20}(\sin\phi), \quad \text{where } P_{20}(x) = \frac{1}{2}(3x^2 - 1), \tag{2}$$

$$V_{22} = \frac{\mu}{r}\left(\frac{R}{r}\right)^2 C_{22} P_{22}(\sin\phi)\cos(2(\lambda' - \lambda_{22})), \quad \text{where } P_{22}(x) = 3(1 - x^2), \tag{3}$$

where R is the equatorial radius of the Moon ($R \approx 1{,}738$ km); P_{20} and P_{22} being the Legendre Associated Functions. We can partially translate their argument $(\sin\phi)$ into Delaunay variables by the way of the spherical trigonometry (see Fig. 1, where the plane of the orbit is at an inclination I): $\sin\phi = \sin I \sin(f + g)$. We then have:

$$V_{20} = +J_2R^2(\mu r^{-3})\left(1 - 3c^2 - 3s^2\cos(2f + 2g)\right)/4. \tag{4}$$

But the coefficient C_{22} makes the longitude λ' to appear in addition to the latitude ϕ. The spherical harmonics being defined with respect to the main axis of inertia of the attracting body, we had to define λ_{22} as the longitude of the lunar longest meridian (minimum inertia). This angle makes the Hamiltonian to be time-dependent, since $\lambda_{22} = \lambda_{\oplus}$ travels at the rate of the synchronous rotation which is $\dot{\lambda}_{\oplus} = n_{☾}$. In order to

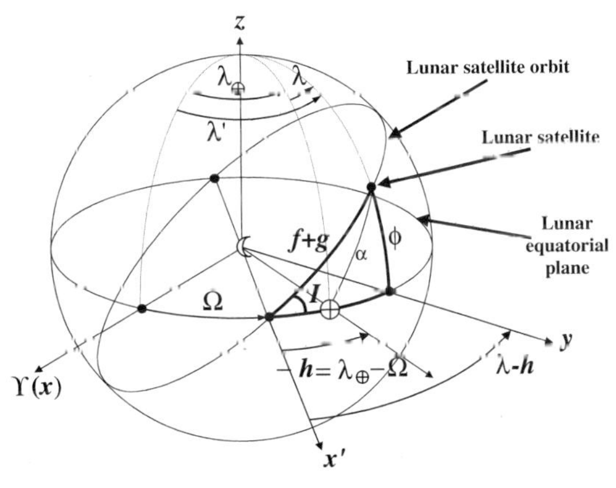

Fig. 1 Simplified selenocentric sphere. The center of the Moon is taken as the origin; the lunar equatorial plane is taken as the (x, y) plane and $\lambda_{\oplus}$ is the longitude of the Earth

eliminate this dependency, we will work in a rotating system whose x' axis now passes through the Earth; we then define new longitudes with respect to $\lambda_\oplus$: $\lambda = \lambda' - \lambda_\oplus$ and we redefine also $h = \Omega - \lambda_\oplus$ (the angles always appear in that combination). A term $(-n_☾ H)$ has to be added to the Hamiltonian in order to take this rotation into account.

With that definition of λ, we have also now $V_{22} = +C_{22}R^2\mu r^{-3}P_{22}(\sin\phi)\cos(2\lambda)$. Once again, the factor $\cos(2\lambda)$ can be partially translated into Delaunay variables by the same way of the spherical trigonometry, which gives finally:

$$\begin{aligned} V_{22} &= +C_{22}R^2(\mu r^{-3})3\{2s^2\cos(2h) + (c+1)^2\cos(2f+2g+2h) \\ &\quad + (c-1)^2\cos(2f+2g-2h)\}/4. \end{aligned} \tag{5}$$

At this stage, there remains in (4) and (5) only r and f to be expressed as a function of (l,g,h) in order to be able to apply a perturbation method. It turns out that the functions $r = r(l,g,h)$ and $f = f(l,g,h)$ cannot be expressed in a closed form, and that one usually falls back at this point into series development in the eccentricity. We would like to avoid this, at least for the following reasons: the results would be much less compact, hence a lack of ease to interpret the results; moreover they would no longer be valid for higher values of the eccentricity.

So we prefer to use the following set of auxiliary variables $(\xi, f, g, h, a, n, e, \eta, s, c)$ in order to describe the position of the lunar satellite:

$$\begin{aligned} &\xi = \tfrac{a}{r} = \frac{1+e\cos f}{1-e^2} = \frac{1}{1-e\cos u}, && f, \\ &a = \frac{L^2}{\mu}, && n = \frac{\mu^2}{L^3}, \\ &e = \sqrt{1-\left(\frac{G}{L}\right)^2} && \eta = \sqrt{1-e^2} = \frac{G}{L}, \\ &s = \sin I = \sqrt{1-\left(\frac{H}{G}\right)^2} && c = \cos I = \frac{H}{G}, \\ &g && h, \end{aligned} \tag{6}$$

where f is the true anomaly.

This set has a major advantage: it leads to formulae in closed form with respect to the eccentricity and inclination. The only two drawbacks are (1) that it is redundant $(e^2+\eta^2=1; c^2+s^2=1)$ and (2) that we need to perform partial derivatives of them with respect to the canonical variables; but it is not too heavy a task. We have for example: $\frac{dA}{dl} = \frac{\partial A}{\partial l} + \frac{\partial A}{\partial \xi}\frac{\partial \xi}{\partial l} + \frac{\partial A}{\partial f}\frac{\partial f}{\partial l}$. This choice of variables and all the partial derivatives of them with respect to the canonical variables (l,g,h,L,G,H) have already been described in De Saedeleer and Henrard (2005). The computation of the partial derivatives themselves requires some caution, but is not too complicated; use has to be made of the Kepler equation $(l = E - e\sin E)$, which links the anomalies. We have for example $\partial f/\partial G = -\sin f(1+\xi\eta^2)/(\eta n a^2 e)$ and also $\partial f/\partial l = \xi^2\eta$, a quantity which plays an important role, since it will allow to switch the integration from l to f.

We can then rewrite the complete Hamiltonian in this set of variables (6). Note that the factor (μr^{-3}) appearing in (4) and (5) is simply written $\xi^3 n^2$. The unperturbed potential is $\mathcal{H}_0^{(0)} = -\mu^2L^{-2}/2$, while we sort the four perturbations by their order of magnitude. The mean motion of the Moon $n_☾$ is about 0.23 rad/day ($n_☾ = 2\pi/T_☾$ with the sidereal rotation period of the Moon $T_☾ \approx 27.32$ days). For a typical lunar orbit (altitude around 500 km), the lunar satellite has a period of 2.64 hours; n is then about 57.12 rad/day. If we choose that frequency as a unit, $n_☾$ is about 4×10^{-3}, while the J_2 and the C_{22} terms are of order 10^{-4} and 10^{-5}, respectively. Additionally, we have the relationship $\gamma = -\mu_\oplus a_\oplus^{-3} = -n_☾^2 M_\oplus/(M_\oplus + M_☾) \approx -n_☾^2$, so that γ is indeed quite very exactly of second-order with respect to $n_☾$.

In summary, we may put the biggest perturbation ($n_☾$) at first-order, and the other lower ones all at second-order, which gives the following final arrangement:

$$\mathcal{H}^{(0)} = \mathcal{H}_0^{(0)} + \mathcal{H}_1^{(0)} + \epsilon\, \mathcal{H}_2^{(0)} + \delta\, \mathcal{HB}_2^{(0)} + \gamma\, \mathcal{HE}_2^{(0)} \tag{7}$$

with:

$$\mathcal{H}_1^{(0)} = -n_☾ H, \tag{8}$$

$$\mathcal{H}_2^{(0)} = \xi^3 n^2 \left(1 - 3c^2 - 3s^2 \cos(2f + 2g)\right)/4, \tag{9}$$

$$\begin{aligned}\mathcal{HB}_2^{(0)} = {} & 3\xi^3 n^2 \{2s^2 \cos(2h) + (c+1)^2 \cos(2f + 2g + 2h) \\ & + (c-1)^2 \cos(2f + 2g - 2h)\}/4,\end{aligned} \tag{10}$$

$$\mathcal{HE}_2^{(0)} = a_\oplus^3 r_\oplus^{-3} a^2 \xi^{-2} P_{20}(\cos\alpha) \tag{11}$$

and with $\epsilon = J_2 R^2, \delta = -C_{22} R^2, \gamma = -\mu_\oplus a_\oplus^{-3}$, and where α is the angle between the Earth and the lunar satellite.

The computation of $\cos\alpha = \vec{r} \cdot \vec{r}_\oplus/(r r_\oplus)$ requires the knowledge of the direction of the Earth from the Moon $\vec{A}_\oplus = (A_\oplus, B_\oplus, C_\oplus)$. For this, we use the lunar theory ELP2000 (Chapront-Touzé and Chapront 1991), which gives the opposite direction, in spherical coordinates. In that theory, the position of the Moon is described by a series of periodic functions mainly of the fundamental arguments L^*, D, l', l^*, F; from which we take the leading terms. Let's recall that L^* is the secular part of the mean longitude of the Moon referred to the mean dynamical ecliptic and equinox of date, D is the secular part of the difference between the mean longitude of the Moon and the geocentric mean longitude of the Sun, l' is the secular part of the geocentric mean anomaly of the Sun, l^* is the secular part of the mean anomaly of the Moon, F is the secular part of the difference between the mean longitude of the Moon and of the longitude of its ascending node on the mean ecliptic of date.

As already mentioned, the development of these perturbations have already been described elsewhere in deeper details (see De Saedeleer 2004 for ϵ and δ, and De Saedeleer and Henrard 2005 for γ and $n_☾$). We just give here in Table 1 the very first terms (|Coefficient| > 0.1) of the second-order perturbations.

In that table, we immediately recognize (9) and (10), while we can rewrite the part corresponding to (11) in full:

$$\begin{aligned}\mathcal{HE}_2^{(0)} = {} & a^2 \xi^{-2} \Big[-0.12466(1 - 3c^2) + 0.37225\, s^2 \cos(2(h - L^*)) \\ & + 0.37397\, s^2 \cos(2(f+g)) + 0.18612\Big\{(1+c)^2 \cos(2(f + g + h - L^*)) \\ & + (1-c)^2 \cos(2(f + g - h + L^*))\Big\}\Big].\end{aligned} \tag{12}$$

We use then the Lie Transform (Deprit 1969) as canonical perturbation method, with the four parameters $(n_☾, \epsilon, \delta, \gamma)$, all gathered in the Lie triangle (see Fig. 2), which is filled by the recursive formula $\mathcal{H}_i^{(j)} = \mathcal{H}_{i+1}^{(j-1)} + \sum_{k=0}^{i} C_i^k \left(\mathcal{H}_{i-k}^{(j-1)}; \mathcal{W}_{k+1}\right)$; note

Table 1 The $\epsilon\, \mathcal{H}_0^{(2)} + \delta\, \mathcal{HB}_0^{(2)} + \gamma\, \mathcal{HE}_0^{(2)}$ series (12 terms)

	f	g	h	L^*	ξ	a	n	c	s	δ	ϵ	γ	$1+c$	$1-c$	Coefficient
cos	0	0	0	0	3	0	2	0	0	0	1	0	0	0	0.25000D+00
cos	0	0	0	0	3	0	2	2	0	0	1	0	0	0	−0.75000D+00
cos	2	2	0	0	3	0	2	0	2	0	1	0	0	0	−0.75000D+00
cos	0	0	0	0	−2	2	0	0	0	0	0	1	0	0	−0.12466D+00
cos	0	0	0	0	−2	2	0	2	0	0	0	1	0	0	0.37398D+00
cos	0	0	2	−2	−2	2	0	0	2	0	0	1	0	0	0.37225D+00
cos	2	2	0	0	−2	2	0	0	2	0	0	1	0	0	0.37397D+00
cos	2	2	2	−2	−2	2	0	0	0	0	0	1	2	0	0.18612D+00
cos	2	2	−2	2	−2	2	0	0	0	0	0	1	0	2	0.18612D+00
cos	0	0	2	0	3	0	2	0	2	1	0	0	0	0	0.15000D+01
cos	2	2	2	0	3	0	2	0	0	1	0	0	2	0	0.75000D+00
cos	2	2	−2	0	3	0	2	0	0	1	0	0	0	2	0.75000D+00

$$
\begin{array}{llll}
\mathcal{H}_0^{(0)} = -\frac{\mu^2}{2L^2} & & & \\
\mathcal{H}_1^{(0)} = -n_{☾} H & \mathcal{H}_0^{(1)} & & \\
\epsilon\, \mathcal{H}_2^{(0)} + \delta\, \mathcal{HB}_2^{(0)} + \gamma\, \mathcal{HE}_2^{(0)} & \mathcal{H}_1^{(1)} & \mathcal{H}_0^{(2)} & \\
0 & \mathcal{H}_2^{(1)} & \mathcal{H}_1^{(2)} & \mathcal{H}_0^{(3)} \\
\vdots & \vdots & \vdots & \vdots \quad \ddots
\end{array}
$$

Fig. 2 Our specific Lie triangle, with the first ($n_{☾}$) and second (ϵ, δ, γ) order perturbations

that an appropriate scaling is done to the perturbations in order to fulfill the scheme $\mathcal{H} = \sum_{i \geq 0} \frac{\epsilon^i}{i!} \mathcal{H}_i^{(0)}$. We write $\mathcal{H}_0^{(i)}$ as $\overline{\mathcal{H}}_0^{(i)}$ in order to remember that the fast angle l has been eliminated; we always put the periodic part in the generator $\mathcal{W}_i$.

3 Averaged Hamiltonian and symbolic manipulation software MM

In order to make the symbolic computations of the averaged theory, we used a specific FORTRAN code called the MM, standing for "Moon's series Manipulator", which has been developed at our university, and which is dedicated to algebraic manipulations. In this tool, each expression is given by a series of linear trigonometric functions, with polynomial coefficients. The property of linearity will make the integrations very straightforward. An example of such a series has been given in Table 1. The computations are done in double precision but we display only five digits, which is sufficient for the purposes of this article.

It is of course impossible to give a comprehensive view of all the results in the scope of this paper, since the series may contain a lot of terms, but we give however explicitly some of them here, and we comment the others. For the first-order, as $\mathcal{H}_1^{(0)}$ is already independent of l, we have $\overline{\mathcal{H}}_0^{(1)} = \mathcal{H}_1^{(0)} = -n_{☾} H$ and $\mathcal{W}_1 = 0$. For the second-order,

Table 2 The $\epsilon\overline{\mathcal{H}}_0^{\omega} = \overline{\mathcal{H}}_0^{(1)} + \epsilon\,\mathcal{B}_0^{(2)} + \delta\,\overline{\mathcal{HB}}_0^{(2)} + \gamma\,\overline{\mathcal{HE}}_0^{(2)}$ series (13 terms)

	$\bar{g}$	$\bar{h}$	L^*	$\bar{a}$	$\bar{n}$	$\bar{e}$	$\bar{\eta}$	$\bar{c}$	$\bar{s}$	δ	$n_{☾}$	ϵ	γ	$1+\bar{c}$	$\bar{H}$	$1-\bar{c}$	Coefficient
cos	0	0	0	0	0	0	0	0	0	0	1	0	0	0	1	0	−0.10000D + 01
cos	0	0	0	0	2	0	−3	0	0	0	0	1	0	0	0	0	0.25000D + 00
cos	0	0	0	0	2	0	−3	2	0	0	0	1	0	0	0	0	−0.75000D + 00
cos	0	0	0	2	0	0	0	0	0	0	0	0	1	0	0	0	−0.12466D + 00
cos	0	0	0	2	0	0	0	2	0	0	0	0	1	0	0	0	0.37398D + 00
cos	0	0	0	2	0	2	0	0	0	0	0	0	1	0	0	0	−0.18699D + 00
cos	0	0	0	2	0	2	0	2	0	0	0	0	1	0	0	0	0.56096D + 00
cos	0	2	−2	2	0	0	0	0	2	0	0	0	1	0	0	0	0.37225D + 00
cos	0	2	−2	2	0	2	0	0	2	0	0	0	1	0	0	0	0.55837D + 00
cos	2	0	0	2	0	2	0	0	2	0	0	0	1	0	0	0	0.93493D + 00
cos	2	2	−2	2	0	2	0	0	0	0	0	0	1	2	0	0	0.46531D + 00
cos	2	−2	2	2	0	2	0	0	0	0	0	0	1	0	0	2	0.46531D + 00
cos	0	2	0	0	2	0	−3	0	2	1	0	0	0	0	0	0	0.15000D + 01

we have:

$$\epsilon\mathcal{H}_0^{(2)} + \delta\mathcal{H}\mathcal{B}_0^{(2)} + \gamma\mathcal{H}\mathcal{E}_0^{(2)} = \epsilon\mathcal{H}_2^{(0)} + \delta\mathcal{H}\mathcal{B}_2^{(0)} + \gamma\mathcal{H}\mathcal{E}_2^{(0)} + \left(\mathcal{H}_0^{(0)};\mathcal{W}_2\right) \tag{13}$$

and we choose:

$$\epsilon\overline{\mathcal{H}}_0^{(2)} + \delta\overline{\mathcal{HB}}_0^{(2)} + \gamma\overline{\mathcal{HE}}_0^{(2)} = \frac{1}{2\pi}\int_0^{2\pi}\left(\epsilon\mathcal{H}_2^{(0)} + \delta\mathcal{H}\mathcal{B}_2^{(0)} + \gamma\mathcal{H}\mathcal{E}_2^{(0)}\right)\mathrm{d}l \tag{14}$$

while $(\mathcal{H}_0^{(0)};\mathcal{W}_2)$ reduces to $n\frac{\partial\mathcal{W}_2}{\partial l}$, which has then to be integrated with respect to l.

The integration of the terms in (ϵ,δ,γ), which is in fact rather a first-order averaging, may be performed in closed form quite easily by using techniques described in De Saedeleer (2004) for (ϵ,δ) using ξ and f, and in Jefferys (1971) for γ, which uses additionally u. Higher orders may be achieved by the same way, provided we are able to compute the integrals. The third-order would contain the combinations of perturbation parameters $(\epsilon n_{☾},\delta n_{☾},\gamma n_{☾})$ and the fourth-order $(\epsilon^2,\delta^2,\gamma^2,\epsilon\delta,\epsilon\gamma,\delta\gamma,\epsilon n_{☾}^2,\delta n_{☾}^2,\gamma n_{☾}^2)$.

In this article, we mainly focus on the first-order effects (ϵ,δ,γ), while some higher order effects (like $\epsilon^2,\delta^2,\epsilon\delta$) have already been described in De Saedeleer (2004). The second-order averaged Hamiltonian (in ϵ,δ,γ) is given in Table 2, from which we can derive the averaged equations of motion.

It can be rewritten in full as follows:

$$\begin{aligned}
\overline{\mathcal{H}} \;=\; & \overline{\mathcal{H}}_0^{(1)} + \epsilon\overline{\mathcal{H}}_0^{(2)} + \delta\overline{\mathcal{HB}}_0^{(2)} + \gamma\overline{\mathcal{HE}}_0^{(2)} = -n_{☾}\bar{H} + \epsilon\frac{\bar{n}^2}{4\bar{\eta}^3}(1-3\bar{c}^2) \\
& + \delta\frac{3\bar{n}^2}{2\bar{\eta}^3}\left(\bar{s}^2\cos(2\bar{h})\right) + \gamma\bar{a}^2\Big[(1-3\bar{c}^2)(-0.12466-0.18699\bar{e}^2) \\
& + 0.37225\,\bar{s}^2\cos(2(\bar{h}-\bar{L}^*)) + 0.55837\bar{s}^2\bar{e}^2\cos(2(\bar{h}-\bar{L}^*)) + 0.93493\,\bar{s}^2\bar{e}^2\cos(2\bar{g}) \\
& + 0.46531\,\bar{e}^2\left\{(1+\bar{c})^2\cos(2(\bar{g}+\bar{h}-\bar{L}^*)) + (1-\bar{c})^2\cos(2(\bar{g}-\bar{h}+\bar{L}^*))\right\}\Big].
\end{aligned} \tag{15}$$

Several validations have been carried out in some previous papers, mainly the effect of J_2. It is not the purpose of this paper to give these validations in details, but we give here however an overview. For the first-order (ϵ): the averaged Hamiltonian and the generator are the same as the results of Brouwer (1959). For the second-order (ϵ^2): the averaged Hamiltonian is the same as (Brouwer 1959), while the generator is equivalent to the generator S_2 given by Eq. 3.2 of Kozai (1962), as it has been shown in De Saedeleer and Henrard (2005), which uses the relationships of Shniad (1970) for the correspondence between generators of von Zeipel (S_i) and the ones of Lie ($\mathcal{W}_i$). We just remind here the expression of the averaged Hamiltonian in ϵ^2:

$$\epsilon^2 \overline{\mathcal{H}}_0^{(4)} = \frac{\epsilon^2 3n^2}{128a^2\eta^7}\Big[5(s^4 - 8c^4) - 4\eta(1-3c^2)^2 - \eta^2(5s^4 - 8c^2) - 2e^2s^2(1-15c^2)\cos(2g)\Big]. \quad (16)$$

4 Numerical integrations

In the following sections, we will investigate numerically the several effects gradually in order to see more clearly the effect of each additional perturbation: ϵ alone in Sect. 5, $(\epsilon+\delta)$ in Sect. 6, $(\epsilon+\delta+n_☾)$ and $(\epsilon+\delta+n_☾+\gamma)$ in Sect. 7. The averaged equations of motion deduced from (15) were integrated numerically; an improved version of the Burlish–Stoer subroutine (Press et al. 1986) has been used. The following set of numerical values for the averaged initial conditions has been chosen:

$$\bar{l}_0 = 10\,\text{rad}, \quad \bar{g}_0 = 1\,\text{rad}, \quad \bar{h}_0 = 2\,\text{rad}, \quad \bar{a}_0 = 3{,}000\,\text{km}, \quad \bar{e}_0 = 0.2, \quad \bar{i}_0 = 30\,\text{deg}. \quad (17)$$

We also took $\mu_☾ = 3.66 \times 10^{13}\,\text{km}^3/\text{day}^2$ and $\mu_\oplus = 81.3\mu_☾$; the period of the satellite is about 4.1 hours for $a = 3{,}000\,\text{km}$. For the perturbation parameters, we took: $\epsilon = 613.573\,\text{km}^2$; $\delta = -67.496\,\text{km}^2$; $n_☾ = 0.230\,\text{rad/day}$; $\gamma = -0.05214\,\text{rad/day}^2$. We can easily select an isolated effect by putting the other parameters to zero.

5 Effect of J_2 alone

The effect of J_2 alone (first- and second-order) is shown in Fig. 3. At first order, (a, e, i) remain constant while the angles g and h do precess, with periods of approximately 3 and 5 years, respectively. These rates are consistent with the two well-known classical formula, given, i.e. in Szebehely (1989); Roy (1968); Jupp (1988):

$$\dot{\omega} = (3n/2)J_2(R/p)^2(2-(5/2)\sin^2 i), \quad (18)$$

$$\dot{\Omega} = (-3n/2)J_2(R/p)^2\cos i, \quad (19)$$

with $p = a(1-e^2)$. The associated peculiar value of the inclination which makes $\dot{\omega}$ to vanish, known as the *critical inclination* $I_c = 63°26'$, is quite famous (Szebehely 1989). Note that the rate of precession of the elements of the orbit of a lunar satellite is much lower than in the case of artificial satellites of the Earth, since the J_2 of the Moon is lower.

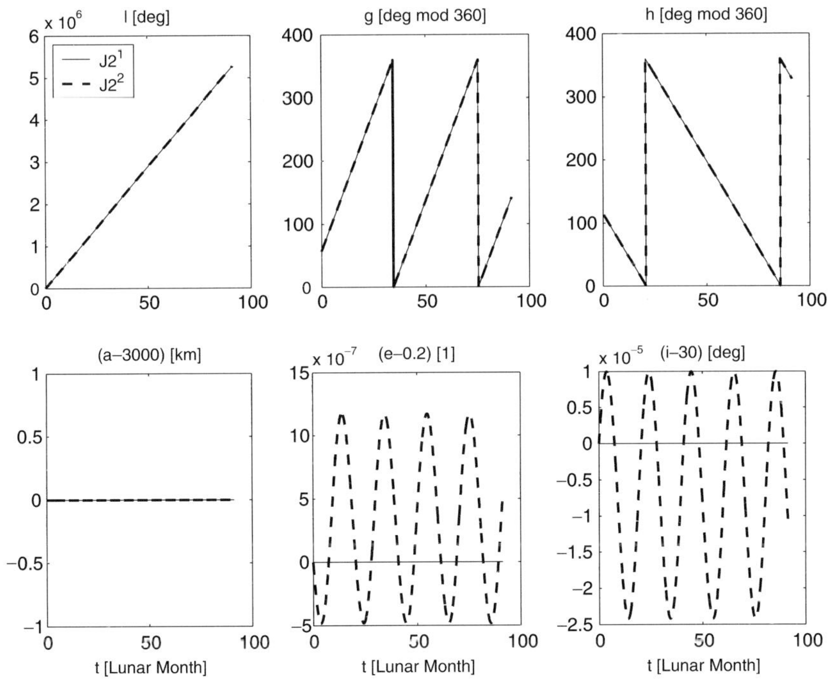

Fig. 3 The effect of J_2 alone ($\epsilon \neq 0, \delta = 0 = n_☾ = \gamma$): integration of the averaged models (15) and (16) for the first- and second-order effect, respectively; in both cases the initial conditions are (17)

At second-order in J_2 (integration of the averaged equations of motion deduced from (16)), e and i start to oscillate, since the averaged Hamiltonian contains a factor like $\cos(2g)$, hence $\dot{G} \neq 0$. The period is then half of 3 years, (around 18 lunar monthes, well noticeable in Fig. 3), but the amplitude of the oscillations are however, small: 1.654×10^{-6} for e, and 3.420×10^{-5} deg for i.

6 Combined effect of J_2 and C_{22}

We come back to the first-order in J_2 now, where (a, e, i) were constant. If we add the perturbation in C_{22}, the angle h enters the game, by a factor like $\cos(2h)$ this time, so that now $\dot{H} \neq 0$, hence i start to oscillate (but still not e, since $\dot{G} = 0$). Now the amplitudes are very significant, since it is a first-order effect; in our numerical example (plotted in Fig. 4), i oscillates roughly from 29 to 37 deg. The period is half of 5 years (around 35 lunar monthes, well noticeable in Fig. 4).

The introduction of C_{22} has another consequence: it modifies quite significantly the classical critical inclination $I_c = 63°26'$ to new critical inclinations I_c^*, as has been shown in De Saedeleer and Henrard (2006). In the case of the Moon, I_c^* may lie in the range 58–72 deg.

7 Additional effect of $n_☾$ and of the Earth

The effect of $n_☾$ and of the Earth is shown in Fig. 5. Let's first look at the dashed curves, labelled "without Earth". This case corresponds to the effect of the perturbations

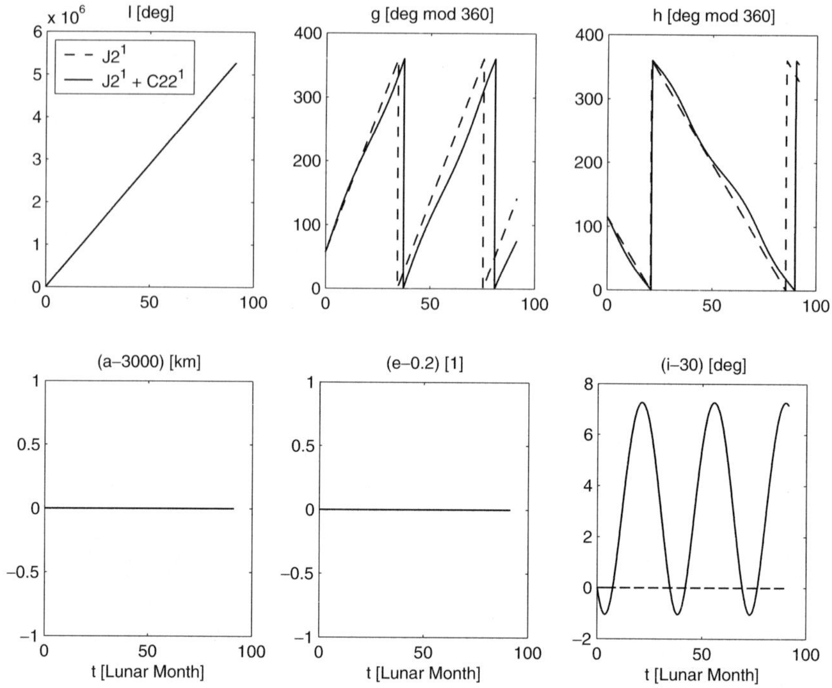

Fig. 4 The combined effect of J_2 and C_{22} ($\epsilon \neq 0, \delta \neq 0, n_{☾} = 0 = \gamma$): integration of the averaged model (15) for the first-order effects; the initial conditions are (17)

($J_2 + C_{22} + n_{☾}$). So, in a first step, only $n_{☾}$ has been added with respect to Sect. 6: the consequence is that the angle h now rotates more quickly: the period is the month (the synchronous rotation) instead of 5 years, hence the inclination also vary, now with a half-month period; the amplitude is quite small: about 0.05 deg.

We then added the effect of the Earth, by considering in a first approximation only a few terms of the third body perturbation (see Table 1). We see that the inclination is now modulated by a period of about 1.2 years with larger amplitude (0.5 deg), coming from a factor like $\sin(2g)$ in $\dot{G}$, with a period of 2.4 years for g. More significant is the variation of the eccentricity, which was constant until now. The eccentricity starts to oscillate, with a fourth month period and quite small amplitudes; but the same long-term modulation as for i also appears (a period of about 1.2 years with larger amplitudes of about 0.02).

It is nowadays known how the stability of a lunar satellite can be strongly affected by the presence of the Earth, especially for higher orbits, while the J_2 effect is stabilizing. The fact that higher orbits are more unstable than lower ones is quite counterintuitive and can lead to surprises. On the other hand, very low orbits are even surprising, since they may also become unstable under the influence of other (odd) gravity harmonics (Knežević and Milani 1998), as was learned the hard way in the past with the crash of Apollo 16 subsatellite only 35 days after its release (Konopliv et al. 1993).

Of course, the dynamics is still strongly dependent on the initial conditions. The eccentricity may sometimes become so high that the satellite crashes on the Moon, as it is the case for polar orbiters. We made a parametric study of the lifetimes of lunar

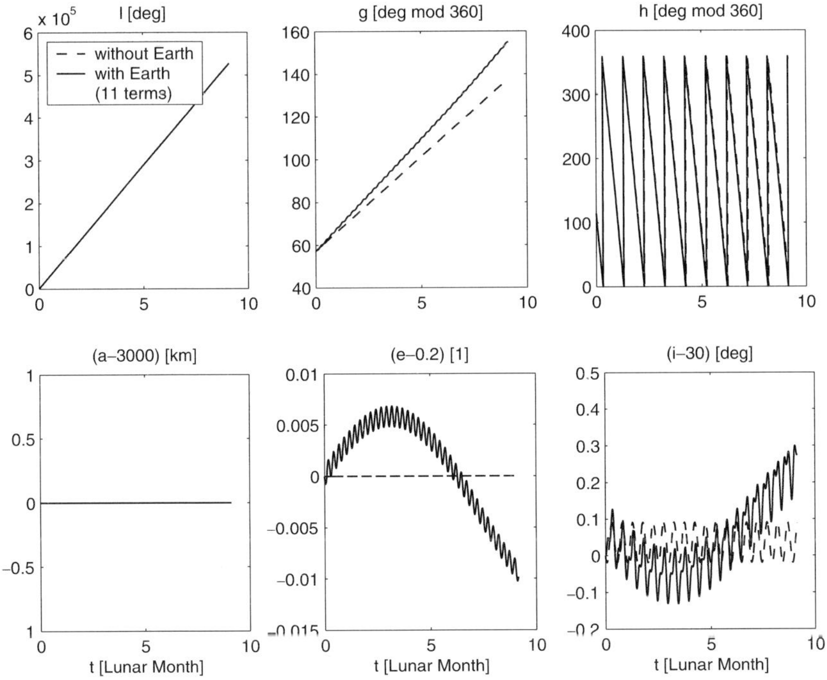

Fig. 5 The additional effect of $n_☾$ and of the Earth: integration of the averaged model (15) for the first-order effects; we took in both cases ($\epsilon \neq 0, \delta \neq 0, n_☾ \neq 0$) and the initial conditions (17). Then we took once $\gamma = 0$ (*dashed line*: without the effect of the Earth) and once $\gamma \neq 0$ (*solid line*: with the effect of the Earth)

polar orbiters and the results were in agreement with (Steichen 1998b; Liu and Wang 2000). The present theory, when it will be completely averaged, will allow that kind of very rapid mission analysis for a wide range of initial conditions.

The effect of the number of terms taken for the Earth is shown in Fig. 6. We integrated the averaged model (15) for the first-order effects, with ($\epsilon \neq 0, \delta \neq 0, n_☾ \neq 0, \gamma \neq 0$) and the initial conditions (17). The terrestrial perturbation contained (1) once 11 terms (accuracy 10^{-6}) and (2) once 350 terms (accuracy 10^{-9}); we then plot the difference (1)–(2) in each of the elements, which is of order 10^{-3}. One conclude that the main trend was already given by the leading terms that were given in Table 1, but that considering more terms can give a somewhat more accurate description.

8 Osculating versus averaged initial conditions

In this section, we present a qualitative validation of the averaging process: we compare the averaged motion (integration of the averaged model (15)) to the osculating one (integration of the osculating model (7)). For illustration purposes, we take again our simple example of Sect. 5: first-order effect in J_2 alone ($\epsilon \neq 0, \delta = 0 = n_☾ = \gamma$), where averaged (a, e, i) were constant.

If we do not pay attention, some surprises can arise. For instance, if we take the same initial conditions for both the osculating motion and the averaged motion, the

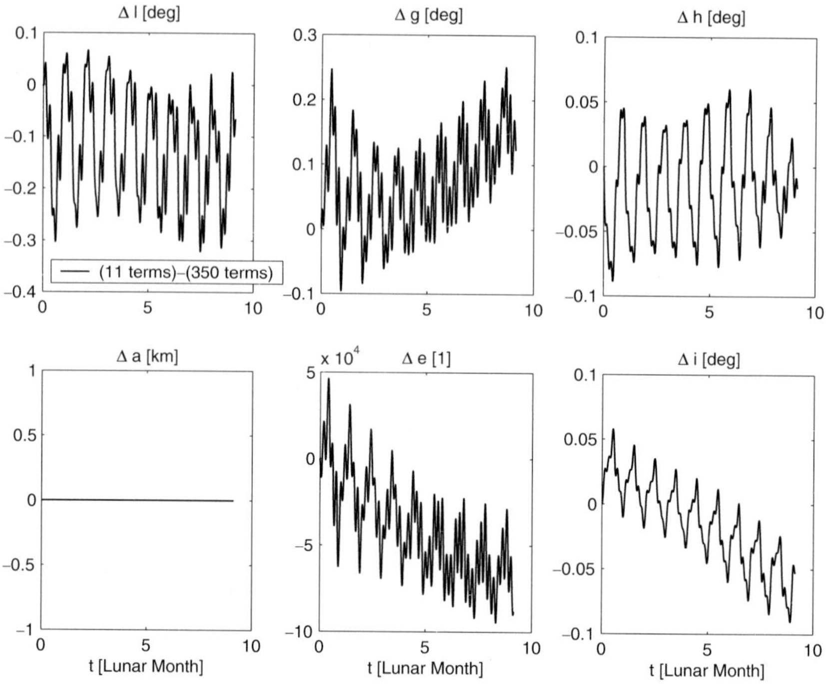

Fig. 6 The effect of the number of terms taken for the Earth: integration of the averaged model (15) for the first-order effects, with ($\epsilon \neq 0, \delta \neq 0, n_{☾} \neq 0, \gamma \neq 0$) and the initial conditions (17). The terrestrial perturbation contained (1) once 11 terms (accuracy 10^{-6}) and (2) once 350 terms (accuracy 10^{-9}); we then plot the difference (1)–(2) in each of the elements, which is of order 10^{-3}

averaged value seems not correct with respect to the osculating one (see Fig. 7, top). In our particular choice, the $\bar{a}$ corresponds to a minimum rather than to the mean value; it should not be the case of course. Note that we can clearly see the period of the satellite on the osculating motion, which is about 4.1 hour for $a = 3{,}000$ km.

The solution is to adapt the initial conditions, using the same transformation which has been used for averaging the Hamiltonian. One has to be careful that the result is inverted: if we use the direct algorithm of the Lie triangle, then we will have the function $\bar{a} = a + \cdots$ If we rather need the function $a = \bar{a} + \cdots$, then we will have to use the algorithm of the inverse (Henrard 1973) as soon as the second-order is considered. A detailed example of such a transformation is given in Appendix, where we used $\beta = (1 - \eta)/e = e/(1 + \eta)$. If we adapt the initial conditions (l, g, h, a, e, i) in such a way, the correspondence between the averaged value and the osculating is then correct (at the order considered for the transformation), as can be seen in Fig. 7 (bottom).

The goal of our work being to build an averaged theory for mission analysis purposes, transformations from averaged to osculating quantities may be very useful in that context; moreover, this transformation is necessary in order to validate numerically the averaged theory. The transformation can be done exactly since the Lie generators are available and since not any real tracking data of lunar satellite are considered.

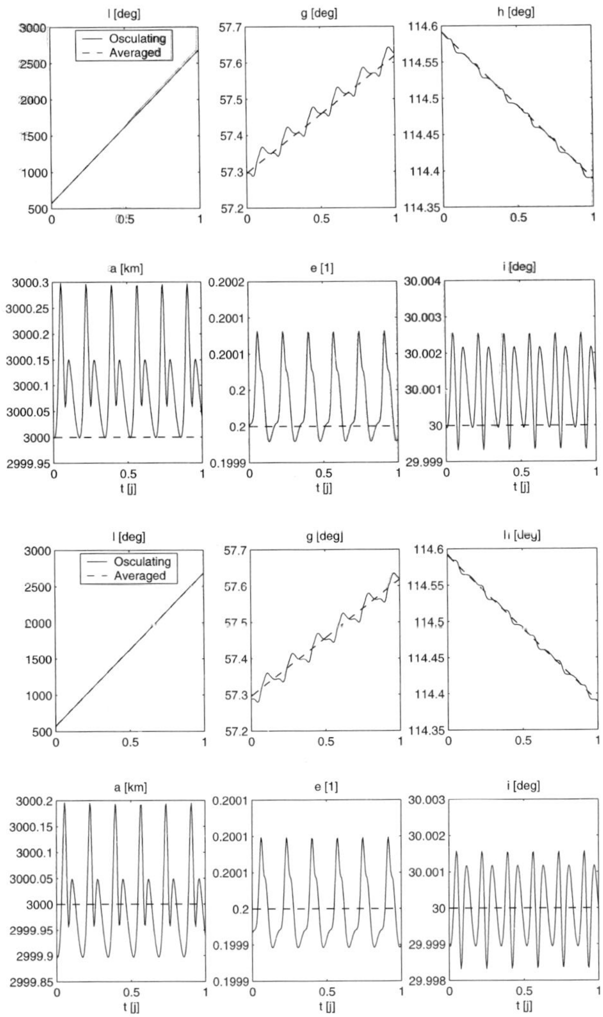

$$\left\{\begin{aligned} \bar{l}_0 &= l_0 &= 10 \text{ rad}\\ \bar{g}_0 &= g_0 &= 1 \text{ rad}\\ \bar{h}_0 &= h_0 &= 2 \text{ rad}\\ \bar{a}_0 &= a_0 &= 3000 \text{ km}\\ \bar{e}_0 &= e_0 &= 0.2\\ \bar{i}_0 &= i_0 &= 30 \text{ deg}\end{aligned}\right.$$

$$\left\{\begin{aligned} \bar{l}_0 &= 0.10000000 \times 10^2 \text{ rad}\\ \bar{g}_0 &= 1.00000000 \times 10^0 \text{ rad}\\ \bar{h}_0 &= 0.20000000 \times 10^1 \text{ rad}\\ \bar{a}_0 &= 0.30000000 \times 10^4 \text{ km}\\ \bar{e}_0 &= 0.20000000 \times 10^0\\ \bar{i}_0 &= 0.30000000 \times 10^2 \text{ deg}\end{aligned}\right.$$

$$\left\{\begin{aligned} l_0 &= 0.10000092 \times 10^2 \text{ rad}\\ g_0 &= 0.99986288 \times 10^0 \text{ rad}\\ h_0 &= 0.20000347 \times 10^1 \text{ rad}\\ a_0 &= 0.29998983 \times 10^4 \text{ km}\\ e_0 &= 0.19996697 \times 10^0\\ i_0 &= 0.29999000 \times 10^2 \text{ deg}\end{aligned}\right.$$

Fig. 7 Comparison of averaged motion (*dashed line*, integration of (15)) with osculating motion (*solid line*, integration of (7)), for the first-order effect of J_2 alone ($\epsilon \neq 0, \delta = 0 = n_{☾} = \gamma$): without adapting the initial conditions (*top*); with adapted initial conditions (*bottom*)

9 Conclusions

We have shown how we built an analytical theory of an artificial satellite of the Moon, by eliminating the short-period terms. We gave some explicit series of the problem, obtained by our home-made algebraic Manipulator software. We performed numerical integrations in order to validate our analytical theory. The effect of each perturbation has been presented progressively and separately as far as possible, in order to achieve a better understanding of the underlying mechanisms. As we could expect, the effect of the Earth plays a major influence, pumping up the eccentricity of the lunar satellite; its role can be modelled already by a few terms only. We stressed and illustrated the importance of adapting the initial conditions from averaged to osculating values in the frame of using an averaged model for mission analysis purposes.

Although not presented in this particular paper, we made other extensive checks (averaged Hamiltonians, generators, liftetimes) with several published works; the results are in good agreement. The results presented here capture the first-order effects of the perturbations by averaging up to order 2, but a full closed-form second-order theory (averaging up to order 4) is also currently being developed and is intended to be published in a forthcoming paper. The third-order will contain the combinations of perturbation parameters ($\epsilon n_{☾}, \delta n_{☾}, \gamma n_{☾}$) and the fourth-order ($\epsilon^2, \delta^2, \gamma^2, \epsilon\delta, \epsilon\gamma, \delta\gamma, \epsilon n_{☾}^2, \delta n_{☾}^2, \gamma n_{☾}^2$).

It is intended to make a more quantitative analysis of the quality of the averaging process, by the same kind of accuracy test as in Knežević and Milani (1995).

Acknowledgements The author is pleased to acknowledge professor Jacques Henrard for fruitful discussions, and for having extended the powerful tool MM ("Moon's series Manipulator") designed by Michèle Moons. A special thank also to Nicolas Rambaux for his help on the numerical integration side.

Appendix: The series giving $a = \bar{a} + \cdots$

Table 3 The a series for (96 terms)

	$\bar{f}$	$\bar{g}$	$\bar{h}$	L^*	$\bar{u}$	$\bar{\xi}$	$\bar{a}$	$\bar{n}$	$\bar{e}$	$\bar{\eta}$	$\bar{c}$	$\bar{s}$	δ	ϵ	γ	$1+\bar{c}$	$\bar{\beta}$	$1-\bar{c}$	Coefficient
cos	0	0	0	0	0	0	1	0	0	0	0	0	0	0	0	0	0	0	0.10000D + 01
cos	0	0	0	0	0	0	1	−2	0	0	0	0	0	0	1	0	0	0	−0.24932D + 00
cos	0	0	0	0	0	0	1	−2	0	0	2	0	0	0	1	0	0	0	0.74795D + 00
cos	0	0	0	0	0	0	1	−2	2	0	0	0	0	0	1	0	0	0	−0.37399D + 00
cos	0	0	0	0	0	0	1	−2	2	0	2	0	0	0	1	0	0	0	0.11219D + 01
cos	0	0	0	0	0	1	1	−2	0	0	0	0	0	0	1	0	0	0	0.24932D + 00
cos	0	0	0	0	0	1	1	−2	0	0	2	0	0	0	1	0	0	0	−0.74795D + 00
cos	0	0	0	0	0	1	1	−2	2	0	0	0	0	0	1	0	0	0	0.37399D + 00
cos	0	0	0	0	0	1	1	−2	2	0	2	0	0	0	1	0	0	0	−0.11219D + 01
cos	0	0	0	0	1	1	1	−2	1	0	0	0	0	0	1	0	0	0	−0.74797D + 00
cos	0	0	0	0	1	1	1	−2	1	0	2	0	0	0	1	0	0	0	0.22439D + 01
cos	0	0	0	0	1	1	1	−2	3	0	0	0	0	0	1	0	0	0	−0.18699D + 00
cos	0	0	0	0	1	1	1	−2	3	0	2	0	0	0	1	0	0	0	0.56096D + 00
cos	0	0	0	0	2	1	1	−2	2	0	0	0	0	0	1	0	0	0	0.37399D + 00
cos	0	0	0	0	2	1	1	−2	2	0	2	0	0	0	1	0	0	0	−0.11219D + 01
cos	0	0	0	0	3	1	1	−2	3	0	0	0	0	0	1	0	0	0	−0.62331D − 01
cos	0	0	0	0	3	1	1	−2	3	0	2	0	0	0	1	0	0	0	0.18699D + 00
cos	0	0	2	−2	0	0	1	−2	0	0	0	2	0	0	1	0	0	0	0.74450D + 00
cos	0	0	2	−2	0	0	1	−2	2	0	0	2	0	0	1	0	0	0	0.11167D + 01
cos	0	0	2	−2	0	1	1	−2	0	0	0	2	0	0	1	0	0	0	−0.74450D + 00
cos	0	0	2	−2	0	1	1	−2	2	0	0	2	0	0	1	0	0	0	−0.11167D + 01
cos	0	0	2	−2	1	1	1	−2	1	0	0	2	0	0	1	0	0	0	0.11167D + 01
cos	0	0	2	−2	1	1	1	−2	3	0	0	2	0	0	1	0	0	0	0.27919D + 00
cos	0	0	2	−2	2	1	1	−2	2	0	0	2	0	0	1	0	0	0	−0.55837D + 00
cos	0	0	2	−2	3	1	1	−2	3	0	0	2	0	0	1	0	0	0	0.93062D − 01
cos	0	0	2	−2	−1	1	1	−2	1	0	0	2	0	0	1	0	0	0	0.11167D + 01
cos	0	0	2	−2	−1	1	1	−2	3	0	0	2	0	0	1	0	0	0	0.27919D + 00
cos	0	0	2	−2	−2	1	1	−2	2	0	0	2	0	0	1	0	0	0	−0.55837D + 00
cos	0	0	2	−2	−3	1	1	−2	3	0	0	2	0	0	1	0	0	0	0.93062D − 01
cos	0	2	0	0	0	0	1	−2	2	0	0	2	0	0	1	0	0	0	0.18699D + 01

Table 3 continued

	$\bar{f}$	$\bar{g}$	$\bar{h}$	L^*	$\bar{u}$	$\bar{\xi}$	$\bar{a}$	$\bar{n}$	$\bar{e}$	$\bar{\eta}$	$\bar{c}$	$\bar{s}$	δ	ϵ	γ	$1+\bar{c}$	$\bar{\beta}$	$1-\bar{c}$	Coefficient
cos	0	2	0	0	0	1	1	−2	2	0	0	2	0	0	1	0	0	0	−0.18699D+01
cos	0	2	0	0	1	1	1	−2	2	0	0	2	0	0	1	0	−1	0	0.93493D+00
cos	0	2	0	0	1	1	1	−2	3	0	0	2	0	0	1	0	0	0	0.46747D+00
cos	0	2	0	0	2	1	1	−2	1	0	0	2	0	0	1	0	−1	0	−0.37397D+00
cos	0	2	0	0	2	1	1	−2	2	0	0	2	0	0	1	0	0	0	−0.18699D+00
cos	0	2	0	0	2	1	1	−2	2	1	0	2	0	0	1	0	0	0	−0.37397D+00
cos	0	2	0	0	3	1	1	−2	2	0	0	2	0	0	1	0	1	0	0.18699D+00
cos	0	2	0	0	3	1	1	−2	3	0	0	2	0	0	1	0	0	0	−0.93493D−01
cos	0	2	0	0	−1	1	1	−2	2	0	0	2	0	0	1	0	1	0	0.93493D+00
cos	0	2	0	0	−1	1	1	−2	3	0	0	2	0	0	1	0	0	0	0.46747D+00
cos	0	2	0	0	−2	1	1	−2	1	0	0	2	0	0	1	0	1	0	−0.37397D+00
cos	0	2	0	0	−2	1	1	−2	2	0	0	2	0	0	1	0	0	0	−0.18699D+00
cos	0	2	0	0	−2	1	1	−2	2	1	0	2	0	0	1	0	0	0	0.37397D+00
cos	0	2	0	0	−3	1	1	−2	2	0	0	2	0	0	1	0	1	0	0.18699D+00
cos	0	2	0	0	−3	1	1	−2	3	0	0	2	0	0	1	0	0	0	−0.93493D−01
cos	0	2	2	−2	0	0	1	−2	2	0	0	0	0	0	1	2	0	0	0.93062D+00
cos	0	2	2	−2	0	1	1	−2	2	0	0	0	0	0	1	2	0	0	−0.93062D+00
cos	0	2	2	−2	1	1	1	−2	2	0	0	0	0	0	1	2	−1	0	0.46531D+00
cos	0	2	2	−2	1	1	1	−2	3	0	0	0	0	0	1	2	0	0	0.23266D+00
cos	0	2	2	−2	2	1	1	−2	1	0	0	0	0	0	1	2	−1	0	−0.18612D+00
cos	0	2	2	−2	2	1	1	−2	2	0	0	0	0	0	1	2	0	0	−0.93062D−01
cos	0	2	2	−2	2	1	1	−2	2	1	0	0	0	0	1	2	0	0	−0.18612D+00
cos	0	2	2	−2	3	1	1	−2	2	0	0	0	0	0	1	2	−1	0	0.93062D−01
cos	0	2	2	−2	3	1	1	−2	3	0	0	0	0	0	1	2	0	0	−0.46531D−01
cos	0	2	2	−2	−1	1	1	−2	2	0	0	0	0	0	1	2	1	0	0.46531D+00
cos	0	2	2	−2	−1	1	1	−2	3	0	0	0	0	0	1	2	0	0	0.23266D+00
cos	0	2	2	−2	−2	1	1	−2	1	0	0	0	0	0	1	2	1	0	−0.18612D+00
cos	0	2	2	−2	−2	1	1	−2	2	0	0	0	0	0	1	2	0	0	−0.93062D−01
cos	0	2	2	−2	−2	1	1	−2	2	1	0	0	0	0	1	2	0	0	0.18612D+00
cos	0	2	2	−2	−3	1	1	−2	2	0	0	0	0	0	1	2	1	0	0.93062D−01
cos	0	2	2	−2	−3	1	1	−2	3	0	0	0	0	0	1	2	0	0	−0.46531D−01
cos	0	2	−2	2	0	0	1	−2	2	0	0	0	0	0	1	0	0	2	0.93062D+00
cos	0	2	−2	2	0	1	1	−2	2	0	0	0	0	0	1	0	0	2	−0.93062D+00
cos	0	2	−2	2	1	1	1	−2	2	0	0	0	0	0	1	0	−1	2	0.46531D+00
cos	0	2	−2	2	1	1	1	−2	3	0	0	0	0	0	1	0	0	2	0.23266D+00
cos	0	2	−2	2	2	1	1	−2	1	0	0	0	0	0	1	0	−1	2	−0.18612D+00
cos	0	2	−2	2	2	1	1	−2	2	0	0	0	0	0	1	0	0	2	−0.93062D−01
cos	0	2	−2	2	2	1	1	−2	2	1	0	0	0	0	1	0	0	2	−0.18612D+00
cos	0	2	−2	2	3	1	1	−2	2	0	0	0	0	0	1	0	−1	2	0.93062D−01
cos	0	2	−2	2	3	1	1	−2	3	0	0	0	0	0	1	0	0	2	−0.46531D−01
cos	0	2	−2	2	−1	1	1	−2	2	0	0	0	0	0	1	0	1	2	0.46531D+00
cos	0	2	−2	2	−1	1	1	−2	3	0	0	0	0	0	1	0	0	2	0.23266D+00
cos	0	2	−2	2	−2	1	1	2	1	0	0	0	0	0	1	0	1	2	−0.18612D+00
cos	0	2	−2	2	−2	1	1	−2	2	0	0	0	0	0	1	0	0	2	−0.93062D−01
cos	0	2	−2	2	−2	1	1	−2	2	1	0	0	0	0	1	0	0	2	0.18612D+00
cos	0	2	−2	2	−3	1	1	−2	2	0	0	0	0	0	1	0	1	2	0.93062D−01
cos	0	2	−2	2	−3	1	1	−2	3	0	0	0	0	0	1	0	0	2	−0.46531D−01
cos	0	0	0	0	0	0	−1	0	0	−3	0	0	0	1	0	0	0	0	0.50000D+00
cos	0	0	0	0	0	0	−1	0	0	−3	2	0	0	1	0	0	0	0	−0.15000D+01
cos	0	0	0	0	0	2	−1	0	0	2	0	0	0	1	0	0	0	0	−0.50000D+00
cos	0	0	0	0	0	2	−1	0	0	−2	2	0	0	1	0	0	0	0	0.15000D+01
cos	1	0	0	0	0	2	−1	0	1	−2	0	0	0	1	0	0	0	0	−0.50000D+00
cos	1	0	0	0	0	2	−1	0	1	−2	2	0	0	1	0	0	0	0	0.15000D+01
cos	1	2	0	0	0	2	−1	0	1	−2	0	2	0	1	0	0	0	0	0.75000D+00
cos	2	2	0	0	0	2	−1	0	0	−2	0	2	0	1	0	0	0	0	0.15000D+01
cos	3	2	0	0	0	2	−1	0	1	−2	0	2	0	1	0	0	0	0	0.75000D+00

Table 3 continued

	$\bar{f}$	$\bar{g}$	$\bar{h}$	L^*	$\bar{u}$	$\bar{\xi}$	$\bar{a}$	$\bar{n}$	$\bar{e}$	$\bar{\eta}$	$\bar{c}$	$\bar{s}$	δ	ϵ	γ	$1+\bar{c}$	$\bar{\beta}$	$1-\bar{c}$	Coefficient
cos	0	0	2	0	0	0	−1	0	0	−3	0	2	1	0	0	0	0	0	0.30000D + 01
cos	0	0	2	0	0	2	−1	0	0	−2	0	2	1	0	0	0	0	0	−0.30000D + 01
cos	1	0	2	0	0	2	−1	0	1	−2	0	2	1	0	0	0	0	0	−0.15000D + 01
cos	1	0	−2	0	0	2	−1	0	1	−2	0	2	1	0	0	0	0	0	−0.15000D + 01
cos	1	2	2	0	0	2	−1	0	1	−2	0	0	1	0	0	2	0	0	−0.75000D + 00
cos	1	2	−2	0	0	2	−1	0	1	−2	0	0	1	0	0	0	0	2	−0.75000D + 00
cos	2	2	2	0	0	2	−1	0	0	−2	0	0	1	0	0	2	0	0	−0.15000D + 01
cos	2	2	−2	0	0	2	−1	0	0	−2	0	0	1	0	0	0	0	2	−0.15000D + 01
cos	3	2	2	0	0	2	−1	0	1	−2	0	0	1	0	0	2	0	0	−0.75000D + 00
cos	3	2	−2	0	0	2	−1	0	1	−2	0	0	1	0	0	0	0	2	−0.75000D + 00

References

Brouwer, D.: Solution of the problem of artificial satellite theory without air drag. Astron. J. **64**, 378–397 (1959)

Chapront-Touzé, M., Chapront, J.: Lunar Tables and Programs 4000 BC to AD 8000. Willmann- Bell. (1991)

De Saedeleer, B., Henrard, J.: Orbit of a lunar artificial satellite: analytical theory of perturbations. *IAU Colloq. 196: Transits of Venus: New Views of the Solar System and Galaxy.* pp. 254–262 (2005)

De Saedeleer, B., Henrard, J.: The combined effect of J_2 and C_{22} on the critical inclination of a lunar orbiter'. Adv. Space Res. **37**(1): 80–87 (2006). The Moon and Near-Earth Objects. Also available as http://dx.doi.org/10.1016/j.asr.2005.06.052

De Saedeleer, B.: Analytical theory of an artificial satellite of the Moon. In: Belbruno, E., Gurfil, P. (eds.) Astrodynamics, Space Missions, and Chaos, of the Annals of the New York Academy of Sciences. *Proceedings of the Conference New Trends in Astrodynamics and Applications, January 20-22, 2003*, Vol. 1017, pp. 434–449. Washington (2004)

De Saedeleer, B.: Complete zonal problem of the artificial satellite: generic compact analytic first order in closed form. Celest. Mech. Dynam. Astron. **91**, 239–268 (2005)

Deprit, A.: Canonical transformations depending on a small parameter. Celest. Mech. **1**, 12–30 (1969)

Henrard, J.: The algorithm of the inverse for Lie transform. In: Szebehely, V., Tapley, B. (eds.) *ASSL: Recent Advances in Dynamical Astronomy*, Vol. 39, pp. 248–257. Dordrecht (1973)

Jefferys, W.: Automated, closed form integration of formulas in elliptic motion. Celest. Mech. **3**, 390–394 (1971)

Jupp, A.: The critical inclination problem: 30 years of progress. Celest. Mech. **43**, 127–138 (1988)

Knežević, Z., Milani, A.: Perturbation theory for low satellites: an application. Bull. Astron. Belgrade **152**, 35–48 (1995)

Knežević, Z., Milani, A.: Orbit maintenance of a lunar polar orbiter. Planet. Space Sci. **46**, 1605–1611 (1998)

Konopliv, A. S., Sjogren, W.L., Wimberly, R.N., Cook, R.A., Vijayaraghavan, A.: A high resolution lunar gravity field and predicted orbit behavior. In *Advances in the Astronautical Sciences, Astrodynamics (AAS/AIAA Astrodynamics Specialist Conference, Pap. # AAS 93–622, Victoria, B.C.)*, Vol. 85, pp. 1275–1295 (1993)

Kozai, Y.: Second-order solution of artificial satellite theory without air drag. Astron. J. **67**, 446–461 (1962)

Liu, L., Wang, X.: On the orbital lifetime of high-altitude satellites. Chinese Astron, Astrophys. **24**, 284–288 (2000)

Milani, A., Knežević, Z.: *Selenocentric proper elements: A tool for lunar satellite mission analysis.* Final Report of a study conducted for ESA, ESTEC, Noordwijk (1995)

Oesterwinter, C.: The motion of a lunar satellite. Celest. Mech. **1**, 368–436 (1970)

Press, W. H., Teukolsky, S. A., Vetterling, W. T., Flannery, B. P.: Numerical Recipes in Fortran 77—The Art of Scientific Computing. Cambridge University Press, Cambridge (1986)

Roy, A.: The theory of the motion of an artificial lunar satellite I. Development of the disturbing function. Icarus **9**, 82–132 (1968)

Shniad, H.: The equivalence of von Zeipel mappings and Lie transforms. Celest. Mech. **2**, 114–120 (1970)
Steichen, D.: An averaging method to study the motion of lunar artificial satellites—I: Disturbing function. Celest. Mech. Dynam. Astron. **68**, 205–224 (1998a)
Steichen, D.: An averaging method to study the motion of lunar artificial satellites—II: Averaging and applications. Celest. Mech. Dynam. Astron. **68**, 225–247 (1998b)
Szebehely, V.: Adventures in Celestial Mechanics. University of Texas Press, Texas (1989)

Celestial Mechanics and Dynamical Astronomy (2006) 95:425–437
DOI 10.1007/s10569-006-9014-0

ORIGINAL ARTICLE

Precise measurement of the solar gravitational red shift

**Alessandro Cacciani · Runa Briguglio ·
Fabrizio Massa · Paolo Rapex**

Received: 13 December 2005 / Revised: 3 March 2006 /
Accepted: 16 March 2006 / Published online: 15 August 2006

Abstract We present the concept and the status of a multi-year project based on a new method to measure the Gravitational Red Shift of the Solar Spectrum with high precision. This project is aimed to conduct experimental verifications of the effect that the Einstein theory of General Relativity predicts for the frequencies of the Fraunhofer lines, that is, the light spectrum emitted by the Sun in its strong gravitational field. Previous determinations of such effect is limited to a precision of 2%. In order to discriminate between classical and relativistic explanations, we need to be sensitive to one part per million of the predicted effect. We have developed a new powerful technique, the Magneto-Optical Filter, that is able to provide far better precision and, for the future, possible space instrumentations able to extend our test to the second-order effect of the relativistic equivalence principle, never done before. The present paper is intended to describe the instrumentation, the procedure and the first encouraging results.

Keywords Tests of general relativity · Gravitational red-shift · Fraunhofer spectrum · Magneto-optical filter

1 Introduction

The reasons why many projects are being conceived concerning sensitive tests of Fundamental Physics, especially General Relativity, are due to both theoretical and experimental aspects. From the theoretical side, the attempt to unify General Relativity and Quantum Theory appears to imply some basic revision of current physical laws; at the same time, exciting progress on precise instrumentations make it possible to plan new measurements able to accomplish new tests and even to open the door to unexpected results. Therefore, there is a worldwide effort to push for new technologies, both from the space and on the ground, with the hope to better clarify Quantum Physics and General Relativity or to explore new discoveries.

A. Cacciani (✉) · R. Briguglio · F. Massa · P. Rapex
Physics Department University "LA SAPIENZA",
Piazzale Aldo Moro 2, 00185 Rome, Italy
e-mail: alessandro.cacciani@roma1.infn.it

During the last decades, we have been developing a particular technology worldwide adopted and known with the name of Magneto-Optical Filter (MOF) (Cacciani et al. 1990, 1994; Cacciani and Moretti 1993). The MOF is able to measure spectral line shifts with unsurpassed precision and absolute reference stability. Therefore, we can focus our interest on a new measurement of the relativistic Gravitational Red Shift on the surface of the Sun that is waiting for more precise determination since the last figure given by LoPresto et al. (1991) The Sun is a massive body, much larger than the Earth. It also displays a resolved image to the observer: so that it is possible to study carefully various competing effects otherwise averaged out on other stars. The status of our project is now at the instrumentation completion level and the first engineering tests. These reveal that we have already reached the precision so far achieved by other authors.

2 Historical review

Following the General Theory of Relativity, the gravitational potential of a celestial body affects the physical time, slowing down any periodic phenomenon assumed as the physical clock. In order to verify this statement, the behavior of different types of clocks, located in different gravitational potentials, must be compared.

The first experiment is the famous Pound–Rebka–Snider experiment (1960) using the Mossbauer effect and measuring the frequency shift of γ-rays ascending a 22.6 m tower at Harvard. The accuracy was of the order of 10%. Another experiment conducted by Vessot et al. (1980) compared the frequency standard of two Hydrogen Masers, one on the ground and the other at 10,000 km on a spacecraft, was able to verify the theoretical prediction at the level of about $2 \cdot 10^{-4}$, at the limit imposed by their Maser stability. Later on (Krisher et al. 1993), a more recent measurement of GRS was performed at radio wavelength using a clock on board of Galileo. They set a limit of 1% of what was expected. A complete review of past and recent experimental works is included in the book of Ohanian (2001). So far the precision of Vessot et al., if true, remains unsurpassed.

3 The solar case

A different kind of possible experiments is provided by the solar spectral lines. The solar gravitational potential shifts their wavelengths relative to laboratory lines on Earth by the amount of $\Delta\lambda = 2.1 \cdot 10^{-6} \cdot \lambda$ towards the red. In terms of velocity, this is equivalent to a Doppler shift of 636.5 m/s outside the Earth's gravitational field and 633.5 m/s at the Earth's surface. The solar spectrum is attractive in this context because the Sun is a massive body monitoring the GRS to values far larger than what is possible on Earth (dynamic range). We can mention here that some unified theories [see the review given by Fujii (1991) "Theoretical background of the fifth force"] suggest the existence of deviations from classical gravity.

The sun is the Earth's closest star, practically at rest, whose mass is well known. This greatly simplifies the problem of disentangling relativistic effects from other frequency shifts. Possible measurement of the gravitational shift on extrasolar systems (ex binary systems) face a series of difficulties discussed in Kopeikin and Ozernoy (1999) or Lindgren and Dravins (2003). Nevertheless, any attempt to measure the solar GRS is complicated by other effects that can change λ, depending on the line formation layer in the atmosphere of the sun: e.g., lower layers are affected by convective upward gas motions to a large extent (about 12 mÅ equivalent Doppler blue shift), while higher layers, at the chromospheric levels, appear less affected. A review article on this is given by Dravins (1982). Also, most of the solar lines are not symmetrical, displaying complex bisectors. Therefore, care must be taken in order to identify the various sources of wavelength shifts. Chromospheric lines like the Sodium doublet, show good symmetry and could be good candidates for the measurement; however they are blended, especially D2, by telluric water vapor lines. The Potassium line at 770 nm appears to be fairly symmetrical and free from water vapor blends but is affected by a poorly known convective blue shift. For this reason LoPresto et al., decided to use chromospheric Oxigen lines in emission, that are formed well above the main convective velocity field, reaching the precision of about 2%. We need to consider here that different heights of line formation sense different gravitational potentials as pointed out in Sect. 3.1 of the article by Lindegren and Dravins (2003). Indeed, at the end of Sect. 5 of the present article, we are proposing a two line measurement as a sensitivity test for our method. Beckers (1977) measured a gravitational redshift of 614 m/s in sunspot umbrae where the convection, and hence the convective blue shift, is suppressed. Within the measurement errors the 3% difference is consistent with the prediction. Potassium and Sodium lines have been used to measure the GRS, on selected points on the solar disk, by Roddier (1965) and Snider (1972), reaching the precision of about 5%. An important remark is that the velocity field on the solar surface is time-dependent (with oscillatory behavior) making the above measurements very uncertain. In order to overcome this difficulty, our experiment is planned to span one or more orbits of the Earth around the Sun (at the observing cadence of one per minute): so that, in addition to a large statistics (see Fig. 1) we could take advantage of the +500 m/s well-known Doppler velocities resulting from the elliptical orbital motion of the Earth for a precise calibration. Moreover, this strategy could be useful to set, in principle, an observational limit to the space isotropy, that is, the gravity constancy versus space directions.

As far as the effect of the oscillatory component of the Doppler signal, it causes a wider Gaussian fluctuation of the measurements while its central value depends only on the total number of experimental points (Fig. 1).

The present project deals with the gravitational effect of the solar mass on its spectral lines with the aim to verify the Einstein theoretical prediction. Our starting point is an experiment from the ground to get eventually the precision of an order of magnitude better than the

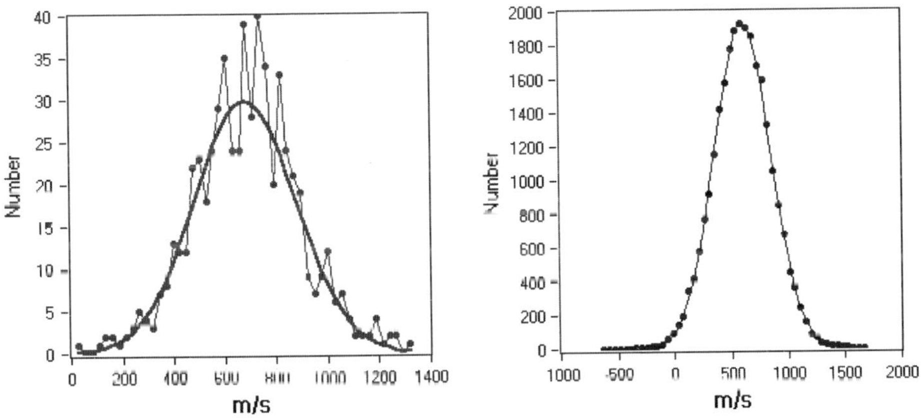

Fig. 1 Effect of the analysis described in Sect. 6 using many points out of a given number of solar Doppler images. The data point fluctuations follow a normal distribution so that a large number of measurements can reduce the oscillatory error at will

previous available figure of 2%. After this accomplishment, we can propose a space mission able to reach the final goal of discriminating classical explanations from General Relativity. This will requires a precision of $\Delta\lambda = 2.1 \cdot 10^{-10} \cdot \lambda$, that is, 1 mm/s Doppler equivalent.

4 The MOF technology

In what follows we show selected pictures of the Sun taken with our MOF. This is an easy way to illustrate the instrument performances. For a comprehensive theoretical and experimental study of the MOF, please refer to Cacciani et al. (1990, 1994), and Cacciani and Moretti (1993); some other recent astrophysical researches accomplished with it can be found in Cacciani et al. (2001), Finsterle et al. (2004a, b), Rhodes et al. (1988, 1990). Here we give brief outlines of its working principle.

Figure 2 shows our instrumental setup as composed by three parts (a telescope, the MOF filter, and an image sensor). Figure 3 shows its practical realization on the roof of our Institute in Rome. Indeed, the MOF is the core of the system that makes this experiment unique. Here let be enough to say that it is a very stable and narrow double band filter. The two bands could be as narrow as 50 milliAngstroms (see Fig. 4), achieving unsurpassed performances as far as its central wavelength reference stability and symmetrical tuning in the wings of the solar lines (Red and Blue sides). Each band can be selected separately at will so that a computer comparison (difference) between the transmitted images can produce the wanted Doppler image of Fig. 6, while their sum gives an intensity image like in Fig. 5.

As a filter, the MOF will be located between the telescope and the image sensor. In this manner, we are able to reject all the other wavelengths of the solar spectrum, but the wanted line, so that we can definitely say that the MOF produces an artificial night, which is the necessary condition to detect faint signals. As an example, in Fig. 5 (which is an image tuned in the core of the Sodium D lines) an intensity reversal (faint emission) is visible in the narrow MOF band-pass wherever magnetic field emerges from below at photospheric levels. Indeed, the magnetic field excites the emitting atoms and modifies the line profiles: the consequent slight increase of its central luminosity appears well visible in our MOF image. This effect should be taken in due consideration to obtain a precise determination of the line shift because it creates macroscopic distortions of the velocity measurement, as shown in Fig. 9: magnetic regions should therefore be avoided, excluding them from the analysis. In Fig. 6 the predominant rotation signal (from the East limb to the West limb) amounts to about 4 km/s. The signal originated by the GRS amounts to 633 m/s (Doppler equivalent). We have been able to measure Doppler signals as low as 1 cm/s by integration over the whole solar disk. Figure 7 shows a plot of the oscillatory signal due to solar p-modes in the 5 min band (peak-to-peak amplitudes ~1 m/s, integration time = 30 s, telescope aperture 2 cm; JPL facility, Pasadena, CA, USA); therefore, our instrument has the potential capability to improve considerably the

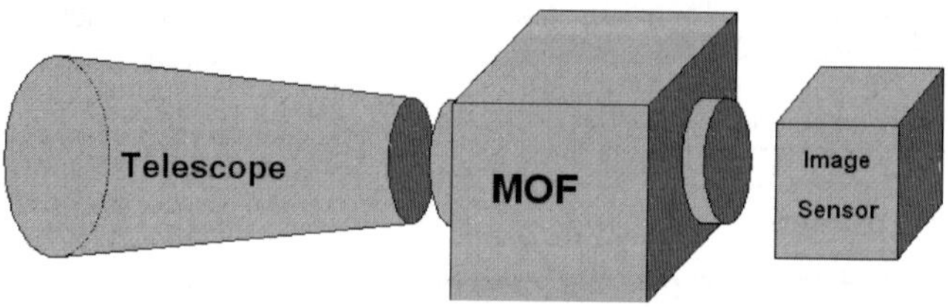

Fig. 2 The MOF setup as composed by three parts: telescope, MOF itself, and image sensor

Fig. 3 Practical realization of the MOF installed on the roof of our institute in Rome

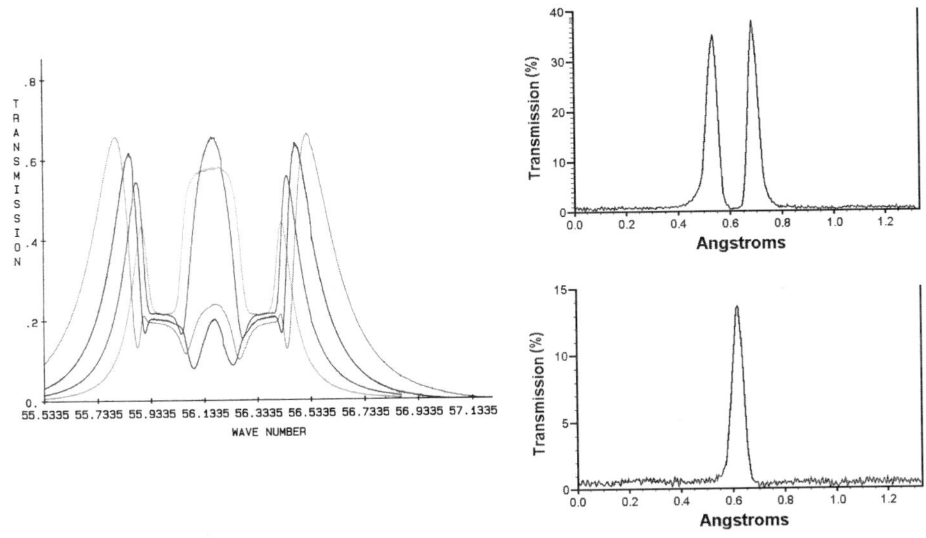

Fig. 4 Typical MOF transmission profiles: we use the double-band transmission of the top right panel

accuracy beyond the few percent so far achieved. Our final goal is to reach the precision of at least one part per mil.

The MOF's weight (about 1 kg) and dimensions ($10 \times 10 \times 30\,\mathrm{cm}^3$) are very attractive for space applications. In this context, we could imagine a space project aimed to test the second order gravitational time dilation (a real test of the General Relativity metrics) going close to the Sun. A comparison between the solar signals from two satellites, one near the Earth and the other close to the Sun, will cancel all the unwanted effects listed in Sect. 4, but the GRS alone. The second-order effect is proportional to ΔU^2 (where U is the gravitational

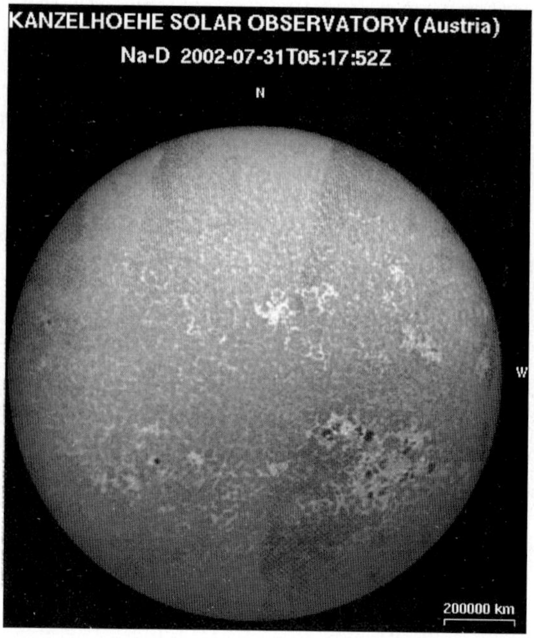

Fig. 5 An intensity image

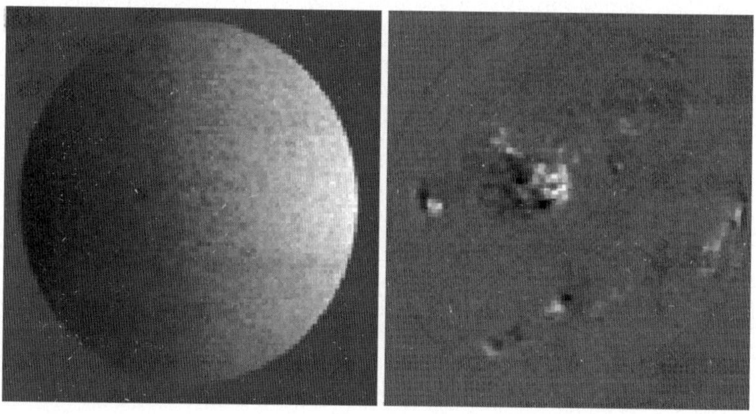

Fig. 6 Left: An example of Doppler image in the K line. Right: An example of Magnetic image

potential and ΔU is its difference between the source and the detector) and is in the range of a mm/s (Doppler equivalent), well within the capability of our MOF (see Fig. 7).

A space mission is also desirable to avoid spurious effects from the Earth's atmosphere; however, due to the high costs and other difficulties of space projects (as, e.g., the spacecraft motions), it is preferable to perform our first measurements from the ground, taking advantage of the absolute wavelength reference of the MOF and the well known Sun–Earth relative Doppler shifts at any time.

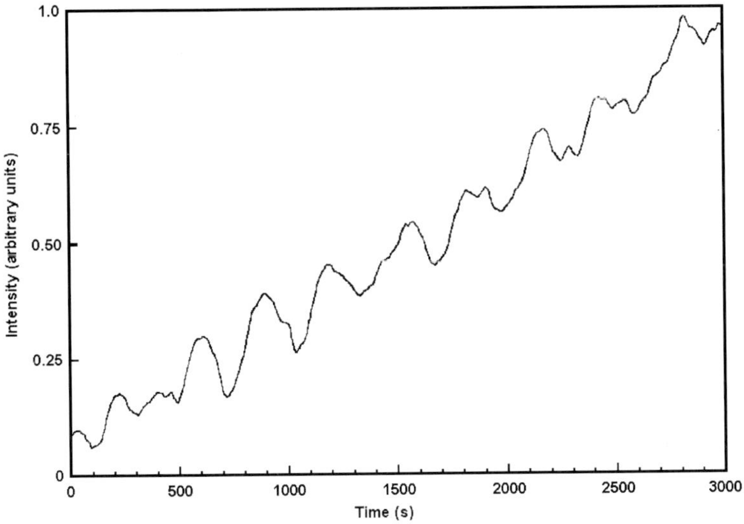

Fig. 7 This figure shows the global oscillations of the sun as a star (no image)

5 The envisioned strategy for the solar GRS measurement

As stated in the previous section, the MOF is able to measure very small Doppler oscillatory signals. However, the present project deals with a far more difficult problem, that is, the evaluation of a constant additive bias among a number of other competing effects. This is the main challenge of the aimed measurement. In the following we list such effects and how the MOF can help disentangling them. What we are showing in this paper are preliminary results and instrumental tests in preparation of the full experiment. The above Fig. 7 shows the most important test run of our instrumental setup. The global oscillations of the Sun as a Star (no image) are displayed with unsurpassed signal-to-noise ratio. After calibration, a signal of 1 m/s (peak-to-peak amplitude) of solar origin is clearly visible on top of the diurnal trend caused by the Earth rotation. It is also visible the beating phenomenon caused by different frequencies present in the solar pulsations.

As far as the GRS however, we need to discuss in detail all the contributions affecting the low-frequency components of the measurement and, particularly, any constant bias. The Sun is a gaseous sphere displaying magnetic features and darkening toward the limbs. The GRS is measured looking at the wavelength position of a solar spectral line respect to a laboratory line. This requires a careful study of their intensity profiles (which are functions of position and time) and the availability of high-resolution spectrographs. Moreover, the average wavelength position is also function of time and depends on the selected point on the solar disk. Indeed, the Sun is an inhomogeneous rotating sphere and its projected disk displays limb foreshortening, convective, and oscillatory signals amounting to as much as 1 km/s (see the mottling in Fig. 6). Therefore, a measurement technique that neglects the above effects, averaging them in a single shot, is definitely not a suitable procedure. A better way of doing would be to measure the wavelength position in all the points at once, in a time interval shorter than the evolution times of the competing signals (possibly few seconds), and proceed to a careful data analysis avoiding magnetic disturbances and correct for convection, rotation, and center/limb variation of the solar line profile. This cannot be achieved with a spectrograph but only with an imaging Doppler–Magnetograph, particularly, the MOF. Figure 9 shows an example of Doppler image obtained to this purpose with our MOF instrument.

The MOF is, by definition, a stable instrument with intrinsic wavelength reference: nevertheless, we need to verify to what extent its stability is maintained during long runs and what is the level of asymmetry between the red and blue transmission bands shown in Fig. 4. We need to check their behavior for each observation using the theory and profiles given in Cacciani et al. (1990, 1994). Moreover, since the MOF works under about 20 Torr internal pressure, we need to take into account also for spectral pressure shifts. Different solar lines suffer from different convective disturbances. They produce an average blue-shift that depends on the chosen spectral line. Chromospheric lines, like the Sodium D_2, are less affected than photospheric ones, although never free from random-like velocities. An additional chromospheric line, very useful in this context, would be the Calcium resonance at 422.7 nm: this is a wide line with a narrow central portion almost entirely chromospheric. At the solar limbs the convective velocities are orthogonal to the line of sight and should become undetectable for all the spectral lines; on the other end, the line profiles undergo modifications (the so called "limb effect"), becoming wider and shallower; moreover, they appear red shifted because of limb disappearance of the convective blue-shift. Also, tangential movements cannot be excluded. For all the above reasons we plan to observe multiple lines simultaneously: the Sodium doublet, the Potassium resonance line at 770 nm and the Calcium line at 422.7 nm. Apart from the method described below in Eq. 1, a comparison among the behaviors of different lines, all over the solar disk, will help disentangling the convective disturbances or identifying the amount of uncertainty produced by this effect. The scientific literature is full of fine observational works regarding this problem in general (see, e.g. the book of Bray et al. 1984). We can mention here that the Potassium line and the Sodium line are formed at two different levels, 300 km apart in height (see also the recent work made by our group at the South Pole, Finsterle et al. (2004a, b), so probing different gravitational potentials: should we be able to see such different effects, this would result in a signal amounting to about $5 \cdot 10^{-4}$ the total effect. The Earth's atmosphere can interfere with the measure, adding blends to the pure solar profile. Figure 8 shows the Sodium solar doublet and, separately, the lines originate in the Earth's atmosphere (mainly H_2O at the bottom of the figure). Their occurrence and equivalent widths are functions of time (slowly changing day by day) and location (dryness or humidity of the observing station). We have been able to model this effect imposing that the diurnal trend (part of which is visible in Fig. 7) be perfectly sinusoidal: any difference is due to the above-mentioned blends and is accounted for by a second-order additive polynomial. In order to monitor and correct for this effect we definitely need continuous data acquisition from more than one station, at least two, at different longitudes. We plan to contact and propose agreements with several observatories around the world that we know are available to host our instrumentation. So far, we have the agreement of SanFernando Observatory (California) and Locarno Observatory (Switzerland).

6 Conclusions and first result

In conclusion, our strategy is to avoid working with a single observing station and few shots at discrete times. The final objective is to accomplish prolonged observing runs for one or more years in order to demonstrate their long-term consistency by fitting the Earth's orbital motion. This will also provide a precise calibration signal, beyond the diurnal and oscillatory shifts. Also, we plan to use more than one solar line, from the photosphere to the chromosphere that

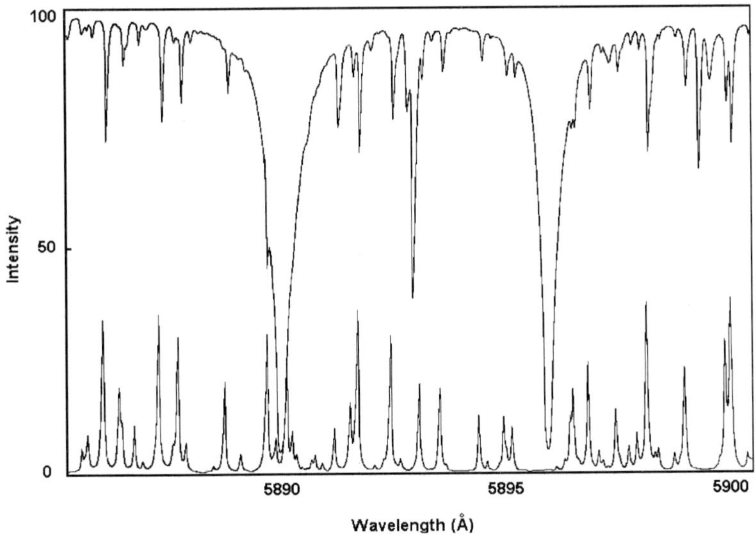

Fig. 8 Spectrum of the Na doublet from the Sun. Several H_2O lines belonging to the Earth's atmosphere are also shown on the bottom of the figure (courtesy of NSO KittPeak Observatory)

can tell us how convection decays with height toward the chromosphere. Finally, it is important to stress again that the MOF imaging capability and its stable and absolute wavelength reference, will allow us to take into account for such effects as the magnetic contamination, the center-to-limb line profile variations and the local oscillatory or convective velocities.

In order to give details about the procedure to identify the GRS, we can refer to a Doppler image like Fig. 9. The analysis can be performed looking at the constant velocity contours that follow, in average, the differential rotation curves. In particular, the zero velocity points follow a winding path around the local convective and oscillatory velocities. The crucial point is that, after correction for all the other possible known biases (the relative Sun–Earth motion and instrumental biases), the zero velocity line does not coincide with the solar rotation axis but appears to be displaced eastwards on the Sun because of the gravitational red shift, convective blue shift and limb effect. The distance between the rotation axis and the zero velocity line at the equator is indeed a measure of them all together.

Several sources of error affect the measurement of such a distance relative to the solar diameter. The following Eq. (1) links all the velocity components along the line of sight between the Sun and the Earth. For each pixel of Fig. 9, we can write:

$$V_{\mathrm{obs}} = V_{\mathrm{GRS}} + V_{\mathrm{BS}} + V_{\mathrm{LE}} + V_{\mathrm{dR}} + V_{\mathrm{EOS}} + V_{\mathrm{IB}} + V_{\approx} \tag{1}$$

where:

V_{Obs} is the local observed velocity;
V_{GRS} is the velocity equivalent to the Gravitational Red Shift, the aim of our research;
V_{BS} is the convective Blue Shift[1] (in this term we include any other contribution, like the solar pressure shift, that vary as $\mu = \cos\Phi$, where Φ is the angle between the outward normal and the direction of the observed ray);
V_{LE} is the Limb Effect[1];

[1] V_{BS} and V_{LE} depend on the chosen spectral line.

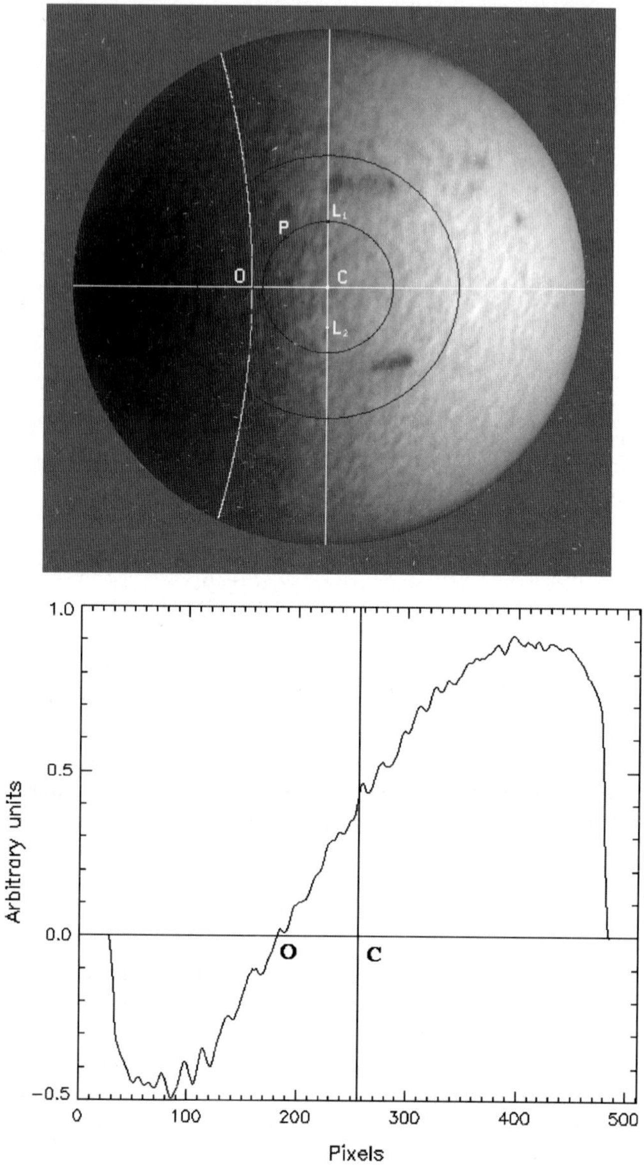

Fig. 9 Velocity image geometry and velocity signal in its equatorial section

V_{dR} is the velocity signal due to the solar differential rotation;
V_{EOS} is the Earth Orbital and Spin velocity;
V_{IB} is the velocity equivalent to any Instrumental Bias; it can be evaluated through laboratory tests;
$V_{\approx}$ is the periodic, oscillatory signal used in Helioseimology; it is a strong signal that will be averaged out using a large statistics.

Now, just few comments about the advantages of using full-disk Doppler images:

(a) It is possible to identify and avoid the magnetic regions during the analysis (see the dark patches in Fig. 9);
(b) Each pixel can be normalized to the local intensity (that is not homogeneous over the solar disk). This could prevent any cross-talk between Doppler and Intensity in the case we had very high-image resolution, practically unachievable. Indeed, included in each resolution element coexist microscopic rising and falling elements, the first being hotter and brighter than the descending ones, so that a net blue shift results when the image resolution is poor;
(c) It is possible, pixel by pixel, to account for the fact that the above blue shift, $V_{\rm BS}$ is maximum at the solar center and vanishes at the limb as $\cos\Phi$;
(d) The solar rotation signal, $V_{\rm dR}$, although differential in latitude, appears always null at the axis where it is perpendicular to the line of sight;
(e) Finally, although the effect of $V_{\rm LS}$ (the center to limb variation of the solar line profile) is more difficult to be accounted for, it is possible to evaluate its contribution assuming a linear behavior with the distance from the disk center, within a limited region (inside the larger circle in Fig. 9).

Each Doppler image will thus offer many points where the full signal can be computed (redundant system of equations). The Ephemeris data will help determining the values for $V_{\rm dR}$ and $V_{\rm EOS}$ to a very high degree of accuracy, while the two quantities $V_{\rm BS}$ and $V_{\rm LS}$ will be obtained measuring distances in pixels (as both are assumed to be linear functions of distance, inside the limited central region). With a large time series of Doppler images we are confident to achieve our result with an error within the 10^{-3} limit. Indeed, our software procedure to identify image centers and diameters has shown that the time series of those quantities is normally distributed, allowing controlling the degree of accuracy as a function of length of the time series.

6.1 First result

The full project is quite expensive and several funding proposals have been submitted. While waiting for the necessary financial support, we have analyzed a few MOF Doppler images, including some taken previously during our 1999 expedition at the Antarctic Italian site of Baia Terranova.

It is remarkable the fact that just looking at Fig. 9, the zero velocity line appears crossing the equator at about one quarter radius eastward which corresponds to about −500 m/s taking into account the line of sight component of the solar rotation only. Assuming, as usual, receding velocities as positive and the following rough figures at the crossing point ($V_{\rm Obs} = 0$):

$$V_{\rm BS} \approx -250\,{\rm m/s}^{2};$$
$$V_{\rm LE} \approx \text{assumed negligible}^{3};$$
$$V_{\rm EOS} \approx +175\,{\rm m/s}^{4};$$

[2] Averaged from "The Sun" by Michael Stix, Springer, II edn. 2004, p. 262, Fig. 6.13.

[3] See Phys. Rev. Lett. 28, 853–856 (1972).

[4] The image was recorded on January 1999 in Antarctica, Terra Nova Bay. The Earth's revolution contribution was ≈+250 m/s, while the contribution from its rotation was ≈−75 m/s.

V_{IB} $\approx$ not measured, assumed negligible;
$V_{\approx}$ $\approx$ assumed negligible averaged over many images;
V_{dR} $\approx -500\,\mathrm{m/s}$.[5]

Eq. (1) gives

$$V_{GRS} \approx 625\,\mathrm{m/s}.$$

The above result is already at the level of 1% internal error; however, it is only indicative being affected by errors, unknown biases and long-term instabilities that our project will be able to characterize very precisely.

Details about the procedure can be found in the following Sect. 6.2.

6.2 Data reduction

Let us write again Eq. (1):

$$V_{Obs} = V_{GRS} + V_{BS} + V_{LE} + V_{dR} + V_{EOS} + V_{IB} + V_{\approx}.$$

Referring to Fig. 9, we shall limit the analysis within a given circle around the center, such as that the two signals V_{BS} (which is maximum at the center) and V_{LE} (which is null at the center and maximum at the limb) vary linearly with the distance r from C:

$$V_{BS} = (1-r)V_{BS,C},$$
$$V_{LE} = r \cdot V_{LE,Max}.$$

For the center C and all the points L_i along the rotation axis, $V_{dR} = 0$ and Eq. (1) becomes

$$\begin{aligned} V_{Obs} &= V_{GRS} + V_{BS,C} + (V_{LE,Max} - V_{BS,C}) \cdot r + V_{EOS} + V_{IB} + V_{\approx} \\ &= V_{GRS} + V_{BS,C} + \gamma \cdot r + V_{EOS} + V_{IB} + V_{\approx}, \end{aligned}$$

where $\gamma = (V_{LE,Max} - V_{BS,C})$ is a constant dependent on the chosen spectral line. Using this equation for the center C and another point L_i, we can determine the value of γ:

$$V_{Obs,C} - V_{Obs,L_i} = -\gamma \cdot r_{L_i} + \Delta V_{\approx} \rightarrow \gamma = (V_{Obs,L_i} - V_{Obs,C})/r_{L_i} + \Delta V_{\approx}$$

(the term $\Delta V_{\approx}$ can be made negligible just smoothing the signal along the rotation axis and taking its average from a long-time series of images). For a rough signal calibration, we can use Eq. (1) for a point L1 on the axis and a point P off axis, but at the same distance from C. We get $V_{Obs,P} - V_{Obs,L_i} = V_{dR,P}$, which is well known. A better calibration will come from the orbital motion of the Earth during a full year of data.

Acknowledgements We thank James LoPresto, Jacques Beckers for their comments and the Observatories of SanFernando (California) and Locarno (Switzerland) for their availability to host our instrumentation for the future accomplishment of the project. We thank the italian PROGETTO ANTARTIDE for the use of Baia Terranova site and financial support.

[5] Assuming the east limb velocity $\approx -2,000\,\mathrm{m/s}$.

References

Beckers, J.M.: Material motions in sunspot umbrae. Ap. J. **213**, 900–905 (1977)

Bray, R.J., Loughhead, R.E., Durrant, C.J.: The solar granulation, 2nd edn. CUP, Cambridge (1984)

Cacciani, A., Rosati, P., Ricci, D., Egidi, A., Apporchaux, T., Marquedant, R., Smith, E.: Theoretical and experimental study of the Magneto-Optical Filter. JPL Document # D11900, Pasadena California (1990 and 1994)

Cacciani, A., Moretti, P.F., Magneto-optical filter (MOF): concept and applications in astronomy. SPIE **2198**, 219–228 (1993)

Cacciani, A., Dolci, M., Moretti, P.F., D'Alessio, F., Giuliani, C., Micolucci, E., Di Cianno, A.: Search for global oscillations on Jupiter with a double cell Sodium Magneto-Optical Filter. A&A **372**: 317–325 (2001)

Dravins, D.: Photospheric Spectrum line asymmetries and wavelength shifts. ARA&A **20**, 61–89 (1982)

Finsterle, W., Jefferies, S.M., Cacciani A., Rapex, P., Giebink, C., Knox, A., and DiMartino, V.: Seismology of the solar atmosphere Solar Phys. **220**, 317–331 (2004a)

Finsterle, W., Jefferies, S.M., Cacciani, A., Rapex, P., McIntosh, S.W.: Helioseismic mapping of the magnetic canopy in the solar chromosphere ApJ. Lett. **613**:L185 (2004b)

Fujii, Y.: Theoretical background of the fifth force. Int. J. Modern Phys. A6 **20**, 3505–3557 (1991)

Kopeikin, S.M., Ozernoy, L.M.: Post-Newtonian theory for precision doppler measurements of binary star orbits. ApJ **523**, 771–785 (1999)

Krisher, T.P., Morabito, D., Anderson, J.D.: The Galileo solar redshift experiment. Phys. Rev. Lett. **70**, 2213–2216 (1993)

Lindegren, L., Dravins, D.: The Fundamenta Definition of "radial velocity". A&A **401**, 1185–1201 (2003)

LoPresto, J.C., Shrader, C., Pierce, A.K.: Solar gravitational red shift from the infrared oxigen triplet. ApJ **367**, 757–760 (1991)

Ohanian, HC.: Special relativity: a modern introduction. Phys. Curric. & Instr. (Edition 2001)

Pound, R.V., Rebka, G.A.: Apparent weigth of photons. Phys. Rev. Lett. **4**, 337 (1960)

Roddier, F.: Etude a haute resolution de quelques raies de Fraunhoffer par observation de la response optique d'un jet atomique. I:Realization d'un spectrograph a jet atomique. Ann. Ap. **28**, 463 (1965)

Rhodes, E., Cacciani, A., Woodard, M., Tomczyk S., Korzennik S., Ulrich R.: On the constancy of intermediate-degree p-mod frequencies during the declining phase of solar cycle 21. ApJ **326**, 479–485 (1988)

Rhodes, E.J., Cacciani, A., Korzennik, S., Tomczyk, S., Ulrich R.: Depth and latitude dependence of the solar internal rotation. ApJ. **351**, 687–700 (1990)

Snider, J.L.: New measurement of the solar gravitational red shift. Phys. Rev. Lett. **28**, 853 (1972)

Vessot et al.: Test of relativistic gravitationwith a space-born hydrogen maser. Phys. Rev. Lett., **45**(26), 2081 (1980)